The IMA Volumes
in Mathematics
and its Applications

Volume 74

Series Editors
Avner Friedman   Willard Miller, Jr.

Institute for Mathematics and
its Applications
IMA

The **Institute for Mathematics and its Applications** was established by a grant from the National Science Foundation to the University of Minnesota in 1982. The IMA seeks to encourage the development and study of fresh mathematical concepts and questions of concern to the other sciences by bringing together mathematicians and scientists from diverse fields in an atmosphere that will stimulate discussion and collaboration.

The IMA Volumes are intended to involve the broader scientific community in this process.

Avner Friedman, Director
Willard Miller, Jr., Associate Director

* * * * * * * * * *

IMA ANNUAL PROGRAMS

1982–1983	Statistical and Continuum Approaches to Phase Transition
1983–1984	Mathematical Models for the Economics of Decentralized Resource Allocation
1984–1985	Continuum Physics and Partial Differential Equations
1985–1986	Stochastic Differential Equations and Their Applications
1986–1987	Scientific Computation
1987–1988	Applied Combinatorics
1988–1989	Nonlinear Waves
1989–1990	Dynamical Systems and Their Applications
1990–1991	Phase Transitions and Free Boundaries
1991–1992	Applied Linear Algebra
1992–1993	Control Theory and its Applications
1993–1994	Emerging Applications of Probability
1994–1995	Waves and Scattering
1995–1996	Mathematical Methods in Material Science

IMA SUMMER PROGRAMS

1987	Robotics
1988	Signal Processing
1989	Robustness, Diagnostics, Computing and Graphics in Statistics
1990	Radar and Sonar (June 18 - June 29)
	New Directions in Time Series Analysis (July 2 - July 27)
1991	Semiconductors
1992	Environmental Studies: Mathematical, Computational, and Statistical Analysis
1993	Modeling, Mesh Generation, and Adaptive Numerical Methods for Partial Differential Equations
1994	Molecular Biology

* * * * * * * * * *

SPRINGER LECTURE NOTES FROM THE IMA:

The Mathematics and Physics of Disordered Media

Editors: Barry Hughes and Barry Ninham
(Lecture Notes in Math., Volume 1035, 1983)

Orienting Polymers

Editor: J.L. Ericksen
(Lecture Notes in Math., Volume 1063, 1984)

New Perspectives in Thermodynamics

Editor: James Serrin
(Springer-Verlag, 1986)

Models of Economic Dynamics

Editor: Hugo Sonnenschein
(Lecture Notes in Econ., Volume 264, 1986)

K.J. Åström   G.C. Goodwin   P.R. Kumar
Editors

Adaptive Control, Filtering, and Signal Processing

With 48 Illustrations

Springer-Verlag

New York Berlin Heidelberg London Paris
Tokyo Hong Kong Barcelona Budapest

K.J. Åström
Department of Automatic Control
Lund Institute of Technology
Lund, Sweden

G.C. Goodwin
Department of Electrical Engineering
 and Computer Science
The University of Newcastle
Rankin Drive
Newcastle, NSW 2308, Australia

P.R. Kumar
Coordinated Science Laboratory
University of Illinois at Urbana-
 Champaign
1308 West Main Street
Urbana, IL 61801-2307 USA

Series Editors:
Avner Friedman
Willard Miller, Jr.
Institute for Mathematics and its
 Applications
University of Minnesota
Minneapolis, MN 55455 USA

Mathematics Subject Classifications (1991): 49N30, 93-06, 93B30, 93C40, 93C10, 93D21, 93E11, 93E12, 93E35

Library of Congress Cataloging-in-Publication Data
Åström, Karl J. (Karl Johan), 1934–.
 Adaptive control, filtering, and signal processing / Karl J.
Åström, G.C. Goodwin, P.R. Kumar.
    p.   cm. — (The IMA volumes in mathematics and its
applications ; v. 74)
   Includes bibliographical references.
   ISBN 0-387-97988-3
   1. Adaptive control systems.   2. Filters (Mathematics)   3. Signal
processing.   I. Goodwin, Graham C. (Graham Clifford), 1945–.
II. Kumar, P. R.   III. Title.   IV. Series.
TJ217.A672   1995
629.8′36 — dc20   95-9906

Printed on acid-free paper.

Production managed by Hal Henglein; manufacturing supervised by Jacqui Ashri.
Camera-ready copy prepared by the IMA.
Printed and bound by Braun-Brumfield, Ann Arbor, MI.
Printed in the United States of America.

9 8 7 6 5 4 3 2 1

ISBN 0-387-97988-3 Springer-Verlag New York Berlin Heidelberg

The IMA Volumes
in Mathematics and its Applications

Current Volumes:

Forthcoming Volumes:

FOREWORD

This IMA Volume in Mathematics and its Applications

ADAPTIVE CONTROL, FILTERING, AND SIGNAL PROCESSING

is based on the proceedings of a workshop that was an integral part of the 1992–93 IMA program on "Control Theory." The area of adaptive systems, which encompasses recursive identification, adaptive control, filtering, and signal processing, has been one of the most active areas of the past decade. Since adaptive controllers are fundamentally nonlinear controllers which are applied to nominally linear, possibly stochastic and time-varying systems, their theoretical analysis is usually very difficult. Nevertheless, over the past decade much fundamental progress has been made on some key questions concerning their stability, convergence, performance, and robustness. Moreover, adaptive controllers have been successfully employed in numerous practical applications, and have even entered the marketplace. The purpose of the meeting was to review the past progress and to focus attention on the fundamental issues that remain.

We thank K.J. Åström, G.C. Goodwin, and P.R. Kumar for organizing the workshop and editing the proceedings. We also take this opportunity to thank the National Science Foundation and the Army Research Office, whose financial support made the workshop possible.

Avner Friedman

Willard Miller, Jr.

PREFACE

This volume is the Proceedings of the Workshop on Adaptive Control, Filtering, and Signal Processing held at IMA, April 12-16, 1993, as part of the year devoted to Control Theory and its Applications.

The Workshop covered topics in the following areas:

(i) Design of adaptive controllers

(ii) Stability of adaptive control systems

(iii) Asymptotic convergence and performance analysis of adaptive systems

(iv) Averaging methods for analysis of adaptive systems

(v) Identification of linear stochastic systems

(vi) Analysis of adaptive filtering algorithms

(vii) Adaptive control of nonlinear systems

(viii) Connections between adaptive systems and learning.

The talks spanned the entire gamut from design of adaptive systems to analysis. The broad spectrum of analytical approaches shows the range of mathematical methods that have been applied to the study of adaptive systems.

We would like to take this opportunity to extend our gratitude to the staff of IMA, Kathy Boyer, Paul Ewing, Joan Felton, Ceil Mcaree, John Pliam, Kathi Polley, Pam Rech, and Mary Saunders. We have fond memories of the extremely warm hospitality in a cool climate. We also thank Professors Avner Friedman and Willard Miller, Jr. for making the Year on Control Theory and its Applications possible, and this Workshop in particular. Their institute inspires all visitors. We thank Patricia V. Brick, Stephan J. Skogerboe, and Kaye Smith for the preparation of the manuscripts.

Finally, we gratefully acknowledge the support of the National Science Foundation and Army Research Office.

K.J. Åström

G.C. Goodwin

P.R. Kumar

CONTENTS

OSCILLATIONS IN SYSTEMS WITH RELAY FEEDBACK

KARL J. ÅSTRÖM*

1. Introduction. Analysis of linear systems with relay feedback is a classical field. The early work on relay feedback was motivated by using relays as amplifiers. These applications became less interesting, because the development of electronic technology made relay amplifiers obsolete. A discussion of relay feedback is found in the classical book [12]. Analysis of systems with relay feedback were given by [27], [10], [11], [22], [23], [24], [25], [7], [16], [18], and [19]. Much of the analysis of relay feedback has been done using the describing function, see [5] and [8]. An interesting discussion of the validity of the describing function approximation is found in [20]. Exact conditions for limit cycle oscillations under relay feedback were developed in the papers by Hamel and Tsypkin. Tsypkin's work is particularly interesting because of its close relation with the approximate methods. An extensive treatment is found in [26], which is an English translation of a book that was first published in Russian in 1974.

In the 1960's it was found that relay feedback could be used in adaptive control. The self-oscillating adaptive controller, developed by Minneapolis Honeywell, used relay feedback in a very clever way. By introducing a relay feedback in the loop there will be a limit cycle oscillation but the propagation of slower signals around the loop will correspond to an amplitude margin $A_m = 2$. The magnitude of the limit cycle oscillation can be adjusted to a desired level by choosing the relay amplitude appropriately. The self-oscilllating adaptive controller, which is described in [21], was tested extensively for flight control systems and it is used in several missiles. The system inspired the development of the dual input describing function, see [8]. The system is analyzed in [14] and [15].

Lately there has been renewed interest in relay feedback because of their use for tuning of simple controllers. This application is based on the idea that a system with relay feedback will oscillate with a frequency that is close to the frequency where the plant has a phase lag of 180°. By relay feedback it is thus possible to determine the cross-over frequency and the magnitude of the transfer function at cross over. With this information it is possible to calculate suitable parameters for simple controllers of the PID type. This idea is described in detail in [2], [3]. Several industrial controllers based on this idea are also available, see [9]. In [17] it is also shown how relay feedback can be used to initialize adaptive controllers.

Use of relay feedback for automatic tuning of controllers has raised questions such as: What are the conditions for stable limit cycle oscillations? Why do the oscillations converge towards the limit cycle so quickly?

---
* Department of Automatic Control, Lund Institute of Technology, Lund, Sweden.

1

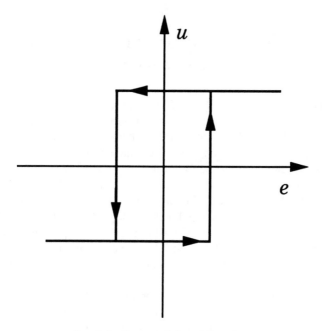

FIG. 1.1. *Characteristics of the relay.*

Is it possible to have several limit cycles depending on the initial conditions? Some of these questions will be discussed in this paper. In Section 2 we present necessary conditions for limit cycles in linear systems with relay feedback. A simple derivation of the classical results is given. A rigorous discussion of the stability of the limit cycles is presented in Section 3. In Section 4 the results are extended to systems with time delays. In Section 5 we give similar results for asymmetric limit cycles and in Section 6 we briefly discuss that very complicated behavior may occur in linear systems with relay feedback.

2. Symmetric oscillations. Consider a linear time invariant system described by

$$(2.1) \qquad \begin{aligned} \frac{dx}{dt} &= Ax + Bu \\ y &= Cx \end{aligned}$$

Let the system be controlled by a relay whose input-output relation is given by

$$(2.2) \qquad u(t) = \begin{cases} d & \text{if } e > \epsilon, \text{ or } e > -\epsilon \text{ and } u(t-) = d \\ -d & \text{if } e < -\epsilon, \text{ or } e < \epsilon \text{ and } u(t-) = -d \end{cases}$$

where d is the relay amplitude, ϵ the hysteresis. The characteristics of the relay is shown in Figure 1.1. The relay is connected to the process in such a way that $e = -y$. For a large class of processes there will be limit cycle oscillations. The limit cycles may have very different characters. We will first consider the case where the oscillations are symmetric. In this case we have the following result.

THEOREM 2.1. *Consider the system (2.1) with the feedback (2.2) where* $e = -y$. *Assume that there exists a symmetric periodic solution with period* $T = 2h$. *Then the following conditions hold.*

$$(2.3) \qquad f(h) = C(I + e^{Ah})^{-1} \int_0^h e^{As} B \, ds = \frac{\epsilon}{d}$$

$$(2.4) \qquad y(t) = C\left(e^{At} a - \int_0^t e^{As} ds \, Bd \right) > -\epsilon \text{ for } 0 \leq t < h$$

Furthermore, the periodic solution is obtained with the initial condition

$$(2.5) \qquad x(0) = a = (I + e^{Ah})^{-1} \int_0^h e^{As} ds \, Bd$$

Proof. Assume that there exists a symmetrical limit cycle with period $T = 2h$ which is obtained with the initial condition a. Let the coordinates be chosen so that $t = 0$ corresponds to a switch where $y = \epsilon$ and $dy/dt > 0$, see Figure 2.1. Integrating Equation (2.1) from time $t = 0$ to time $t = h$ when the next switch occurs we get

$$x(h) = \Phi a - \Gamma d$$

where

$$(2.6) \qquad \Phi = \Phi(h) = e^{Ah} \quad \text{and} \quad \Gamma = \Gamma(h) = \int_0^h e^{As} ds \, B.$$

Since the limit cycle is symmetrical, it follows that $x(h) = -a$. Hence

$$-a = \Phi a - \Gamma d.$$

The initial state is thus given by (2.5). Since the relay switches from d to $-d$ at time h, it follows from Equation (2.2) that

$$\epsilon = Ca = C(I + \Phi)^{-1} \Gamma d$$

This condition is equivalent to (2.3). To have a periodic solution with period $2h$ it must also be required that the relay does not switch before time h. This means that

$$y(t) > -\epsilon$$

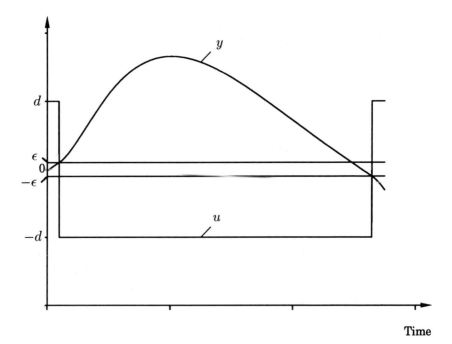

FIG. 2.1. *Input and output signals for a stable symmetric limit cycle.*

for all t in the interval $0 \le t < h$, which gives condition (2.4). □

REMARK 1. *Notice that condition (2.4) implies that the derivative of the output must be negative at time h. By symmetry this implies that it must be positive at time $t = 0$, i.e.*

$$(2.7) \qquad \frac{dy}{dt}(0) = C\frac{dx}{dt}(0) = Cv > 0$$

where v is the velocity of the state vector at time $t = 0-$, i.e.

$$(2.8) \qquad v = \frac{dx}{dt}(0) = Aa + Bd$$

Numerical procedures will be needed to find a value of h that satisfies (2.3). An expression for the derivative of the function is useful to develop efficient algorithms. We have

LEMMA 2.2. *The derivative of the function f given by Equation (2.3)*

is

(2.9) $$\frac{df}{dh} = 2Ce^{Ah}(I + e^{Ah})^{-2}B = \frac{1}{2}\,C\left(\cosh\frac{Ah}{2}\right)^{-2}B$$

If the matrix A is nonsingular it also follows that

$$f(h) = CA^{-1}\left(\tanh\frac{Ah}{2}\right)B$$

Proof. We have

(2.10) $$A\int_0^h e^{As}\,ds = e^{Ah} - I$$

which can be proven by series expansion of the exponential function. Differentiating the function f defined by Equation (2.3) we get

$$\begin{aligned}
\frac{df}{dh} &= -C(I + e^{Ah})^{-2}e^{Ah}A\int_0^h e^{As}B\,ds + C(I + e^{Ah})^{-1}e^{Ah}B \\
&= C(I + e^{Ah})^{-2}e^{Ah}\left(-A\int_0^h e^{As}\,ds + I + e^{Ah}\right)B \\
&= 2C(I + e^{Ah})^{-2}e^{Ah}B = \frac{1}{2}C\left(\cosh\frac{Ah}{2}\right)^{-2}B
\end{aligned}$$

This proves the first part of the lemma. To prove the second part we assume that A is nonsingular. Then notice that it follows from (2.10) that

$$\Gamma = A^{-1}(e^{Ah} - I)$$

and it follows from (2.3) that

(2.11) $$\begin{aligned}
f(h) &= C(e^{Ah} + I)^{-1}A^{-1}(e^{Ah} - I)B \\
&= C\left(e^{\frac{1}{2}Ah} + e^{-\frac{1}{2}Ah}\right)^{-1}A^{-1}\left(e^{\frac{1}{2}Ah} - e^{-\frac{1}{2}Ah}\right)B \\
&= CA^{-1}\left(\tanh\frac{1}{2}Ah\right)B
\end{aligned}$$

Notice that A commutes with e^A. $\square$

2.1. Frequency domain interpretations. The conditions in Theorem 1 have a nice frequency domain interpretation. To show this consider the system given by Equation (2.1). Sampling the system with a zero-order hold and period h we find that the pulse transfer function is

$$H(z) = C(zI - \Phi)^{-1}\Gamma$$

see [4]. This implies that the condition (2.3) can be written as

$$(2.12) \qquad\qquad H(-1) = -\frac{\epsilon}{d}$$

This condition has a nice intuitive interpretation. The response of a linear system to a square-wave input can be computed exactly by sampling the system. In particular the value $H(-1)$ gives the steady-state gain at the sampling instants. The condition (2.12) tells that the values of the input and the output at the sampling instants have opposite signs and the magnitude of the output is ϵ if the input has magnitude d.

Let $G(s)$ be the transfer function of the system (2.1), i.e.

$$G(s) = C(sI - A)^{-1}B$$

The condition (2.7) implies that

$$(2.13) \qquad\qquad H_v(-1) < 0$$

where $H_v(z)$ is the zero-order hold sampling of $sG(s)$, i.e. the transfer functions that relates dy/dt to u.

An alternative expression is obtained by using the following relation between the zero-order hold sampling of a system with transfer function $G(s)$ and its z-transform $H(z)$.

$$H(z) = \sum_{n=-\infty}^{\infty} \frac{1 - z^{-1}}{\log z + 2\pi i n} G\left(\frac{1}{h}\log z + \frac{2\pi i n}{h}\right)$$

See [4]. Introducing $\omega = \pi/h$ we get

$$(2.14)\; H(-1) = \frac{2}{i\pi} \sum_{n=-\infty}^{\infty} \frac{G(i\omega(1 + 2n))}{1 + 2n} = \frac{4}{\pi} \sum_{n=0}^{\infty} \frac{\operatorname{Im} G(i\omega(1 + 2n))}{1 + 2n}$$

Proceeding in the same way we find that

$$(2.15) \qquad\qquad H_v(-1) = \frac{4\omega}{\pi} \sum_{n=0}^{\infty} \operatorname{Re} G(i\omega(1 + 2n))$$

The conditions for oscillation given by Equations (2.3) and (2.7) can thus equivalently be expressed by Equations (2.14) and (2.15). These were the conditions originally introduced by Tsypkin. Introduce the function $\Lambda(s)$ defined as

$$(2.16)\Lambda(i\omega) = \frac{4\omega}{\pi} \sum_{n=0}^{\infty} \operatorname{Re} G(i\omega(1 + 2n)) + i\frac{4}{\pi} \sum_{n=0}^{\infty} \frac{\operatorname{Im} G(i\omega(1 + 2n))}{2n + 1}$$

The frequency of a possible oscillation can thus be determined by plotting the locus of $\Lambda(i\omega)$, called the hodograph or the Tsypkin locus, and finding the frequency where the curve intersects the half line

$$\text{Im } s = -\frac{\epsilon}{d}, \quad \text{Re } s \leq 0$$

Notice that if $|G(i\omega)|$ decreases rapidly with increasing ω, the series can be approximated by their first terms and we have approximately

$$H(-1) = \frac{4}{\pi} \sum_{n=0}^{\infty} \frac{\text{Im } G(i\omega(1+2n))}{1+2n} \approx \frac{4}{\pi}\text{Im } G(i\omega)$$

and

$$H_v(-1) = \frac{4\omega}{\pi} \sum_{n=0}^{\infty} \text{Re } G(i\omega(1+2n)) \approx \frac{4\omega}{\pi}\text{Re } G(i\omega)$$

An approximate version of condition (2.3) is thus

$$\text{Im } G(i\omega) = -\frac{\pi\epsilon}{4d}$$

This is the same result as is obtained with the describing function approximation.

3. Stability of the limit cycle. Local stability of the limit cycle will now be investigated. To do this we will calculate the Jacobian of the Poincaré map. We have the following result.

THEOREM 3.1. *Consider the closed loop system given by (2.1) and (2.2). Assume that there is a symmetric periodic solution. Let a be the initial state that generates the periodic motion. The Jacobian of the Poincaré map is given by*

$$(3.1) \qquad W = \left(I - \frac{vC}{Cv}\right)\Phi$$

where $\Phi = e^{Ah}$ *and* $v = Aa + Bd$. *The limit cycle is locally stable if and only if W has all its eigenvalues inside the unit disk.*

Proof. Consider a trajectory with initial condition $x(0) = a$. Let another solution have initial condition $x(0) = a + \delta a$, where δa is chosen to be on the switching plane, i.e. such that

$$C(a + \delta a) = 0.$$

The solution is

$$x(t) = e^{At}(a + \delta a) + \int_0^t e^{A(t-s)} Bu(s)\, ds.$$

Assume that the solution reaches the switching plane at time $h + \delta h$. Then the control signal has the value $u(t) = -d$ for $0 < t < h + \delta h$. Hence

$$x(h + \delta h) = e^{A(h+\delta h)}(a + \delta a) + \int_0^{h+\delta h} e^{A(h+\delta h - s)} B u(s)\, ds.$$

Making a series expansion in δa and δh we get

$$
\begin{aligned}
(3.2) \quad x(h + \delta h) &= \Phi(I + A\delta h)(a + \delta a) - (I + A\delta h)\Gamma d - B d \delta h + \mathcal{O}(\delta^2) \\
&= \Phi a - \Gamma d + \Phi \delta a + A(\Phi a - \Gamma d)\delta h - B d \delta h + \mathcal{O}(\delta^2) \\
&= -a + \Phi \delta a - (Aa + Bd)\delta h + \mathcal{O}(\delta^2) \\
&= -a + \Phi \delta a - v \delta h + \mathcal{O}(\delta^2)
\end{aligned}
$$

where $-v$ is the velocity at time h. The equation has a simple physical interpretation. The perturbation in the trajectory at time $h + \delta h$ is composed of two terms. The first term is the initial perturbation multiplied by the transition matrix. The second term is the state velocity multiplied by δh. Since $x(h + \delta h)$ is on the switching plane, we have

$$Cx(h + \delta h) = -\epsilon.$$

Neglecting terms of order δ^2 this gives

$$C\Phi\delta a = Cv\delta h$$

Notice that it follows from Theorem 1 that $Cv > 0$ if there is a symmetric limit cycle. Hence

$$\delta h = \frac{C\Phi}{Cv} \delta a$$

and Equation (3.2) then becomes

$$x(h + \delta h) = -a + \left(I - \frac{vC}{Cv}\right)\Phi\delta a + \mathcal{O}(\delta a^2)$$

which proves the theorem. □

REMARK 2. *Notice that it follows from Theorem 1 that $Cv > 0$ if a symmetric limit cycle exists.*

REMARK 3. *Notice that the matrix*

$$P = \left(I - \frac{vC}{Cv}\right)$$

has the property $P^2 = P$. It is thus a projection matrix.

REMARK 4. *Notice that C is a left eigenvector of P with eigenvalue 0.*

3.1. Analysis of a given system. Theorems 1 and 2 provide tools for analyzing if a given system has a stable limit cycle. The procedure is simply as follows.

Step 1 Find h such that $f(h) = \epsilon/d$.

Step 2 Compute a, v, W, and check the conditions $Cv > 0$ and $|\lambda(W)| < 1$.

Step 3 Check the condition $y(t) > -\epsilon$ for $0 < t < h$.

3.2. Examples. Some examples will now be considered.

EXAMPLE 1. *Consider a system with the transfer function*

$$G(s) = (s+1)^{-3}$$

Assume that the relay has $d = 1$ and no hysteresis, i.e. $\epsilon = 0$. Tedious but straightforward calculations give

$$f(h) = \frac{1 + (1 - 2h - h^2)e^{-h} - (1 + 2h - h^2)e^{-2h} - e^{-3h}}{(1 + e^{-h})^3}.$$

The shape of this function is shown in Figure 3.1.

The function has only one zero for positive h. Numerical calculations give the zero at $h = 1.8399$. The limit cycle thus has period $T = 6.68$. To investigate the problem further we introduce the following state-space representation

$$A = \begin{pmatrix} -1 & 1 & 0 \\ 0 & -1 & 1 \\ 0 & 0 & -1 \end{pmatrix}, \qquad B = \begin{pmatrix} 0 \\ 0 \\ 1 \end{pmatrix}, \qquad C = (1 \quad 0 \quad 0)$$

Furthermore we get

$$a = \begin{pmatrix} 0 \\ 0.2906 \\ 0.7259 \end{pmatrix}$$

and

$$Cv = 0.2906$$

The Jacobian of the Poincaré map becomes

$$W = \begin{pmatrix} 0 & 0 & 0 \\ -0.2379 & -0.2788 & -0.1104 \\ -0.1498 & -0.2757 & -0.0947 \end{pmatrix}$$

This matrix has the eigenvalues 0, 0.0104 and -0.3840. We can thus conclude that there is a unique symmetric limit cycle, which is stable. From the magnitude of the eigenvalues we can also conclude that convergence towards the limit cycle is quite rapid. The linearized analysis shows that

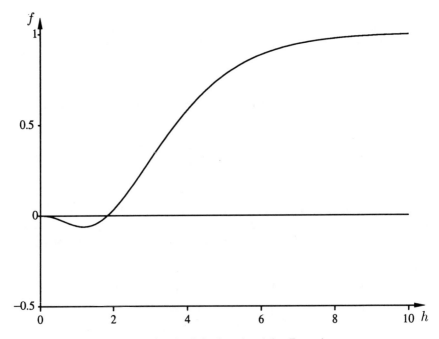

FIG. 3.1. *Graph of the function f for Example 1.*

perturbations are reduced by at least $0.384^4 \approx 0.02$ *after only four switches! In Figure 3.2 we show a simulation of the system under relay feedback, which illustrates the fast convergence.*

Applying the describing function method to Example 1 we obtain the same qualitative result, because the Nyquist curve only intersects the negative real line at one point. This occurs for

$$\arg G(i\omega) = \pi$$

which gives $\omega = \sqrt{3}$. The corresponding period is thus $T = 3.63$, which can be compared with the value $T = 3.68$ obtained by the exact analysis. The next example is a case where there is larger differences between the exact analysis and the describing function analysis.

EXAMPLE 2. *Consider a system with the transfer function*

$$G(s) = \frac{(s + \alpha)^2}{s(s + 1)^2}$$

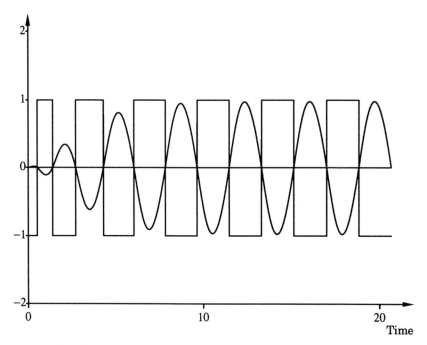

FIG. 3.2. *Results of simulation of the system in Example 1. The relay amplitude is* $d = 1$ *and there is no hysteresis.*

The Nyquist curve of this transfer function intersects the negative real axis at two points if $\alpha > 3 + 2\sqrt{2}$. *The describing function analysis thus gives two possible limit cycles. To investigate the system we introduce the state-space representation*

$$
A = \begin{pmatrix} 0 & 1 & 1 \\ 0 & -1 & \alpha - 1 \\ 0 & 0 & -1 \end{pmatrix}, \quad B = \begin{pmatrix} 1 \\ \alpha - 1 \\ \alpha - 1 \end{pmatrix}, \quad C = \begin{pmatrix} 1 & 0 & 0 \end{pmatrix}
$$

The function $f(h)$ *is shown in Figure 3.3 for* $\alpha = 7$. *The function has zeros for* $h = 0.8076$ *and* $h = 1.9464$. *The velocity condition is satisfied for both solutions but only the second solution corresponds to a stable limit cycle. We can thus conclude that there is only one stable symmetric limit cycle with period* $T = 3.89$.

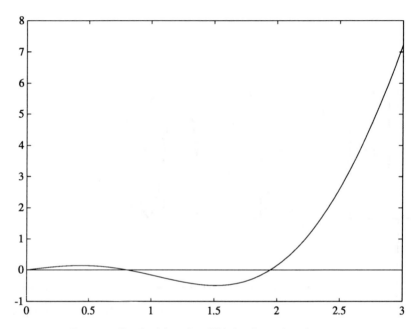

FIG. 3.3. *Graph of function $f(h)$ for Example 2 for $\alpha = 7$.*

EXAMPLE 3. *Consider a system with the transfer function*

$$G(s) = \frac{1}{s^2}$$

Assume that the relay has $d = 1$ and no hysteresis, i.e. $\epsilon = 0$. The pulse transfer function of the system is

$$H(z) = \frac{h^2}{2} \frac{z+1}{(z-1)^2}.$$

Hence

$$f(h) = H(-1) = 0.$$

The function f thus vanishes identically for all values of h. Furthermore we get

$$\Phi = \begin{pmatrix} 1 & h \\ 0 & 1 \end{pmatrix} \quad and \quad \Gamma = \begin{pmatrix} h^2/2 \\ h \end{pmatrix}.$$

Hence

$$a = \begin{pmatrix} 0 \\ h/2 \end{pmatrix} \quad and \quad v = \begin{pmatrix} h/2 \\ 1 \end{pmatrix}.$$

and thus

$$Cv = h/2 \quad and \quad W = \begin{pmatrix} 0 & 0 \\ -2/h & -1 \end{pmatrix}.$$

We thus find that there are limit cycles with any period. The amplitude is arbitrary. The limit cycles are stable but not asymptotically stable.

EXAMPLE 4. *Consider a system with the transfer function*

$$G(s) = \frac{1}{s^2 + 1}.$$

Assume that there is no hysteresis in the relay, i.e. $\epsilon = 0$. Introduce the state-space representation

$$A = \begin{pmatrix} 0 & 1 \\ -1 & 0 \end{pmatrix}, \quad B = \begin{pmatrix} 0 \\ 1 \end{pmatrix}, \quad C = (1\,1 \quad 0).$$

Straightforward calculations give

$$\Phi(h) = \begin{pmatrix} \cos h & \sin h \\ -\sin h & \cos h \end{pmatrix}, \quad \Gamma(h) = \begin{pmatrix} 1 - \cos h \\ \sin h \end{pmatrix}.$$

Hence

$$f(h) = 0.$$

The condition (2.3) thus holds automatically. Furthermore

$$a = \begin{pmatrix} 0 \\ \frac{\sin h}{1 + \cos h} \end{pmatrix} d.$$

This gives

$$y(t) = \left(\frac{\sin h \sin t}{1 + \cos h} - 1 + \cos t \right) d \qquad 0 \le t \le h.$$

For $0 < h < \pi$ we have $y(t) > 0$, which means that condition (2.4) is also satisfied. Notice that $y(t)$ becomes infinite for some $t < h$ if $h \ge \pi$. Furthermore we have

$$W = \begin{pmatrix} 0 & 0 \\ -\frac{1 + \cos h}{\sin h} & -1 \end{pmatrix}.$$

We can thus conclude that there are limit cycles with any period $0 < T < 2\pi$. The limit cycles are stable but not asymptotically stable.

4. Systems with time delays. The results of Sections 2 and 3 will now be extended to systems with time delays. This is of considerable interest, since such models are common. It is assumed that the system is described by

$$(4.1) \qquad \begin{aligned} \frac{dx(t)}{dt} &= Ax(t) + Bu(t - \tau) \\ y(t) &= Cx(t) \end{aligned}$$

This is an infinite dimensional system because of the time delay. The state of the system can be chosen as $x(t)$ and the function $u(s), t - \tau < s \le t$. The state will be simplified significantly, because the control signal is generated by relay feedback. We have the following result.

THEOREM 4.1. *Consider the system (4.1) with the relay (2.2) where* $e = -y$. *Assume that there exists a symmetric limit cycle with period* $T = 2h$. *Let m be the integer part of* τ/h *and* τ_1 *the remainder when dividing* τ *by h. Then the following conditions hold:*

$$(4.2) \qquad (i) \quad f(h) = C(I + \Phi(h))^{-1}(\Gamma_0(h) - \Gamma_1(h))(-1)^m = \frac{\epsilon}{d}$$

$$(4.3) \qquad (ii) \quad y(t) = Cx(t) > -\epsilon \text{ for } 0 \le t < h$$

where

$$\Phi(h) = e^{Ah}$$

$$\Gamma_0(h) = \int_0^{h-\tau_1} e^{As} B \, ds$$

$$\Gamma_1(h) = \int_{h-\tau_1}^h e^{As} B \, ds$$

The periodic solution corresponds to the initial condition

$$(4.4) \qquad x(0) = a = (I + \Phi(h))^{-1}(\Gamma_0(h) - \Gamma_1(h))(-1)^m d$$

Proof. Assume that there exists a limit cycle with period $T = 2h$. Choose time so that $t = 0$ corresponds to the time when the output y reaches ϵ with positive slope. Integrating (4.1) from $t = 0$ to $t = h$ gives

$$x(h) = \Phi(h)a - (\Gamma_0(h) - \Gamma_1(h))(-1)^m d$$

Since the oscillation was symmetric with period 2h, we have $x(h) = -a$, which implies

$$-a = \Phi(h)a - (\Gamma_0(h) - \Gamma_1(h))(-1)^m d$$

Solving this equation with respect to a gives (4.4). Since the output is ϵ at $t = 0$, we have

$$y(0) = Cx(0) = \epsilon$$

which implies (4.2). Since the relay can not switch before $t = h$, the condition (4.3) must also hold. □

REMARK 5. *Let v be the velocity of the state at time $0-$. Then*

$$v = Aa + B(-1)^m d$$

To compute a value of h that satisfies (4.2) it is useful to have analytic expressions for the derivative of the function f. We have:

LEMMA 4.2. *Consider the function f defined by Equation (4.2). Assume that $\tau < h$, then the function and its derivative are given by*

$$f(h) = C(I + \Phi(h))^{-1}(\Gamma_0(h) - \Gamma_1(h))$$
$$\frac{df(h)}{dh} = C(I + \Phi(h))^{-1}\Phi(h)(2e^{-A\tau_1} - I)B$$
$$- C(I + \Phi(h))^{-2}\Phi A(\Gamma_0(h) - \Gamma_1(h))$$

Proof. The proof is obtained by a straightforward calculation. □

4.1. Stability analysis. The analysis of the stability of the limit cycles is complicated for systems with time delays, because the state of the system is infinite dimensioned. The state can, e.g., be chosen as $x(t)$ and $\{u(s), t - \tau \leq s \leq t\}$. The problem is simplified a little because of the relay feedback, which makes the control signal piecewise constant. It is simplified even more if we consider limit cycles with period $T > 2\tau$, because $\{u(s), t - \tau \leq s \leq t\}$ is then constant and can be specified by one real variable. Moreover, with relay feedback this variable is uniquely given by the feedback. In the particular case $T > 2\tau$ the state of the system can thus be represented by $x(t)$. Also notice that in this case we have $m = 0$ and $\tau_1 = \tau$ in Theorem 3. We have the following result.

THEOREM 4.3. *Consider the closed loop system given by (4.1) and (2.2). Assume that there is a symmetric periodic solution whose period is larger than 2τ. Let a be the initial state that generates the periodic motion. The Jacobian of the Poincaré map is given by*

$$(4.5) \qquad\qquad W = \left(I - \frac{vC}{Cv}\right)\Phi$$

where $\Phi = e^{Ah}$ and $v = Aa + Bd$. The limit cycle is locally stable if and only if W has all its eigenvalues inside the unit disk.

Proof. The proof is by a direct calculation. Choose the coordinates so that $t = 0$ corresponds to the time when the relay switches from d to $-d$. Consider a trajectory with initial condition $x(0) = a$. Let another solution have initial condition $x(0) = a + \delta a$, where δa is chosen to be on the switching plane, i.e. such that

$$C(a + \delta a) = 0$$

The solution is

$$x(t) = e^{At}(a + \delta a) + \int_0^t e^{A(t-s)} Bu(s - \tau)\, ds$$

Proceeding in the same way as in the proof of Theorem 2 we find after some calculations:

$$x(h + \delta h) = -a + \Phi \delta a - v \delta h + \mathcal{O}(\delta^2)$$

where $-v$ is the velocity at time h. Since $x(h + \delta h)$ is on the switching plane, we have

$$Cx(h + \delta h) = -\epsilon.$$

This gives

$$C\Phi \delta a = Cv \delta h.$$

Notice that it follows from Theorem 3 that $Cv > 0$ if there is a symmetric limit cycle. Hence

$$\delta h = \frac{C\Phi}{Cv} \delta a$$

and we get

$$x(h + \delta h) = -a + \left(I - \frac{vC}{Cv} \right) \Phi \delta a + \mathcal{O}(\delta a^2)$$

which proves the theorem. $\square$

4.2. Examples. Limit cycles in systems with time delays will now be illustrated by a number of examples.

EXAMPLE 5. *Consider the system*

(4.6) $$\frac{dx(t)}{dt} = \alpha x(t) + \beta u(t - \tau)$$

The limit cycle obtained is illustrated in Figure 4.1, which shows a simulation of a system with the transfer function

$$G(s) = 2 \frac{e^{-s\tau}}{s + 1}$$

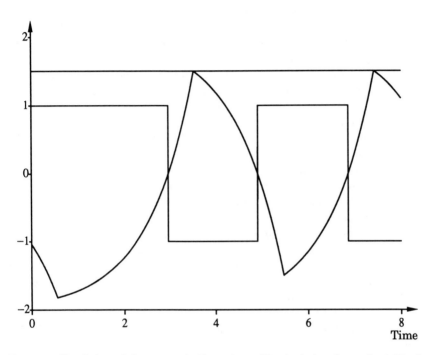

FIG. 4.1. *Simulation of the system in Example 5. The dead time is $\tau = \log 1.75$, the relay amplitude is $d = 1$ and there is no hysteresis.*

where $\tau = \log 1.75 = 0.56$. The relay amplitude is $d = 1$ and there is no hysteresis. Let us investigate limit cycles with periods larger than 2τ. For $h > \tau$ we have

$$\Phi = e^{\alpha h}, \qquad \Gamma_0 = \beta \frac{e^{\alpha(h-\tau)} - 1}{\alpha}, \qquad \Gamma_1 = \beta \frac{e^{\alpha h} - e^{\alpha(h-\tau)}}{\alpha}$$

These expressions also hold for $\alpha = 0$ if the values are interpreted as limits when α goes to zero. Hence

$$f(h) = \beta \frac{2e^{\alpha(h-\tau)} - e^{\alpha h} - 1}{\alpha(1 + e^{\alpha h})}.$$

If the relay has no hysteresis, i.e. $\epsilon = 0$ we find that the function f has a *unique zero if $\alpha\tau < \log 2$. The zero is given by*

(4.7) $$h = \frac{1}{\alpha}\log\frac{1}{(2e^{-\alpha\tau}-1)}$$

for $\alpha \neq 0$ and $\alpha\tau < \log 2$. For $\alpha = 0$ the zero is obtained for $h = 2\tau$. Limit cycle oscillations are thus always possible if the system (4.6) is stable. If the system is unstable, limit cycles can occur only if the dead-time is not too large in relation to the unstable pole. The admissible delay will be smaller if the relay has hysteresis.

The limit cycle corresponds to the initial condition $x = 0$ and $dx/dt = d$. If the system is unstable ($\alpha > 0$) and if the initial condition is too large in magnitude there will not be any limit cycles, because the relay feedback is unable to keep $|x|$ from growing. Since the system is of first order, we have automatically $W = 0$. This means that the limit cycle is locally stable. Since $W = 0$, convergence to the limit cycle will occur after one switch in the linear range. If the initial condition is such that $|x(0)| < 1/\alpha$ then the limit cycle will be obtained after one switch. This is easily seen from Equation (4.7), because at the first switch we have $x = 0$, which corresponds to the limit cycle.

For the simulation shown in Figure 4.1 a limit cycle exists if the initial state is such that $|x(0)| < d = 1$, since $\log\tau = 1.75 < 2$. The peak amplitude of the limit cycle oscillation is $2(e^\tau - 1) = 1.5$. In the simulation we have chosen the initial value as $x(0) = -0.52$. The line corresponding to $x(t) = 1.5$ is also shown. The rapid convergence of the limit cycle is clearly seen in the figure.

Next we will consider a special case of Example 5, which is of particular interest.

EXAMPLE 6. *Consider the special case of Example 5 when $\alpha = 0$. The system then has the transfer function*

$$G(s) = \frac{e^{-s\tau}}{s}$$

It follows from Equation (4.7) that $h = 2\tau$, which implies that the period of the limit cycle is $T = 4\tau$. Since the dynamics is an integrator, the limit cycle is a triangular wave as is shown in Figure 4.2. Notice the rapid convergence towards the limit cycle.

The describing function condition gives the following condition for the frequency of the limit cycle.

$$\arg G(i\omega_{osc}\tau) = -\pi.$$

This gives

$$\omega_{osc}\tau + \frac{\pi}{2} = \pi.$$

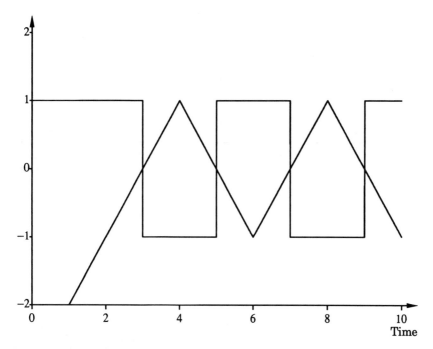

FIG. 4.2. *Simulation of the system in Example 6. The dead time is $\tau = 1$, the relay amplitude is $d = 1$ and there is no hysteresis.*

Hence $\omega_{osc} = \pi/(2\tau)$ *and the period becomes $T = 4\tau$. In the particular case when the dynamics is an integrator with a time delay the describing function analysis will thus give the same result as the exact analysis.*

In Examples 5 and 6 we were only considering limit cycles with periods larger than 2τ. It follows from Theorem 3 that there are many limit cycles with shorter periods. This is illustrated by an additional example.

EXAMPLE 7. *Consider the system*

$$\frac{dy}{dt} = u(t - \tau).$$

Assuming that $\tau = 1.25$ and $h = 0.5$ we get $m = 2$ and $\tau_1 = 0.25$. Furthermore

$$\Phi = 1 \qquad and \qquad \Gamma_0(h) = \Gamma_1(h) = 0.25.$$

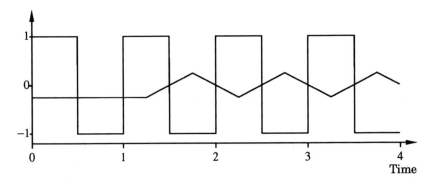

FIG. 4.3. *A limit cycle with period* $2h = 1$ *for a system with the transfer function* $G(s) = \frac{1}{s}e^{-1.25s}$.

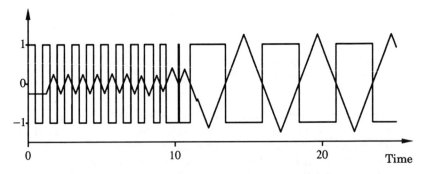

FIG. 4.4. *Simulation of the system in Figure 4.3 over a longer time.*

The condition (4.2) is thus satisfied for a relay without hysteresis. By proper initialization it is also possible to satisfy the condition (4.3) and we thus find that there is a limit cycle with period $2h = 1 < \tau$. Figure 4.3 shows a simulation of the system. In this simulation the system has been initialized at time $t = 1.5$. Past values of the control signal have been chosen to correspond to the limit cycle. The figure shows that there is a limit cycle with period 1, which is shorter than τ. A calculation of the Jacobian of the Poincaré map shows that the limit cycle is unstable. This is illustrated in Figure 4.4, where the simulation in Figure 4.3 has been continued for a longer time. The figure shows that the solution converges to a stable limit cycle with period $4\tau = 5$.

This example illustrates that complicated and unexpected behavior can be obtained for systems with time delay. It is easy to see that we can obtain many limit cycles in this case.

5. Asymmetric oscillations. It is of interest to also consider the case of asymmetric oscillations. Let the relay be characterized by

$$(5.1) \qquad u(t) = \begin{cases} d_1 & \text{if } e > \epsilon \text{ or } (e > -\epsilon \text{ and } u(t-) = d_1) \\ -d_2 & \text{if } e < -\epsilon \text{ or } (e < \epsilon \text{ and } u(t-) = -d_2) \end{cases},$$

where $e = -y$. We have

THEOREM 5.1. *Consider the system (2.1) with the feedback law (5.1), where $e = -y$. Assume that the matrix $\Phi - I$ is regular. Necessary conditions for a limit cycle with period T are*

$$(5.2) \qquad \begin{cases} C(I - \Phi)^{-1}(\Phi_2\Gamma_1 d_1 - \Gamma_2 d_2) = -\epsilon \\ C(I - \Phi)^{-1}(-\Phi_1\Gamma_2 d_2 + \Gamma_1 d_1) = \epsilon \end{cases},$$

where

$$(5.3) \qquad \begin{aligned} \Phi = e^{AT} \quad \Phi_1 = e^{A\tau} \quad \Phi_2 = e^{A(T-\tau)} \\ \Gamma_1 = \int_0^\tau e^{As}\,ds\,B \quad \Gamma_2 = \int_0^{T-\tau} e^{As}\,ds\,B \end{aligned}$$

Proof. Assume that a limit cycle exists, where the relay switches twice per period. Let the switching times be $\{t_k\}$, where even indices correspond to up-crossings of y and odd indices correspond to down-crossings. Integration of the state equations over one period gives

$$(5.4) \qquad \begin{aligned} x(t_{2k+1}) &= \Phi_1 x(t_{2k}) + \Gamma_1 u(t_{2k}) \\ x(t_{2k+2}) &= \Phi_2 x(t_{2k+1}) + \Gamma_2 u(t_{2k+1}) \end{aligned}$$

where the matrices Φ_1, Φ_2, Γ_1, and Γ_2 are given by (5.3). Notice that the matrices Φ_1 and Φ_2 commute. The condition that the relay switches at times t_k can be expressed as

$$\begin{aligned} y(t_{2k}) &= Cx(t_{2k}) = -\epsilon \\ y(t_{2k+1}) &= Cx(t_{2k+1}) = \epsilon \end{aligned}$$

If a limit cycle exists then the state will be a periodic function. Introducing

$$\begin{aligned} x(t_{2k+1}) &= x(t_{2k-1}) \\ x(t_{2k+2}) &= x(t_{2k}) \\ u(t_{2k}) &= d_1 \\ u(t_{2k+1}) &= -d_2 \end{aligned}$$

into (5.4) then gives

$$(5.5) \qquad \begin{aligned} x(t_{2k+1}) &= a_1 = (I - \Phi)^{-1}(-\Phi_1\Gamma_2 d_2 + \Gamma_1 d_1) \\ x(t_{2k+2}) &= a_2 = (I - \Phi)^{-1}(\Phi_2\Gamma_1 d_1 - \Gamma_2 d_2) \end{aligned}$$

and (5.2) after straightforward calculations. □

Notice that the conditions (5.2) is a transcendental equation in the variables T and τ. To find possible oscillations we must thus find the solutions to Equation (5.5). We then have to proceed to check additional conditions.

5.1. Frequency domain interpretations. It is intuitively appealing to reformulate the result in the frequency domain. The key observation is that under limit cycle conditions the behavior of the system (2.1) can be described as a multivariable linear time-invariant discrete time system (the stroboscopic transformation). The inputs are $u(t_{2k})$ and $u(t_{2k+1})$ and the outputs are $y(t_{2k+1})$ and $y(t_{2k+2})$. Introduce

$$z_k = \begin{pmatrix} x(t_{2k-1}) \\ x(t_{2k}) \end{pmatrix}, \quad u_k = \begin{pmatrix} u(t_{2k}) \\ u(t_{2k+1}) \end{pmatrix}, \quad y_k = \begin{pmatrix} Cx(t_{2k-1}) \\ Cx(t_{2k}) \end{pmatrix}$$

Equation (5.4) can then be written as

$$z_{k+1} = \begin{pmatrix} 0 & \Phi_1 \\ 0 & \Phi \end{pmatrix} z_k + \begin{pmatrix} \Gamma_1 & 0 \\ \Phi_2\Gamma_1 & \Gamma_2 \end{pmatrix} u_k$$
$$y_k = \begin{pmatrix} C & 0 \\ 0 & C \end{pmatrix} z_k$$

This is a time-invariant discrete time system. Let the pulse transfer function of the system be

$$(5.6) \qquad H(z) = \begin{pmatrix} C(zI - \Phi)^{-1}\Gamma_1 & z^{-1}C(zI - \Phi)^{-1}\Phi_1\Gamma_2 \\ C(zI - \Phi)^{-1}\Phi_2\Gamma_1 & C(zI - \Phi)^{-1}\Gamma_2 \end{pmatrix}$$

Putting $z = 1$ in (5.6) it follows that the condition (5.2) can be written as

$$(5.7) \qquad H(1) \cdot \begin{pmatrix} d_1 \\ -d_2 \end{pmatrix} = \begin{pmatrix} \epsilon \\ -\epsilon \end{pmatrix}$$

5.2. Stability of the limit cycle. Conditions for local stability of the limit cycle given by Theorem 4 will now be explored. The following result holds.

THEOREM 5.2.   *Consider the system (2.1) with the feedback (5.1). Assume that the matrix $\Phi - I$ is regular and that τ and T are such that (5.2) is satisfied. Let the matrix*

$$(5.8) \qquad W = \left(I - \frac{w_2 C}{Cw_2} \right) \Phi_2 \left(I - \frac{w_1 C}{Cw_1} \right) \Phi_1$$

where

$$\begin{aligned} w_1 &= \Phi_1(Aa_1 + Bd_1) \\ w_2 &= \Phi_2(Aa_2 - Bd_2) \end{aligned}$$

$$a_1 = (I - \Phi)^{-1}(\Phi_2\Gamma_1 d_1 - \Gamma_2 d_2)$$
$$a_2 = (I - \Phi)^{-1}(-\Phi_1\Gamma_2 d_2 + \Gamma_1 d_1)$$

has all its eigenvalues inside the unit disc. The limit cycle is then locally stable.

Proof. It follows from the proof of Theorem 4.3 that

$$x(t_{2k}) = a_1 \quad \text{and} \quad x(t_{2k+1}) = a_2$$

for the limit cycle. Compare with Equation (5.5). Consider a solution to (2.1) and (5.1), where the initial condition is perturbed from a_1 to $a_1 + \delta a_1$. Hence

$$(5.9) \quad x(\tau + \delta\tau) = e^{A(\tau+\delta\tau)}(a_1 + \delta a_1) + \int_0^{\tau+\delta\tau} e^{As} ds Bd_1$$

$$= \Phi_1 a_1 + \Gamma_1 d_1 + \Phi_1 \delta a_1 + \Phi_1(Aa_1 + Bd_1)\delta\tau + \mathcal{O}(\delta^2)$$

where $\mathcal{O}(\delta^2)$ denotes terms of second and higher order in δ. The value of $\delta\tau$, where the control signal switches, is given by

$$y(\tau + \delta\tau) = Cx(\tau + \delta\tau) = \epsilon.$$

Hence

$$C\Phi_1 \delta a_1 + C\Phi_1(Aa_1 + Bd_1)\delta\tau + \mathcal{O}(\delta^2) = 0$$

or

$$\delta\tau = \frac{-C\Phi_1 \delta a_1}{C\Phi_1(Aa_1 + Bd_1)} + \mathcal{O}(\delta^2) = -\frac{C\Phi_1 \delta a_1}{Cw_1} + \mathcal{O}(\delta^2)$$

Equation (5.9) can then be written as

$$x(\tau + \delta\tau) = a_2 + \left(I - \frac{w_1 C}{Cw_1}\right)\Phi_1 \delta a_1 + \mathcal{O}(\delta^2)$$

Repeating the same analysis for the time interval, where $u(t) = -d_2$, we find that the relay switches at $T + \delta T$ and that

$$x(T + \delta T) = x(t_{2k+2}) + W \cdot \delta a_1 + \mathcal{O}(\delta^2)$$

For small perturbations in the initial conditions the changes in the state at the switching instants are thus governed by the difference equation

$$\delta a_{2k+2} = W \cdot \delta a_{2k}$$

The sequence $\{\delta a_{2k}\}$ then converges to zero exponentially to zero, since W is a constant matrix with eigenvalues inside the unit disc. $\square$

6. Conclusions. Relay feedback has played a role in adaptive control for a long time. Lately it has been used for automatic tuning and for initialization of adaptive controllers, see [3] and [17]. In this paper we have analyzed limit cycles and their stability in simple cases. Particular emphasis has been given to symmetric limit cycles.

It should be observed that linear systems with relay feedback can exhibit very complicated behavior. In [1] it is, e.g., shown that a second order oscillatory system with time delay can have different stable limit cycles depending on the initial conditions. In [13] it is shown that very complicated limit cycles can be obtained in similar systems. An example with a limit cycle, where the relay switches 12 times in a period, is given. Cook has found chaotic behavior for an unstable second order linear system, [6]. Another example of chaotic behavior is given in [13].

The cases where complicated behavior has been observed have all been where the linear system has poorly damped or unstable oscillatory poles. This is in strong contrast with nonoscillatory poles, where unique stable limit cycles often are found. In spite of the progress made it appears to be many interesting unsolved research problems for systems with relay feedback.

7. Acknowledgements. The work reported in this paper was partially supported by the Swedish Research Council for Engineering Sciences under contract 91-721. The paper was written during a one month stay at the Institute of Mathematics and Its Applications at University of Minnesota. This support is gratefully acknowledged.

REFERENCES

[1] ARZÉN, K. E., *Realization of Expert System Based Feedback Control*, PhD thesis TFRT-1029, Department of Automatic Control, Lund Institute of Technology, Lund, Sweden (1987).

[2] ÅSTRÖM, K.J. AND T. HÄGGLUND, "Automatic tuning of simple regulators with specifications on phase and amplitude margins," *Automatica*, **20** (1984), pp. 645–651.

[3] ÅSTRÖM, K.J. AND T. HÄGGLUND, *Automatic Tuning of PID Controllers*, Instrument Society of America, Research Triangle Park, North Carolina (1988).

[4] ÅSTRÖM, K.J. AND B. WITTENMARK, *Computer Controlled Systems–Theory and Design*, Prentice-Hall, Englewood Cliffs, New Jersey, second edition (1990).

[5] ATHERTON, D.P., *Nonlinear Control Engineering–Describing Function Analysis and Design*, Van Nostrand Reinhold Co., London, UK (1975).

[6] COOK, P.A., "Simple feedback systems with chaotic behaviour," *Systems and Control Letters*, **6** (1985), pp. 223–227.

[7] ECKMAN, D.P., "Phase-plane analysis. A general method of solution for two-position process control," *Trans. ASME*, **76** (1954), pp. 109–116.

[8] GELB, A. AND W.E.V. VELDE, *Multiple-Input Describing Functions and Nonlinear System Design*, McGraw-Hill, New York (1968).

[9] HÄGGLUND, T. AND K.J. ÅSTRÖM, "Industrial adaptive controllers based on frequency response techniques," *Automatica*, **27** (1991), pp. 599–609.

[10] HAMEL, B., "Étude mathématique des systèmes à plusieurs degrès de liberté décrits

par des équations linéaires avec un terme de commande discontinu," in *Proceedings of the Journées d'Études des Vibrations A.E.R.A.*, Paris, France (1950).

[11] HAMEL, B., "A mathematical study of on-off controlled higher order systems," in *Proceedings of the Symposium on Nonlinear Circuit Analysis, Polytechnic Institute of Brooklyn*, New York, volume 6 (1956), pp. 225–232.

[12] HAZEN, H.L., "Theory of servomechanisms," *J. Franklin Institute*, **218** (1934), pp. 279–330.

[13] HOLMBERG, U., *Relay Feedback of Simple Systems*, PhD thesis TFRT-1034, Department of Automatic Control, Lund Institute of Technology, Lund, Sweden (1991).

[14] HOROWITZ, I.M., "Comparison of linear feedback systems with self-oscillating adaptive systems," *IEEE Transactiosn on Automatic Control*, **AC-9** (1964), pp. 386–392.

[15] HOROWITZ, I.M., J.W. SMAY, AND A. SHAPIRO, "A synthesis theory for self-oscillating adaptive systems (soas)," *Automatica*, **10** (1974), pp. 381–392.

[16] LOZIER, J.C., "Carrier-controlled relay servos," *Elec. Eng.*, **69** (1950), pp. 1052–1056.

[17] LUNDH, M. AND K.J. ÅSTRÖM, "Automatic initialization of robust adaptive controllers," in *Preprints 4th IFAC Symposium on Adaptive Systems in Control and Signal Processing*, Grenoble (1992), pp. 439–444.

[18] MAEDA, H. AND S. KODAMA, "Stability of a relay servosystem," *IEEE Transactions on Automatic Control*, **14** (1969), pp. 555–558.

[19] MAEDA, H. AND S. KODAMA, "A new stability criterion of a relay servosystem," *IEEE Transactions on Automatic Control*, **15** (1970), pp. 275–276.

[20] MEES, A.I., *Dynamics of Feedback Systems*, John Wiley & Sons, New York (1981).

[21] SCHUCK, O.H., "Honeywell's history and philosophy in the adaptive control field," in *Proc. Self Adaptive Flight Control Symposium*, Wright Patterson AFB, Ohio. Wright Air Development Center, P.C. Gregory, ed. (1959).

[22] TSYPKIN, YA.Z., "Theory of intermittent control," *Avtomatika i Telemekhanika*, Vol. **10** (1949) Part I, pp. 189–224; Part II, pp. 342–361; Vol. 11 (1950) Part III, pp. 300–331.

[23] TSYPKIN, YA.Z., "Stability and self-oscillations in relay systems of automatic control," *Trudy LKVVIA*, No. **32** (1950), pp. 38–66.

[24] TSYPKIN, YA.Z., "On the stability of periodic modes in relay systems of automatic control," *Avtomatika i Telemekhanika*, **14:5** (1953), pp. 638–646.

[25] TSYPKIN, YA.Z., "On the determination of steady-state oscillations of on-off feedback systems," *Trans. IRE Circuit Theory*, **CT-9** (1962), p. 279.

[26] TSYPKIN, YA.Z., *Relay Control Systems*, Cambridge University Press, Cambridge, UK (1984).

[27] WEISS, H.K., "Analysis of relay servomechanisms," *J. Aeronautical Sciences*, **13** (1946), pp. 364–376.

COMPATIBILITY OF STOCHASTIC AND WORST CASE SYSTEM IDENTIFICATION: LEAST SQUARES, MAXIMUM LIKELIHOOD AND GENERAL CASES*

ER-WEI BAI[†] AND MARK S. ANDERSLAND[†]

Abstract. Stochastic and worst case system identification are different and are usually treated separately. We believe that under certain assumptions there exist estimates of unknown systems that are near optimal from both the stochastic and worst case points of view. This paper studies some algorithms that produce such estimates. The algorithms combine a classical least squares or maximum likelihood estimate with a projection. It is shown that the modified estimates are closer to the true system than the least squares and maximum likelihood estimates, and that they are convergent and near optimal in the worst case setting. It is also shown that these results extend to more general cases.

1. Introduction. There is a vast literature studying system identification problems from a stochastic point of view (see [6]). In this setting, measurement data is assumed to be contaminated by random noise and certain statistical information about the "unknown" noise is required to be available a priori. Under these assumptions, an estimate is derived which possesses nice statistical properties. Simply put, the derived model is a good estimate of the true system "on average." Moreover, for a particular experiment, it is "likely to be good" with some confidence. However, no explicit error bounds can be derived. In fact, the actual error between the estimate and the true system could be unbounded in the worst case, although this is unlikely.

An alternative approach, called worst case identification, is a deterministic one (see [2]). In this setting, the noise is considered to be deterministic and an upper bound on the noise is assumed to be known a priori. This worst case approach allows for some modeling uncertainties in identification, and is based on the assumption that the noise is very malicious and that whatever identification algorithm is used, the worst possible noise will occur. Hence, the model derived by this approach is conservative compared with that of the stochastic approach. However, it guards against the most difficult case, which even if unlikely, can happen. Thus a worst case approach commits, on average, an error that is likely to be larger than that of a stochastic one, despite the fact that the error produced by the worst case approach is guaranteed to be within some known bounds. On the other hand, there exists some noise for which the error produced by a stochastic approach would be much larger than that of a worst case approach.

---

* The authors would like to thank Professor G. Gu of LSU and Professor R. Tempo of CNR for their comments on the paper. This research supported in part by NSF Grant ECS-9011359, and DARPA Grant N00174-91-C-0116.

† Department of Electrical and Computer Engineering, The University of Iowa, Iowa City, IA 52242-1595. phone: (319) 335-5198.

Moreover, this error could be unbounded in the worst case scenario.

These two approaches are obviously based on different philosophies. In the literature they are usually treated differently and separately. Motivated by the work of [7,8] we believe that these two approaches are not necessarily incompatible but are instead complementary. In general, an optimal estimate in the stochastic sense is not worst case optimal and may in fact diverge. Similar comments can be made about worst case optimal estimates. We believe, however, that

> *Under certain assumptions, there exist estimates that are near-optimal from both the stochastic and worst case points of view.*

Roughly speaking, under certain assumptions there exist identification algorithms that produce estimates that are "at least as good" as the ones derived by a stochastic approach and at the same time are guaranteed to satisfy error bounds for the worst case scenario.

In this paper, we study such algorithms by investigating projections of simple least squares and maximum likelihood estimates. The idea is as follows. Given experimental data, we first use either a classical least squares or a maximum likelihood estimator to determine an estimate. This estimate is optimal in some stochastic sense, however no explicit error bound can be given and the actual error can be unbounded in worst case scenario. We then modify this estimate based on some prior information. It is shown that the modified estimate is "closer to the true system" than the unmodified estimate. Next, we study the modified estimate in a worst case setting. It is shown that it is convergent in the worst case scenario. In other words, the final estimate derived by modifying the one obtained from the least squares method or the maximum likelihood estimator is "closer to the true system", and is also convergent and near-optimal in a worst case setting. We conclude the paper by showing that these results extend to more general cases.

2. Problem formulation. The class of systems under consideration consists of stable single-input-single-output (SISO) time invariant discrete time systems described by the convolution $y = h * u$, where u, y, and h are, respectively, the system input, output, and impulse response. The system transfer function $G(z)$ is defined through its impulse response sequence $H = (h(0), h(1), \ldots)^T$, i.e.,

$$(2.1) \qquad G(z) = \sum_{i=0}^{\infty} h(i) z^{-i} .$$

In the paper, we identify each system either by its impulse response H or by its transfer function $G(z)$. Thus, with a slight abuse of notation, we have $H = G(z)$. Let S be the collection of all exponentially stable systems of the form (2.1). Let $B_{\rho,M}$,

$$B_{\rho,M} = \left\{ G(z) = \sum_{i=0}^{\infty} h(i)z^{-i} \mid |h(i)| \leq M\rho^i, \ M < \infty, \ 0 < \rho < 1 \right\} ,$$
(2.2)

denote a subset of S.

The experimental data consists of a finite set of measurements of the corrupted system frequency response,

(2.3) $D^N(H,\eta) = \left\{ d(k) = G(e^{-j\frac{2\pi}{N}k}) + \eta(k) \mid , \ k = 0, \ldots, N-1 \right\} ,$

where $G(e^{-j\frac{2\pi}{N}k})$ is the frequency response of the system at the normalized frequency $\frac{K}{N} \to e$ and $\eta(k)$ models the effects of noise.

An identification algorithm $\mathcal{A}$ can be regarded as a mapping from the data $D^N(H,\eta)$ to a finite order approximation $\widehat{H}_m$ of H or $\widehat{G}_m(z)$ of $G(z)$, i.e.,

(2.4) $$\mathcal{A}(D^N(H,\eta)) = \widehat{G}_m(z) = \sum_{i=0}^{m-1} \widehat{h}(i)z^{-i} ,$$

where $m \leq N$ is the order of the approximation (dependent on N) and $\widehat{H}_m = (\widehat{h}(0), \widehat{h}(1), \ldots, \widehat{h}(m-1))^T$ are the estimates of the first m components of H, i.e. $H_m = (h(0), \ldots, h(m-1))^T$. In a stochastic setting, the noise $\eta(k)$ is assumed to be a random variable. Some statistical properties of $\eta(k)$ are assumed to be available and an estimate is derived in some optimal sense based on these statistical properties. No explicit error bounds can be derived. In a worst case setting, $\eta(k)$ is assumed to be a deterministic noise that is bounded by some known level ξ but is otherwise unknown. An estimate is derived together with an explicit worst case error bound. The estimate is said to be convergent if this worst case error bound converges to zero as the noise level ξ goes to zero and the number of data points N goes to infinity.

3. Least squares estimators

3.1. Least square methods. The least squares method is a classical stochastic approach to determining $\widehat{H}_m = (\widehat{h}(0), \ldots, \widehat{h}(m-1))^T$, and consequently, the estimate $\widehat{G}_m(z)$. To use the least squares method it is assumed that the unknown system is a member of $B_{\rho,M}$ of (2.2) for some ρ and M. The noise $\eta(k)$, $k = 0, 1, \ldots, N-1$, is assumed to consist of N independent, identically distributed, complex random variables with zero mean and variance σ^2. Under these assumptions, the data points in $D^N(H,\eta)$ can be written as

$$d(k) = h(0)e^{-j\frac{2\pi}{N}k\cdot 0} + h(1)e^{-j\frac{2\pi}{N}k\cdot 1} + \cdots + h(m-1)e^{-j\frac{2\pi}{N}k\cdot(m-1)}$$

(3.1) $$+ \sum_{i=m}^{\infty} h(i)e^{-j\frac{2\pi}{N}k\cdot i} + \eta(k) \qquad k = 0, 1, \ldots, N-1 ,$$

i.e.,

(3.2) $$D = AH_M + R_N + \eta_N$$

where

$$D = \begin{bmatrix} d(0) \\ \vdots \\ d(N-1) \end{bmatrix},$$

(3.3) $$A = \begin{bmatrix} 1 & 1 & 1 & \cdots & 1 \\ 1 & e^{-j\frac{2\pi}{N}} & e^{-j\frac{2\pi}{N}2} & \cdots & e^{-j\frac{2\pi}{N}(m-1)} \\ \vdots & \vdots & \ddots & \ddots & \vdots \\ 1 & e^{-j\frac{2\pi}{N}(N-1)} & \cdots & \cdots & e^{-j\frac{2\pi}{N}(N-1)(m-1)} \end{bmatrix},$$

(3.4) $$H_m = \begin{bmatrix} h(0) \\ \vdots \\ h(m-1) \end{bmatrix}, \quad \eta_N = \begin{bmatrix} \eta(0) \\ \vdots \\ \eta(N-1) \end{bmatrix},$$

and

(3.5) $$R_N = \begin{bmatrix} r(0) \\ \vdots \\ r(N-1) \end{bmatrix} = \begin{bmatrix} \sum_{i=m}^{\infty} h(i)e^{-j\frac{2\pi}{N}0i} \\ \vdots \\ \sum_{i=m}^{\infty} h(i)e^{-j\frac{2\pi}{N}(N-1)i} \end{bmatrix}.$$

The objective of the least squares method is to find an estimate $\widehat{H}_m$ of H_m that minimizes $(D - A\widehat{H}_m)^*(D - A\widehat{H}_m)$, where $*$ denotes conjugate transpose. It is well known that the solution of the least square method is given by

(3.6) $$\widehat{H}_m = \begin{bmatrix} \widehat{h}(0) \\ \vdots \\ \widehat{h}(m-1) \end{bmatrix} = (A^*A)^{-1}A^*D$$

when $N \geq m$. It can be easily verified that $(A^*A)^{-1} = \frac{1}{N}I$ so that

(3.7) $$\widehat{h}(i) = \frac{1}{N} \sum_{k=0}^{N-1} d(k)e^{j\frac{2\pi}{N}ki}, \qquad i = 0, 1, \ldots, m-1.$$

Note that this is precisely the form of an N point discrete Fourier transform (DFT). As such, we see that it is computationally efficient since the essential operation can be implemented as a fast Fourier transform (FFT).

To establish the estimate's statistical properties, note that

$$\widehat{h}(i) \;=\; \frac{1}{N}\sum_{k=0}^{N-1}\left(\sum_{\ell=0}^{m-1} h(\ell)e^{-j\frac{2\pi}{N}k\ell}\right)e^{j\frac{2\pi}{N}ki} + \frac{1}{N}\sum_{k=0}^{N-1}\left(r(k)+\eta(k)\right)e^{j\frac{2\pi}{N}ki}$$

$$(3.8) \quad =\; h(i)+\frac{1}{N}\sum_{k=0}^{N-1}r(k)e^{j\frac{2\pi}{N}ki}+\frac{1}{N}\sum_{k=0}^{N-1}\eta(k)e^{j\frac{2\pi}{N}ki}\;,$$

where $r(k)$ denotes the unmodeled dynamics as in (3.5). Let

$$(3.9) \qquad\qquad \widehat{G}_m(z) = \sum_{i=0}^{m-1}\widehat{h}(i)z^{-i}\;.$$

Because

$$(3.10) \qquad\qquad |r(k)| \le \frac{M\rho^m}{1-\rho} \qquad \text{for each } k\;,$$

it is easy to see that,

$$(3.11) \qquad\qquad \left|E\widehat{H}_m - H_m\right| \le \frac{M\rho^m}{1-\rho}\;,$$

and therefore

$$\left\|E\widehat{G}_m(z)-G(z)\right\|_\infty \le \left\|E\widehat{G}_m(z)-G(z)\right\|_1 \longrightarrow 0\;, \qquad \text{as } N \ge m \to \infty\;,$$
(3.12)
where E is the expectation operator, and $\|\cdot\|_\infty$ and $\|\cdot\|_1$ are the standard infinity and 1 operator norms. Thus the least squares method is asymptotically unbiased in both the parameter estimate $\widehat{H}_m$ and the transfer function estimate $\widehat{G}_m(z)$. Note that the least squares method is only asymptotically unbiased because some model mismatch is allowed in identification—i.e., the assumed order m of the estimate is lower than that of the actual system $G(z)$ which could be of infinity. If m is the order of the actual system $G(z)$, then the least squares method is unbiased.

Now let us study the consistency of the least squares method.

LEMMA 3.1. *Assume that N is chosen such that $\frac{N}{m} \to \infty$ as $m \to \infty$. Then*

$$(3.13) \qquad\qquad E\left\|\widehat{G}_m(z)-G(z)\right\|_\infty \longrightarrow 0 \text{ as } m \to \infty\;.$$

Proof. Note that

$$\left|\widehat{G}_m-G\right|^2 = \left[\sum_{i=0}^{m-1}(\widehat{h}(i)-h(i))e^{-j\omega i}\right]\left[\sum_{i=0}^{m-1}(\widehat{h}(i)-h(i))e^{-j\omega i}\right]^{*}$$

$$- \left[\sum_{i=0}^{m-1} (\widehat{h}(i) - h(i))e^{-j\omega i} \right] \left[\sum_{i=m}^{\infty} h(i)e^{-j\omega i} \right]^{*}$$

$$- \left[\sum_{i=m}^{\infty} h(i)e^{-j\omega i} \right] \left[\sum_{i=0}^{m-1} (\widehat{h}(i) - h(i))e^{-j\omega i} \right]^{*}$$

$$(3.14) \qquad + \left[\sum_{i=m}^{\infty} h(i)e^{-j\omega i} \right] \left[\sum_{i=m}^{\infty} h(i)e^{-j\omega i} \right]^{*} .$$

After tedious but straightforward manipulation, it can be verified that

$$(3.15) \qquad \left| E \left[\sum_{i=0}^{m-1} (\widehat{h}(i) - h(i))e^{-j\omega i} \right] \left[\sum_{i=0}^{m-1} (\widehat{h}(i) - h(i))e^{-j\omega i} \right]^{*} \right| \leq \frac{m}{N}\sigma^2$$

$$+ \frac{M^2 m^2 \rho^{2m}}{(1 - \rho)^2}$$

$$(3.16) \qquad \left| E \left[\sum_{i=0}^{m-1} (\widehat{h}(i) - h(i))e^{-j\omega i} \right] \left[\sum_{i=m}^{\infty} h(i)e^{-j\omega i} \right]^{*} \right| \leq \frac{M^2 m \rho^{2m}}{(1 - \rho)^2}$$

$$(3.17) \qquad \left| E \left[\sum_{i=m}^{\infty} h(i)e^{-j\omega i} \right] \left[\sum_{i=m}^{\infty} h(i)e^{-j\omega i} \right]^{*} \right| \leq \frac{M^2 \rho^{2m}}{(1 - \rho)^2} .$$

Taking the expectation of (3.14), applying the triangle inequality, and using (3.15)–(3.17), it follows that

$$(3.18) \qquad E \left| \widehat{G}_m(e^{j\omega}) - G(e^{j\omega}) \right|^2 \longrightarrow 0 \text{ as } m \to \infty .$$

This completes the proof.   □

Remark 3.2. The least squares method is asymptotically unbised and consistent even in the presence of model uncertainties.

Remark 3.3. When $m = N$, the least squares method is exactly the N-point Lagrange interpolation method [2]. Because N is allowed to be much larger than m here, the statistical properties of the least squares method are superior to those of the Lagrange method. For instance, the average squared error of (3.18) goes to zero as N goes to infinity in the presence of random noise.

The costs of using the least squares method, or more generally any stochastic approach, are that at least some information about the noise statistics must be assumed known, and that no explicit error bounds can be given. Additionally, the least squares method is biased if there is a model uncertainty, although it is asymptotically unbiased. One way to improve the least squares estimate is to increase the order of the estimate $\widehat{G}_m(z)$. Hopefully, as m increases, the quality of the estimate improves. Unfortunately, this is not always the case. If $G(z)$ is infinite order and $\widehat{G}_m(z) = \sum_{i=0}^{m-1} \widehat{h}(i)z^{-i}$ denotes the least squares estimate of $G(z)$ it is known [2] that the hard error bound on $\|\widehat{G}_m(z) - G(z)\|_\infty$ may diverge in the standard H_∞ norm as m goes to infinity, and that the rate of divergence is proportional to $\ln m$. In fact, the error bound is also divergent in other standard norms, e.g., the ℓ_1 norm. Thus, the estimate $\widehat{G}_m(z)$ derived from the least squares method may not be a good one from a worst case point of view.

3.2. A modified least squares estimate. As discussed in the last section, an estimate derived by the least squares method is computationally efficient and has nice statistical properties, but no explicit error bound can be derived. In addition, in the presence of model uncertainty, the actual error could be unbounded in the worst case scenario. In this section we modify the estimate using prior information about ρ and M so that the modified estimate is "closer to the true system" than the unmodified one.

As before, let $G(z) = \sum_{i=0}^{\infty} h(i)z^{-i}$ and $\widehat{G}_m(z) = \sum_{i=0}^{m-1} \widehat{h}(i)z^{-i}$ be, respectively, the true system transfer function and the estimate derived by the least squares method. Assume that $G(z)$ is an element of $B_{\rho,M}$ for some ρ and M. In this section, we assume that ρ and M are known. Clearly, $\widehat{G}_m(z) - G(z)$ may diverge because the estimate $\widehat{G}_m$ need not be in S, let alone $B_{\rho,M}$. To ensure convergence, it suffices to find a $\overline{G}_m(z)$ (or $\overline{H}_m$) in $B_{\rho,M}$ such that $\overline{G}_m(z)$ is "closest" to $\widehat{G}_m(z)$ among all models in $B_{\rho,M}$. More precisely, for a given $\widehat{G}_m(z)$ or $\widehat{H}_m$, it suffices to let $\overline{H}_m = (\overline{h}(0), \ldots, \overline{h}(m-1))^T$ and $\overline{G}_m(z) = \sum_{i=0}^{m-1} \overline{h}(i)z^{-i}$, where

$$(3.19) \qquad \overline{H}_m = \arg \min_{Y \in \{Y \in R^m, |y(i)| \le M\rho^i\}} \left\| Y - \widehat{H}_m \right\|_p$$

for some $p = 1, 2, \ldots, \infty$, and $\| \cdot \|_p$ is the standard p norm in R^m, i.e.,

$$(3.20) \qquad \|Y\|_p = \left[\sum_{i=0}^{m-1} |y(i)|^p \right]^{1/p} \qquad \text{for } p = 1, 2, \ldots < \infty$$

where $Y = (y(0), \ldots, y(m-1))^T$ and

$$(3.21) \qquad \|Y\|_\infty = \max_i |y(i)| .$$

LEMMA 3.4. *The minimization of (3.19) is achieved when* $\overline{H}_m =$ proj $\widehat{H}_m$ *for any* $p = 1, 2, \ldots, \infty$, *where* proj $\widehat{H}_m = (\text{proj} \, \widehat{h}(0), \ldots, \text{proj} \, \widehat{h}(m-1))^T$ *with*

$$(3.22) \quad \text{proj} \, \widehat{h}(i) = \begin{cases} \widehat{h}(i) & |\widehat{h}(i)| \leq M\rho^i \\ M\rho^i & \widehat{h}(i) \geq M\rho^i \\ -M\rho^i & \widehat{h}(i) \leq -M\rho^i \end{cases} \quad i = 0, 1, \ldots, m-1 \ .$$

The corresponding modified estimate proj $\widehat{G}_m(z)$ is defined as

$$(3.23) \qquad \qquad \text{proj} \, \widehat{G}_m(z) = \sum_{i=0}^{m-1} \text{proj} \, \widehat{h}(i) z^{-i} \ .$$

Proof. Obviously, proj $\widehat{G}_m(z) \in B_{\rho,M}$. Now, for $p < \infty$, $\overline{H}_m = \arg\min \|Y - \widehat{H}_m\|_p$ if and only if

$$(3.24) \qquad \left[\left\| \overline{H}_m - \widehat{H}_m \right\|_p \right]^p \leq \left[\left\| Y - \widehat{H}_m \right\|_p \right]^p$$

for every $Y \in \{(y(1), \cdots y(m))^T \in R^m, |y(i)| \leq M\rho^i\}$.

This is fulfilled if $|\overline{h}(i) - \widehat{h}(i)| \leq |y(i) - \widehat{h}(i)|$ for each $i = 1, 2, \ldots, m-1$ and all $|y(i)| \leq M\rho^i$. Clearly, the solution is $\overline{h}(i) = \text{proj} \, \widehat{h}(i)$. When $p = \infty$, $\|\overline{H}_m - \widehat{H}_m\|_\infty \leq \|Y - \widehat{H}_m\|_\infty$ is guaranteed if $|\overline{h}(i) - \widehat{h}(i)| \leq |y(i) - \widehat{h}(i)|$ which is again achieved by setting $\overline{h}(i) = \text{proj} \, \widehat{h}(i)$. This completes the proof.   □

Intutitively, if $\widehat{G}_m(z)$ is in $B_{\rho,M}$, then proj $\widehat{G}_m(z) = \widehat{G}_m(z)$. If $\widehat{G}_m(z)$ is not in $B_{\rho,M}$, then it is a falsified estimate with respect to the prior information on ρ and M. This, however, does not imply that for any unfalsified $\overline{G}_m \in B_{\rho,M}$ the distance between $\overline{G}_m$ and $G(z)$ is smaller than that between $\widehat{G}_m(z)$ and $G(z)$. It is possible that for some $\overline{G}_m(z)$ in $B_{\rho,M}$, $G(z)$ would be closer to $\widehat{G}_m(z)$ than to $\overline{G}_m(z)$, although $\widehat{G}(z)$ could be falsified and $\overline{G}_m$ unfalsified with respect to ρ and M. Fortunately when $\overline{G}_m(z)$ is chosen to be proj $\widehat{G}_m(z)$, proj $\widehat{G}_m(z)$ is guaranteed to be closer to $G(z)$ than $\widehat{G}_m(z)$. More precisely, the following result holds.

LEMMA 3.5.   Let $\|G(z)\|_1 = \sum_{i=0}^{\infty} |h(i)|$ and $\|G\|_2 = [\frac{1}{2\pi} \int_0^{2\pi} |G(e^{j\omega})|^2 d\omega]^{\frac{1}{2}}$ be the standard ℓ_1 and H_2 operator norms. Then for every m and $\widehat{G}_m(z)$ or $\widehat{H}_m$,

$$(3.25) \qquad \left\| \text{proj} \, \widehat{H}_m - H_m \right\|_p \leq \left\| \widehat{H}_m - H_m \right\|_p \qquad p = 1, 2, \ldots, \infty$$

$$(3.26) \qquad \left\| G(z) - \text{proj} \, \widehat{G}_m(z) \right\|_1 \leq \left\| G(z) - \widehat{G}_m(z) \right\|_1$$

$$(3.27) \qquad \left\| G(z) - \text{proj} \, \widehat{G}_m(z) \right\|_2 \leq \left\| G(z) - \widehat{G}_m(z) \right\|_2$$

Proof. Noting that $|h(i) - \text{proj}\,\widehat{h}(i)| \leq |h(i) - \widehat{h}(i)|$ for all i, the result follows directly. □

Remark 3.6. By projection, the modified estimate $\text{proj}\,\widehat{G}_m(z)$ is not only the best in $B_{\rho,M}$, but is also closer to the true system $G(z)$ in both the parameter space and operator norms $\|\cdot\|_p$, and ℓ_1 and H_2. In fact, these properties hold for any experimental data.

Remark 3.7. Let $y_{\text{proj}}(k)$, $y_{\widehat{G}_m}(k)$, and $y(k)$, be the outputs of, respectively, the systems $\text{proj}\,\widehat{G}_m(z)$, $\widehat{G}_m(z)$, and $G(z)$ for some input $u(k)$. Since $\text{proj}\,\widehat{G}_m(z)$ is closer to $G(z)$ than $\widehat{G}_m(z)$ in the ℓ_1 sense, we have

$$(3.28) \quad \sup_k \left| y_{\text{proj}}(k) - y(k) \right| \leq \sup_k \left| y_{\widehat{G}_m}(k) - y(k) \right| \quad \text{for all } k \text{ and } u(\cdot),$$

i.e., the maximum output error is reduced by the projection operator.

Remark 3.8. Lemma 3.5 holds for every stochastic approach. Let $g(z)$ be an estimate obtained by any stochastic approach, then

$$(3.29) \quad \|G(z) - \text{proj}\,g(z)\|_i \leq \|G(z) - g(z)\|_i, \quad i = 1, 2.$$

Remark 3.9. Prior information on ρ and M is required to define the projection operator. However, this does not imply that the exact values of ρ and M are required. Instead only the estimates $\widehat{\rho}$ and $\widehat{M}$ are needed. If $\rho \leq \widehat{\rho} \leq 1$ and $M \leq \widehat{M}$, the results in Lemma 3.1 still hold. If $\widehat{\rho} < \rho$ and/or $\widehat{M} < M$, there will be an error term. This error term is bounded and goes to zero when $\widehat{\rho} - \rho$ and $\widehat{M} - M$ go to zero. For details see [1].

3.3. Worst case performance of the modified estimate. In this section, we study the performance of the modified estimate in a worst case setting. It will be shown that an explicit error bound can be given that is finite, even in the worst case scenario.

The set up of the worst case approach (see [2]) is similar to that of the stochastic approach of Section 2 but there are differences. The unknown noise $\eta(k)$ is assumed to be deterministic and bounded by some constant ξ. Also, the unknown true system is assumed to be $G(z) = \sum_{i=0}^{\infty} h(i)z^{-i}$, although the estimate $\widehat{G}_m^{wor}(z) = \sum_{i=0}^{m-1} \widehat{h}^{wor}(i)z^{-i}$ could be finite order. Thus structural mismatches between the true system and the estimate are also allowed in the worst case setting.

Let $\widehat{G}_m^{wor}(z) = \text{proj}\,\widehat{G}_m(z)$ as in Section 3. $\text{proj}\,\widehat{G}_m(z)$ is a modification of the stochastic estimator obtained by the least squares method that is closer to $G(z)$ than $\widehat{G}_m(z)$ in the ℓ_1 and H_2 norms. However, it is not clear whether $\text{proj}\,\widehat{G}_m(z)$ converges in a worst case setting. This raises the following question: What is the performance of $\text{proj}\,\widehat{G}_m(z)$ in a worst case setting?

We begin with known results from [1] and [2].

LEMMA 3.10. *Consider the least square method of Section 2. Assume that* $G(z) = \sum_{i=0}^{\infty} h(i)z^{-i}$, *with* $|h(i)| \leq M\rho^i$ *for some* M, ρ. *In the absence of noise, i.e., when* $\eta(k) = 0$ *for* $k = 0, 1, \ldots, N-1$, *the least squares estimate* $\widehat{G}_m(z)$ *satisfies*

$$(3.30) \quad \left\| G(z) - \widehat{G}_m(z) \right\|_{\infty} \leq \left\| G(z) - \widehat{G}_m(z) \right\|_1 \leq M_1\rho^m, \quad \text{for all } m \leq N ,$$

where $M_1 = \frac{2M}{(1-\rho)^2}$ *and*

$$(3.31) \qquad \left| h(i) - \widehat{h}(i) \right| \leq \xi , \qquad i = 0, 1, \ldots, m-1 .$$

Where ξ *is the bound on the unknown noise.*

Given Lemma 3.10, the convergence of $\text{proj}\,\widehat{G}_m(z)$ is an immediate consequence of Theorem 2.2 of [3] and Theorem 3.1 of [1].

THEOREM 3.11. *Let* $\text{proj}\,\widehat{G}_m(z)$ *be defined as in (3.23). Then, for all* m,

$$
\begin{aligned}
\sup_{|\eta(k)| \leq \xi,\, G \in B_{\rho, M}} & \left\| \text{proj}\,\widehat{G}_m(z) - G(z) \right\|_{\infty} \\
(3.32) \qquad & \leq \sup_{|\eta(k)| \leq \xi,\, G \in B_{\rho, M}} \left\| \text{proj}\,\widehat{G}_m(z) - G(z) \right\|_1 \\
& \leq M_1\rho^m + \sum_{i=0}^{m-1} \min\{\xi, 2M\rho^i\} \\
& \leq c\xi + M_1\rho^m
\end{aligned}
$$

where

$$(3.33) \quad c = \begin{cases} \frac{1}{1-\rho} & 2M \leq \xi \\ k + \frac{1}{1-\rho} & 2M \geq \xi \end{cases} \qquad \text{For any } k \text{ such that } 2M\rho^k \leq \xi .$$

Thus, in the worst case setting, an explicit error bound on $\text{proj}\,\widehat{G}_m(z) - G(z)$ can be derived, and the bound goes to zero as ξ goes to zero and m goes to infinity.

4. Maximum likelihood estimators. In Section 3, we discussed the projection of a least squares estimator. In this section, we consider the projection of a maximum likelihood estimator. The results are almost identical.

4.1. Maximum likelihood methods. From equation (3.2), we know that

$$(4.1) \qquad D = AH_m + \overset{\bullet}{\eta}_N + R_N .$$

Let $a_i, i = 1, 2, \ldots, N - 1$ denote the ith row of the matrix A. Then we have

(4.2) $\qquad d(k) = a_k H_m + r(k) + \eta(k) \qquad k = 1, 2, \ldots, N - 1$.

The philosophy of the worst case identification is fundamentally non-statistical. However, one may argue that having bounds on the noise $\eta(k)$ is equivalent to considering noise that is uniformly distributed within known bounds. In this case, the complex $\eta(k)$'s can be considered to be independent pairs of random variables uniformly distributed in a circle of radius ξ, i.e. the probability density function of each $\eta(k)$ is given by

(4.3) $\qquad p(x) = \begin{cases} \frac{1}{\pi \xi^2} & |x| \leq \xi \\ 0 & |x| > \xi \end{cases}$.

For a given order m, a maximum likelihood estimate H_m^{ML} of H_m is one which satisfies

(4.4) $\qquad p(D^N | H_m^{ML}) = \max_{\overline{H}_m} p(D^N | \overline{H}_m)$

for every $D^N(H, \eta)$. Clearly $\overline{H}_m = H_m^{ML}$ if and only if $\overline{H}_m$ corresponds to a noise value $\overline{\eta}(k)$ for which the probability density function of $\eta(k)$ at $\overline{\eta}(k)$ assumes its maximum value of $\frac{1}{\pi \xi^2}$ for all $k = 1, 2, \ldots, N - 1$. For arbitray m, this is not always possible, because the random variable $d(k) - a_k H_m = r(k) + \eta(k)$ is uniformly distributed in a circle with radius ξ centered at unknown $r(k)$ and the magnitude of $r(k)$ could be larger than ξ. However, if $G(z)$ is in $B_{\rho, M}$ and $N \geq m$ is large enough, it is possible.

To see this let $N \geq m$ be such that

(4.5) $\qquad \dfrac{M \rho^m}{1 - \rho} = \xi_1 < \xi$.

Choose $\overline{H}_m$ such that

(4.6) $\qquad \left| d(k) - a_k \overline{H}_m \right| \leq \xi - \xi_1 \qquad k = 0, 1, \ldots, N - 1$.

Since $|r(k)| \leq \frac{M \rho^m}{1 - \rho} \leq \xi_1$, this choice of $\overline{H}_m$ results in a value $\overline{\eta}(k)$

$$
\begin{aligned}
|\overline{\eta}(k)| &= \left| d(k) - a_k \overline{H}_m - r(k) \right| \\
&\leq \left| d(k) - a_k \overline{H}_m \right| + |r(k)| \\
&\leq \xi - \xi_1 + \xi_1 = \xi .
\end{aligned}
$$

(4.7)

Thus the probability density function $p(D^N | \overline{H}_m)$ assumes its maximum value, and consequently, $\overline{H}_m$ is a maximum likelihood estimate. Call this estimate $H_m^{ML} = (h_m^{ML}(0), h_m^{ML}(1), \ldots, h_m^{ML}(m - 1))^T$. The corresponding transfer function is defined by

(4.8) $\qquad G_m^{ML}(z) = \sum_{i=0}^{m-1} h_m^{ML}(i) z^{-i}$.

4.2. A modified maximum likelihood estimate and its performace in the worst case setting. There is little we can say about the performance of H_m^{ML} or $G_m^{ML}(z)$ in terms of the operator norms $\|\cdot\|_1$ and $\|\cdot\|_\infty$ because we can not determine an explicit error bound on $G_m^{ML}(z) - G(z)$. In fact, the worst case error bound on $G_m^{ML}(z) - G(z)$ could be divergent. To this end, we modify $G_m^{ML}(z)$ in a manner analogous to that of Section 3. Define

$$(4.9) \qquad \text{proj } G_m^{ML}(z) = \sum_{i=0}^{m-1} \text{proj } h_m^{ML}(i) z^{-i}$$

with

$$(4.10) \qquad \text{proj } h_m^{ML}(i) = \begin{cases} h_m^{ML}(i) & |h_m^{ML}(i)| \leq M\rho^i \\ M\rho^i & h_m^{ML}(i) \geq M\rho^i \\ -M\rho^i & h_m^{ML}(i) \leq -M\rho^i \end{cases}$$
$$i = 0, 1, \ldots, m-1 .$$

Similar to the results of Section 3, we have

LEMMA 4.1. *For every m and $G_m^{ML}(z)$,*

$$(4.11) \qquad \left\| G(z) - \text{proj } G_m^{ML}(z) \right\|_1 \leq \left\| G(z) - G_m^{ML}(z) \right\|_1 ,$$

$$(4.12) \qquad \left\| G(z) - \text{proj } G_m^{ML}(z) \right\|_2 \leq \left\| G(z) - G_m^{ML}(z) \right\|_2 .$$

Thus $\text{proj } G_m^{ML}(z)$ is closer to $G(z)$ than $G_m^{ML}(z)$ for every $G_m^{ML}(z)$. Additionally, we have the following result.

THEOREM 4.2. *Let $\text{proj } G_m^{ML}(z)$ and $G_m^{ML}(z)$ be defined as before. Then for all m such that $\frac{M\rho^m}{1-\rho} = \xi_1 < \xi$,*

$$\sup_{|\eta(k)| \leq \xi, \, G \in B_{\rho, M}} \left\| \text{proj } G_m^{ML}(z) - G(z) \right\|_\infty$$
$$\leq \sup_{|\eta(k)| \leq \xi, \, G \in B_{\rho, M}} \left\| \text{proj } G_m^{ML}(z) - G(z) \right\|_1$$
$$\leq M_1 \rho^m + \sum_{i=0}^{m-1} \min\{2\xi, 2M\rho^i\}$$
$$\leq c\xi + M_1 \rho^m ,$$

(4.13)
where

$$(4.14) \qquad c = \begin{cases} \frac{1}{1-\rho} & M \leq \xi \\ 2(k + \frac{1}{1-\rho}) & M \geq \xi \end{cases}$$

for any k such that $M\rho^k \leq \xi$.

Proof. The proof has two steps. In the first step, we show that

$$(4.15) \qquad \max_i |h_m^{ML}(i) - h(i)| \le 2\xi \qquad i = 1, 2, \ldots, m-1 \ .$$

Let H_m^{ML} be chosen as in equation (4.4). Then

$$(4.16) \qquad |d(k) - a_k H_m^{ML}| \le \xi - \xi_1 \qquad \text{for all } k = 1, 2, \ldots, N-1 \ .$$

Define $H_m^{ML} = H_m + \Delta H_m$ or $h_m^{ML}(i) = h(i) + \Delta h_m(i)$ for $i = 1, 2, \ldots, m-1$. ΔH_m or $\Delta h_m(i)$'s are the parameter errors. Notice that

$$(4.17) \qquad \begin{aligned} |d(k) - a_k H_m^{ML}| &= |d(k) - a_k H_m - a_k \Delta H_m| \\ &\ge |a_k \Delta H_m| - |d(k) - a_k H_m| \ . \end{aligned}$$

Thus

$$(4.18) \qquad \begin{aligned} |a_k \Delta H_m| &\le |d(k) - a_k H_m| + |d(k) - a_k H_m^{ML}| \\ &\le \xi - \xi_1 + |\eta(k) + r(k)| \\ &\le \xi - \xi_1 + \xi + \xi_1 = 2\xi \ . \end{aligned}$$

Because a_k is the kth row of the matrix A, $\|A\Delta H_m\|_\infty \le 2\xi$, and

$$(4.19) \qquad \begin{aligned} \|\Delta H_m\|_\infty &= \|(A^*A)^{-1}A^* A \Delta H_m\|_\infty \\ &\le \frac{1}{N} \|A^*\|_\infty \cdot \|A\Delta H_m\|_\infty \\ &\le \|A\Delta H_m\|_\infty \\ &\le 2\xi \ . \end{aligned}$$

This shows that $\max_i |h_m^{ML}(i) - h(i)| \le 2\xi$.

The second step is similar to the proof of Theorem 3.11, i.e.,

$$(4.20) \qquad \begin{aligned} \sup_{|\eta(k)| \le \xi, \, G \in B_{\rho, M}} \left\| \text{proj} \, G_m^{ML}(z) - G(z) \right\|_1 &\le M_1 \rho^m + \sum_{i=0}^{m-1} \min\{2\xi, 2M\rho^i\} \\ &\le c\xi + M_1 \rho^m \end{aligned}$$

where c is given in equation (4.14). This completes the proof. $\qquad \square$

Thus the modified maximum likelihood estimate $\text{proj} \, G_m^{ML}(z)$ is convergent and also near optimal in the worst case setting.

5. Generalized cases. In this section, we extend the results of Theorems 3.11 and 4.2 to generalized cases.

Let $\mathcal{A}$ be any stochastic identification algorithm that maps the noisy data (2.3) into the identified model (2.4). Assume that

$$\widehat{G}_m^0(z) = \sum_{i=0}^{m-1} \widehat{h}(i) z^{-i}$$

and

$$\widehat{G}_m^\eta(z) = \sum_{i=0}^{m-1} (\widehat{h}(i) + \Delta h(i)) z^{-i}$$

are the models that the algorithm $\mathcal{A}$ produces when, respectively, $\eta(k) \equiv 0$ and $\eta(k) \neq 0$ i.e., in the absence and presence of noise. Note that $\Delta h(i)$ represents the effects of noise $\eta(k)$ on the algorithm $\mathcal{A}$. Analogus to the proof of Theorem 3.11, we have

THEOREM 5.1. *Let $\mathcal{A}$ be any identification algorithm as described above. Assume that $|\Delta h(i)| \leq M \sup_k |\eta(k)| = M\epsilon$ for all i and some $M < \infty$. Then the convergence of $\widehat{G}_m^0(z)$ in the absence of noise implies the convergence of $proj\widehat{G}_m^\eta(z)$ in the presence of noise in the worst case sense, where the operator proj is defined in Section 3.*

Remark 5.2. Theorem 5.1 tells us that any identification algorithm for which the coefficient perturbations $\Delta h(i)$ due to noise are linearly bounded by ϵ, is convergent in the presence of noise in the worst case sense if it is convergent in the absence of noise. This covers a very large class of identification algorithms.

6. Concluding remarks. In this paper it was shown that the estimates $proj\,\widehat{G}_m(z)$ and $proj\,G_m^{ML}(z)$, the modified versions of the least squares estimate $\widehat{G}_m(z)$ and the maximum likelihood estimate $G_m^{ML}(z)$ are, closer to $G(z)$ than $\widehat{G}_m(z)$ and $G_m^{ML}(z)$. In this sense, they are better than either the least squares or the maximum likelihood estimates. It was also shown that these modified estimates and the more general estimates in Section 5 are convergent. Thus, they are near optimal from both the statistical and worst case points of view.

The key to our combination of the stochastic and worst case approachs is the utilization of prior information on M and ρ. When other types of prior information are available, they can also be used to improve our estimates.

Although frequency data were assumed in the paper, the results can be easily extended to identification algorithms using time domain data, e.g., the impulse or step response of a system. The point that we wish to make through this work is that stochastic and worst case system identification are not necessarily incompatible. We hope that this work will initiate further discussions along this line.

REFERENCES

[1] E.W. BAI AND S. RAMAN, *Robust system identification with noisy time/frequency response experimental data: projection and linear algorithms*, Automatica (to appear).

[2] A.J. HELMICKI, C.A. JACOBSON AND C.N. NETT, *Control oriented system identification: a worst case/deterministic approach in* H_∞, IEEE Trans. Automat. Control, vol. AC-36, (1991), pp. 1163–1176.

[3] C.A. JACOBSON, C.N. NETT AND J.R. PARTINGTON, *Worst case system identification in* ℓ_1: *optimal algorithms and error bounds*, Syst. and Contr. Letters, vol. 19, (1992), pp. 419–424.

[4] G. GU AND P.P. KHARGONEKAR, *Linear and nonlinear algorithms for identification in* $\mathcal{H}_\infty$ *with error bounds*, IEEE Trans. Automat. Control, vol. AC-37, (1992), pp. 953–963.

[5] J.M. KRAUSE AND P.P. KHARGONEKAR, *A comparison of classical stochastic and deterministic robust identification*, IEEE Trans. Automat. Control, vol. AC-37, (1992), pp. 994–1000.

[6] L. LJUNG, *System Identification: Theory For The User*, Prentice-Hall, Englewood Cliffs, NJ, 1987.

[7] R. TEMPO AND G.W. WASILKOWSKI, *Maximum likelihood estimators and worst case optimal algorithms for system identification*, Syst. and Contr. Letters, vol. 10, (1988), pp. 265–270.

[8] R. TEMPO, *Robust and optimal algorithms for worst case parametric system identification*, (preprint) 1993.

[9] J.F. TRAUB, G.W. WASILKOWSKI AND H. WOZNIAKOWSKI, *Information Based Complexity*, Academic Press, New York, 1988.

[10] B. WAHLBERG AND L. LJUNG, *Hard frequency-domain model error bounds from least squares like identification techniques*, IEEE Trans. Automat. Control, vol. AC-37, (1992), pp. 900–912.

SOME RESULTS FOR THE ADAPTIVE BOUNDARY CONTROL OF STOCHASTIC LINEAR DISTRIBUTED PARAMETER SYSTEMS[*]

T.E. DUNCAN[†]

1. Introduction. Linear distributed parameter systems are an important family of models for many physical phenomena. A useful subfamily of these models is described by analytic semigroups, [14]. To model some perturbations or inaccuracies in these models it is often reasonable to introduce white noise in these systems to obtain linear stochastic distributed parameter systems. For controlled linear distributed parameter systems it is often natural to consider that the control occurs on the boundary or at discrete points because in many applications it is impractical or not feasible to apply the control throughout the domain.

Since in many control situations there are unknown parameters in these linear, stochastic distributed parameter systems, it is necessary to solve a stochastic adaptive control problem. The control problem is described by an ergodic, quadratic cost functional so that under suitable assumptions the optimal control can be characterized as the solution of an algebraic or stationary Riccati equation. Since there is boundary or point control, the linear transformation for the control in the state equation is an unbounded operator in addition to the unbounded operator that is the infinitesimal generator of the analytic semigroup. The unknown parameters in the model appear affinely in both the infinitesimal generator of the the semigroup and the linear transformation of the control. The solution of such a stochastic adaptive control problem means exhibiting a strongly consistent family of estimates of the unknown parameters and constructing an adaptive control that is self-optimizing, that is, it achieves the optimal ergodic cost if the system were known. A stochastic adaptive control problem with distributed control is solved in [4].

A brief summary of each of the following sections is given now. In Section 2 one adaptive boundary control problem is formulated and its solution is described. An Itô formula is given for smooth functions of the solution of a stochastic distributed parameter system with boundary or point control. A family of least squares estimates of the unknown parameters is shown to be strongly consistent. A certainty equivalence adaptive control is shown to be self-optimizing, that is, the family of average costs converges to the optimal ergodic cost. The proofs of these results are given in [7]. Some examples are given that satisfy the assumptions that are used. These examples include stochastic parabolic problems with boundary control and a structurally damped plate with random loading and point control.

---

[*] Research partially supported by NSF Grants ECS-9102714 and DMS-9305936.

[†] Department of Mathematics, University of Kansas, Lawrence, KS 66045.

In Section 3 a second adaptive boundary control problem is formulated and solved. The unknown parameters appear in more linear operators on the boundary but in a more restricted way in the infinitesimal generator than in the model given in Section 2. Strong consistency of a family of least squares estimates is verified using a diminishingly exciting input. A certainty equivalence control is shown to be self-tuning and self-optimizing. The proofs of these results are given in [8].

In Section 4 some other distributed parameter models are briefly introduced that are linear but not described by analytic semigroups or that are not linear. Specifically the adaptive control of some linear stochastic hyperbolic systems is described as well as the adaptive control of a stochastic semilinear system.

2. Adaptive Boundary Control for the First Model. The unknown linear stochastic distributed parameter system with boundary or point control is formally described by the following stochastic differential equation

(2.1) $$dX(t; \alpha) = (A(\alpha)X(t; \alpha) + B(\alpha)U(t))dt + \phi dW(t)$$
$$X(0; \alpha) = X_0$$

where $X(t; \alpha) \in H$, H is a real, separable, infinite dimensional Hilbert space, $(W(t), t \geq 0)$ is a cylindrical Wiener process on H, $\phi \in \mathcal{L}(H)$, $\alpha = (\alpha^1, ..., \alpha^q)$ and $t \geq 0$.

The probability space is denoted $(\Omega, \mathcal{F}, P)$ where P is a probability measure that is induced from the cylindrical Wiener measure and $\mathcal{F}$ is the P-completion of the Borel σ-algebra on Ω. Let $(\mathcal{F}_t, t \geq 0)$ be an increasing P-complete family of sub-σ-algebras of $\mathcal{F}$ such that X_t is $\mathcal{F}_t$-measurable for $t \geq 0$ and $(\langle \ell, W(t) \rangle, \mathcal{F}_t, t \geq 0)$ is a martingale for each $\ell \in H$. $A(\alpha)$ is the infinitesimal generator of an analytic semigroup on H. For some $\beta \geq 0$ the operator $-A(\alpha) + \beta I$ is strictly positive so that the fractional powers $(-A(\alpha) + \beta I)^\gamma$ and $(-A(\alpha)^* + \beta I)^\gamma$ and the spaces $D_{A(\alpha)}^\gamma = \mathcal{D}((-A(\alpha) + \beta I)^\gamma)$ and $D_{A^*(\alpha)}^\gamma = \mathcal{D}((-A^*(\alpha) + \beta I)^\gamma)$ with the graph norm topology for $\gamma \in \mathbb{R}$ can be defined. It is assumed that $B(\alpha) \in \mathcal{L}(H_1, D_{A(\alpha)}^{\epsilon-1})$ where H_1 is a real, separable Hilbert space and $\epsilon \in (0, 1)$ (cf. (A4)). For the solution of (2.1) on $[0, T]$ the control $(U(t), t \in [0, T])$ is an element of $M_W^p(0, T, H_1)$ where $M_W^p(0, T, H_1) = \{u : [0, T] \times \Omega \to H_1$ u if $(\mathcal{F}_t)$-nonanticaptive and $E \int_0^T |u(t)|^p dt < \infty\}$, and $p > \max(2, 1/\epsilon)$ is fixed.

A selection of the following assumptions is used subsequently in this section:

(A1) The family of unknown parameters are the elements of a compact set $\mathcal{K}$.

(A2) For $\alpha \in \mathcal{K}$ the operator $\phi^*(-A^*(\alpha) + \beta I)^{-1/2+\delta}$ is Hilbert-Schmidt for some $\delta \in (0, \frac{1}{2})$.

(A3) There are real numbers $M > 0$ and $\omega > 0$ such that for $t > 0$ and $\alpha \in \mathcal{K}$

$$|S(t;\alpha)|_{\mathcal{L}(H)} \le Me^{-\omega t}$$

and

$$|A(\alpha)S(t;\alpha)|_{\mathcal{L}(H)} \le Mt^{-1}e^{-\omega t}$$

where $(S(t;\alpha), t \ge 0)$ is the analytic semigroup generated by $A(\alpha)$.

(A4) For all $\alpha_1, \alpha_2 \in \mathcal{K}, \mathcal{D}(A(\alpha_1)) = \mathcal{D}(A(\alpha_2))$, $D^\delta_{A(\alpha_1)} = D^\delta_{A(\alpha_2)}$ and $D^\delta_{A^\bullet(\alpha_1)} = D^\delta_{A^\bullet(\alpha_2)}$ for $\delta \in \mathbb{R}$.

(A5) For each $\alpha \in \mathcal{K}$ and $x \in H$ there is a control $u_{\alpha,x} \in L^2(\mathbb{R}_+, H_1)$ such that $y(\cdot) = S(\cdot;\alpha)x + \int_0 S(\cdot - t;\alpha)B(\alpha)u_{\alpha,x}(t)dt \in L^2(\mathbb{R}_+, H)$.

(A6) The operator $A(\alpha)$ has the form

$$A(\alpha) = F_0 + \sum_{i=1}^{q} \alpha^i F_i$$

where F_i is a linear, densely defined operator on H for $i = 0, 1, ..., q$ such that $\cap_{i=0}^{q} \mathcal{D}(F_i^*)$ is dense in H.

It is well known that the strong solution of (2.1) may not exist so usually the mild solution of (2.1) is used, that is,

$$X(t;\alpha) = S(t;\alpha)X_0 + \int_0^t S(t-r;\alpha)B(\alpha)U(r)dr + \int_0^t S(t-r;\alpha)\phi dW(r)$$

(2.2)

where $S(t;\alpha) = e^{tA(\alpha)}$. The mild solution is equivalent to the following inner product equation: for each $y \in \mathcal{D}(A^*(\alpha))$

(2.3)
$$\langle y, X(t;\alpha)\rangle = \langle y, X(0)\rangle + \int_0^t \langle A^*(\alpha)y, X(s;\alpha)\rangle ds$$
$$+ \int_0^t \langle \psi(\alpha)y, U(s)\rangle ds + \langle \phi^* y, W(t)\rangle$$

where $\psi(\alpha) = B^*(\alpha) \in \mathcal{L}(D^{1-\epsilon}_{A^*(\alpha)}, H_1)$. The following lemma verifies that $(X(t;\alpha), t \in [0,T])$ is a well defined process in $M_W^p(0, T, H)$.

LEMMA 2.1. *Assume that (A2) is satisfied. For $T > 0$ and $\alpha \in \mathcal{K}$ the process $(Z(t;\alpha), t \in [0,T])$ and $(\hat{Z}(t;\alpha), t \in [0,T])$ given by the equations*

(2.4)
$$Z(t;\alpha) = \int_0^t S(t-r;\alpha)\phi \, dW(r)$$

$$(2.5) \qquad \hat{Z}(t;\alpha) = \int_0^t S(t-r;\alpha)B(\alpha)U(r)dr$$

for $U \in M_W^p(0,T,H_1)$ are elements of $M_W^p(0,T,H)$ with versions that have continuous sample paths.

If $A(\alpha) = A^*(\alpha)$ and $(A(\alpha) - \beta I)^{-1}$ is compact then the assumption (A2) is equivalent to the assumption that

$$\int_0^T t^{-2\delta}|S(t;\alpha)\phi|_{HS}^2 dt < \infty$$

for $T > 0$.

Consider the quadratic cost functional

$$(2.6) \quad J(X_0,U,\alpha,T) = \int_0^T [< QX(s), X(s) > + < PU(s), U(s) >]ds$$

where $T \in (0,\infty], X(0) = X_0, Q \in \mathcal{L}(H), P \in \mathcal{L}(H_1)$ are self adjoint operators satisfying

$$(2.7) \qquad\qquad\qquad \langle Qx, x \rangle \geq r_1|x|^2$$
$$(2.8) \qquad\qquad\qquad \langle Py, y \rangle \geq r_2|y|^2$$

for $x \in H$, $y \in H_1$ and constants $r_1 > 0$ and $r_2 > 0$. For the deterministic control problem for (2.1) with $\phi \equiv 0$ and the cost functional (2.6) with $T = +\infty$ assuming (A5), the optimal cost is $\langle V(\alpha)X_0, X_0 \rangle$ [3,11,12] where V satisfies the formal stationary Riccati equation

$$(2.9) \quad A^*(\alpha)V(\alpha) + V(\alpha)A(\alpha) - V(\alpha)B(\alpha)P^{-1}\psi(\alpha)V(\alpha) + Q = 0$$

and $\psi(\alpha) = B^*(\alpha)$.

The equation (2.9) can be modified to a meaningful inner product equation as

$$(2.10)\langle A(\alpha)x, Vy \rangle + \langle Vx, A(\alpha)y \rangle - \langle P^{-1}\psi(\alpha)Vx, \psi(\alpha)Vy \rangle + \langle Qx, y \rangle = 0$$

for $x, y \in \mathcal{D}(A(\alpha))$. It has been shown [3,11, 12] that if (A5) is satisfied then V is the unique nonnegative, self adjoint solution of (2.10) and $V \in \mathcal{L}(H, D_{A^*}^{1-\epsilon})$. The solution of (2.9) is understood to be the solution of (2.10).

For adaptive control, the control policies $(U(t), t \geq 0)$ that are considered are linear feedback controls, that is,

$$(2.11) \qquad\qquad\qquad U(t) = K(t)X(t)$$

where $(K(t), t \geq 0)$ is an $\mathcal{L}(H, H_1)$-valued process that is uniformly bounded almost surely by a constant $R > 0$. Let $\Delta > 0$ be fixed. It is assumed that the $\mathcal{L}(H, H_1)$-valued process $(K(t), t \geq 0)$ has the property that $K(t)$ is adapted to $\sigma(X(u), u \leq t - \Delta)$ for each $t \geq \Delta$ and it is assumed that $(K(t), t \in [0, \Delta])$ is a deterministic, operator-valued function. For such an admissible adaptive control there is a unique solution of (2.1) with $K(t) = \tilde{K}(X(s), 0 \leq s \leq t - \Delta)$. If $\Delta = 0$ then (2.1) may not have a unique solution. Furthermore, the delay $\Delta > 0$ accounts for some time that is required to compute the adaptive feedback control law from the observation of the solution of (2.1).

Two more assumptions ((A7) and (A8)) are given that are used for the verification of the strong consistency of a family of least squares estimates of the unknown parameter vector α. Define $\mathbb{K} \subset \mathcal{L}(H, H_1)$ as

$$\mathbb{K} = \{K \in \mathcal{L}(H, H_1) : |K|_{\mathcal{L}(H, H_1)} \leq R\}$$

where R is given above.

Assume that $B(\alpha)$ is either independent of $\alpha \in \mathcal{K}$ or has the form

$$(2.12) \qquad\qquad B(\alpha) = \psi^*(\alpha)$$

where $\psi(\alpha) = \hat{B}^* A^*(\alpha) \in \mathcal{L}(D^{1-\epsilon}_{A^*(\alpha)}, H_1)$ and the operator $\hat{B} \in \mathcal{L}(H_1, D^{\epsilon}_{A(\alpha)})$ is given.

(A7) There is a finite dimensional projection $\tilde{P}$ on H with range in $\cap^q_{i=1}$ $\mathcal{D}(F^*_i)$ such that $i_{\tilde{P}} \phi \phi^* i^*_{\tilde{P}} > 0$ where $i_{\tilde{P}} : H \to \tilde{P}(H)$ is the projection map and $B(\alpha)$ is either independent of α or has the form (2.12). In the latter case there is a finite dimensional projection $\hat{P}$ on H and a constant $c > 0$ such that

$$|\hat{P}(I + K^* \hat{B}^*)F^* \tilde{P}|_{\mathcal{L}(H)} > c$$

is satisfied for all $F \in \{F_1, ..., F_q\}$ and $K \in \mathbb{K}$.

It is easy to verify that if H is infinite dimensional, $\hat{B} \in \mathcal{L}(H_1, H)$ is compact and $(F^*_i)^{-1} \in \mathcal{L}(H)$ for $i = 1, 2, ..., q$ then (A7) is satisfied.

Let $(U(t), t \geq 0)$ be an admissible control, denoted generically as

$$U(t) = K(t)X(t)$$

where $(X(t), t \geq 0)$ is the (unique) mild solution of (2.1) using the above admissible control. Let

$$(2.13) \qquad\qquad \mathcal{A}(t) = (a_{ij}(t))$$

and

$$(2.14) \qquad\qquad \tilde{\mathcal{A}}(t) = (\tilde{a}_{ij}(t))$$

where

$$(2.15a) \qquad\qquad a_{ij}(t) = \int_0^t \langle \tilde{P}F_i X(s), \tilde{P}F_j X(s) \rangle ds$$

if B does not depend on α or

(2.15b) $$a_{ij}(t) = \int_0^t \langle \tilde{P}(F_i + F_i \hat{B} K(s)) X(s), \tilde{P}(F_j + F_j \hat{B} K(s)) X(s) \rangle ds$$

if $B(\alpha)$ has the form (2.12) and

(2.16) $$\tilde{a}_{ij}(t) = \frac{a_{ij}(t)}{a_{ii}(t)}$$

It is easy to verify that the integrations in (2.15a) and (2.15b) are well defined.

For the verification of the strong consistency of a family of least squares estimates of the unknown parameter vector, the following assumption is used.

(A8) For each admissible adaptive control law, $(\tilde{\mathcal{A}}(t), t \geq 0)$ satisfies

$$\liminf_{t \to \infty} |\det \tilde{\mathcal{A}}(t)| > 0 \qquad a.s.$$

The following result is an Itô formula for a smooth function of the solution of (2.1).

PROPOSITION 2.2. *Assume that (A2) is satisfied. Let* $V \in C^{1,2}([0,T] \times H)$ *be such that* $V_x(t,x) \in D_{A^*}^{1-\epsilon}, V_x(t,\cdot) : H \to D_{A^*}^{1-\epsilon}$ *is continuous,* $\langle Ax, V_x(t,x) \rangle$ *for* $x \in \mathcal{D}(A)$ *can be extended to a continuous function* $h : [0,T] \times H \to \mathbb{R}$ *and*

$$|h(t,x)| + |V(t,x)| + |V_x(t,x)|_{D_{A^*}^{1-\epsilon}} + |V_{xx}(t,x)|_{\mathcal{L}(H)} + |V_t(t,x)| \leq k(1 + |x|^p)$$

for $(t,x) \in [0,T) \times H$ *and* $p > 0, k > 0$. *Assume that one of the following three conditions is satisfied.*

i) ϕ *is Hilbert-Schmidt.*

ii) $V_{xx}(t,x)$ *is nuclear,* $V_{xx}(t,\cdot)$ *is continuous in the norm* $|\cdot|_1$ *of nuclear operators and* $|V_{xx}(t,x)|_1 \leq k(1 + |x|^p)$ *for* $(t,x) \in (0,T) \times H$ *where* $k > 0$ *and* $p > 0$.

iii) $V_{xx}(t,x) \in \mathcal{L}(D_A^{\delta-(1/2)}, D_{A^*}^{(1/2)-\delta})$ *for* $(t,x) \in [0,T] \times H$, *the function* $L(\cdot) = (R^*(\beta))^{-(1/2)+\delta} V_{xx}(t,\cdot)(R(\beta))^{-(1/2)+\delta} : H \to \mathcal{L}(H)$ *is continuous and* $|L(x)|_{\mathcal{L}(H)} \leq k(1 + |x|^p)$ *is satisfied for* $t > 0$ *and* $x \in H$.
Then

$$V(t, X(t)) - V(\tau, X(\tau)) = \int_\tau^t [h(s, X(s)) + V_s(s, X(s))$$

(2.17) $$+ \langle u(s), \psi V_x(s, X(s)) \rangle + \frac{1}{2}\pi(s, X(s))] ds$$

$$+ \int_\tau^t \langle \phi^* V_x(s, X(s)), dW(s) \rangle$$

where $0 \leq \tau \leq t \leq T$, $\psi = B^*$ *and* $(X(t), t \in [0, T])$ *satisfies (2.1) and for
i) and ii),* $\pi(t, x) = TrV_{xx}(t, x)\phi\phi^*$ *and for iii)* $\pi(t, x) = Tr(R^*(\beta))^{\delta - (1/2)}$
$V_{xx}(t, x)\phi\phi^*(R^*(\beta))^{(1/2) - \delta}$.

For use of this Itô formula in the adaptive control problem it is useful
to state explicitly the case where $v(x) = \langle Vx, x \rangle$ where $V \in \mathcal{L}(H)$ is a
self-adjoint operator.

COROLLARY 2.3. *Let* $V \in \mathcal{L}(H)$ *be self adjoint such that* $V \in \mathcal{L}(H, D_{A^*}^{1 - \epsilon})$
and $|\langle Vx, Ax \rangle| \leq k|x|^2$ *for* $x \in \mathcal{D}(A)$ *where* $k > 0$. *Assume that one of the
following conditions is satisfied:*

i) ϕ *is Hilbert-Schmidt*
ii) V *is nuclear*
iii) $V \in \mathcal{L}(D_A^{\delta - (1/2)}, D_{A^*}^{(1/2) - \delta})$
Then for all $0 \leq \tau \leq t \leq T$

$$\langle VX(t), X(t) \rangle - \langle VX(\tau), X(\tau) \rangle$$
$$= \int_\tau^t [h(X(s)) + 2\langle U(s), \Psi V X(s) \rangle + \pi(V)]ds + 2\int_\tau^t \langle \phi^* V X(s), dW(s) \rangle \text{ a.s.}$$

(2.18)

where h *is the continuous extension of* $2\langle Vx, Ax \rangle$ *on* H, *and for i) and ii)*
$\Pi(V) = TrV\phi\phi^*$ *and for iii)* $\Pi(V) = Tr(R^*(\beta))^{\delta - (1/2)}V\phi\phi^*(R^*(\beta))^{(1/2) - \delta}$.

For the identification of the unknown parameters in the linear stochastic distributed parameter system (2.1), a family of least squares estimates
is formed. It is assumed for the identification that $\beta = 0$, that is, $-A(\alpha)$
is strictly positive. Let $\tilde{P}$ be the projection given in (A7). The estimate
of the unknown parameter vector at time t, $\hat{\alpha}(t)$, is the minimizer of the
quadratic functional of α, $L(t; \alpha)$, given by

(2.19)
$$L(t; \alpha) = -\int_0^t \langle \tilde{P}(A(\alpha) + B(\alpha)K(s))X(s), d\tilde{P}X(s) \rangle$$
$$+ \frac{1}{2}\int_0^t |\tilde{P}(A(\alpha) + B(\alpha)K(s))X(s)|^2 ds$$

where $U(s) = K(s)X(s)$ is an admissible adaptive control.

THEOREM 2.4. *Let* $(K(t), t \geq 0)$ *be an admissible feedback control
law. Assume that (A2, A6, A7, A8) are satisfied and* $\alpha_0 \in \mathcal{K}^0$. *Then the
family of least squares estimates* $(\hat{\alpha}(t), t > 0)$, *where* $\hat{\alpha}(t)$ *is the minimizer
of (2.19), is strongly consistent, that is,*

(2.20)
$$P_{\alpha_0}(\lim_{t \to \infty} \hat{\alpha}(t) = \alpha_0) = 1$$

where α_0 *is the true parameter vector.*

For the applications of identification and adaptive control it is important to have recursive estimators of the unknown parameters. Let

$\langle \tilde{F}(s)x, y \rangle$ be the vector whose i^{th} component is $\langle \tilde{P}F_i(I + \hat{B}K(s))x, y \rangle$. We have

$$(2.21) \qquad \hat{\alpha}(t) = \mathcal{A}^{-1}(t) \int_0^t \langle \tilde{F}(s)X(s), d\tilde{P}X(s) - \tilde{P}F_0 X(s)ds \rangle$$

Since $\mathcal{A}^{-1}(t)$ satisfies the differential equation

$$d\mathcal{A}^{-1}(t) = -\mathcal{A}^{-1}(t)d\mathcal{A}(t)\mathcal{A}^{-1}(t)$$

the differential of (2.21) satisfies

$$(2.22)\ d\hat{\alpha}(t) = \mathcal{A}^{-1}(t)\langle \tilde{F}(t)X(t), d\tilde{P}X(s) - \tilde{P}A(\hat{\alpha}(t))(I + \hat{B}K(t))X(t)dt \rangle$$

Now the certainty equivalence, optimal ergodic control law is shown to be self-tuning and self-optimizing. The self-tuning property is obtained by using the continuity of the solution of a stationary Riccati equation with respect to parameters in the topology induced by a suitable operator norm. Since the unbounded operator $B(\alpha)$ appears in the linear transformation of the control in (2.1), this operator topology is more restrictive than for bounded linear transformations on the Hilbert space. This continuity property is also used to show that the certainty equivalence control stabilizes the unknown system in a suitable sense. The self-optimizing property is verified for this adaptive control.

The solution V of the stationary Riccati equation (2.9) satisfies the assumptions of Corollary 2.1 if one of the following three conditions is satisfied: (i) ϕ is Hilbert-Schmidt, (ii) V is nuclear or (iii) A is strictly negative. By (A5) $V \in \mathcal{L}(H, D_{A^*}^{1-\epsilon})$ [11, 12] and (2.10) implies that

$$|\langle Ax, Vx \rangle| = |\langle Rx, x \rangle| \le k|x|^2$$

for some $R \in \mathcal{L}(H)$. If A is strictly negative then it easily follows that $V \in \mathcal{L}(D_A^{\delta-(1/2)}, D_{A^*}^{(1/2)-\delta})$. Moreover, if (A2) is satisfied with $\phi = I$ then V is nuclear because from (Theorems 1 and 2 [12]) it follows that $P_a = (-A^* + \beta I)^a V \in \mathcal{L}(H)$ for each $a \in (0, 1)$. Thus $V = P_a^*(-A + \beta I)^{-a}$ is nuclear because $(-A + \beta I)^{-a}$ is nuclear for $a = 1 - 2\delta$ by (A2, A5).

If an adaptive control is self-tuning and some stability properties are satisfied for the solution of (2.1), then this adaptive control is self-optimizing.

PROPOSITION 2.5. *Assume that (A2, A5) are satisfied, that the solution V of (2.10) satisfies the assumptions of Corollary 2.1 and that*

$$(2.23) \qquad\qquad \lim_{t \to \infty} \frac{1}{t}\langle VX(t), X(t) \rangle = 0 \qquad a.s.$$

$$(2.24) \qquad\qquad \limsup_{t \to \infty} \frac{1}{t}\int_0^t |X(s)|^2 ds < \infty \qquad a.s.$$

where $(X(t), t \geq 0)$ is the solution of (2.1) with $\alpha_0 \in \mathcal{K}$ and the control $U \in \cap_{T>0} M_W^p(0, T, H_1)$. Then

$$(2.25) \qquad \liminf_{T \to \infty} \frac{1}{T} J(X_0, U, \alpha_0, T) \geq \Pi(V) \qquad a.s.$$

where V is the solution of (2.10) with $\alpha = \alpha_0$. Furthermore, if U is an admissible control $U(t) = K(t)X(t)$ such that

$$(2.26) \qquad \lim_{t \to \infty} K(t) = k_0 \qquad a.s.$$

in the uniform $\mathcal{L}(H, H_1)$ topology where $k_0 = -P^{-1}\psi V$ then

$$(2.27) \qquad \lim_{T \to \infty} \frac{1}{T} J(X_0, U, \alpha_0, T) = \Pi(V). \qquad a.s.$$

Now it is shown that the stability conditions (2.23-2.24) are satisfied for an admissible, self-tuning adaptive control.

PROPOSITION 2.6. *Assume that (A2, A5) are satisfied. Let the solution V of (2.10) satisfy the assumptions of Corollary 2.1. If $(X(t), t \geq 0)$ is the solution of (2.1) with $\alpha_0 \in \mathcal{K}$ and an adaptive control law $(K(t), t \geq 0)$ that satisfies (2.26) then (2.23-2.24) are satisfied.*

The self-optimizing property is now verified for a self-tuning adaptive control.

THEOREM 2.7. *Assume that (A1-A4, A6-A8) are satisfied. Let $(\hat{\alpha}(t),$ $t \geq 0)$ be the family of least squares estimates where $\hat{\alpha}(t)$ is the minimizer of (2.19). Let $(K(t), t \geq 0)$ be an admissible adaptive control law such that*

$$(2.28) \qquad K(t) = -P^{-1}\psi(\hat{\alpha}(t - \Delta))V(\hat{\alpha}(t - \Delta))$$

where $\psi(\alpha) = B^*(\alpha)$ and $V(\alpha)$ is the solution of (2.10) for $\alpha \in \mathcal{K}$. Then the family of estimates $(\hat{\alpha}(t), t \geq 0)$ is strongly consistent,

$$(2.29) \qquad \lim_{t \to \infty} K(t) = k_0 \qquad a.s.$$

in $\mathcal{L}(H, H_1)$ where $k_0 = -P^{-1}\psi(\alpha_0)V(\alpha_0)$ and

$$(2.30) \qquad \lim_{T \to \infty} \frac{1}{T} J(X_0, U, \alpha_0, T) = Tr\Pi(V(\alpha_0)) \qquad a.s.$$

where $U(t) = K(t)X(t)$ and $\Pi(V)$ is given in Corollary 2.1.

Now some examples are given that satisfy the assumptions for the adaptive boundary control problem. Consider

$$(2.31) \qquad \frac{\partial y}{\partial t}(t, x) = \alpha L(x, D)y(t, x) + \eta(t, x), \qquad (t, x) \in \mathbb{R}_+ \times G$$

$$(2.32) \qquad \frac{\partial y}{\partial v} = u(t, x), \qquad (t, x) \in \mathbb{R}_+ \times \partial G$$

$$y(0, x) = y_0(x)$$

where

$$L(x, D)f = \sum_{i,j=1}^{n} D_i a_{ij}(x) D_j f + \sum_{i=1}^{n} [b_i(x) D_i f + D_i(d_i(x)f)] + c(s)f$$

the coefficients a_{ij}, b_i, d_i and c are elements of $C^\infty(G), (a_{ij})$ is symmetric, positive definite, G is a bounded, open domain in $\mathbb{R}^n$ with C^∞-boundary ∂G with G locally on one side of ∂G. $(\eta(t, x), t \in \mathbb{R}_+, x \in G)$ is a space dependent white noise, $u \in L^2(0, T, L^2(\partial G))$ for each $T > 0$. $\alpha \in \mathcal{K} = [\alpha_1, \alpha_2]$ is a scalar parameter for $0 < \alpha_1 < \alpha_2$. Assume that the operator A corresponding to $L(x, D)$ is strictly negative and (A2) is satisfied. Using the semigroup model we have that

(2.33)
$$y(t; \alpha) = S(t; \alpha, A)y_0 + \alpha \int_0^t S(t - r; \alpha) BU(r)dr$$

$$+ \int_0^t S(t - r; \alpha) \phi dW(r)$$

where $S(t; \alpha) = e^{t\alpha A}$, $B = [\hat{B}^* A^*]^* \in \mathcal{L}(D_{A^*}^{1-\epsilon}, H_1)$ and $\hat{B} \in \mathcal{L}(H_1, D_A^\epsilon)$ solves the elliptic problem

$$L(x, D)(\hat{B}g) = 0 \qquad \text{on} \quad G$$
$$\frac{\partial}{\partial v}(\hat{B}g) = -g \qquad \text{on} \quad \partial G.$$

The assumptions (A1), (A3-A6) are now trivially satisfied because $A(\alpha) = \alpha A, \alpha \in [\alpha_1, \alpha_2], \alpha_1 > 0$ and A is strictly negative. We have that $(A^*)^{-1} \in \mathcal{L}(H), \hat{B} \in \mathcal{L}(H_1, D_A^\epsilon)$ and the embedding $D_A^\epsilon \to H$ is compact so (A7) is satisfied. Since the parameter is scalar, (A8) is trivially satisfied. Thus by Theorem 2.4 the family of least squares estimates given in its statement is strongly consistent for $\alpha_0 \in (\alpha_1, \alpha_2)$. For any strongly consistent family of estimators $(\hat{\alpha}(t), t \geq 0)$, the cost functional (2.6) with a uniformly positive $Q \in \mathcal{L}(L^2(G))$ and $P \in \mathcal{L}(L^2(\partial G))$, the system (2.1) with $A(\alpha_0) = \alpha_0 A, B$ as above, $\beta = 0$ and the adaptive control

$$U(t) = -P^{-1}\psi(\hat{\alpha}(t - \Delta))V(\hat{\alpha}(t - \Delta))X(t)$$

has the self-optimizing property (2.30) by Theorem 2.7.

An elementary example of a boundary control problem with a vector parameter α is described that satisfies (A8). Let $H = L^2([0, 1], \mathbb{R})$ and F_1 and F_2 be the linear operators

$$F_1 = \frac{d^2}{dx^2}$$

$$F_2 = \frac{d}{dx}$$

and let $A(\alpha)$ be

$$A(\alpha) = \alpha_1 F_1 + \alpha_2 F_2$$

For the infinitesimal generator it can be verified that (A8) is satisfied.

Now consider an example of a structurally damped plate with random loading and point control. Consider the following model of a plate in the deflection w.

(2.34) $w_{tt}(t, x) + \Delta^2 w(t, x) - \alpha \Delta w(t, x) = \delta(x - x_0) u(t) + \eta(t, x)$

for $(t, x) \in \mathbb{R}_+ \times G$

(2.35) $w(0, \cdot) = w_0 \qquad w_t(0, \cdot) = w_1$

(2.36) $w|_{\mathbb{R}_+ \times \partial G} = \Delta w|_{\mathbb{R}_+ \times \partial G} = 0$

where $\alpha > 0$ is an unknown constant, $\eta(t, x)$ formally represents a space dependent Gaussian white noise on the open, bounded, smooth domain $G \subset \mathbb{R}^n$ for $n \le 3$ and $\delta(x - x_0)$ is the Dirac distribution at $x_0 \in G$. The cost functional is

(2.37) $J(w_0, w_1, u, \alpha, T) = \displaystyle\int_0^T (|w(t)|^2_{H^2(G)} + |w_t(t)|^2_{L^2(G)} + |u(t)|^2) dt.$

For a mathematical treatment of the deterministic problem (2.34-2.37) where $\eta \equiv 0$ the reader is referred to [2,13] and the references given therein. Define the linear operator $\mathcal{A}$ by the equation $\mathcal{A}h = \Delta^2 h$ where $\mathcal{D}(\mathcal{A}) = \{h \in H^4(G) : h|_{\partial G} = \Delta h|_{\partial G} = 0\}$. Following [2,13] the equations (2.34-2.37) are rewritten in the form (2.1, 2.6) where $H = \mathcal{D}(\mathcal{A}^{1/2}) \times L^2(G) = (H^2(G) \cap H^1_0(G)) \times L^2(G)$, $H_1 = \mathbb{R}$

$$A(\alpha) = \begin{bmatrix} 0 & I \\ -\mathcal{A} & -\alpha \mathcal{A}^{1/2} \end{bmatrix}$$

$$Bu = \begin{bmatrix} 0 \\ \delta(x - x_0) u \end{bmatrix}$$

$$\phi = \begin{bmatrix} 0 & 0 \\ 0 & \phi_1 \end{bmatrix}$$

where $\phi_1 \in \mathcal{L}(L^2(G))$ is a Hilbert-Schmidt operator and $\phi_1 \phi_1^* > 0$, $Q = I$, $P = I$ and $(W(t), t \ge 0)$ in (2.1) is a cylindrical Wiener process on H. It is known [2] that $A(\alpha)$ generates a stable analytic semigroup, $(S(t; \alpha), t \ge 0)$, and that $B \in \mathcal{L}(H_1, D^{\epsilon-1}_{A(\alpha)})$ for $\epsilon \in (0, 1 - \frac{n}{4})$ which is possible for $n \le 3$ (cf. [13]). Suppose that the unknown parameter $\alpha \in K = [a_0, a_1]$ where $0 < a_0 < a_1$. The assumptions (A1), (A2), (A4), (A5) and (A6) are clearly satisfied. Since B does not depend on $\alpha \in \mathcal{K}$ the assumption (A7) is satisfied with a finite dimensional projection $\tilde{P} : H \rightarrow \tilde{P}(H)$ of the form

$$\tilde{P} = \begin{bmatrix} 0 & 0 \\ 0 & \tilde{P}_1 \end{bmatrix}$$

where $\tilde{P}_1 : L^2(G) \rightarrow H^2(G)$ and $\tilde{P}_1 \neq 0$. The assumption (A8) is trivially satisfied because the parameter α is scalar. The assumptions of the uniform analyticity and the exponential stability of the semigroup $(S(t; \alpha), t \geq 0)$ and the continuous dependence of this semigroup on α, can be verified by the explicit spectral expansions of $A(\alpha)$ and $(S(t; \alpha), t \geq 0)$ ([2], Theorem A3). Therefore by Theorem 2.4 the family of least squares estimates given in its statement is strongly consistent for $\alpha_0 \in (a_0, a_1)$. For any strongly consistent family of estimators, $(\hat{\alpha}(t), t \geq 0)$ the system (2.1) with $A(\alpha_0), B, \phi$ as above, $\beta = 0$ and the adaptive control $(U(t), t \geq 0)$ given by

$$U(t) = -\psi(\hat{\alpha}(t - \Delta))V(\hat{\alpha}(t - \Delta))X(t)$$

has the self-optimizing property (2.30) by Theorem 2.7.

3. Adaptive Boundary Control for the Second Model. Before describing the model it is useful to highlight some of the differences between this model and the model in Section 2. In this model the unknowns appear in the linear transformation of the control in a more general way than in the first model. However, in this model the "highest order" operator in the infinitesimal generator is assumed to be known and only the "lower order" operators contain unknown parameters. For the identification of the unknown parameters that occur in the linear operator acting on the control, it is necessary to ensure that there is sufficient excitation. This is accomplished by a diminishing excitation that has no effect on the ergodic cost functional but ensures sufficient excitation for strong consistency. This approach was motivated by [1]. The control at time t is required to be measurable with respect to the past of the state process until time $t - \Delta$ where $\Delta > 0$ is arbitrary but fixed. This assumption can account for some delay in processing the information for the construction of the control. No boundedness assumptions are made on the range of the unknown parameters. The conditions for strong consistency of a family of least squares estimates in this section seem to be often more readily verifiable than those in Section 2.

The stochastic system is described by the stochastic evolution equation

$$(3.1) \qquad \begin{aligned} dX(t; \alpha) &= [A_0 + A_1(\alpha) + A_0 BC(\alpha)]X(t; \alpha)dt \\ &+ A_0 BD(\alpha)U(t)dt + G \, dW(t) \\ X(0; \alpha) &= x \end{aligned}$$

in a separable Hilbert space H with inner product $\langle \cdot, \cdot \rangle$ where A_0 is the infinitesimal generator of an exponentially stable analytic semigroup $(S_0(t),$

$t \geq 0$) on H, $A_0 = A_0^*$, $\alpha \in \mathcal{K} \subset \mathbb{R}^q$. Let D_A^γ for $\gamma \in \mathbb{R}$ be the domain of the fractional power $(-A_0)^\gamma$ with the $(-A_0)^\gamma$ graph norm. Let $B \in \mathcal{L}(H_1, D_A^\epsilon)$ for some $\epsilon \in (0,1)$, $A_1^*(\alpha) \in \mathcal{L}(D_A^\eta, H)$ for some $\eta \in [0,1)$, $C(\alpha) \in \mathcal{L}(H, H_1)$ and $D(\alpha) \in \mathcal{L}(H_2, H_1)$ for each value of $\alpha \in \mathbb{R}^m$ where H_1 and H_2 are separable Hilbert spaces. The formal process $(W(t), t \geq 0)$ is a cylindrical Wiener process with the incremental covariance the identity, $I \in \mathcal{L}(H)$, that is defined on a probability space $(\Omega, \mathcal{F}, P)$ with a filtration $(\mathcal{F}_1, t \geq 0)$. For $p \geq 2$ let $M_W^p(H_2) = \cap_{T>0} M_W^p(0, T, H_2)$ where

$$
\text{(3.2)} \qquad
\begin{aligned}
M_W^P(0, T, H_2) = \{U | U : [0, T] \times \Omega \to H_2, (U(t), t \geq 0) \\
\text{is } (\mathcal{F}_t) \text{ adapted and } E \int_0^T |U(t)|^P dt < \infty\}.
\end{aligned}
$$

The control process $(U(t), t \geq 0)$ in (3.1) is assumed to belong to the space $M_W^p(H_2)$ for some fixed $p > \max(\frac{1}{\epsilon}, \frac{1}{1-\eta})$ and $p \geq 2$.

For the control problem, the following ergodic, quadratic cost functional is used

$$
\text{(3.3)} \qquad J(\alpha, U) = \limsup_{t \to \infty} \frac{1}{t} J(t, x, \alpha, U)
$$

where

$$
J(t, x, \alpha, U) = \int_0^t [\langle Q_1 X(s, \alpha), X(s; \alpha) \rangle + \langle Q_2 U(s), U(s) \rangle] ds
$$

and $Q_1 = Q_1^* \in \mathcal{L}(H)$, $Q_1 \geq 0$, $Q_2 = Q_2^* \in \mathcal{L}(H_2)$ and $Q_2 \geq cI, c > 0$.

The following conditions are used:

(C1) $(-A_0)^{-\delta}G$ is Hilbert-Schmidt for some $0 \leq \delta < \frac{1}{2}$.

(C2) (Compactness of the resolvent) A_0^{-1} is compact.

(C3) (Continuous dependence on parameters) $A_1^*(\cdot), C(\cdot), D(\cdot)$ are continuous functions from the parameter set $\mathcal{K} \subset \mathbb{R}^q$ into $\mathcal{L}(D_A^\eta, H)$, $\mathcal{L}(H, H_1)$ and $\mathcal{L}(H_2, H_1)$ respectively.

(C4) (Uniform detectability and stabilizability) There are linear operators $F \in \mathcal{L}(H, H_2)$ and $K \in \mathcal{L}(H)$ and constants $c > 0$ and $\rho > 0$ such that
1. $|\exp[t(A_0 + A_1^*(\alpha) + C^*(\alpha)\psi + QK)|_{\mathcal{L}(H)} \leq ce^{-\rho t}$
2. $|\exp[t(A_0 + A_1^*(\alpha) + C^*(\alpha)\psi + F^*D^*(\alpha)\psi)]|_{\mathcal{L}(H)} \leq ce^{-\rho t}$ for all $t \geq 0$ and $\alpha \in \mathcal{K}$.

If (C4) is satisfied then there is a unique, nonnegative, self-adjoint linear operator V on H such that $V \in \mathcal{L}(H, D_A^\gamma)$ for all $\gamma \in (0, 1)$ and

$$
\begin{aligned}
\text{(3.4)} \qquad &\langle (A_0 + C(\alpha))x, Vy \rangle + \langle (A_0 + C(\alpha))y, Vx \rangle + \langle Q_1 x, y \rangle \\
&- \langle Q_2^{-1} B^*(\alpha)Vx, B^*(\alpha)Vy \rangle = 0.
\end{aligned}
$$

PROPOSITION 3.1. *If (C2-C4) are satisfied then*

$$(3.5) \qquad \lim_{\alpha \to \alpha_0} |V(\alpha) - V(\alpha_0)|_{\mathcal{L}(H, D_A^{1-\epsilon})} = 0$$

where $V(\cdot)$ is the solution of (3.4).

From Theorem 5.3 [13] it is clear that (C2) can be weakened e.g., (C2) can be replaced by the assumptions of stability of $(S(t; \alpha), t \geq 0, \alpha \in \mathcal{K})$ and the positivity of Q_1. However, (C2) is not restrictive for the examples that are described subsequently.

To estimate the parameters of the unknown system (3.1) a family of least squares estimates is given that is shown to be strongly consistent. Some additional conditions are introduced.

(C5) The semigroup generated by $A_0 + \mathcal{C}(\alpha)$ is stable for each α where
$\mathcal{C}(\alpha) = A_1(\alpha) + [C^*(\alpha)\psi]^*$.

(C6) The linear operators $A_1(\alpha), C(\alpha)$ and $D(\alpha)$ have the following form:

$$A_1(\alpha) = A_{10} + \sum_{i=1}^{q_1} \alpha^i A_{1i}$$

$$C(\alpha) = C_0 + \sum_{i=1}^{q_1} \alpha^i C_i$$

$$D(\alpha) = D_0 + \sum_{i=q_1+1}^{q} \alpha^i D_i$$

where $A_{1i}^* \in \mathcal{L}(D_A^\eta, H)$, $C_i \in \mathcal{L}(H, H_1)$ for $i = 0, ..., q_1$ and $D_i \in \mathcal{L}(H_2, H_1)$ for $i = 0, q_1 + 1, ..., q$. Define the linear operators $\mathcal{C}_i$ and $\mathcal{B}_i$ as follows

$$\mathcal{C}_i = A_{1i} + [C_i^*\psi]^*$$

for $i = 0, ..., q_1$ and

$$\mathcal{B}_i = [D_i^*\psi]^*$$

for $i = 0, q_1 + 1, ..., q$. Clearly $\mathcal{C}_i \in \mathcal{L}(H, D_A^{-\gamma})$ for $i = 0, ..., q_1$. Where $\gamma = \max(1 - \epsilon, \eta)$ and $\mathcal{B}_i \in \mathcal{L}(H_2, D_A^{\epsilon-1})$.

(C7) There is a finite dimensional projection $\tilde{P} : D_A^{-1} \to \tilde{P}(D_A^{-1}) \subset H$ and $(\tilde{P}\mathcal{B}_i, i = q_1 + 1, ...q)$ are linearly independent and for each nonzero $\beta \in \mathbb{R}^{q_1}$

$$tr \sum_{i=1}^{q_1} \beta_i(\tilde{P}(\mathcal{C}_i)) \int_0^{\Delta} S(r; \alpha_0) GG^* S^*(r; \alpha_0) dr \sum_{i=1}^{q_1} \beta_i(\tilde{P}(\mathcal{C}_i))^* > 0$$

where $(S(t; \alpha_0), t \geq 0)$ is the C_0-semigroup with the infinitesimal generator $A_0 + \mathcal{C}(\alpha_0)$.

Let $(\Omega, \mathcal{F}, P)$ denote a probability space for (3.1) where P includes a measure induced from the cylindrical Wiener process and a family of independent random variables for a diminishingly excited control introduced subsequently. $\mathcal{F}$ is the P-completion of an appropriate σ-algebra on Ω and $(\mathcal{F}_t, t \geq 0)$ is a filtration so that the cylindrical Wiener process $(W(t), t \geq 0)$, the solution $(X(t), t \geq 0)$ of (3.1) and the diminishingly excited control are adapted to $(\mathcal{F}_t, t \geq 0)$.

For the adaptive control problem it is convenient to enlarge the class of controls to $\tilde{M}_W^p(H_2) = \cap_{T>0} \tilde{M}_W^p(0, T, H_2)$ where

$$\tilde{M}_W^p(0, T, H_2) = \{U | U : [0, T] \times \Omega \to H_2,$$
$$(U(t), t \geq 0) \text{ is } (\mathcal{F}_t) \text{ adapted and}$$
$$\int_0^T |U(s)|^p ds < \infty \qquad a.s.\}$$

It is elementary to verify that the regularity properties of the sample paths of the solution of (3.1) with $U \in M_W^p(H_2)$ carry over to $U \in \tilde{M}_W^p(H_2)$.

Define $\varphi \times \varphi$ and $\varphi \times \beta$ where $\beta \in \mathbb{R}^q$ by the equations

$$\varphi \times \varphi = ((\langle \varphi_i, \varphi_j \rangle))$$
$$\varphi \times \beta = ((\langle \varphi_i, \beta \rangle)).$$

More generally if a is an ℓ-tuple of $\mathbb{R}^k$ vectors and b is an m-tuple of $\mathbb{R}^k$ vectors then define $a \times b \in \mathcal{L}(\mathbb{R}^\ell, \mathbb{R}^m)$ as $a \times b = ((\langle a_i, b_j \rangle))$. If $F \in \mathcal{L}(\tilde{P}D_A^{-1})$ then define $\tilde{F}\varphi$ by the equation

$$\tilde{F}\varphi = (F\varphi_i)$$

The stochastic differential equation for $(\tilde{P}X(t), t \geq 0)$ can be expressed as

$$d\tilde{P}X(t) = [\tilde{P}(A_0 + C_0)]X(t)dt$$

(3.6)
$$+ \tilde{P}B_0 U(t)dt + \varphi(t) \cdot \alpha dt + \tilde{P}GdW(t).$$

Fix $a > 0$ and define the $\mathcal{L}(\tilde{P}D_A^{-1})$-valued process $(\Gamma(t), t \geq 0)$ as

(3.7)
$$\Gamma(t) = \left(\int_0^t \varphi(s) \times \varphi(s)ds + a^{-1}I \right)^{-1}.$$

A family of least squares estimates $(\hat{\alpha}(t), t \geq 0)$ of the true parameter vector α_0 is defined as the solution of the following affine stochastic differential equation

(3.8)
$$d\hat{\alpha}(t) = \Gamma(t)[\varphi(t) \times (d\tilde{P}X(t) - \tilde{P}(A_0 + C_0)X(t)dt$$
$$- \tilde{P}B_0 U(t)dt - \varphi(t) \cdot \hat{\alpha}(t)dt]$$
$$\hat{\alpha}(0) = \alpha(0).$$

where $U \in \tilde{M}_W^p(H_2)$.

Let $\tilde{\alpha}(t) = \alpha_0 - \hat{\alpha}(t)$ for $t \geq 0$. The process $(\tilde{\alpha}(t), t \geq 0)$ satisfies the following stochastic differential equation

$$(3.9) \qquad d\tilde{\alpha}(t) = -\Gamma(t)[\varphi(t) \times (\varphi(t) \cdot \tilde{\alpha}(t)dt + \tilde{P}GdW(t))]$$
$$\tilde{\alpha}(0) = \alpha_0 - \alpha(0).$$

Since

$$(3.10) \qquad \frac{d\Gamma}{dt} = -\Gamma(t)[\varphi(t) \times \varphi(t)]\Gamma(t)$$
$$\Gamma(0) = aI$$

we have that the solution of (3.9) is

$$(3.11) \qquad \tilde{\alpha}(t) = -\Gamma(t)\Gamma^{-1}(0)\tilde{\alpha}(0) - \Gamma(t)\int_0^t (\varphi(s) \times \tilde{P}GdW(s)).$$

The control is a sum of a desired (adaptive) control and a diminishing excitation control. Let $(Z_n, n \in \mathbb{N})$ be a sequence of H_2-valued, independent, identically distributed, random variables that is independent of the cylindrical Wiener process $(W(t), t \geq 0)$. It is assumed that $EZ_n = 0$ and the covariance of Z_n is Λ for all n where Λ is positive and nuclear and there is a $\sigma > 0$ such that $|Z_n|^p \leq \sigma$ a.s. Choose $\tilde{\epsilon} \in (0, \frac{1}{2})$ and fix it. Define the H_2-valued process $(V(t), t \geq 0)$ as

$$(3.12) \qquad V(t) = \sum_{n=0}^{[t/\Delta]} \frac{Z_n}{n^{\tilde{\epsilon}/2}} 1_{[n\Delta,(n+1)\Delta)}(t).$$

Clearly we have that

$$(3.13) \qquad \lim_{t\to\infty} |V(t)| = 0 \qquad a.s.$$

and for each $\ell_1, \ell_2 \in H_2^* = H_2$

$$\lim_{t\to\infty} \frac{1}{t^{1-\tilde{\epsilon}}} \int_0^t \langle \ell_1, V(s) \rangle \langle \ell_2, V(s) \rangle ds$$

$$(3.14)$$
$$= \lim_{t\to\infty} \frac{1}{t^{1-\tilde{\epsilon}}} \sum_{i=1}^{[t/\Delta]} \frac{\langle \ell_1, Z_i \rangle \langle \ell_2, Z_i \rangle}{i^{\tilde{\epsilon}}} \Delta + o(1)$$

$$= \Delta^{\tilde{\epsilon}}(1 - \tilde{\epsilon})^{-1} \langle \Lambda \ell_1, \ell_2 \rangle \qquad a.s.$$

It is assumed that $Z_n \in \mathcal{F}_{n\Delta}$ and Z_n is independent of $\mathcal{F}_s$ for $s < n\Delta$ for all $n \in \mathbb{N}$.

The diminishingly excited control is

$$(3.15) \qquad\qquad U(t) = U^d(t) + V(t)$$

for all $t \geq 0$.

LEMMA 3.2. *Let $\tilde{\alpha}(t)$ be given by (3.11) for $t \geq 0$ and let $\lambda_{\min}(t)$ and $\lambda_{\max}(t)$ be the minimum eigenvalue and the maximum eigenvalue respectively of $\Gamma(t)$. The following inequality is satisfied*

$$(3.16) \qquad\qquad |\tilde{\alpha}(t)|^2 \leq 0 \left(\frac{\log \lambda_{\max}(t)}{\lambda_{\min}(t)} \right) \qquad a.s.$$

as $t \to \infty$.

THEOREM 3.3. *Let $\tilde{\epsilon} \in (0, \frac{1}{2})$ be determined from the definition of $(V(t), t \geq 0)$ in (3.12). If (C1-C7) are satisfied and the control process $(U(t), t \geq 0)$ for (3.1) is given by (3.15) where $U^d(t) \in \mathcal{F}((t-\Delta) \vee 0)$ for $t \geq 0, U^d \in \tilde{M}_W^p(H_2)$ and*

$$(3.17) \qquad\qquad \limsup_{t \to \infty} \frac{1}{t^{1+\delta}} \int_0^t |U^d(s)|^2 ds < \infty \qquad a.s.$$

for some $\delta \in [0, 1-2\tilde{\epsilon})$, then

$$(3.18) \qquad\qquad |\alpha_0 - \hat{\alpha}(t)|^2 = 0 \left(\frac{\log t}{t^\beta} \right) \qquad a.s.$$

as $t \to \infty$ for each $\beta \in (\frac{1+\delta}{2}, 1-\tilde{\epsilon})$ and $(\hat{\alpha}(t), t \geq 0)$ satisfies (3.8).

A self-optimizing adaptive control is constructed for the unknown linear stochastic system (3.1) with the ergodic quadratic cost functional (3.3) using the family of least squares estimates $(\hat{\alpha}(t), t \geq 0)$ that satisfies (3.8).

The family of admissible controls $\mathcal{U}(\Delta)$ for the minimization of (3.3) is

$$(3.19) \quad \mathcal{U}(\Delta) = \{ U : U(t) = U^d(t) + U^1(t), U^d(t) \in \mathcal{F}((t-\Delta) \vee 0)$$
$$\text{and } U^1(t) \in \sigma(V(s), (t-\Delta) \vee 0 \leq s \leq t) \text{ for all } t \geq 0,$$
$$U \in \tilde{M}_W^p(H_2), \limsup_{t \to \infty} \frac{|X(t)|^2}{t} = 0 \quad a.s., \quad \text{and}$$
$$\limsup_{t \to \infty} \frac{1}{t} \int_0^t (|X(s)|^2 + |U(s)|^2) ds < \infty \quad a.s.\}.$$

Since $A_0 + \mathcal{C}(\alpha_0)$ is the infinitesimal generator of a stable analytic semigroup it is known that for the deterministic infinite time boundary control problem with $G = 0$ there is a solution P of the algebraic Riccati equation that is formally expressed as

$$(3.20) \qquad\qquad A^*P + PA - P\tilde{B}Q_2^{-1}\tilde{B}^*P + Q_1 = 0$$

where $A = A_0 + \mathcal{C}(\alpha_0)$ and $\tilde{B} = \mathcal{B}(\alpha_0)$. This formal Riccati equation can be expressed as a precise inner product equation

$$(3.21) \qquad \langle Ax, Py \rangle + \langle Px, Ay \rangle - \langle Q_2^{-1}\tilde{B}^* Px, \tilde{B}^* Py \rangle + \langle Q_1 x, y \rangle = 0.$$

where $x, y \in \mathcal{D}(A)$. This solution P is the strong limit of the family of solutions of the differential Riccati equations as the final time tends to infinity. This solution is called the minimal solution of (3.20) or (3.21).

The solution P of the Riccati equation (3.21) satisfies the hypotheses for the Itô formula because $P \in \mathcal{L}(H, D_A^{1-\epsilon})$ from the results for the infinite time deterministic control problem [3,10,12], and $P \in \mathcal{L}(D_A^{-\delta}, D_A^{-\delta})$ because A is strictly negative. Thus we can apply the Itô formula to $(\langle PX(t(, X(t)), t \geq 0)$ and use (3.21) to obtain

$$
\begin{aligned}
\langle PX(t), X(t) \rangle - \langle Px, x \rangle - & \int_0^t [2\langle U(s), \tilde{B}^* PX(s) \rangle \\
(3.22) \qquad + \langle Q_2^{-1}\tilde{B}^* PX(s), \tilde{B}^* PX(s) \rangle & - \langle Q_1 X(s), X(s) \rangle] ds \\
+ t Tr(-A_0)^\delta PGG^*(-A_0)^{-\delta} & + 2 \int_0^t \langle PX(s), GdW(s) \rangle.
\end{aligned}
$$

Rewriting (4.4) we have

$$
\begin{aligned}
\langle PX(t), X(T) \rangle - \langle Px, x \rangle + & \int_0^t \langle Q_1 X(s), X(s) \rangle + \langle Q_2 U(s), U(s) \rangle ds \\
(3.23) \quad = & \int_0^t \langle U(s) + Q_2^{-1}\tilde{B}^* PX(s), Q_2(U(s) + Q_2^{-1}\tilde{B}^* PX(s) \rangle ds \\
+ t Tr(-A_0)^\delta PGG^*(-A_0)^{-\delta} & + 2 \int_0^t \langle PX(s), GdW(s) \rangle.
\end{aligned}
$$

Define the H-valued process $(\hat{X}(t), t \geq \Delta)$ by the equation

$$(3.24) \qquad \hat{X}(t) = S(\Delta; \alpha_0)X(t - \Delta) + \int_{t-\Delta}^t S(t - s; \alpha_0)\mathcal{B}(\alpha_0)V(s)ds.$$

Clearly for $t \geq \Delta$

$$X(t) = \hat{X}(t) + \int_{t-\Delta}^t S(t - s; \alpha_0)\mathcal{B}(\alpha_0)U(s)ds + \int_{t-\Delta}^t S(t - s; \alpha_0)GdW(s)$$

where $(X(t), t \geq 0)$ satisfies (3.1) and the input or control in (3.1) is a sum of V and $U \in \mathcal{U}(\Delta)$.

By Lemma 3.2 and (3.23) we have that for any $U \in \mathcal{U}(\Delta)$

$$(3.25) \quad \limsup_{t \to \infty} \frac{1}{t} J(x, U, \alpha_0, t) = Tr(-A_0)^\delta PGG^*(-A_0)^{-\delta}$$

$$+ \limsup_{t \to \infty} \frac{1}{t} \int_\Delta^t \langle U(s) + Q_2^{-1} \tilde{B}^* P \hat{X}(s)$$

$$+ Q_2^{-1} \tilde{B}^* P \left(\int_{s-\Delta}^s S(s - r; \alpha_0) \tilde{B} U(r) dr + \int_{s-\Delta}^s S(s - r; \alpha_0) GdW(r) \right),$$

$$Q_2(U(s) + Q_2^{-1} \tilde{B}^* P \hat{X}(s) + Q_2^{-1} \tilde{B}^* P \left(\int_{s-\Delta}^s S(s - r; \alpha_0) \tilde{B} U(r) dr \right.$$

$$+ \left. \int_{s-\Delta}^s S(s - r; \alpha_0) GdW(r) \right) \rangle ds$$

$$\geq Tr(-A_0)^\delta PGG^*(-A_0)^{-\delta} + Tr\tilde{B}^* PR(\Delta) P \tilde{B} Q_2^{-1} \qquad \qquad a.s.$$

where J is given by (3.3) and $R(\Delta)$ satisfies

$$R(\Delta) = \int_0^\Delta S(r; \alpha_0) GG^* S^*(r; \alpha_0) dr.$$

By Lemma 3.2 and (3.25) it is clear that $-Q_2^{-1} \tilde{B}^* P \hat{X}(\cdot) - \int_{\cdot - \Delta} Q_2^{-1} \tilde{B}^* P \hat{X}$
$(s) ds \in \mathcal{U}(\Delta)$ and it minimizes the ergodic cost functional (3.3) for the family off controls $\mathcal{U}(\Delta)$.

Define the H_2-valued (control) process $(U^0(t), t \geq \Delta)$ by the equation

$$\begin{aligned} U^0(t) &= -Q_2^{-1} \tilde{B}^*(t - \Delta) P(t - \Delta)(S(\Delta; t - \Delta) X(t - \Delta) \\ (3.26) \quad &+ \int_{t-\Delta}^t S(t - s; t - \Delta) \tilde{B}(t - \Delta) U^d(s) ds) \end{aligned}$$

where $\tilde{B}^*(t) = (\mathcal{B}^*(\hat{\alpha}(t)))^*$, $S(\tau; t) = e^{\tau A(t)}$ and $A(t)$ is defined as

$$(3.27) \quad A(t) = \begin{cases} A_0 + \mathcal{C}(\hat{\alpha}(t)) & \text{if } A_0 + \mathcal{C}(\hat{\alpha}(t)) \text{ is stable} \\ \tilde{A} & \text{otherwise} \end{cases}$$

and $\tilde{A}$ is a fixed stable infinitesimal generator (that is, the associated semigroup is stable) such that $\tilde{A} = A_0 + \mathcal{C}(\alpha_1)$ for some parameter vector α_1, $P(t)$ is the minimal solution of (3.21) using $A(t)$ and $\tilde{B}^*(t)$. It will be clear by the construction of U^d that $U^0 \in \mathcal{U}(\Delta)$.

Define two sequences of stopping times $(\sigma_n, n = 0, 1, ...)$ and $(\tau_n, n = 1, 2, ..)$ as follows:

$$\alpha_0 \equiv 0$$

$$(3.28) \qquad \sigma_n = \sup\left\{ t \geq \tau_n : \int_0^s |U^0(r)|^p dr \leq \tau_n^\delta s \text{ for all } s \in [\tau_n, t) \right\}$$

$$(3.29) \qquad \tau_n = \inf\left\{ t > \sigma_{n-1} + 1 : \int_0^t |U^0(r)|^p dr \leq t^{1+\delta/2} \right.$$

$$\left. \text{and } |X(t - \Delta)|^2 \leq t^{1+\delta/2} \right\}.$$

where $\delta > 0$ is fixed and $\frac{1+\delta}{2} < 1 - \epsilon$. It is clear that $(\tau_n - \sigma_{n-1}) \geq 1$ on $\{\sigma_{n-1} < \infty\}$ for all $n \geq 1$.

Define the adaptive control $(U^*(t), t \geq 0)$ by the equation

$$(3.30) \qquad\qquad U^*(t) = U^d(t) + V(t)$$

for $t \geq 0$ where

$$(3.31) \qquad U^d(t) = \begin{cases} 0 & \text{if } t \in [\sigma_n, \tau_{n+1}) \text{ for some } n \geq 0 \\ U^0(t) & \text{if } t \in [\tau_n, \sigma_n) \text{ for some } n \geq 1 \end{cases}$$

and $U^0(t), V(t)$ satisfy (3.26), (3.12) respectively. It is clear that $U^d \in \tilde{M}_W^p(H_2)$.

THEOREM 3.4. *If (C1-C7) are satisfied then the adaptive control $(U^*(t), t \geq 0)$ for (3.1) given by (3.30) is an element of $\mathcal{U}(\Delta)$ and is self-optimizing, that is,*

$$(3.32) \inf_{U \in \mathcal{U}(\Delta)} \limsup_{t \to \infty} \frac{1}{t} J(x, U, \alpha_0, t) = \lim_{t \to \infty} \frac{1}{t} J(x, U^*, \alpha_0, t)$$

$$= Tr(-A_0)^\delta PGG^*(-A_0)^{-\delta} + Tr\tilde{B}^* PR(\Delta)P\tilde{B}Q_2^{-1} \qquad a.s.$$

where J is given by (3.3).

4. Adaptive Control for some other Models. Another important family of linear distributed parameter systems with boundary or point control that is not included in analytic semigroups is hyperbolic systems. Hyperbolic systems include examples from beam, plate and wave equations. Since the hyperbolic systems lack the smoothing properties of the analytic semigroups, the optimal control problem for hyperbolic systems is more subtle than for analytic semigroups, e.g., the optimal control in general lacks any smoothing properties that occur with analytic semigroups. For the adaptive boundary control of a linear stochastic hyperbolic system the continuity of the solution of an algebraic Riccati equation with respect to parameters and the self-tuning of a certainty equivalence adaptive control are more difficult than for the adaptive control problems in Sections 2 and 3. Nonetheless some adaptive control problems for hyperbolic systems have been solved [5,6].

An unknown linear stochastic hyperbolic system with boundary or point control can be formally described by the following stochastic differential equation

$$dX(; \alpha) = (A(\alpha)X(t; \alpha) + B(\alpha)U(t))dt + CdW(t)$$
$$X(0; \alpha) = x$$

where $A(\alpha)$ is the generator of a C_0-semigroup of bounded linear operators on a separable Hilbert space H and $(W(t), .t \geq 0)$ is a cylindrical Wiener process. With suitable assumptions this stochastic differential equation can describe many stochastic hyperbolic systems and an adaptive control problem can be solved for an ergodic quadratic cost functional.

Since distributed parameter phenomena cannot always be effectively modelled by linear systems it is natural to investigate some nonlinear distributed parameter systems. A useful subfamily of nonlinear systems is semilinear systems. Semilinear equations are effectively influenced by the linear part of the equation. For adaptive control of semilinear systems a number of important questions arise. For invariant measures there are the questions of existence, uniqueness and continuous dependence on parameters. For adaptive control it is necessary to have some good information about the optimal control to establish self-tuning and self-optimality of an adaptive control. An adaptive control problem for a stochastic semilinear system is formulated and solved in [9]. Specifically the unknown stochastic system is described by the following stochastic differential equation

$$dX(t; \alpha) = (A(\alpha)X(t; \alpha) + F(\alpha, X(t; \alpha)) + U(t))dt + CdW(t)$$

where $A(\alpha)$ is the generator of a C_0-semigroup of bounded linear operators on a separable Hilbert space H and $(W(t), t \geq 0)$ is a cylindrical Wiener measure. The cost functional for the ergodic control problem is

$$J(x, U, \alpha) = \limsup_{T \to \infty} \frac{1}{T} E_x \left[\int_0^T g(X(t; \alpha)) + \frac{1}{2} |U(t)|^2 dt \right]$$

under suitable assumptions on the systems, the admissible control and the cost functional an adaptive control problem can be solved.

REFERENCES

[1] H.F. Chen, T.E. Duncan and B. Pasik-Duncan, *Stochastic adaptive control for continuous time linear systems with quadratic cost*, to appear in Appl. Math. Optim.

[2] S. Chen and R. Triggiani, *Proof of extensions of two conjectures on structural damping for elastic systems*, Pacific J. Math. 136 (1989), pp. 15–55.

[3] G. DaPrato and A. Ichikawa, *Riccati equations with unbounded coefficients*, Ann. Mat. Pura Appl. 140 (1985), pp. 209–221.

[4] T.E. Duncan, B. Goldys and B. Pasik-Duncan, *Adaptive control of linear stochastic evolution systems*, Stochastics and Stochastic Reports 35 (1991), pp. 129–142.

[5] T.E. Duncan, I. Lasiecka and B. Pasik-Duncan, *Some aspects of the adaptive boundary control of stochastic linear hyperbolic systems*, Proc. 32nd Conf. on Decision and Control (1993), pp. 2430–2434.

[6] T.E. Duncan, I. Lasiecka and B. Pasik-Duncan, *Adaptive boundary and point control of a stochastic linear hyperbolic systems*, in preparation.

[7] T.E. Duncan, B. Maslowski and B. Pasik-Duncan, *Adaptive boundary control of linear stochastic distributed parameter systems*, SIAM J. Control Optim., 32 (1994), pp. 648–672.

[8] T.E. Duncan, B. Maslowski and B. Pasik-Duncan, *Adaptive boundary control of linear distributed parameter systems described by analytic semigroups*, to appear in Appl. Math. Optim.

[9] T.E. Duncan and B. Pasik-Duncan, *Adaptive control of some stochastic semilinear equations*, Proc. 32nd Conf. on Decision and Control, (1993), pp. 2419–2423.

[10] F. Flandoli, *Direct solution of a Riccati equation arising in a stochastic control problem with control and observations on the boundary*, J. Appl. Math. Optim. 14 (1986), pp. 107–129.

[11] F. Flandoli, *Algebraic Riccati equation arising in boundary control problems*, SIAM J. Control Optim. 25 (1987), 612–636.

[12] I. Lasiecka and R. Triggiani, *The regulator problem for parabolic equations with Dirichlet boundary control I*, Appl. Math. Optim. 16 (1987), 147–168.

[13] I. Lasiecka and R. Triggiani, *Numerical approximations of algebraic Riccati equations modelled by analytic semigroups and applications*, Math. Computation 57 (1991), 639–662 and 513–537.

[14] A. Pazy, *Semigroups of Linear Operators and Applications to Partial Differential Equations*, Springer-Verlag, New York, 1983.

LMS IS H^∞ OPTIMAL*

BABAK HASSIBI[†], ALI H. SAYED[‡], AND THOMAS KAILATH[§]

Abstract. We show that the celebrated LMS (Least-Mean Squares) adaptive algorithm is H^∞ optimal. In other words, the LMS algorithm, which has long been regarded as an approximate least-mean squares solution, is in fact an exact minimizer of a certain so-called H^∞ error norm. In particular, the LMS minimizes the energy gain from the disturbances to the *predicted* errors, while the so-called normalized LMS minimizes the energy gain from the disturbances to the *filtered* errors. Moreover, since these algorithms are *central* H^∞ filters, they minimize a certain exponential cost function and are thus also risk-sensitive optimal (in the sense of Whittle). We discuss the various implications of these results, and show how they provide theoretical justification for the widely observed excellent robustness properties of the LMS filter.

1. Introduction. The LMS algorithm was originally conceived as an approximate recursive procedure that solves the following adaptive problem [1,2]: given a sequence of $1 \times M$ input row vectors $\{h_i\}$, and a corresponding sequence of desired responses $\{d(i)\}$, find an estimate of an $M \times 1$ column vector of weights w such that the sum squared error $\sum_{i=0}^{N} |d(i) - h_i w|^2$ is minimized. The LMS solution recursively updates estimates of the weight vector along the direction of the instantaneous gradient of the squared error. The introduction of the LMS adaptive filter by Widrow and Hoff in 1960 came as a significant development for a broad range of engineering applications since the LMS adaptive linear-estimation procedure requires essentially no advance knowledge of the signal statistics. Despite the name, however, we should note that the LMS algorithm does not minimize the sum of squared errors, and has long been thought to be an approximate minimizing solution.

Algorithms that exactly minimize the sum of squared errors, for every value of N are known: they are the well-known recursive least squares (RLS) algorithms (see, e.g., [3]). They have better convergence properties, but are computationally more complex, and are less robust than the sim-

* This work was supported in part by the Air Force Office of Scientific Research, Air Force Systems Command under Contract AFOSR91-0060 and by the Army Research Office under contract DAAL03-89-K-0109. This manuscript is submitted for publication with the understanding that the US Government is authorized to reproduce and distribute reprints for Government purposes notwithstanding any copyright notation thereon. The views and conclusions contained in this document are those of the authors and should not be interpreted as necessarily representing the official policies or endorsements, either express or implied, of the Air Force Office of Scientific Research or the U.S. Government.

† Contact author: Information Systems Laboratory, Stanford University, Stanford CA 94305. Phone (415) 723-1538 Fax (415) 723-8473 E-mail: hassibi@rascals.stanford.edu

‡ Department of Electrical and Computer Engineering, University of California, Santa Barbara, CA 93106.

§ Information Systems Laboratory, Stanford University, Stanford, CA 94305.

ple LMS algorithm. For example, it has been observed that the LMS has better tracking capabilities than the RLS algorithm in the presence of nonstationary inputs [3]. We show here that the superior robustness properties of the LMS algorithm are due to the fact that it is a minimax algorithm, or more specifically an H^∞ optimal algorithm. We shall define precisely what this means in Section 2, here we note only that the H^∞ criterion was introduced to address the fact that in many applications one is often faced with model uncertainties and lack of statistical information on the exogeneous signals. The great recent interest in H^∞ filtering may be seen from [4,5,6,7,8,9,10,11] and the many other references therein.

In this paper, we shall use some of the well known results in H^∞ estimation theory in order to show that the LMS algorithm is the so-called central *a priori* H^∞-optimal filter, while the so-called normalized LMS algorithm is the central *a posteriori* H^∞-optimal filter. This provides LMS with a rigorous basis and furnishes a minimization criterion that has long been missing. Morever, since LMS and normalized LMS are shown to be *central* filters they also minimize a certain exponential cost function, and are thus risk-sensitive optimal [16].

The remainder of the paper is organized as follows. In Section 2 we review the H^∞ estimation problem as one that minimizes the energy gain from the disturbances to the estimation error. We consider the a posteriori and a priori cases which correspond to filtered and predicted estimation errors, respectively. Section 3 gives the expressions for the H^∞ a posteriori and a priori filters, as well as their full parametrization, since such filters are not unique. In Section 4, we formulate the H^∞ adaptive filtering problem. Section 5 shows that the normalized LMS algorithm is the central a posteriori H^∞ optimal adaptive filter, and that if the learning rate is chosen appropriately, LMS is the central a priori H^∞ optimal adaptive filter. We then consider a simple example that demonstrates the robustness of LMS compared to RLS, and in Section 5.4 present a discussion on the merit of the different H^∞ optimal algorithms. With this in mind, we develop the full parametrization of all H^∞ optimal adaptive filters in Section 6, and in Section 7 show that LMS and normalized LMS have the additional property of being risk-sensitive optimal. This provides LMS and normalized LMS with a stochastic interpretation in the special case of disturbances that are independent Gaussian random variables. Section 8 offers a very brief summary.

We find it ironic that the LMS algorithm is not H^2 optimal, contrary to what its name suggests, but that it rather satisfies a minimax criterion. Moreover, in most H^∞ problems, the optimum solution has not been determined in closed form - what is usually determined is a certain type of suboptimal solution. We show, however, that for the adaptive problem at hand, the optimum solution can be determined.

2. The H^∞ problem. We first give a brief review of some of the results in H^∞ estimation theory using the notation of the companion papers [12,13]. The reader is also referred to [4,5,6,7,8,9,10,11] and the references therein for earlier results and alternative approaches.

We begin with the definition of the H^∞ norm of a transfer operator. As will presently become apparent, the motivation for introducing the H^∞ norm is to capture the worst case behaviour of a system.

Let h_2 denote the vector space of square-summable complex-valued causal sequences $\{f_k,\ 0 \le k < \infty\}$, viz.,

$$h_2 = \{\text{set of sequences } \{f_k\} \text{ such that } \sum_{k=0}^{\infty} f_k^* f_k < \infty\}$$

with inner product $< \{f_k\}, \{g_k\} > = \sum_{k=0}^{\infty} f_k^* g_k$, where $*$ denotes complex conjugation. Let T be a transfer operator that maps a causal input sequence $\{u_i\}$ to a causal output sequence $\{y_i\}$. Then the H^∞ norm of T is given by

$$\|T\|_\infty = \sup_{u \in h_2, u \ne 0} \frac{\|y\|_2}{\|u\|_2},$$

where the notation $\|u\|_2$ denotes the h_2-norm of the causal sequence $\{u_k\}$, viz.,

$$\|u\|_2^2 = \sum_{k=0}^{\infty} u_k^* u_k.$$

The H^∞ norm may be thus regarded as the maximum *energy gain* from the input u to the output y.

2.1. Formulation of the H^∞ problem. We now consider a state-space model of the form

$$
\begin{aligned}
x_{i+1} &= F_i x_i + G_i u_i, \quad x_0 \\
y_i &= H_i x_i + v_i
\end{aligned}
$$

(2.1)

where x_0, $\{u_i\}$, and $\{v_i\}$ are *unknown* quantities and y_i is the measured output. We can regard v_i as a measurement noise and u_i as a process noise or driving disturbance. Let z_i be linearly related to the state x_i via a given matrix L_i, viz.,

$$z_i = L_i x_i$$

We shall be interested in the following two cases. Let $\hat{z}_{i|i} = \mathcal{F}_f(y_0, y_1, \ldots, y_i)$ denote the estimate of z_i given observations $\{y_j\}$ from time 0 up to and including time i, and $\hat{z}_i = \mathcal{F}_p(y_0, y_1, \ldots, y_{i-1})$ denote the estimate of z_i given

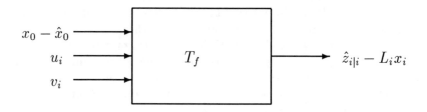

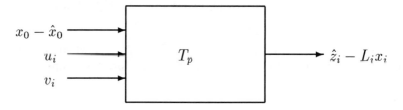

FIG. 2.1. *Transfer matrices from disturbances to filtered and predicted estimation error*

observations $\{y_j\}$ from time 0 to time $i - 1$. We then have the following two estimation errors: the *filtered* error

$$(2.2) \qquad e_{f,i} = \hat{z}_{i|i} - L_i x_i,$$

and the *predicted* error

$$(2.3) \qquad e_{p,i} = \hat{z}_i - L_i x_i.$$

Let T_f (T_p) denote the transfer operator that maps the unknown disturbances $\{x_0 - \hat{x}_0, u_i, v_i\}$ (where $\hat{x}_0$ denotes an initial guess of x_0) to the filtered (predicted) error $e_{f,i}$ ($e_{p,i}$). See Figure 2.1. The H^∞ estimation problem can now be stated as follows.

PROBLEM 1 (OPTIMAL H^∞ PROBLEM). *Find H^∞-optimal estimation strategies $\hat{z}_{i|i} = \mathcal{F}_f(y_0, y_1, \ldots, y_i)$ and $\hat{z}_i = \mathcal{F}_p(y_0, y_1, \ldots, y_{i-1})$ that respectively minimize $\| T_f \|_\infty$ and $\| T_p \|_\infty$, and obtain the resulting*

$$(2.4) \qquad \gamma_{f,o}^2 = \inf_{\mathcal{F}_f} \| T_f \|_\infty^2 = \inf_{\mathcal{F}_f} \sup_{x_0, u \in h_2, v \in h_2} \frac{\| e_f \|_2^2}{(x_0 - \hat{x}_0)^* \Pi_0^{-1} (x_0 - \hat{x}_0) + \| u \|_2^2 + \| v \|_2^2}$$

and

$$(2.5) \qquad \gamma_{p,o}^2 = \inf_{\mathcal{F}_p} \| T_p \|_\infty^2 = \inf_{\mathcal{F}_p} \sup_{x_0, u \in h_2, v \in h_2} \frac{\| e_p \|_2^2}{(x_0 - \hat{x}_0)^* \Pi_0^{-1} (x_0 - \hat{x}_0) + \| u \|_2^2 + \| v \|_2^2}$$

where Π_0 *is a positive definite matrix that reflects a priori knowledge as to how close* x_0 *is to the initial guess* $\hat{x}_0$.

Note that the infimum in (2.5) is taken over all *strictly* causal estimators $\mathcal{F}_p$, whereas in (2.4) the estimators $\mathcal{F}_f$ are causal since they have additional access to y_i. This is relevant since the solution to the H^∞ problem, as we shall see, depends on the structure of the information available to the estimator.

The above problem formulation shows that H^∞ optimal estimators guarantee the smallest estimation error energy over all possible disturbances of fixed energy. H^∞ estimators are thus over conservative, which reflects in a better robust behaviour to disturbance variation.

A closed form solution of the optimal H^∞ problem is available only for some special cases (one of which is the adaptive filtering problem to be studied), and a simpler problem results if one relaxes the minimization condition and settles for a suboptimal solution.

PROBLEM 2 (SUB-OPTIMAL H^∞ PROBLEM). *Given scalars* $\gamma_f > 0$ *and* $\gamma_p > 0$, *find estimation strategies* $\hat{z}_{i|i} = \mathcal{F}_f(y_0, y_1, \ldots, y_i)$ *and* $\hat{z}_i = \mathcal{F}_p(y_0, y_1, \ldots, y_{i-1})$ *that respectively achieve* $\| T_f \|_\infty \leq \gamma_f$ *and* $\| T_p \|_\infty \leq \gamma_p$. *This clearly requires checking whether* $\gamma_f \geq \gamma_{f,o}$ *and* $\gamma_p \geq \gamma_{p,o}$.

To guarantee $\| T_f \|_\infty \leq \gamma_f$ we shall proceed as follows. Let $T_{f,i}$ be the transfer operator that maps the disturbances $\left\{ x_0 - \hat{x}_0, \{u_j\}_{j=0}^i, \{v_j\}_{j=0}^i \right\}$ to the filtered errors $\left\{ \{e_{f,j}\}_{j=0}^i \right\}$. We shall find a γ_f that ensures $\| T_{f,i} \|_\infty < \gamma_f$ for all i. Likewise we shall find a γ_p that ensures for $\| T_{p,i} \|_\infty < \gamma_p$ for all i.

3. The H^∞ filters. We now briefly review some of the results on H^∞ filters using the notation of [12,13].

THEOREM 3.1 (THE H^∞ A POSTERIORI FILTER). *For a given* $\gamma > 0$, *if the* F_i *are nonsingular then an estimator with* $\|T_{f,i}\|_\infty < \gamma$ *exists if, and only if,*

$$(3.1) \qquad P_j^{-1} + H_j^* H_j - \gamma^{-2} L_j^* L_j > 0, \qquad j = 0, \ldots, i$$

where $P_0 = \Pi_0$ *and* P_j *satisfies the Riccati recursion*

$$(3.2) \qquad P_{j+1} = F_j P_j F_j^* + G_j G_j^* -$$

$$F_j P_j \begin{bmatrix} L_j^* & H_j^* \end{bmatrix} \left\{ \begin{bmatrix} -\gamma^2 I & 0 \\ 0 & I \end{bmatrix} + \begin{bmatrix} L_j \\ H_j \end{bmatrix} P_j \begin{bmatrix} L_j^* & H_j^* \end{bmatrix} \right\}^{-1} \begin{bmatrix} L_j \\ H_j \end{bmatrix} P_j F_j^*$$

If this is the case, then one possible H_∞ filter with level γ is given by

$$\hat{z}_{j|j} = L_j \hat{x}_{j|j}$$

where $\hat{x}_{j|j}$ is recursively computed as

$$(3.3) \qquad \hat{x}_{j+1|j+1} = F_j \hat{x}_{j|j} + K_{f,j}(y_{j+1} - H_{j+1} F_j \hat{x}_{j|j}), \qquad \hat{x}_{-1|-1}$$

and

$$(3.4) \qquad K_{f,j} = P_{j+1}H_{j+1}^*(I + H_{j+1}P_{j+1}H_{j+1}^*)^{-1}$$

THEOREM 3.2 (THE H^∞ A PRIORI FILTER). *For a given $\gamma > 0$, if the F_i are nonsingular then an estimator with $\|T_{p,i}\|_\infty < \gamma$ exists if, and only if,*

$$(3.5) \qquad \tilde{P}_j^{-1} = P_j^{-1} - \gamma^{-2}L_j^*L_j > 0, \qquad j = 0, \ldots, i$$

where P_j is the same as in Theorem 3.1. If this is the case, then one possible H_∞ filter with level γ is given by

$$(3.6) \qquad \hat{z}_j = L_j\hat{x}_j$$

$$(3.7) \qquad \hat{x}_{j+1} = F_j\hat{x}_j + K_{p,j}(y_j - H_j\hat{x}_j), \qquad \hat{x}_0$$

where

$$(3.8) \qquad K_{p,j} = F_j\tilde{P}_jH_j^*(I + H_j\tilde{P}_jH_j^*)^{-1}$$

Note that the above two estimators bear a striking resemblance to the celebrated Kalman filter:

$$(3.9) \quad \begin{aligned} \hat{x}_{j+1} &= F_j\hat{x}_j + F_jP_jH_j^*(I + H_jP_jH_j^*)^{-1}(y_j - H_j\hat{x}_j) \\ P_{j+1} &= F_jP_jF_j^* + G_jG_j^* - F_jP_j(I + H_jP_jH_j^*)^{-1}P_jF_j^* \end{aligned}$$

and that the only difference is that the P_j of equation (3.4), and $\tilde{P}_j$ of equation (3.8), satisfy Riccati recursions that differ with (3.9). However, as $\gamma \to \infty$, the Riccati recursion (3.2) collapses to the Kalman filter recursion (3.9). This suggests that the H^∞ norm of the Kalman filter may be quite large, indicating that it may have poor robustness properties.

It is also interesting that the structure of the H^∞ estimators depends, via the Riccati recursion (3.2), on the linear combination of the states that we intend to estimate (*i.e.* the L_i). This is as opposed to the Kalman filter, where the estimate of any linear combination of the state is given by that linear combination of the state estimate. Intuitively, this means that the H^∞ filters are specifically tuned towards the linear combination L_ix_i.

Note also that condition (3.5) is more stringent than condition (3.1), indicating that the existence of an a priori filter of level γ implies the existence of an a posteriori filter of level γ, but not necessarily vice versa.

We further remark that the filter of Theorem 3.1 (and Theorem 3.2) is one of many possible filters with level γ. A full parametrization of all estimators of level γ are given by the following Theorems. (For proofs see [13]).

THEOREM 3.3 (ALL H^∞ A POSTERIORI ESTIMATORS). *All H^∞ a posteriori estimators that achieve a level γ (assuming they exist) are given by*

$$(3.10) \qquad \hat{z}_{i|i} = L_i \hat{x}_{i|i} + [I - L_i(P_i^{-1} + H_i^* H_i)^{-1} L_i^*]^{\frac{1}{2}} \times$$

$$\mathcal{S}_i \left((I + H_i P_i H_i^*)^{\frac{1}{2}} (y_i - H_i \hat{x}_{i|i}), \ldots, (I + H_0 P_0 H_0^*)^{\frac{1}{2}} (y_0 - H_0 \hat{x}_{0|0}) \right)$$

where $\hat{x}_{i|i}$ is given by Theorem 3.1, and

$$\mathcal{S}(a_i, \ldots, a_0) = \begin{bmatrix} \mathcal{S}_0(a_0) \\ \mathcal{S}_1(a_1, a_0) \\ \vdots \\ \mathcal{S}_i(a_i, \ldots, a_0) \end{bmatrix}$$

is any (possibly nonlinear) contractive causal mapping, i.e.,

$$\sum_{j=0}^{i} |\mathcal{S}_j(a_j, \ldots, a_0)|^2 \leq \sum_{j=0}^{i} |a_j|^2$$

THEOREM 3.4 (ALL H^∞ A PRIORI ESTIMATORS). *All H^∞ a priori estimators that achieve a level γ (assuming they exist) are given by*

$$(3.11) \qquad \hat{z}_i = L_i \hat{x}_i + [I - L_i P_i L_i^*]^{\frac{1}{2}} \times$$

$$\mathcal{S}_i \left((I + H_{i-1} \tilde{P}_{i-1} H_{i-1}^*)^{-\frac{1}{2}} (y_{i-1} - H_{i-1} \hat{x}_{i-1}), \ldots, (I + H_0 \tilde{P}_0 H_0^*)^{-\frac{1}{2}} (y_0 - H_0 \hat{x}_0) \right)$$

where $\hat{x}_i$ and $\tilde{P}_i$ are given by Theorem 3.2, and $\mathcal{S}$ is any (possibly nonlinear) contractive causal mapping.

Note that although the filters obtained in Theorems 3.1 and 3.2 are linear, the full parametrization of all H^∞ filters with level γ are given by a *nonlinear* causal contractive mapping $\mathcal{S}$. The filters of Theorems 3.1 and 3.2 are known as the *central* filters and correspond to $\mathcal{S} = 0$.

4. Formulation of the H^∞ adaptive problem. Suppose we observe an output sequence $\{d_i\}$ that obeys the following model:

$$(4.1) \qquad d_i = h_i w + v_i$$

where $h_i = \begin{bmatrix} h_1(i) & h_2(i) & \ldots & h_M(i) \end{bmatrix}$ is a known $1 \times M$ vector whose elements are the inputs from M input channels ($h_k(i)$ denotes the input at time i to the kth channel), $w = \begin{bmatrix} w_1 & w_2 & \ldots & w_M \end{bmatrix}^T$ is an unknown $M \times 1$ weight vector, and v_i is an unknown disturbance, which may also include modelling errors. We shall not make any assumptions on the noise sequence $\{v_i\}$ (such as stationarity, whiteness, normal distributed, etc.).

Note that equation (4.1) can be restated in the following state-space form:

$$x_{i+1} = x_i, \qquad x_0 = w$$

(4.2)
$$d_i = h_i x_i + v_i$$

This is a relevant step since it reduces the adaptive filtering problem to an equivalent state-space estimation problem. For example, the RLS algorithm follows if one applies the Kalman filter to (4.2). Here we shall show that the LMS and normalized LMS algorithms follow from applying the H^∞ theory to (4.2).

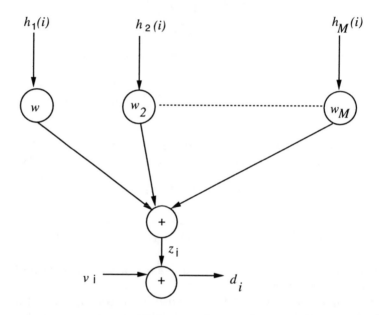

FIG. 4.1. *Signal model*

Consider the uncorrupted output $z_i = h_i x_i$ of (4.2). Let $\hat{z}_{i|i} = \mathcal{F}_f(d_0, d_1, \ldots, d_i)$ denote the estimate of z_i using the noisy measurements $\{d_j\}$ and the input vectors $\{h_j\}$ from time 0 up to and including i. Likewise, let $\hat{z}_i = \mathcal{F}_p(d_0, d_1, \ldots, d_{i-1})$ denote the estimate of z_i using the noisy measurements $\{d_j\}$ and the input vectors $\{h_j\}$ from time 0 to time $i-1$. As before, we have the *filtered* error

(4.3)
$$e_{f,i} = \hat{z}_{i|i} - h_i x_i,$$

and the *predicted* error

(4.4)
$$e_{p,i} = \hat{z}_i - h_i x_i.$$

Let T_f (T_p) denote the transfer operator that maps the unknown disturbances $\{w - \hat{w}_{|-1}, v_i\}$ (where $\hat{w}_{|-1}$ is an initial guess of w) to the filtered (predicted) error $e_{f,i}$ ($e_{p,i}$). The H^∞ adaptive filtering problem can then be stated as follows.

PROBLEM 3 (OPTIMAL H^∞ ADAPTIVE PROBLEM). *Find H^∞-optimal estimation strategies $\hat{z}_{i|i} = \mathcal{F}_f(d_0, d_1, \ldots, d_i)$ and $\hat{z}_i = \mathcal{F}_p(d_0, d_1, \ldots, d_{i-1})$ that respectively minimize $\| T_f \|_\infty$ and $\| T_p \|_\infty$, and obtain the resulting*

$$(4.5)\ \gamma_{f,o}^2 = \inf_{\mathcal{F}_f} \| T_f \|_\infty^2 = \inf_{\mathcal{F}_f} \sup_{w,v \in h_2} \frac{\| e_f \|_2^2}{(w - \hat{w}_{|-1})^* \Pi_0^{-1} (w - \hat{w}_{|-1}) + \| v \|_2^2}$$

and

$$(4.6)\ \gamma_{p,o}^2 = \inf_{\mathcal{F}_p} \| T_p \|_\infty^2 = \inf_{\mathcal{F}_p} \sup_{w,v \in h_2} \frac{\| e_p \|_2^2}{(w - \hat{w}_{|-1})^* \Pi_0^{-1} (w - \hat{w}_{|-1}) + \| v \|_2^2}$$

where Π_0 is a positive definite matrix that reflects a priori knowledge as to how close w is to the initial guess $\hat{w}_{|-1}$.

From now on we shall assume, without loss of generality, that Π_0 has the special form $\Pi_0 = \mu I$, where μ is a positive constant.

Before closing this section we should remark that the conventional H^2 (or least squares) criterion recursively minimizes the following cost function:

$$(4.7) \qquad \min_w\ (w - \hat{w}_{|-1})^* \Pi_0^{-1} (w - \hat{w}_{|-1}) + \sum_{j=0}^{i} |d(j) - h_j w|^2$$

When w and the $\{v_j\}$ are independent Gaussian random variables with variances Π_0 and I respectively, the above criterion yields the maximum likelihood estimate of w. The recursive solution in its most most natural form, involves propagating a Riccati variable, yielding the so-called RLS algorithm. It has long been thought that LMS is an approximate algorithm where the Riccati variable is set equal to a constant matrix (most commonly, a multiple of the identity matrix), which leads to a simpler and faster algorithm.

However, we shall presently see that the LMS algorithm, does in fact exactly minimize a *different* criterion, namely the H^∞ criterion.

Note that the H^∞ optimal adaptive filters guarantee the smallest estimation energy over all possible disturbances of fixed energy, and therefore will have better robust behaviour to disturbance variation. Moreover, in the special case when w and the $\{v_j\}$ are independent Gaussian random variables with variances Π_0 and I, respectively, we shall obtain an additional interpretation of the LMS algorithm, *viz.* it is an optimal risk-sensitive solution in the sense of Whittle.

5. Main result. At this point we need one more definition.

DEFINITION 5.1 (EXCITING INPUTS). *The input vectors h_i are called exciting if, and only if,*

$$\lim_{N \to \infty} \sum_{i=0}^{N} h_i h_i^* = \infty$$

5.1. The normalized LMS algorithm. We first consider the a posteriori filter and show that it collapses to the normalized LMS algorithm.

THEOREM 5.2 (NORMALIZED LMS ALGORITHM). *Consider the state-space model (4.2), and suppose we want to minimize the H^∞ norm of the transfer operator $T_{f,i}$ from the unknowns w and $\{v_j\}_{j=0}^{i}$ to the filtered error $\{e_{f,j} = \hat{z}_{j|j} - h_j w\}_{j=0}^{i}$. If the input data $\{h_j\}$ is exciting, then the minimum H^∞ norm is*

$$\gamma_{opt} = 1.$$

In this case the central optimal H^∞ a posteriori filter is

$$\hat{z}_{j|j} = h_j \hat{w}_{|j}$$

where $\hat{w}_{|j}$ is given by the normalized LMS algorithm with parameter μ:

$$(5.1) \qquad \hat{w}_{|j+1} = \hat{w}_{|j} + \frac{\mu h_{j+1}^*}{1 + \mu h_{j+1} h_{j+1}^*}(d_{j+1} - h_{j+1}\hat{w}_{|j}) , \qquad \hat{w}_{|-1}$$

Intuitively it is not hard to convince oneself that γ_{opt} cannot be less than one. To this end suppose that the estimator has chosen some initial guess $\hat{w}_{|-1}$. Then one may conceive of a disturbance that yields an observation that coincides with the output expected from $\hat{w}_{|-1}$, i.e.,

$$h_i \hat{w}_{|-1} = h_i w + v_i = d_i$$

In this case one expects that the estimator will not change its estimate of w, so that $\hat{w}_{|i} = \hat{w}_{|-1}$ for all i. Thus the filtered error is

$$e_{f,i} = h_i \hat{w}_{|i} - h_i w = h_i \hat{w}_{|-1} - h_i w = v_i$$

and the ratio in (4.5) can be made arbitrarily close to one.

The surprising fact though is that γ_{opt} is exactly one and that the normalized LMS algorithm achieves it. What this means is that normalized LMS guarantees that the energy of the filtered error will never exceed the energy of the disturbances. This is not true for other estimators. For example, in the case of the recursive least-squares (RLS) algorithm, one can come up with a disturbance of small energy that will yield a filtered error of large energy.

Proof of Theorem 5.2: We shall use the a posteriori filter of Theorem 3.1 with $F_i = I$, $G_i = 0$, $H_i = h_i$, and $L_i = h_i$. Thus the Riccati recursion simplifies to

$$P_{i+1} = P_i - P_i \begin{bmatrix} h_i^* & h_i^* \end{bmatrix} \left\{ \begin{bmatrix} -\gamma^2 I & 0 \\ 0 & I \end{bmatrix} + \begin{bmatrix} h_j \\ h_j \end{bmatrix} P_i \begin{bmatrix} h_i^* & h_i^* \end{bmatrix} \right\}^{-1} \begin{bmatrix} h_i \\ h_i \end{bmatrix} P_i$$

which, using the matrix inversion lemma, implies that

$$\begin{aligned} P_{i+1}^{-1} &= P_i^{-1} + \begin{bmatrix} h_i^* & h_i^* \end{bmatrix} \begin{bmatrix} -\gamma^{-2} I & 0 \\ 0 & I \end{bmatrix} \begin{bmatrix} h_i \\ h_i \end{bmatrix} \\ &= P_i^{-1} + (1 - \gamma^{-2}) h_i^* h_i \end{aligned}$$

Consequently, starting with $P_0^{-1} = \mu^{-1} I$, we get

$$(5.2) \qquad P_{i+1}^{-1} = \mu^{-1} I + (1 - \gamma^{-2}) \sum_{j=0}^{i} h_j^* h_j$$

Now we need to check the existence condition (3.1) and find the optimum γ_{opt}. It follows from the above expression for P_{i+1}^{-1} that we have

$$(5.3) \quad P_{i+1}^{-1} + H_{i+1}^* H_{i+1} - \gamma^{-2} L_{i+1}^* L_{i+1} = \mu^{-1} I + (1 - \gamma^{-2}) \sum_{j=0}^{i+1} h_j^* h_j$$

Suppose $\gamma < 1$ so that $1 - \gamma^{-2} < 0$. Since the $\{h_j\}$ are exciting, we conclude that for some k, and for large enough i, we must have

$$\sum_{j=0}^{i+1} |h_k(j)|^2 > \frac{\mu^{-1}}{\gamma^{-2} - 1}$$

This implies that the kth diagonal entry of the matrix on the right hand side of (5.3) is negative, viz.,

$$\mu^{-1} + (1 - \gamma^{-2}) \sum_{j=0}^{i+1} |h_k(j)|^2 < 0$$

Consequently, $P_{i+1}^{-1} + H_{i+1}^* H_{i+1} - \gamma^{-2} L_{i+1}^* L_{i+1}$ cannot be positive-definite. Therefore, $\gamma_{opt} \geq 1$. We now verify that γ_{opt} is indeed 1. For this purpose, we note that if we consider $\gamma = 1$ then from equation (5.2) we have $P_i = \mu I > 0$ for all i and the existence condition is satisfied. If we now write the a posteriori filter for $\gamma_{opt} = 1$, with $P_i = \mu I$, we get the desired so-called normalized LMS algorithm.

$$\square$$

5.2. The LMS algorithm. We now apply the a priori H^∞-filter and show that it collapses to the LMS algorithm.

THEOREM 5.3 (LMS ALGORITHM). *Consider the state-space model (4.2), and suppose we want to minimize the H^∞ norm of the transfer operator $T_{p,i}$ from the unknowns w and $\{v_j\}_{j=0}^i$ to the predicted error $\{e_{p,j} = \hat{z}_j - h_j w\}_{j=0}^i$. If the input data $\{h_j\}$ is exciting, and*

$$(5.4) \qquad\qquad 0 < \mu < \inf_i \frac{1}{h_i h_i^*}$$

then the minimum H^∞ norm is

$$\gamma_{opt} = 1.$$

In this case the central optimal a priori H^∞ filter is

$$\hat{z}_j = h_i \hat{w}_{|j-1}$$

where $\hat{w}_{|j-1}$ is given by the LMS algorithm with learning rate μ, viz.,

$$(5.5) \qquad \hat{w}_{|j} = \hat{w}_{|j-1} + \mu h_j^* (d_j - h_j \hat{w}_{|j-1}), \qquad \hat{w}_{|-1}$$

Proof: The proof is similar to that for the normalized LMS case. For $\gamma < 1$ the matrix $\tilde{P}_i$ of Theorem 3.2 cannot be positive-definite. For $\gamma = 1$ we get $P_i = \mu I > 0$ for all i, and

$$\begin{aligned} \tilde{P}_i^{-1} &= P_i^{-1} - L_i^* L_i \\ &= \mu^{-1} I - h_i^* h_i \end{aligned}$$

It is straightforward to see that the the eigenvalues of $\tilde{P}_i^{-1}$ are

$$\{\mu^{-1}, \mu^{-1}, ..., \mu^{-1}, \mu^{-1} - h_i h_i^*\}$$

Thus $\tilde{P}_i^{-1}$ is positive definite if, and only if, (5.4) is satisfied, which leads to $\gamma_{opt} = 1$. Writing the H^∞ a priori filter equations for $\gamma = 1$ yields

$$\begin{aligned} \hat{w}_{|i} &= \hat{w}_{|i-1} + \tilde{P}_i h_i^* (I + h_i \tilde{P}_i h_i^*)^{-1}(d_i - h_i \hat{w}_{|i-1}) \\ &= \hat{w}_{|i-1} + \tilde{P}_i (I + h_i^* h_i \tilde{P}_i)^{-1} h_i^* (d_i - h_i \hat{w}_{|i-1}) \\ &= \hat{w}_{|i-1} + (\tilde{P}_i^{-1} + h_i^* h_i)^{-1} h_i^* (d_i - h_i \hat{w}_{|i-1}) \\ &= \hat{w}_{|i-1} + \mu h_i^* (d_i - h_i \hat{w}_{|i-1}) \end{aligned}$$

$\square$

The above result indicates that if the learning rate μ is chosen according to (5.4), then LMS ensures that the energy of the predicted error will never exceed the energy of the disturbances. It is interesting that we have obtained an upper bound on the learning rate μ that guarantees this H^∞

optimality, since it is a well known fact that LMS behaves poorly if the learning rate is chosen too large. It is also interesting to compare the bound in (5.4) with the bound studied in [2] and [21].

We further note that if the input data is not exciting, then $\sum_{i=0}^{\infty} h_i^* h_i$ will have a finite limit, and the minimum H^∞ norm of the a posteriori and a priori filters will be the smallest γ that ensures

$$\mu^{-1} I + (1 - \gamma^{-2}) \sum_{i=0}^{\infty} h_i^* h_i > 0$$

This will in general yield $\gamma_{opt} < 1$, and Theorems 3.1 and 3.2 can be used to write the optimal filters for this γ_{opt}. In this case the LMS and normalized LMS algorithms will still correspond to $\gamma = 1$, but will now be suboptimal.

5.3. Example. To illustrate the robustness of the LMS algorithm we consider a special case of model (4.2) where h_i is now a scalar that randomly takes on the values $+1$ and -1.

$$\begin{aligned} x_{i+1} &= x_i, & x_0 = w \\ d_i &= h_i x_i + v_i \end{aligned}$$

(5.6)

Assuming we have observed N points of data, we can then use the LMS algorithm to write the transform operator $T_{lms,N}(\mu)$ that maps the disturbances $\{\mu^{-\frac{1}{2}} x_0, v_i\}$ to the $\{e_{p,i}\}$.

(5.7)

$$\begin{bmatrix} e_{p,0} \\ e_{p,1} \\ \vdots \\ e_{p,N-1} \end{bmatrix} =$$

$$\underbrace{\begin{bmatrix} \mu^{\frac{1}{2}} h_0 & 0 & 0 & \cdots & 0 \\ \mu^{\frac{1}{2}}(1-\mu)h_1 & -\mu h_1 h_0 & 0 & \cdots & 0 \\ \mu^{\frac{1}{2}}(1-\mu)^2 h_2 & -\mu(1-\mu)h_2 h_0 & -\mu h_2 h_1 & \cdots & 0 \\ \vdots & \vdots & \vdots & \ddots & \vdots \\ \mu^{\frac{1}{2}}(1-\mu)^{N-1} h_{N-1} & -\mu(1-\mu)^{N-2} h_{N-1} h_0 & -\mu(1-\mu)^{N-3} h_{N-1} h_1 & \cdots & -\mu h_{N-1} h_{N-2} \end{bmatrix}}_{T_{lms,N}(\mu)}$$

$$\begin{bmatrix} \mu^{-\frac{1}{2}} x_0 \\ v_0 \\ \vdots \\ v_{N-2} \end{bmatrix}$$

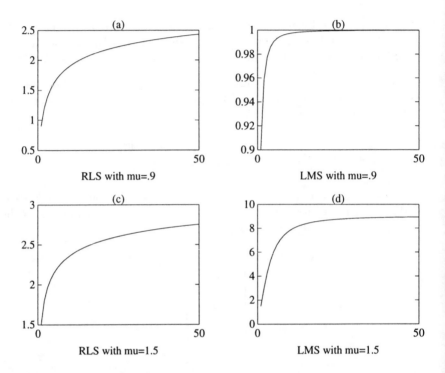

FIG. 5.1. *Maximum singular value of transfer operators* $T_{lms,N}(\mu)$ *and* $T_{rls,N}(\mu)$ *as a function of* N *for the values* $\mu = .9$ *and* $\mu = 1.5$.

Suppose now we use the RLS algorithm (*viz.* the Kalman filter) to estimate the states in (5.6), *i.e.*,

$$\hat{x}_{i+1} = \hat{x}_i + k_{p,i}(d_i - h_i\hat{x}_i)$$

where

$$k_{p,i} = \frac{p_i h_i^*}{1 + p_i|h_i|^2} \, ,$$

and

(5.8) $$p_{i+1} = p_i - \frac{|h_i|^2 p_i^2}{1 + p_i|h_i|^2} = p_i - \frac{p_i^2}{1 + p_i} = \frac{p_i}{1 + p_i} \quad , \quad p_0 = \mu$$

then we can write the transfer operator $T_{rls,N}$ that maps the disturbances to the predicted errors as follows:

(5.9)
$$\begin{bmatrix} e'_{p,0} \\ e_{p,1} \\ \vdots \\ e'_{p,N-1} \end{bmatrix} =$$

$$\underbrace{\begin{bmatrix} \mu^{\frac{1}{2}} h_0 & 0 & 0 & \cdots & 0 \\ \mu^{\frac{1}{2}} \frac{h_1}{1+\mu} & -\mu\frac{h_1 h_0}{1+\mu} & 0 & \cdots & 0 \\ \mu^{\frac{1}{2}} \frac{h_2}{1+2\mu} & -\mu\frac{h_2 h_0}{1+2\mu} & -\mu\frac{h_2 h_1}{1+2\mu} & \cdots & 0 \\ \vdots & \vdots & \vdots & \ddots & \vdots \\ \mu^{\frac{1}{2}} \frac{h_{N-1}}{1+(N-1)\mu} & -\mu\frac{h_{N-1} h_0}{1+(N-1)\mu} & -\mu\frac{h_{N-1} h_1}{1+(N-1)\mu} & \cdots & -\mu\frac{h_{N-1} h_{N-2}}{1+(N-1)\mu} \end{bmatrix}}_{T_{rls,N}(\mu)}$$

$$\begin{bmatrix} \mu^{-\frac{1}{2}} x_0 \\ v_0 \\ \vdots \\ v_{N-2} \end{bmatrix}$$

We now study the maximum singular values of $T_{lms,N}(\mu)$ and $T_{rls,N}(\mu)$ as a function of μ and N. Note that in this special problem, condition (5.4) implies that μ must be less than one to guarantee the H^∞ optimality of LMS. Therefore we chose the two values $\mu = .9$ and $\mu = 1.5$ (one greater and one less than $\mu = 1$). The results are illustrated in Figure 5.1 where the maximum singular values of $T_{lms,N}(\mu)$ and $T_{rls,N}(\mu)$ are plotted against the number of observations N. As expected, for $\mu = .9$ the maximum singular value of $T_{lms,N}(\mu)$ remains constant at one, whereas the maximum singular value of $T_{rls,N}(\mu)$ is greater than one and increases with N. For $\mu = 1.5$

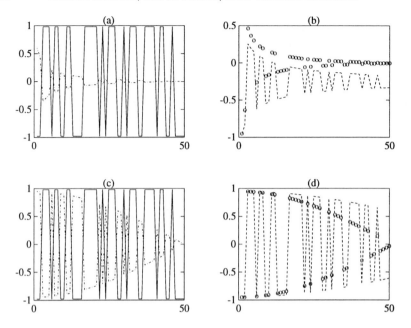

FIG. 5.2. *Worst case disturbances and the corresponding predicted errors for RLS and LMS. (a) The solid line represents the uncorrupted output $h_i x_i$ and the dashed line represents the worst case RLS disturbance. (b) The dashed line and the dotted line represent the RLS and LMS predicted errors, respectively, for the worst case RLS disturbance. (c) The solid line represents the uncorrupted output $h_i x_i$ and the dashed line represents the worst case LMS disturbance. (d) The dashed line and the dotted line represent the RLS and LMS predicted errors, respectively, for the worst case LMS disturbance.*

both RLS and LMS display maximum singular values greater than one, with the performance of LMS being significantly worse.

Figure 5.2 shows the worst case disturbance signals for the RLS and LMS algorithms in the $\mu = .9$ case, and the corresponding predicted errors. These worst case disturbances are found by computing the maximum singular vectors of $T_{rls,50}(.9)$ and $T_{lms,50}(.9)$, respectively. The worst case RLS disturbance, and the uncorrupted output $h_i x_i$, are depicted in Figure 5.2a. As can be seen from Figure 5.2b the corresponding RLS predicted error does not go to zero (it is actually biased), whereas the LMS predicted error does. The worst case LMS disturbance signal is given in Figure 5.2c, and as before, the LMS predicted error tends to zero, while the RLS predicted error does not. The form of the worst case disturbances (especially for RLS) are quite interesting; they compete with the true output early on, and then go to zero.

The disturbance signals considered in this example are rather contrived and may not happen in practice. However, they serve to illustrate the fact

that the RLS algorithm may have poor performance even if the disturbance signals have small energy. On the other hand, LMS will have robust
performance over a wide range of disturbance signals.

5.4. Discussion. In Section 5.1 we motivated the $\gamma_{opt} = 1$ result
for normalized LMS by considering a disturbance strategy that made the
observed output d_i coincide with the expected output $h_i \hat{w}_{|-1}$. It is now
illuminating to consider the *dual* strategy for the estimator.

Recall that in the a posteriori adaptive filtering problem the estimator
has access to observations $d_0, d_1, \ldots, d_i$ and is required to construct an
estimate $\hat{z}_{i|i}$ of the uncorrupted output $z_i = h_i x_i$. The dual to the above
mentioned disturbance strategy would be to construct an estimate that
coincides with the observed output, *viz.*,

$$(5.10) \qquad\qquad \hat{z}_{i|i} = d_i$$

The corresponding filtered error is:

$$e_{f,i} = \hat{z}_{i|i} - h_i x_i = d_i - h_i x_i = v_i$$

Thus the ratio in (4.5) can be made arbitrarily close to one, and the estimator (5.10) will achieve the same $\gamma_{opt} = 1$ that the normalized LMS
algorithm does.

Formally, the estimator (5.10) may be obtained from the normalized
LMS algorithm (5.1) by letting $\mu \to \infty$. However, (5.10) will achieve
$\gamma_{opt} = 1$ for any value of μ.

The fact that the simplistic estimator (5.10) (which is obviously of no
practical use) is an optimal H^∞ a posteriori filter seems to question the
very merit of being H^∞ optimal. A first indication towards this direction
may be the fact that the H^∞ estimators that achieve a certain level γ are
nonunique. In our opinion the property of being H^∞ optimal (i.e. of minimizing the energy gain from the disturbances to the errors) is a desirable
property in itself. The sensitivity of the RLS algorithm to different disturbance signals, as illustrated in the example of Section 5.3, clearly indicates
the desirability of the H^∞ optimality property. However, different estimators in the set of all H^∞ optimal estimators may have drastically different
behaviour with respect to other *desirable* performance measures.

In Section 6 we shall develop the full parametrization of all H^∞ optimal
a posteriori and a priori adaptive filters, and show how to obtain (5.10) as
a special case of this parametrization. As indicated in Theorems 5.2 and
5.3, the LMS and normalized LMS algorithms correspond to the so-called
central filters. These central filters have other desirable properties that we
shall discuss in Section 7: they are risk-sensitive optimal (i.e. they optimize
a certain exponential cost criterion) and can also be shown to be maximum
entropy.

The main problem with the estimator (5.10) is that it makes absolutely no use of the state-space model (4.2). We should note that it is not possible to come up with such a simple minded estimator in the a priori case: indeed as we shall see in the next section, the a priori estimator corresponding to (5.10) is highly nontrivial. The reason seems to be that since in the a priori case one deals with predicted error energy, it is inevitable that one must make use of the state-space model (4.2) in order to construct an optimal prediction of the *next* output. Thus in the a priori case, the problems arising from such unreasonable estimators such as (5.10) are avoided.

6. All H^∞ adaptive filters. In Section 5.4 we came up with an alternative optimal H^∞ a posteriori filter. We shall presently use the results of Theorems 3.3 and 3.4 to parametrize all optimal H^∞ a priori and a posteriori filters.

THEOREM 6.1 (ALL H^∞ A POSTERIORI ADAPTIVE FILTERS). *All H^∞ optimal a posteriori adaptive filters that achieve $\gamma_{opt} = 1$ are given by*

$$(6.1) \quad \hat{z}_{i|i} = h_i \hat{x}_{i|i} + (I + \mu h_i h_i^*)^{-\frac{1}{2}}$$
$$\mathcal{S}_i \left((I + \mu h_i h_i^*)^{\frac{1}{2}} (d_i - h_i \hat{x}_{i|i}), \ldots, (I + \mu h_0 h_0^*)^{\frac{1}{2}} (d_0 - h_0 \hat{x}_{0|0}) \right)$$

where $\hat{x}_{i|i}$ is the estimated state of the normalized LMS algorithm with parameter μ, and

$$\mathcal{S}(a_i, \ldots, a_0) = \begin{bmatrix} \mathcal{S}_0(a_0) \\ \mathcal{S}_1(a_1, a_0) \\ \vdots \\ \mathcal{S}_i(a_i, \ldots, a_0) \end{bmatrix}$$

is any (possibly nonlinear) contractive causal mapping, i.e.,

$$\sum_{j=0}^{i} |\mathcal{S}_j(a_j, \ldots, a_0)|^2 \leq \sum_{j=0}^{i} |a_j|^2$$

Proof: Using the result of Theorem 3.3 with $H_i = h_i$ and $L_i = h_i$, the full parametrization of all H^∞ a posteriori adaptive filters is given by

$$(6.2) \quad \hat{z}_{i|i} = h_i \hat{x}_{i|i} + [I - h_i(P_i^{-1} + h_i^* h_i)^{-1} h_i^*]^{\frac{1}{2}}$$
$$\mathcal{S}_i \left((I + h_i P_i h_i^*)^{\frac{1}{2}} (d_i - h_i \hat{x}_{i|i}), \ldots, (I + h_0 P_0 h_0^*)^{\frac{1}{2}} (d_0 - h_0 \hat{x}_{0|0}) \right)$$

Now from the proof of Theorem 5.2 we know that for all a posteriori filters that achieve $\gamma_{opt} = 1$, we have $P_i = \mu I$. Moreover we have the identity

$$I - h_i(P_i^{-1} + h_i^* h_i)^{-1} h_i^* = (I + h_i P_i h_i^*)^{-1}$$

Replacing the above expression along with $P_i = \mu I$ into (6.2) yields the desired result.

☐

At this point we should note the significance of some special choices for the causal contraction $\mathcal{S}$.

- $\mathcal{S} = 0$: This yields the normalized LMS algorithm.
- $\mathcal{S} = I$: This yields

$$\hat{z}_{i|i} = h_i \hat{x}_{i|i} + (I + \mu h_i h_i^*)^{-\frac{1}{2}}(I + \mu h_i h_i^*)^{\frac{1}{2}}(d_i - h_i \hat{x}_{i|i}) = d_i$$

which is the simple minded estimator of Section 5.4.

- $\mathcal{S} = -I$: This yields

$$\hat{z}_{i|i} = h_i \hat{x}_{i|i} - (I + \mu h_i h_i^*)^{-\frac{1}{2}}(I + \mu h_i h_i^*)^{\frac{1}{2}}(d_i - h_i \hat{x}_{i|i}) = 2h_i \hat{x}_{i|i} - d_i$$

Thus it is quite obvious that the different H^∞ optimal a posteriori adaptive filters may have quite different behaviour with respect to other desirable criteria.

THEOREM 6.2 (ALL H^∞ A PRIORI ADAPTIVE FILTERS). *If the input data $\{h_i\}$ is exciting, and*

$$0 < \mu < \inf_i \frac{1}{h_i h_i^*}$$

then all H^∞ optimal a priori adaptive filters are given by

(6.3)
$$\hat{z}_i = h_i \hat{x}_i + (I - \mu h_i h_i^*)^{\frac{1}{2}}$$

$$\mathcal{S}_i\Big((I - \mu h_{i-1} h_{i-1}^*)^{\frac{1}{2}}(d_{i-1} - h_{i-1}\hat{x}_{i-1}), \ldots, (I - \mu h_0 h_0^*)^{\frac{1}{2}}(d_0 - h_0 \hat{x}_0)\Big)$$

where $\hat{x}_i$ is the state estimate of the LMS algorithm with learning rate μ, and $\mathcal{S}$ is any (possibly nonlinear) contractive causal mapping.

Proof: Using the result of Theorem 3.4 with $H_i = h_i$ and $L_i = h_i$, the full parametrization of all H^∞ a priori adaptive filters is given by

(6.4)
$$\hat{z}_i = h_i \hat{x}_i + [I - h_i P_i h_i^*]^{\frac{1}{2}}$$

$$\mathcal{S}_i\Big((I + h_{i-1}\tilde{P}_{i-1}h_{i-1}^*)^{-\frac{1}{2}}(d_{i-1} - h_{i-1}\hat{x}_{i-1}), \ldots, (I + h_0 \tilde{P}_0 h_0^*)^{-\frac{1}{2}}(d_0 - h_0 \hat{x}_0)\Big)$$

where $\tilde{P}_i = (P_i^{-1} - h_i^* h_i)^{-1}$. Now from the proof of Theorem 5.3 we know that for all a priori filters that achieve $\gamma_{opt} = 1$, we have $P_i = \mu I$. Moreover we have the identity

$$I + h_i \tilde{P}_i h_i^* = I + h_i (P_i^{-1} - h_i^* h_i)^{-1} h_i = (I - h_i P_i h_i^*)^{-1}$$

Replacing the above expression along with $P_i = \mu I$ into (6.4) yields the desired result.

☐

It is once more interesting to note the consequences of some special choices of the causal contraction $\mathcal{S}$.

- $S = 0$: This yields the LMS algorithm.
- $S = I$: This yields

$$\hat{z}_i = h_i \hat{x}_i + (I - \mu h_i h_i^*)^{\frac{1}{2}} (I - \mu h_{i-1} h_{i-1}^*)^{\frac{1}{2}} (d_{i-1} - h_{i-1} \hat{x}_{i-1})$$

which is the a priori adaptive filter that corresponds to the simple minded estimator of Section 5.4. Note that in this case the filter is highly nontrivial.

- $S = -I$: This yields

$$\hat{z}_i = h_i \hat{x}_i - (I - \mu h_i h_i^*)^{\frac{1}{2}} (I - \mu h_{i-1} h_{i-1}^*)^{\frac{1}{2}} (d_{i-1} - h_{i-1} \hat{x}_i)$$

Note that it does not seem possible to obtain a simplistic a priori estimator that achieves optimal performance.

7. Risk-sensitive optimality. In this section we shall focus on a certain property of the central H^∞ filters, namely the fact that they are risk-sensitive optimal filters. This will give further insight into the LMS and normalized LMS algorithms, and in particular will provide a stochastic interpretation in the special case of disturbances that are independent Gaussian random variables.

The risk-sensitive (or exponential cost) criterion was introduced in [14] and further studied in [15,16,17]. We begin with a brief introduction to the risk-sensitive criterion. For much more on this subject consult the recent textbook [16].

7.1. The exponential cost function. Although it is straightforward to consider the risk-sensitive criterion in the full generality of the state-space model of Section 2, we shall only deal with the special case of our interest. To this end, consider the state-space model corresponding to the adaptive filtering problem we have been studying:

$$\begin{aligned}
x_{i+1} &= x_i, & x_0 = w \\
(7.1) \qquad d_i &= h_i x_i + v_i
\end{aligned}$$

where w and the $\{v_i\}$ are now independent Gaussian random variables with covariances Π_0 and I, respectively. Moreover, w is assumed to have mean $\hat{w}_{|-1}$, and the $\{v_i\}$ are assumed to be zero mean. As before, we are interested in the filtered and predicted estimates $\hat{z}_{i|i} = \mathcal{F}_f(d_0, d_1, \ldots, d_i)$ and $\hat{z}_i = \mathcal{F}_f(d_0, d_1, \ldots, d_{i-1})$ of the uncorrupted output $z_i = h_i x_i$. The corresponding filtered and predicted errors are given by $e_{f,i} = \hat{z}_{i|i} - z_i$ and $e_{p,i} = \hat{z}_i - z_i$. The conventional Kalman filter is an estimator that performs the following minimization (see e.g. [22,23,20]):

$$(7.2) \qquad \min_{\{\hat{z}_j\}} E \sum_{j=0}^{i} e_{p,j}^* e_{p,j} = \min_{\{\hat{z}_j\}} E \parallel e_p \parallel^2$$

where the expectation is taken over the Gaussian random variables w and $\{v_j\}$ whose joint conditional distribution is given by:

$$p(w, v_0, \ldots, v_i | d_0, \ldots, d_i) \propto$$

$$exp\left[-\frac{1}{2}\left((w - \hat{w}_{|-1})^* \Pi_0^{-1}(w - \hat{w}_{|-1}) + \sum_{j=0}^{i}(d_j - h_j x_j)^*(d_j - h_j x_j)\right)\right]$$

and where the symbol $\propto$ stands for 'proportional to'. In the terminology of [16], the filter that minimizes (7.2) is known as the *risk-neutral* filter.

An alternative criterion that is risk-sensitive has been extensively studied in [14,15,16,17] and corresponds to the following minimization problem

$$(7.3a) \qquad \min_{\{\hat{z}_{j|j}\}} \mu_{f,i}(\theta) = \min_{\{\hat{z}_{j|j}\}}\left(-\frac{2}{\theta}log\left[E exp(-\frac{\theta}{2}\mathbf{C}_{f,i})\right]\right)$$

or

$$(7.3b) \qquad \min_{\{\hat{z}_j\}} \mu_{p,i}(\theta) = \min_{\{\hat{z}_j\}}\left(-\frac{2}{\theta}log\left[E exp(-\frac{\theta}{2}\mathbf{C}_{p,i})\right]\right)$$

where $\mathbf{C}_{f,i} = \sum_{j=0}^{i} e_{f,i}^* e_{f,i}$ and $\mathbf{C}_{p,i} = \sum_{j=0}^{i} e_{p,i}^* e_{p,i}$. The criteria in (7.3a) and (7.3b) are known as the a posteriori and a priori *exponential cost functions*, and any filters that minimize $\mu_{f,i}(\theta)$ and $\mu_{p,i}(\theta)$ are referred to as a posteriori and a priori risk-sensitive filters, respectively. The scalar parameter θ is correspondingly called the *risk-sensitivity* parameter. Some intuition concerning the nature of this modified criterion is obtained by expanding $\mu_i(\theta)$ (where we have dropped the subscripts f and p since the argument follows for both filtered and predicted estimates) in terms of θ and writing,

$$\mu_i(\theta) = E(\mathbf{C}_i) - \frac{\theta}{4}Var(\mathbf{C}_i) + O(\theta^2)$$

The above equation shows that for $\theta = 0$, we have the risk-neutral case (*i.e.*, the conventional Kalman filter). When $\theta > 0$, we seek to maximize $E exp(-\frac{\theta}{2}\mathbf{C}_i)$, which is convex and decreasing in $\mathbf{C}_i$. Such a criterion is termed *risk-seeking* (or optimistic) since larger weights are on small values of $\mathbf{C}_i$, and hence we are more concerned with the frequent occurrence of moderate values of $\mathbf{C}_i$ than with the occasional large values. When $\theta < 0$, we seek to minimize $E exp(-\frac{\theta}{2}\mathbf{C}_i)$, which is convex and increasing in $\mathbf{C}_i$. Such a criterion is termed *risk-averse* (or pessimistic) since large weights are on large values of $\mathbf{C}_i$, and hence we are more concerned with the occasional occurrence of large values than with the frequent occurrence of moderate ones.

The relationship between the risk-sensitive criterion and the H^∞ criterion was first noted in [24] and has been further discussed in [16,13]. It may be formally stated as follows: *In the risk-averse case $\theta < 0$, the risk-sensitive optimal filter with parameter θ is given by the central H^∞ filter with level $\gamma = -\theta^{-\frac{1}{2}}$.* In particular, there is a certain *smallest* value of the risk-sensitivity parameter $\bar{\theta}$, after which the minimizing property of $\mu_i(\theta)$ breaks down, and it is this value that yields the optimal central H^∞ filter with $\gamma_{opt} = -\bar{\theta}^{-\frac{1}{2}}$.

7.2. Risk-sensitive adaptive filtering. Using the discussion of Section 7.1, we are now in a position to state the risk-sensitive results for LMS and normalized LMS.

THEOREM 7.1 (NORMALIZED LMS AND RISK-SENSITIVITY). *Consider the state-space model (7.1) where the w and $\{v_j\}$ are independent Gaussian random variables with means $\hat{w}_{|-1}$ and 0, and variances μI and I, respectively. The solution to the following minimization problem*

$$(7.4) \qquad \min_{\{\hat{z}_{j|j}\}} \mu_{f,i}(\theta) = \min_{\{\hat{z}_{j|j}\}} \left(2log \left[Eexp(\frac{1}{2}\mathbf{C}_{f,i}) \right] \right)$$

where $\mathbf{C}_{f,i} = \sum_{j=0}^{i} e_{f,i}^* e_{f,i}$, and the expectation is taken over w and $\{v_j\}$ subject to observing $\{d_0, d_1, \ldots, d_i\}$, is given by the normalized LMS algorithm

$$\hat{z}_{i|i} = h_i \hat{w}_{|i}$$

and

$$(7.5) \qquad \hat{w}_{|i+1} = \hat{w}_{|i} + \frac{\mu h_{i+1}^*}{1 + \mu h_{i+1} h_{i+1}^*}(d_{i+1} - h_{i+1}\hat{w}_{|i}) , \qquad \hat{w}_{|-1}$$

THEOREM 7.2 (LMS AND RISK-SENSITIVITY). *Consider the state-space model (7.1) where the w and $\{v_j\}$ are independent Gaussian random variables with means $\hat{w}_{|-1}$ and 0, and variances μI and I, respectively. Suppose moreover, that the $\{h_i\}$ are exciting, and that*

$$0 < \mu < \inf_i \frac{1}{h_i h_i^*}$$

Then the solution to the following minimization problem

$$(7.6) \qquad \min_{\{\hat{z}_j\}} \mu_{p,i}(\theta) = \min_{\{\hat{z}_j\}} \left(2log \left[Eexp(\frac{1}{2}\mathbf{C}_{p,i}) \right] \right)$$

where $\mathbf{C}_{p,i} = \sum_{j=0}^{i} e_{p,i}^* e_{p,i}$, and the expectation is taken over w and $\{v_j\}$ subject to observing $\{d_0, d_1, \ldots, d_{i-1}\}$, is given by the LMS algorithm

$$\hat{z}_i = h_i \hat{w}_{i-1}$$

and

(7.7) $$\hat{w}_{|i} = \hat{w}_{|i-1} + \mu h_i^*(d_i - h_i \hat{w}_{|i-1}) , \qquad \hat{w}_{|-1}$$

Thus LMS and normalized LMS are risk-averse filters that avoid the occasional occurrence of large estimation error energies at the price of tolerating the frequent occurrence of moderate ones. This fact is in agreement with the intuition we have gained from the H^∞ optimality of these algorithms.

Before closing this section we should remark that the central H^∞ filters possess other properties in addition to the one described above. In the game theoretic formulation of H^∞ estimation, the central filter corresponds to the *solution* of the game. Morever, among all H^∞ estimators that achieve a certain level γ, the central solution can be shown to be the maximum entropy [18] solution. However, we shall not pursue these directions here.

8. Conclusion. We have demonstrated that the LMS algorithm is H^∞ optimal. This result solves a long standing issue of finding a rigorous basis for the LMS algorithm, and also confirms its robustness. We find it quite interesting that despite the fact that there has only been recent interest in the field of H^∞ estimation, there has existed an H^∞ optimal estimation algorithm that has been widely used in practice for the past three decades.

Acknowledgement. The first author would like to thank Prof. L. Ljung for contributing to the discussion in Section 5.4.

REFERENCES

[1] B. Widrow and M.E. Hoff, Jr. Adaptive switching circuits. *IRE WESCON Conv. Rec.*, pages 96–104, 1960. Pt. 4.

[2] B. Widrow and S. D. Stearns. *Adaptive Signal Processing*. Prentice-Hall, Inc., Englewood Cliffs, NJ, 1985.

[3] S. Haykin. *Adaptive Filter Theory*. Prentice-Hall, Englewood Cliffs, NJ, second edition, 1991.

[4] H. Kwakernaak. A polynomial approach to mimimax frequency domain optimization of multivariable feedback systems. *Int. J. of Control*, 44:117–156, 1986.

[5] J.C. Doyle, K. Glover, P. Khargonekar, and B. Francis. State-space solutions to standard H_2 and H_∞ control problems. *IEEE Transactions on Automatic Control*, 34(8):831–847, August 1989.

[6] P.P. Khargonekar and K.M. Nagpal. Filtering and smoothing in an $H^\infty-$ setting. *IEEE Trans. on Automatic Control*, AC-36:151–166, 1991.

[7] T. Basar. Optimum performance levels for minimax filters, predictors and smoothers. *Systems and Control Letters*, 16:309–317, 1991.

[8] D.J. Limebeer and U. Shaked. New results in h^∞-filtering. In *Proc. Int. Symp. on MTNS*, pages 317–322, June 1991.

[9] U. Shaked and Y. Theodor. $H^\infty-$optimal estimation: A tutorial. In *Proc. IEEE Conference on Decision and Control*, pages 2278–2286, Tucson, AZ, Dec. 1992.

[10] I. Yaesh and U. Shaked. $H^\infty-$optimal estimation: The discrete time case. In *Proc. Inter. Symp. on MTNS*, pages 261–267, Kobe, Japan, June 1991.

[11] M.J. Grimble. Polynomial matrix solution of the H^∞ filtering problem and the relationship to Riccati equation state-space results. *IEEE Trans. on Signal Processing*, 41(1):67–81, January 1993.

[12] B. Hassibi, A.H. Sayed, and T. Kailath. Recursive linear estimation in Krein spaces - part I: Theory. In *Proc. IEEE Conference on Decision and Control*, San Antonio, TX, Dec. 1993, pp. 3489–3495.

[13] B. Hassibi, A.H. Sayed, and T. Kailath. Recursive linear estimation in Krein spaces - Part II: Applications. In *Proc. IEEE Conference on Decision and Control*, San Antonio, TX, Dec. 1993, pp. 3495–3501.

[14] D.H. Jacobson. Optimal stochastic linear systems with exponential performance criteria and their relation to deterministic games. *IEEE Trans. Automatic Control*, 18(2), April 1973.

[15] J. Speyer, J. Deyst, and D.H. Jacobson. Optimization of stochastic linear systems with additive measurement and process noise using exponential performance criteria. *IEEE Trans. Automatic Control*, 19:358–366, August 1974.

[16] P. Whittle. *Risk Sensitive Optimal Control*. John Wiley and Sons, New York, 1990.

[17] J.L. Speyer, C. Fan, and R.N. Banavar. Optimal stochastic estimation with exponential cost criteria. In *Proc. IEEE Conference on Decision and Control*, pages 2293–2298, Tucson, Arizona, December 1992.

[18] K. Glover and D. Mustafa. Derivation of the maximum entropy H^∞ controller and a state space formula for its entropy. *Int. J. Control*, 50:899–916, 1989.

[19] T. Kailath. *Linear Systems*. Prentice-Hall, Englewood Cliffs, NJ, 1980.

[20] A. H. Sayed and T. Kailath. A state-space approach to adaptive filtering. In *Proc. IEEE International Conference on Acoustics, Speech, and Signal Processing*, Minneapolis, MN, 1993.

[21] B. Widrow, et al. Stationary and nonstationary learning characteristics of the LMS adaptive filter. *Proceedings IEEE*, 64(8):1151–1162, August 1976.

[22] A.H. Jazwinski. *Stochastic Processes and Filtering Theory*, volume 64 of *Mathematics in Science and Engineering*. Academic Press, New York, 1970.

[23] B.D.O. Anderson and J.B. Moore. *Optimal Filtering*. Prentice-Hall Inc., Englewood Cliffs, NJ, 1979.

[24] K. Glover and J.C. Doyle. State-space formulae for all stabilizing controllers that satisfy an H^∞-norm bound and relations to risk sensitivity. *System and Control Letters*, 11:167–172, 1988.

ADAPTIVE CONTROL OF NONLINEAR SYSTEMS: A TUTORIAL

IOANNIS KANELLAKOPOULOS*

Abstract. We present a tutorial overview of recent Lyapunov-based results in adaptive control of nonlinear systems. Design tools like backstepping, overparametrization, and tuning functions are illustrated via simple examples. We also discuss the recently introduced "κ-terms", which endow the resulting adaptive systems with new properties, such as boundedness without adaptation and systematic transient performance improvement through trajectory initialization.

Key words. Adaptive control, nonlinear systems, Lyapunov designs, backstepping.

1. Introduction. Over the last few years, adaptive control of nonlinear systems has emerged as an exciting new research area, which has witnessed rapid and impressive developments [19,25,21,1,26,6,20] leading to global stability and tracking results for large classes of nonlinear systems [7,8,3,9,10,11,12,15,16,17,22,27]. The latter results are all Lyapunov-based, i.e., the design procedure achieves the desired objectives by constructing a suitable Lyapunov function and rendering its derivative nonpositive. Estimation-based adaptive controllers for similar classes of nonlinear systems have also been recently developed, and are presented in a companion paper in this volume [14].

In this paper we provide a tutorial overview of Lyapunov-based designs developed by our research group. These results are based on several design tools such as adaptive backstepping, observer backstepping, overparametrization, tuning functions, and κ-terms, which are used as building blocks for the construction of systematic design procedures. First, in Section 2, we state the general results without proving them, and then proceed to illustrate the corresponding design procedures by several simple examples in the following sections. Section 3 deals with overparametrized schemes, while tuning functions are introduced in Section 4. Output-feedback schemes for nonlinear systems are illustrated in Section 5, and their application to linear systems in Section 6. In Section 7 we discuss the recently introduced κ-terms, and, finally, in Section 8, we show how these terms and trajectory initialization can be combined with existing design procedures to yield adaptive controllers which guarantee boundedness without adaptation and without knowledge of bounds on the unknown parameters, and systematic improvement of transient tracking performance with the choice of design parameters.

---

* Department of Electrical Engineering, UCLA, Los Angeles, CA 90024-1594.
This work was supported in part by NSF through Grant RIA ECS-9309402, in part by UCLA through the SEAS Dean's Fund, and in part by the Institute for Mathematics and its Applications with funds provided by the National Science Foundation.

2. General results. The design procedures illustrated by the examples in this paper have been generalized [7–13] to classes of nonlinear systems characterized by *canonical forms*: every member of a class can be transformed into the corresponding canonical form via a diffeomorphism. Most of these classes can be characterized in a *coordinate-free* fashion through differential geometric conditions which are necessary and sufficient for the existence of the appropriate diffeomorphism. The discussion of these geometric characterizations is beyond the scope of this tutorial, and thus here we assume that the system at hand has already been transformed into one of the canonical forms.

Once in the canonical form, there are several underlying assumptions which are common to all the results stated in this paper:

- The unknown parameters enter linearly into the system equations, or they can be reparametrized to yield a linear parametrization.
- The nonlinearitics satisfy the strict-feedback condition, i.e., they depend only on state-variables which are "fed back", but are otherwise not restricted by any growth conditions.
- The nonlinearities can depend only on measured variables.
- The relative degree ρ of the system is known.
- The dynamic order of the system n is also known.[1]
- The zero dynamics subsystem is bounded-input-bounded-state (BIBS) stable, where any states (other than the subsystem states) entering the equations of the subsystem are treated as inputs. If the states of the zero dynamics are not measured, then the zero dynamics must be linear and exponentially stable.
- The sign of the leading coefficient of the control u is known.[2]

Under these assumptions, the following results can be obtained:

1. Global stability and tracking under full-state feedback for systems in the *strict-feedback canonical form*:

$$
\begin{aligned}
\dot{x}_1 &= x_2 + \theta^{\mathrm{T}} \varphi_1(x_1) \\
\dot{x}_2 &= x_3 + \theta^{\mathrm{T}} \varphi_2(x_1, x_2) \\
&\ \ \vdots \\
\dot{x}_{n-1} &= x_n + \theta^{\mathrm{T}} \varphi_{n-1}(x_1, \ldots, x_{n-1}) \\
\dot{x}_n &= \sigma(x) u + \varphi_0(x) + \theta^{\mathrm{T}} \varphi_n(x),
\end{aligned}
$$

(2.1)

where $\sigma(x) \neq 0$ for all $x \in \mathrm{I\!R}$ and θ is the vector of unknown parameters.

2. Global stability and tracking under full-state feedback for systems in

---

[1] If the system is given in the canonical form, then this assumption can be relaxed to require knowledge of only an upper bound on the system order. However, if one has to verify geometric conditions and perform a diffeomorphism, the system order must be known.

[2] This is the nonlinear analog of the known high-frequency gain assumption in adaptive control of linear systems.

the *partially strict-feedback canonical form*:

$$
\begin{aligned}
\dot{x}_1 &= x_2 + \theta^{\mathrm{T}} \varphi_1(x_1) \\
\dot{x}_2 &= x_3 + \theta^{\mathrm{T}} \varphi_2(x_1, x_2) \\
&\ \ \vdots \\
\dot{x}_{q-1} &= x_q + \theta^{\mathrm{T}} \varphi_{q-1}(x_1, \ldots, x_{q-1}) \\
\dot{x}_q &= x_{q+1} + \theta^{\mathrm{T}} \varphi_q(x_1, \ldots, x_q, x^r) \\
&\ \ \vdots \\
\dot{x}_{\rho-1} &= x_\rho + \theta^{\mathrm{T}} \varphi_{\rho-1}(x_1, \ldots, x_{\rho-1}, x^r) \\
\dot{x}_\rho &= x_{\rho+1} + \varphi_0(x) + \theta^{\mathrm{T}} \varphi_\rho(x) + b_{n-\rho} \sigma(x) u \\
\dot{x}^r &= \Phi_0(x_1, \ldots, x_q, x^r) + \theta^{\mathrm{T}} \Phi(x_1, \ldots, x_q, x^r) \\
y &= x_1 ,
\end{aligned}
$$

(2.2)

where θ and $b_{n-\rho}$ are unknown, the sign of $b_{n-\rho}$ is known, $\sigma(x) \neq 0$ for all $x \in \mathbb{R}$, and the x^r-subsystem is BIBS stable with respect to $x_1, \ldots, x_q$ as its inputs.

3. Regional stability and tracking under full-state feedback for systems in the *pure-feedback canonical form*:

$$
\begin{aligned}
\dot{x}_1 &= x_2 + \theta^{\mathrm{T}} \varphi_1(x_1, x_2) \\
\dot{x}_2 &= x_3 + \theta^{\mathrm{T}} \varphi_2(x_1, x_2, x_3) \\
&\ \ \vdots \\
\dot{x}_{n-1} &= x_n + \theta^{\mathrm{T}} \varphi_{n-1}(x_1, \ldots, x_n) \\
\dot{x}_n &= \sigma(x) u + \varphi_0(x) + \theta^{\mathrm{T}} \varphi_n(x) .
\end{aligned}
$$

(2.3)

This form is clearly more general than the strict-feedback form (2.1), since the nonlinearities are allowed to depend on one more state variable in each equation. The price paid for this enlargement of the class of nonlinear systems is the loss of globality: stability and tracking can now be guaranteed only in a region around the origin.

4. Global stability and tracking under output feedback for systems in the *output-feedback canonical form*:

$$
\begin{aligned}
\dot{x}_1 &= x_2 + \varphi_{0,1}(y) + \theta^{\mathrm{T}} \varphi_1(y) \\
&\ \ \vdots \\
\dot{x}_{\rho-1} &= x_\rho + \varphi_{0,\rho-1}(y) + \theta^{\mathrm{T}} \varphi_{\rho-1}(y) \\
\dot{x}_\rho &= x_{\rho+1} + \varphi_{0,\rho}(y) + \theta^{\mathrm{T}} \varphi_\rho(y) + b_{n-\rho} \sigma(y) u \\
&\ \ \vdots \\
\dot{x}_n &= \varphi_{0,n}(y) + \theta^{\mathrm{T}} \varphi_n(y) + b_0 \sigma(y) u \\
y &= x_1 ,
\end{aligned}
$$

(2.4)

where θ and $b_0, \ldots, b_{n-\rho}$ are unknown, $\sigma(y) \neq 0$ for all $y \in \mathbb{R}$, the sign of $b_{n-\rho}$ is known, the polynomial $B(s) = b_{n-\rho} s^{n-\rho} + \cdots + b_1 s + b_0$ is Hurwitz, and the only measured variable is the output y.

5. Global stability and tracking under partial-state feedback for systems in the *partial-feedback canonical form*:

$$\dot{x}_1 = x_2 + \theta^{\mathrm{T}}\varphi_1(x_1)$$
$$\dot{x}_2 = x_3 + \theta^{\mathrm{T}}\varphi_2(x_1, x_2)$$

$$\vdots$$

$$\dot{x}_{q-1} = x_q + \theta^{\mathrm{T}}\varphi_{p-1}(x_1, \ldots, x_{q-1})$$
$$\dot{x}_q = x_{q+1} + \varphi_{0,q}(x_1, \ldots, x_q) + \theta^{\mathrm{T}}\varphi_q(x_1, \ldots, x_q)$$

(2.5)
$$\vdots$$

$$\dot{x}_{p-1} = x_p + \varphi_{0,p-1}(x_1, \ldots, x_q) + \theta^{\mathrm{T}}\varphi_{p-1}(x_1, \ldots, x_q)$$
$$\dot{x}_p = x_{p+1} + \varphi_{0,p}(x_1, \ldots, x_q)$$
$$\qquad + \theta^{\mathrm{T}}\varphi_p(x_1, \ldots, x_q) + b_{n-p}\sigma(x_1, \ldots, x_q)u$$

$$\vdots$$

$$\dot{x}_n = \varphi_{0,n}(x_1, \ldots, x_q) + \theta^{\mathrm{T}}\varphi_n(x_1, \ldots, x_q) + b_0\sigma(x_1, \ldots, x_q)u,$$

where θ and $b_0, \ldots, b_{n-p}$ are unknown, $\sigma(x_1, \ldots, x_q) \neq 0$ for all $(x_1, \ldots, x_q)$ $\in \mathrm{IR}^q$, the sign of b_{n-p} is known, the polynomial $B(s) = b_{n-p}s^{n-p} + \cdots + b_1 s + b_0$ is Hurwitz, and the only measured variables are $x_1, \ldots, x_q$.

3. Overparametrized schemes. Let us now illustrate some of the design procedures that led to the results stated in the previous section. We start with the overparametrized design procedure of [7], which is applied to the system:

(3.1)
$$\dot{x}_1 = x_2 + \theta^{\mathrm{T}}\varphi_1(x_1)$$
$$\dot{x}_2 = x_3 + \theta^{\mathrm{T}}\varphi_2(x_1, x_2)$$
$$\dot{x}_3 = u + \theta^{\mathrm{T}}\varphi_3(x_1, x_2, x_3),$$

where θ is an unknown constant parameter vector. The system (3.1) is in the strict-feedback canonical form, as a comparison with (2.1) reveals. The reason for denoting this form as "strict-feedback" is evident from Figure 3.1: in the block diagram representation of the system (3.1), the nonlinearities are placed on feedback paths only.

Our objective is to design an adaptive state-feedback controller that guarantees global stability and asymptotic convergence of x_1 to zero.

Step 1. In the first step of the design procedure we define the error variable

(3.2)
$$z_1 = x_1,$$

and write its derivative as

(3.3)
$$\dot{z}_1 = \dot{x}_1 = x_2 + \theta^{\mathrm{T}}\varphi_1(x_1).$$

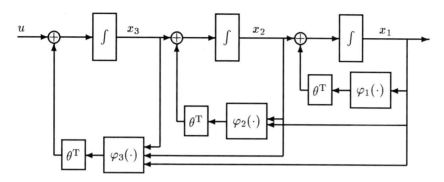

FIG. 3.1. *Block diagram of the strict-feedback system (3.1). The nonlinearities depend only on variables that are "fed back".*

In (3.3) we view x_2 as a "virtual control" and design for it a *stabilizing function* α_1. The difference between the actual value of x_2 and its "desired" value α_1 is defined to be the second error variable:

$$(3.4) \qquad\qquad z_2 = x_2 - \alpha_1 .$$

Since θ is unknown, α_1 employs a parameter estimate ϑ_1:

$$(3.5) \qquad\qquad \alpha_1 = -c_1 z_1 - \vartheta_1^{\mathrm{T}} \varphi_1 ,$$

where $c_1 > 0$ is a positive design constant. Substituting (3.4) and (3.5) into (3.3), we obtain

$$(3.6) \qquad\qquad \dot{z}_1 = -c_1 z_1 + z_2 + (\theta - \vartheta_1)^{\mathrm{T}} \varphi_1 .$$

To design the update law for the parameter estimate ϑ_1, we form the partial Lyapunov function

$$(3.7) \qquad V_1(z_1, \vartheta_1) = \frac{1}{2} z_1^2 + \frac{1}{2} (\theta - \vartheta_1)^{\mathrm{T}} \Gamma^{-1} (\theta - \vartheta_1) ,$$

where $\Gamma > 0$ is the adaptation gain matrix. The derivative of V_1 is

$$
\begin{aligned}
\dot{V}_1 &= z_1 \dot{z}_1 - (\theta - \vartheta_1)^{\mathrm{T}} \Gamma^{-1} \dot{\vartheta}_1 \\
(3.8) \qquad &= z_1 z_2 - c_1 z_1^2 + (\theta - \vartheta_1)^{\mathrm{T}} \left(\varphi_1 z_1 - \Gamma^{-1} \dot{\vartheta}_1 \right) .
\end{aligned}
$$

To cancel the last cross-term in the above derivative, we choose the update law

$$(3.9) \qquad\qquad \dot{\vartheta}_1 = \Gamma \varphi_1 z_1 ,$$

which yields

$$(3.10) \qquad\qquad \dot{V}_1 = z_1 z_2 - c_1 z_1^2 .$$

The equations (3.6) and (3.9) form the first error subsystem $\mathcal{S}_1$:

(3.11) $\mathcal{S}_1$:
$$\begin{aligned}
\dot{z}_1 &= -c_1 z_1 + z_2 + (\theta - \vartheta_1)^{\mathrm{T}} \varphi_1 \\
\dot{\vartheta}_1 &= \Gamma \varphi_1 z_1 .
\end{aligned}$$

Step 2. The derivative of z_2 is now expressed as

$$\begin{aligned}
\dot{z}_2 &= \dot{x}_2 - \dot{\alpha}_1 \\
&= x_3 + \theta^{\mathrm{T}} \varphi_2 - \frac{\partial \alpha_1}{\partial x_1} \dot{x}_1 - \frac{\partial \alpha_1}{\partial \vartheta_1} \dot{\vartheta}_1 .
\end{aligned}$$

Substituting the $\dot{x}_1$ equation from (3.1) and the update law (3.9) results in:

$$\dot{z}_2 \quad - \quad x_3 + \theta^{\mathrm{T}} \varphi_2 \quad \frac{\partial \alpha_1}{\partial x_1} \left(x_2 + \theta^{\mathrm{T}} \varphi_1 \right) - \frac{\partial \alpha_1}{\partial \vartheta_1} \Gamma \varphi_1 z_1$$

(3.12) $$= \quad x_3 - \frac{\partial \alpha_1}{\partial x_1} x_2 - \frac{\partial \alpha_1}{\partial \vartheta_1} \Gamma \varphi_1 z_1 + \theta^{\mathrm{T}} \left(\varphi_2 - \frac{\partial \alpha_1}{\partial x_1} \varphi_1 \right) .$$

The variable x_3 is now viewed as the virtual control in (3.12). Accordingly, the new error variable z_3 is defined as

(3.13) $z_3 = x_3 - \alpha_2 .$

The stabilizing function α_2 should now be designed to render nonpositive the derivative of a Lyapunov function. A first attempt is based on the augmented partial Lyapunov function

$$V_2(z_1, z_2, \vartheta_1) = V_1(z_1, \vartheta_1) + \frac{1}{2} z_2^2 ,$$

whose derivative, using (3.10) and (3.12), is

$$\begin{aligned}
\dot{V}_2 &= \dot{V}_1 + z_2 \dot{z}_2 \\
&= z_1 z_2 - c_1 z_1^2 \\
&\quad + z_2 \left[z_3 + \alpha_2 - \frac{\partial \alpha_1}{\partial x_1} x_2 - \frac{\partial \alpha_1}{\partial \vartheta_1} \Gamma \varphi_1 z_1 + \theta^{\mathrm{T}} \left(\varphi_2 - \frac{\partial \alpha_1}{\partial x_1} \varphi_1 \right) \right] \\
&= z_2 z_3 - c_1 z_1^2 \\
&\quad + z_2 \left[z_1 + \alpha_2 - \frac{\partial \alpha_1}{\partial x_1} x_2 - \frac{\partial \alpha_1}{\partial \vartheta_1} \Gamma \varphi_1 z_1 + \theta^{\mathrm{T}} \left(\varphi_2 - \frac{\partial \alpha_1}{\partial x_1} \varphi_1 \right) \right] .
\end{aligned}$$

Note that in the last equality, the term $z_1 z_2$ has been intentionally grouped together with α_2, since it is going to be dealt with in this design step. In contrast, the term $z_2 z_3$ has been separated from the rest, since it is going to be dealt with in the next step. The designer of α_2 is now called upon to cancel as many cross-terms in the above derivative as possible. To counteract the terms containing the unknown parameter θ, α_2 must use a parameter estimate. Let us suppose that the existing estimate ϑ_1 were to be used again:

$$\alpha_2 = -z_1 - c_2 z_2 + \frac{\partial \alpha_1}{\partial x_1} x_2 + \frac{\partial \alpha_1}{\partial \vartheta_1} \Gamma \varphi_1 z_1 - \vartheta_1^{\mathrm{T}} \left(\varphi_2 - \frac{\partial \alpha_1}{\partial x_1} \varphi_1 \right).$$

Then the derivative of V_2 would be

$$\dot{V}_2 = z_2 z_3 - c_1 z_1^2 - c_2 z_2^2 + (\theta - \vartheta_1)^{\mathrm{T}} \left(\varphi_2 - \frac{\partial \alpha_1}{\partial x_1} \varphi_1 \right) z_2.$$

In the above equation there is nothing we can do to cancel the $(\theta - \vartheta_1)$-term, since the update law for ϑ_1 has already been assigned in (3.9). To overcome this difficulty, we replace ϑ_1 in the expression for α_2 with a *new* estimate ϑ_2:

$$(3.14) \quad \alpha_2 = -z_1 - c_2 z_2 + \frac{\partial \alpha_1}{\partial x_1} x_2 + \frac{\partial \alpha_1}{\partial \vartheta_1} \Gamma \varphi_1 z_1 - \vartheta_2^{\mathrm{T}} \left(\varphi_2 - \frac{\partial \alpha_1}{\partial x_1} \varphi_1 \right).$$

Since ϑ_1 and ϑ_2 are both estimates of θ, this scheme is using multiple estimates of the same parameters, and is therefore called an *overparametrized* scheme.

With the choice (3.14), the $\dot{z}_2$-equation becomes

$$(3.15) \quad \dot{z}_2 = -c_2 z_2 - z_1 + z_3 + (\theta - \vartheta_2)^{\mathrm{T}} \left(\varphi_2 - \frac{\partial \alpha_1}{\partial x_1} \varphi_1 \right).$$

The presence of the new parameter estimate ϑ_2 dictates the following augmentation of the Lyapunov function:

$$(3.16) \quad V_2(z_1, z_2, \vartheta_1, \vartheta_2) = V_1 + \frac{1}{2} z_2^2 + \frac{1}{2} (\theta - \vartheta_2)^{\mathrm{T}} \Gamma^{-1} (\theta - \vartheta_2),$$

whose derivative is:

$$
\begin{aligned}
\dot{V}_2 &= \dot{V}_1 + z_2 \dot{z}_2 - (\theta - \vartheta_2)^{\mathrm{T}} \Gamma^{-1} \dot{\vartheta}_2 \\
&= z_1 z_2 - c_1 z_1^2 + z_2 \left[-c_2 z_2 - z_1 + z_3 \right. \\
&\quad \left. + (\theta - \vartheta_2)^{\mathrm{T}} \left(\varphi_2 - \frac{\partial \alpha_1}{\partial x_1} \varphi_1 \right) \right] - (\theta - \vartheta_2)^{\mathrm{T}} \Gamma^{-1} \dot{\vartheta}_2 \\
(3.17) \quad &= z_2 z_3 - c_1 z_1^2 - c_2 z_2^2 + (\theta - \vartheta_2)^{\mathrm{T}} \left(\varphi_2 - \frac{\partial \alpha_1}{\partial x_1} \varphi_1 - \Gamma^{-1} \dot{\vartheta}_2 \right).
\end{aligned}
$$

The $(\theta - \vartheta_2)$-term can be eliminated with the update law

$$(3.18) \quad \dot{\vartheta}_2 = \Gamma \left(\varphi_2 - \frac{\partial \alpha_1}{\partial x_1} \varphi_1 \right) z_2,$$

which yields

(3.19) $$\dot{V}_2 = z_2 z_3 - c_1 z_1^2 - c_2 z_2^2 .$$

The equations (3.15) and (3.18) along with the first error subsystem (3.11) form the second error subsystem $\mathcal{S}_2$:

(3.20) $\mathcal{S}_2$:
$$\begin{aligned}
\dot{z}_1 &= -c_1 z_1 + z_2 + (\theta - \vartheta_1)^{\mathrm{T}} \varphi_1 \\
\dot{z}_2 &= -c_2 z_2 - z_1 + z_3 + (\theta - \vartheta_2)^{\mathrm{T}} \left(\varphi_2 - \tfrac{\partial \alpha_1}{\partial x_1} \varphi_1 \right) \\
\dot{\vartheta}_1 &= \Gamma \varphi_1 z_1 \\
\dot{\vartheta}_2 &= \Gamma \left(\varphi_2 - \tfrac{\partial \alpha_1}{\partial x_1} \varphi_1 \right) z_2 .
\end{aligned}$$

Step 3. This is the final design step, since the actual control u appears in the derivative of z_3:

$$\begin{aligned}
\dot{z}_3 &= \dot{x}_3 - \dot{\alpha}_2 \\
&= u + \theta^{\mathrm{T}} \varphi_3 - \frac{\partial \alpha_2}{\partial x_1} \dot{x}_1 - \frac{\partial \alpha_2}{\partial x_2} \dot{x}_2 - \frac{\partial \alpha_2}{\partial \vartheta_1} \dot{\vartheta}_1 - \frac{\partial \alpha_2}{\partial \vartheta_2} \dot{\vartheta}_2 \\
&= u + \theta^{\mathrm{T}} \varphi_3 - \frac{\partial \alpha_2}{\partial x_1} (x_2 + \theta^{\mathrm{T}} \varphi_1) - \frac{\partial \alpha_2}{\partial x_2} (x_3 + \theta^{\mathrm{T}} \varphi_2) \\
&\quad - \frac{\partial \alpha_2}{\partial \vartheta_1} \Gamma \varphi_1 z_1 - \frac{\partial \alpha_2}{\partial \vartheta_2} \Gamma \left(\varphi_2 - \frac{\partial \alpha_1}{\partial x_1} \varphi_1 \right) z_2 \\
&= u - \frac{\partial \alpha_2}{\partial x_1} x_2 - \frac{\partial \alpha_2}{\partial x_2} x_3 - \frac{\partial \alpha_2}{\partial \vartheta_1} \Gamma \varphi_1 z_1 - \frac{\partial \alpha_2}{\partial \vartheta_2} \Gamma \left(\varphi_2 - \frac{\partial \alpha_1}{\partial x_1} \varphi_1 \right) z_2 \\
\end{aligned}$$
(3.21) $$\quad + \theta^{\mathrm{T}} \left(\varphi_3 - \frac{\partial \alpha_2}{\partial x_2} \varphi_2 - \frac{\partial \alpha_2}{\partial x_1} \varphi_1 \right) .$$

Following the same route as in Step 2, we arrive at the conclusion that since the update laws for ϑ_1 and ϑ_2 have already been assigned, yet another new estimate ϑ_3 is needed. The control u and the update law for ϑ_3 are designed to render nonpositive the derivative of the full Lyapunov function

$$\begin{aligned}
V_3(z, \vartheta_1, \vartheta_2, \vartheta_3) &= V_2(z_1, z_2, \vartheta_1, \vartheta_2) + \frac{1}{2} z_3^2 + \frac{1}{2} (\theta - \vartheta_3)^{\mathrm{T}} \Gamma^{-1} (\theta - \vartheta_3) \\
&= \frac{1}{2} (z_1^2 + z_2^2 + z_3^2) + \frac{1}{2} (\theta - \vartheta_1)^{\mathrm{T}} \Gamma^{-1} (\theta - \vartheta_1) \\
\end{aligned}$$
(3.22) $$\quad + \frac{1}{2} (\theta - \vartheta_2)^{\mathrm{T}} \Gamma^{-1} (\theta - \vartheta_2) + \frac{1}{2} (\theta - \vartheta_3)^{\mathrm{T}} \Gamma^{-1} (\theta - \vartheta_3) .$$

Using (3.19) and (3.21), the derivative of V_3 is computed as

$$
\begin{aligned}
\dot{V}_3 &= \dot{V}_2 + z_3\dot{z}_3 - (\theta - \vartheta_3)^{\mathrm{T}}\Gamma^{-1}\dot{\vartheta}_3 \\
&= -c_1 z_1^2 - c_2 z_2^2 + z_3 \left[u + z_2 - \frac{\partial \alpha_2}{\partial x_1}x_2 - \frac{\partial \alpha_2}{\partial x_2}x_3 \right. \\
&\quad - \frac{\partial \alpha_2}{\partial \vartheta_1}\Gamma\varphi_1 z_1 - \frac{\partial \alpha_2}{\partial \vartheta_2}\Gamma\left(\varphi_2 - \frac{\partial \alpha_1}{\partial x_1}\varphi_1\right)z_2 \\
&\quad \left. +\theta^{\mathrm{T}}\left(\varphi_3 - \frac{\partial \alpha_2}{\partial x_2}\varphi_2 - \frac{\partial \alpha_2}{\partial x_1}\varphi_1\right) \right] - (\theta - \vartheta_3)^{\mathrm{T}}\Gamma^{-1}\dot{\vartheta}_3 .
\end{aligned}
$$

(3.23)

The choice of control

$$
\begin{aligned}
u &= -z_2 - c_3 z_3 + \frac{\partial \alpha_2}{\partial x_1}x_2 + \frac{\partial \alpha_2}{\partial x_2}x_3 + \frac{\partial \alpha_2}{\partial \vartheta_1}\Gamma\varphi_1 z_1 \\
&\quad + \frac{\partial \alpha_2}{\partial \vartheta_2}\Gamma\left(\varphi_2 - \frac{\partial \alpha_1}{\partial x_1}\varphi_1\right)z_2 - \vartheta_3^{\mathrm{T}}\left(\varphi_3 - \frac{\partial \alpha_2}{\partial x_2}\varphi_2 - \frac{\partial \alpha_2}{\partial x_1}\varphi_1\right),
\end{aligned}
$$

(3.24)

when substituted into (3.23), results in

$$
\begin{aligned}
\dot{V}_3 &= -c_1 z_1^2 - c_2 z_2^2 - c_3 z_3^2 \\
&\quad + (\theta - \vartheta_3)^{\mathrm{T}}\left(\varphi_3 - \frac{\partial \alpha_2}{\partial x_2}\varphi_2 - \frac{\partial \alpha_2}{\partial x_1}\varphi_1 - \Gamma^{-1}\dot{\vartheta}_3\right).
\end{aligned}
$$

(3.25)

The $(\theta - \vartheta_3)$-term is now eliminated with the update law

$$
\dot{\vartheta}_3 = \Gamma\left(\varphi_3 - \frac{\partial \alpha_2}{\partial x_2}\varphi_2 - \frac{\partial \alpha_2}{\partial x_1}\varphi_1\right)z_3 ,
$$

(3.26)

which yields

$$
\dot{V}_3 = -c_1 z_1^2 - c_2 z_2^2 - c_3 z_3^2 .
$$

(3.27)

With the choices (3.24) and (3.26) the complete error system $\mathcal{S}_3$ becomes:

(3.28) $\mathcal{S}_3$:
$$
\begin{aligned}
\dot{z}_1 &= -c_1 z_1 + z_2 + (\theta - \vartheta_1)^{\mathrm{T}}\varphi_1 \\
\dot{z}_2 &= -c_2 z_2 - z_1 + z_3 + (\theta - \vartheta_2)^{\mathrm{T}}\left(\varphi_2 - \frac{\partial \alpha_1}{\partial x_1}\varphi_1\right) \\
\dot{z}_3 &= -c_3 z_3 - z_2 + (\theta - \vartheta_3)^{\mathrm{T}}\left(\varphi_3 - \frac{\partial \alpha_2}{\partial x_2}\varphi_2 - \frac{\partial \alpha_2}{\partial x_1}\varphi_1\right) \\
\dot{\vartheta}_1 &= \Gamma\varphi_1 z_1 \\
\dot{\vartheta}_2 &= \Gamma\left(\varphi_2 - \frac{\partial \alpha_1}{\partial x_1}\varphi_1\right)z_2 \\
\dot{\vartheta}_3 &= \Gamma\left(\varphi_3 - \frac{\partial \alpha_2}{\partial x_2}\varphi_2 - \frac{\partial \alpha_2}{\partial x_1}\varphi_1\right)z_3 .
\end{aligned}
$$

Let us now rewrite this error system in the more informative matrix form:

$$
\frac{d}{dt}\begin{bmatrix} z_1 \\ z_2 \\ z_3 \end{bmatrix} = \begin{bmatrix} -c_1 & 1 & 0 \\ -1 & -c_2 & 1 \\ 0 & -1 & -c_3 \end{bmatrix}\begin{bmatrix} z_1 \\ z_2 \\ z_3 \end{bmatrix}
$$

(3.29)
$$
+ \begin{bmatrix} \varphi_1 & 0 & 0 \\ 0 & \varphi_2 - \frac{\partial \alpha_1}{\partial x_1}\varphi_1 & 0 \\ 0 & 0 & \varphi_3 - \frac{\partial \alpha_2}{\partial x_2}\varphi_2 - \frac{\partial \alpha_2}{\partial x_1}\varphi_1 \end{bmatrix}^{\mathrm{T}} \begin{bmatrix} \theta - \vartheta_1 \\ \theta - \vartheta_2 \\ \theta - \vartheta_3 \end{bmatrix}
$$

$$
\frac{d}{dt}\begin{bmatrix} \vartheta_1 \\ \vartheta_2 \\ \vartheta_3 \end{bmatrix} = \Gamma \begin{bmatrix} \varphi_1 & 0 & 0 \\ 0 & \varphi_2 - \frac{\partial \alpha_1}{\partial x_1}\varphi_1 & 0 \\ 0 & 0 & \varphi_3 - \frac{\partial \alpha_2}{\partial x_2}\varphi_2 - \frac{\partial \alpha_2}{\partial x_1}\varphi_1 \end{bmatrix} \begin{bmatrix} z_1 \\ z_2 \\ z_3 \end{bmatrix}
$$

In the above equation, it is important to note that

- the constant system matrix has negative terms along its diagonal, and its off-diagonal terms are skew-symmetric, and
- the matrix that multiplies the parameter errors in the $\dot{z}$-equation is used in the update laws for the parameter estimates.

Due to the structure of the closed-loop error system (3.29), it is possible to use the simple Lyapunov function (3.22) to conclude the stability of all the variables. Indeed, combining (3.22) with (3.27) we conclude that V_3 is bounded, and therefore z_1, z_2, z_3 and $\vartheta_1, \vartheta_2, \vartheta_3$ are bounded. Since $z_1 = x_1$, we see that x_1 is bounded also. The boundedness of x_2 then follows from the boundedness of α_1 (defined in (3.5)) and the fact that $x_2 = z_2 + \alpha_1$. Similarly, the boundedness of x_3 follows from (3.13) and (3.14). Using (3.24) we conclude that the control u is also bounded. Finally, the fact that V_3 is integrable and $\dot{V}_3$ is bounded (since the derivatives of all the variables are also bounded), leads to $z \to 0$ as $t \to \infty$. In particular, this implies that $x_1 = z_1$ goes to zero asymptotically, but does *not* imply the regulation of x_2 and x_3. For example, let us consider x_2: from (3.4) and (3.5) we see that $x_2 + \vartheta_1\varphi_1(0)$ will converge to zero. Thus, x_2 is not guaranteed to converge to zero unless $\varphi_1(0) = 0$.

The design procedure illustrated in the above example can be generalized to systems of arbitrary order which can be transformed into the strict-feedback (2.1) or the partially strict-feedback (2.2) canonical form. The number of design steps required is equal to the relative degree ρ of the system. At each step, a new stabilizing function α_i and a new parameter estimate ϑ_i is generated. As a result, if a system contains p unknown parameters, the overparametrized adaptive controller may employ as many as $p\rho$ parameter estimates. A schematic representation of this design procedure is given in Figure 3.2.

4. Tuning functions. The increase in the number of parameter estimates caused by overparametrization can be an undesirable feature, since it rapidly increases the dynamic order of the resulting adaptive controller.

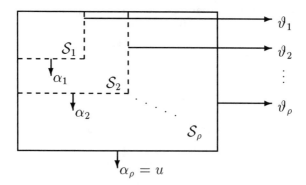

FIG. 3.2. *The design procedure for overparametrized schemes.*

Soon after the design procedure of [7] was developed, the number of parameter estimates required was reduced in half [3]. Subsequently, the need for overparametrization was eliminated with the introduction of *tuning functions* [12], which require only as many estimates as there are unknown parameters.

To illustrate the design procedure with tuning functions, let us consider again the system (3.1). Since there are several similarities with the design procedure presented in the previous section, the exposition will focus more on the differences that arise from the elimination of multiple parameter estimates.

Step 1. With $z_1 = x_1$ and x_2 viewed as the virtual control in the $\dot{z}_1$-equation, we define the first stabilizing function α_1 as in (3.5):

$$(4.1) \qquad \alpha_1 = -c_1 z_1 - \hat{\theta}^T \varphi_1 \,.$$

Comparing (4.1) with (3.5), we see that the parameter estimate ϑ_1 has been replaced by the parameter estimate $\hat{\theta}$. The difference in notation indicates that in this design procedure only one estimate per unknown parameter will be used.

The partial Lyapunov function is now chosen as

$$(4.2) \qquad V_1(z_1, \hat{\theta}) = \frac{1}{2} z_1^2 + \frac{1}{2} \tilde{\theta}^T \Gamma^{-1} \tilde{\theta} \,,$$

where

$$(4.3) \qquad \tilde{\theta} = \theta - \hat{\theta}$$

is the parameter error, and $\Gamma > 0$ is the adaptation gain matrix. The derivative of V_1 is ($z_2 = x_2 - \alpha_1$)

$$(4.4) \qquad \dot{V}_1 = z_1 z_2 - c_1 z_1^2 + \tilde{\theta}^T \left(\varphi_1 z_1 - \Gamma^{-1} \dot{\hat{\theta}} \right) \,.$$

In contrast to the overparametrized scheme, here we do not employ the update law (3.9) to cancel the $\tilde{\theta}$-term, because it was exactly that choice that made the use of multiple estimates necessary. Instead, we define the first *tuning function* as

$$(4.5) \qquad\qquad \tau_1 = \varphi_1 z_1 ,$$

and note that the $\tilde{\theta}$-term in $\dot{V}_1$ would have been eliminated with the choice of update law $\dot{\hat{\theta}} = \Gamma \tau_1$. Since this is not the last design step, we postpone the choice of update law, and rewrite $\dot{V}_1$ as:

$$(4.6) \qquad\qquad \dot{V}_1 = z_1 z_2 - c_1 z_1^2 + \tilde{\theta}^{\mathrm{T}} \left(\tau_1 - \Gamma^{-1} \dot{\hat{\theta}} \right).$$

The first error subsystem $\mathcal{S}_1$ becomes:

$$(4.7) \qquad\qquad \mathcal{S}_1 : \qquad
\begin{aligned}
\dot{z}_1 &= -c_1 z_1 + z_2 + \tilde{\theta}^{\mathrm{T}} \varphi_1 \\
\tau_1 &= \varphi_1 z_1 .
\end{aligned}$$

Step 2. The derivative of $z_2 = x_2 - \alpha_1$ is

$$
\begin{aligned}
\dot{z}_2 &= \dot{x}_2 - \frac{\partial \alpha_1}{\partial x_1} \dot{x}_1 - \frac{\partial \alpha_1}{\partial \hat{\theta}} \dot{\hat{\theta}} \\
&= x_3 + \theta^{\mathrm{T}} \varphi_2 - \frac{\partial \alpha_1}{\partial x_1}(x_2 + \theta^{\mathrm{T}} \varphi_1) - \frac{\partial \alpha_1}{\partial \hat{\theta}} \dot{\hat{\theta}} \\
(4.8) \qquad &= x_3 - \frac{\partial \alpha_1}{\partial x_1} x_2 + \theta^{\mathrm{T}} \left(\varphi_2 - \frac{\partial \alpha_1}{\partial x_1} \varphi_1 \right) - \frac{\partial \alpha_1}{\partial \hat{\theta}} \dot{\hat{\theta}} .
\end{aligned}
$$

The virtual control here is x_3, so $z_3 = x_3 - \alpha_2$. Using this definition and the identity $\theta = \hat{\theta} + \tilde{\theta}$, we rewrite the $\dot{z}_2$-equation as:

$$\dot{z}_2 = z_3 + \alpha_2 - \frac{\partial \alpha_1}{\partial x_1} x_2 + \hat{\theta}^{\mathrm{T}} \left(\varphi_2 - \frac{\partial \alpha_1}{\partial x_1} \varphi_1 \right) + \tilde{\theta}^{\mathrm{T}} \left(\varphi_2 - \frac{\partial \alpha_1}{\partial x_1} \varphi_1 \right) - \frac{\partial \alpha_1}{\partial \hat{\theta}} \dot{\hat{\theta}} .$$
$$(4.9)$$

To design the stabilizing function α_2, we consider the the augmented partial Lyapunov function

$$(4.10) \qquad\qquad V_2 = V_1 + \frac{1}{2} z_2^2 .$$

The only difference between (4.10) and (3.16) is the absence of the new parameter error $(\theta - \vartheta_2)$ in (4.10). The derivative of $\dot{V}_2$ is

$$
\begin{aligned}
\dot{V}_2 &= \dot{V}_1 + z_2 \dot{z}_2 \\
&= z_1 z_2 - c_1 z_1^2 + \tilde{\theta}^{\mathrm{T}} \left(\tau_1 - \Gamma^{-1} \dot{\hat{\theta}} \right) + z_2 \left[z_3 + \alpha_2 - \frac{\partial \alpha_1}{\partial x_1} x_2 \right. \\
&\quad \left. + \hat{\theta}^{\mathrm{T}} \left(\varphi_2 - \frac{\partial \alpha_1}{\partial x_1} \varphi_1 \right) + \tilde{\theta}^{\mathrm{T}} \left(\varphi_2 - \frac{\partial \alpha_1}{\partial x_1} \varphi_1 \right) - \frac{\partial \alpha_1}{\partial \hat{\theta}} \dot{\hat{\theta}} \right] \\
&= z_2 z_3 - c_1 z_1^2 + \tilde{\theta}^{\mathrm{T}} \left[\tau_1 + z_2 \left(\varphi_2 - \frac{\partial \alpha_1}{\partial x_1} \varphi_1 \right) - \Gamma^{-1} \dot{\hat{\theta}} \right]
\end{aligned}
$$

$$
(4.11) \qquad + z_2 \left[z_1 + \alpha_2 - \frac{\partial \alpha_1}{\partial x_1} x_2 + \hat{\theta}^{\mathrm{T}} \left(\varphi_2 - \frac{\partial \alpha_1}{\partial x_1} \varphi_1 \right) - \frac{\partial \alpha_1}{\partial \hat{\theta}} \dot{\hat{\theta}} \right] .
$$

In the last equation, the term $z_1 z_2$ has been grouped together with α_2, since it is going to be dealt with via α_2 in this design step. In contrast, the term $z_2 z_3$ has been separated from the rest, since it is going to be dealt with at the next step. Moreover, all the terms containing $\tilde{\theta}$ have been grouped together. If we were to choose an update law now, the choice $\dot{\hat{\theta}} = \Gamma \tau_2$, where τ_2 is the second tuning function

$$
(4.12) \qquad \tau_2 = \tau_1 + z_2 \left(\varphi_2 - \frac{\partial \alpha_1}{\partial x_1} \varphi_1 \right),
$$

would eliminate the $\tilde{\theta}$-terms in (4.11). Then, the last term in $\dot{V}_2$ above would be rendered equal to $-c_2 z_2^2$ with the choice

$$
(4.13) \qquad \alpha_2 = -z_1 - c_2 z_2 + \frac{\partial \alpha_1}{\partial x_1} x_2 - \hat{\theta}^{\mathrm{T}} \left(\varphi_2 - \frac{\partial \alpha_1}{\partial x_1} \varphi_1 \right) + \frac{\partial \alpha_1}{\partial \hat{\theta}} \Gamma \tau_2 .
$$

However, since the design procedure continues, the choice of update law is again postponed. Substituting the expressions for τ_2 and α_2 into (4.11) we obtain:

$$
(4.14) \quad \dot{V}_2 = z_2 z_3 - c_1 z_1^2 - c_2 z_2^2 + \tilde{\theta}^{\mathrm{T}} \left[\tau_2 - \Gamma^{-1} \dot{\hat{\theta}} \right] - z_2 \frac{\partial \alpha_1}{\partial \hat{\theta}} \left(\dot{\hat{\theta}} - \Gamma \tau_2 \right) .
$$

Comparing (4.13) with (3.14), we see that instead of including the actual update law in the stabilizing function α_2 (as we did for $\dot{\vartheta}_1$ in (3.14)), in (4.13) we replaced $\dot{\hat{\theta}}$ by $\Gamma \tau_2$. The intuition behind this substitution is clear: since the update law has not been assigned yet, we replace it in α_2 with our best estimate of it available at this step, i.e., $\Gamma \tau_2$. The last term in (4.14) represents the mismatch between the actual update law and the estimate we used for it in this step. This term, which will be dealt with in the next step, was not present in the overparametrized scheme, where update laws were assigned in each design step and could therefore be exactly accounted for in subsequent steps.

With the choices (4.13) and (4.12), the second error subsystem S_2 becomes:

(4.15) $S_2 :$
$$\dot{z}_1 = -c_1 z_1 + z_2 + \tilde{\theta}^{\mathrm{T}} \varphi_1$$
$$\dot{z}_2 = -c_2 z_2 - z_1 + z_3$$
$$+ \tilde{\theta}^{\mathrm{T}} \left(\varphi_2 - \frac{\partial \alpha_1}{\partial x_1} \varphi_1 \right) - \frac{\partial \alpha_1}{\partial \hat{\theta}} \left(\dot{\hat{\theta}} - \Gamma \tau_2 \right)$$
$$\tau_2 = \varphi_1 z_1 + \left(\varphi_2 - \frac{\partial \alpha_1}{\partial x_1} \varphi_1 \right) z_2 .$$

Step 3. This is the final design step, since the actual control u appears in the derivative of z_3:

$$\dot{z}_3 = u + \theta^{\mathrm{T}} \varphi_3 - \frac{\partial \alpha_2}{\partial x_1} (x_2 + \theta^{\mathrm{T}} \varphi_1) - \frac{\partial \alpha_2}{\partial x_2} (x_3 + \theta^{\mathrm{T}} \varphi_2) - \frac{\partial \alpha_2}{\partial \hat{\theta}} \dot{\hat{\theta}}$$

$$= u - \frac{\partial \alpha_2}{\partial x_1} x_2 - \frac{\partial \alpha_2}{\partial x_2} x_3 + \theta^{\mathrm{T}} \left(\varphi_3 - \frac{\partial \alpha_2}{\partial x_1} \varphi_1 - \frac{\partial \alpha_2}{\partial x_2} \varphi_2 \right) - \frac{\partial \alpha_2}{\partial \hat{\theta}} \dot{\hat{\theta}}$$

$$= u - \frac{\partial \alpha_2}{\partial x_1} x_2 - \frac{\partial \alpha_2}{\partial x_2} x_3 + \hat{\theta}^{\mathrm{T}} \left(\varphi_3 - \frac{\partial \alpha_2}{\partial x_1} \varphi_1 - \frac{\partial \alpha_2}{\partial x_2} \varphi_2 \right)$$

(4.16) $$+ \tilde{\theta}^{\mathrm{T}} \left(\varphi_3 - \frac{\partial \alpha_2}{\partial x_1} \varphi_1 - \frac{\partial \alpha_2}{\partial x_2} \varphi_2 \right) - \frac{\partial \alpha_2}{\partial \hat{\theta}} \dot{\hat{\theta}} ,$$

Since this is the final step, we will design the actual control law and the actual update law based on the full Lyapunov function

(4.17) $$V_3 = V_2 + \frac{1}{2} z_3^2 = \frac{1}{2} \left(z_1^2 + z_2^2 + z_3^2 \right) + \frac{1}{2} \tilde{\theta}^{\mathrm{T}} \Gamma^{-1} \tilde{\theta} ,$$

whose derivative is:

$$\dot{V}_3 = \dot{V}_2 + z_3 \dot{z}_3$$

$$= z_2 z_3 - c_1 z_1^2 - c_2 z_2^2 + \tilde{\theta}^{\mathrm{T}} \left[\tau_2 - \Gamma^{-1} \dot{\hat{\theta}} \right] - z_2 \frac{\partial \alpha_1}{\partial \hat{\theta}} \left(\dot{\hat{\theta}} - \Gamma \tau_2 \right)$$

$$+ z_3 \left[u - \frac{\partial \alpha_2}{\partial x_1} x_2 - \frac{\partial \alpha_2}{\partial x_2} x_3 + \hat{\theta}^{\mathrm{T}} \left(\varphi_3 - \frac{\partial \alpha_2}{\partial x_1} \varphi_1 - \frac{\partial \alpha_2}{\partial x_2} \varphi_2 \right) \right.$$

$$\left. + \tilde{\theta}^{\mathrm{T}} \left(\varphi_3 - \frac{\partial \alpha_2}{\partial x_1} \varphi_1 - \frac{\partial \alpha_2}{\partial x_2} \varphi_2 \right) - \frac{\partial \alpha_2}{\partial \hat{\theta}} \dot{\hat{\theta}} \right]$$

$$= -c_1 z_1^2 - c_2 z_2^2 + \tilde{\theta}^{\mathrm{T}} \left[\tau_2 + z_3 \left(\varphi_3 - \sum_{j=1}^{2} \frac{\partial \alpha_2}{\partial x_j} \varphi_j \right) - \Gamma^{-1} \dot{\hat{\theta}} \right]$$

$$+ z_3 \left[u + z_2 - \sum_{j=1}^{2} \frac{\partial \alpha_2}{\partial x_j} x_{j+1} + \hat{\theta}^{\mathrm{T}} \left(\varphi_3 - \sum_{j=1}^{2} \frac{\partial \alpha_2}{\partial x_j} \varphi_j \right) - \frac{\partial \alpha_2}{\partial \hat{\theta}} \dot{\hat{\theta}} \right]$$

(4.18) $$- z_2 \frac{\partial \alpha_1}{\partial \hat{\theta}} \left(\dot{\hat{\theta}} - \Gamma \tau_2 \right) .$$

The update law is now chosen to eliminate the $\tilde{\theta}$-term in (4.18):

$$(4.19) \qquad \dot{\hat{\theta}} = \Gamma \tau_3 \,,$$

where τ_3 is the last tuning function:

$$(4.20) \qquad \tau_3 = \tau_2 + z_3 \left(\varphi_3 - \sum_{j=1}^{2} \frac{\partial \alpha_2}{\partial x_j} \varphi_j \right).$$

With this definition, $\dot{V}_3$ can be rewritten as

$$
\begin{aligned}
\dot{V}_3 &= -c_1 z_1^2 - c_2 z_2^2 + z_3 \left[u + z_2 - \sum_{j=1}^{2} \frac{\partial \alpha_2}{\partial x_j} x_{j+1} \right. \\
(4.21) \qquad &\left. + \hat{\theta}^{\mathrm{T}} \left(\varphi_3 - \sum_{j=1}^{2} \frac{\partial \alpha_2}{\partial x_j} \varphi_j \right) - \frac{\partial \alpha_2}{\partial \hat{\theta}} \Gamma \tau_3 \right] - z_2 \frac{\partial \alpha_1}{\partial \hat{\theta}} \Gamma (\tau_3 - \tau_2) \,.
\end{aligned}
$$

As we already mentioned, the last term in (4.21) represents the mismatch between the actual update law and its estimate $\Gamma \tau_2$ that was used in the definition of α_2. Since the difference $\tau_3 - \tau_2$ contains z_3 as a factor (see (4.20)), we can group this term together with the other z_3-terms and cancel it with u:

$$
\begin{aligned}
\dot{V}_3 &= -c_1 z_1^2 - c_2 z_2^2 + z_3 \left[u + z_2 - \sum_{j=1}^{2} \frac{\partial \alpha_2}{\partial x_j} x_{j+1} + \hat{\theta}^{\mathrm{T}} \left(\varphi_3 - \sum_{j=1}^{2} \frac{\partial \alpha_2}{\partial x_j} \varphi_j \right) \right. \\
(4.22) \qquad &\left. - \frac{\partial \alpha_2}{\partial \hat{\theta}} \Gamma \tau_3 - z_2 \frac{\partial \alpha_1}{\partial \hat{\theta}} \Gamma \left(\varphi_3 - \sum_{j=1}^{2} \frac{\partial \alpha_2}{\partial x_j} \varphi_j \right) \right].
\end{aligned}
$$

Then, the choice

$$
\begin{aligned}
u &= -c_3 z_3 - z_2 + \sum_{j=1}^{2} \frac{\partial \alpha_2}{\partial x_j} x_{j+1} - \hat{\theta}^{\mathrm{T}} \left(\varphi_3 - \sum_{j=1}^{2} \frac{\partial \alpha_2}{\partial x_j} \varphi_j \right) \\
(4.23) \qquad &+ \frac{\partial \alpha_2}{\partial \hat{\theta}} \Gamma \tau_3 + z_2 \frac{\partial \alpha_1}{\partial \hat{\theta}} \Gamma \left(\varphi_3 - \sum_{j=1}^{2} \frac{\partial \alpha_2}{\partial x_j} \varphi_j \right),
\end{aligned}
$$

results in

$$(4.24) \qquad \dot{V}_3 = -c_1 z_1^2 - c_2 z_2^2 - c_3 z_3^2 \,.$$

With the choices (4.20) and (4.23) the complete error system $\mathcal{S}_3$ becomes:

$$
\begin{aligned}
\dot{z}_1 &= -c_1 z_1 + z_2 + \tilde{\theta}^{\mathrm{T}} \varphi_1 \\
\dot{z}_2 &= -c_2 z_2 - z_1 + z_3 + \tilde{\theta}^{\mathrm{T}} \left(\varphi_2 - \tfrac{\partial \alpha_1}{\partial x_1} \varphi_1 \right) \\
&\quad - \tfrac{\partial \alpha_1}{\partial \theta} \Gamma \left(\varphi_3 - \textstyle\sum_{j=1}^2 \tfrac{\partial \alpha_2}{\partial x_j} \varphi_j \right) z_3 \\
\dot{z}_3 &= -c_3 z_3 - z_2 + \tilde{\theta}^{\mathrm{T}} \left(\varphi_3 - \textstyle\sum_{j=1}^2 \tfrac{\partial \alpha_2}{\partial x_j} \varphi_j \right) \\
&\quad + \tfrac{\partial \alpha_1}{\partial \theta} \Gamma \left(\varphi_3 - \textstyle\sum_{j=1}^2 \tfrac{\partial \alpha_2}{\partial x_j} \varphi_j \right) z_2 \\
\tau_3 &= \varphi_1 z_1 + \left(\varphi_2 - \tfrac{\partial \alpha_1}{\partial x_1} \varphi_1 \right) z_2 + \left(\varphi_3 - \textstyle\sum_{j=1}^2 \tfrac{\partial \alpha_2}{\partial x_j} \varphi_j \right) z_3 \,.
\end{aligned}
$$

(4.25) $\mathcal{S}_3$:

Once again, the matrix form of this error system is more informative:

$$
\frac{d}{dt} \begin{bmatrix} z_1 \\ z_2 \\ z_3 \end{bmatrix} = \begin{bmatrix} -c_1 & 1 & 0 \\ -1 & -c_2 & 1 - \tfrac{\partial \alpha_1}{\partial \theta} \Gamma \phi_3 \\ 0 & -1 + \tfrac{\partial \alpha_1}{\partial \theta} \Gamma \phi_3 & -c_3 \end{bmatrix} \begin{bmatrix} z_1 \\ z_2 \\ z_3 \end{bmatrix} + \begin{bmatrix} \phi_1^{\mathrm{T}} \\ \phi_2^{\mathrm{T}} \\ \phi_3^{\mathrm{T}} \end{bmatrix} \tilde{\theta}
$$

$$
(4.26) \quad \tau_3 = \begin{bmatrix} \phi_1 & \phi_2 & \phi_3 \end{bmatrix} \begin{bmatrix} z_1 \\ z_2 \\ z_3 \end{bmatrix} ,
$$

where we have used the convenient notation

$$
(4.27) \quad \phi_1 = \varphi_1, \quad \phi_2 = \varphi_2 - \frac{\partial \alpha_1}{\partial x_1} \varphi_1, \quad \phi_3 = \varphi_3 - \sum_{j=1}^2 \frac{\partial \alpha_2}{\partial x_j} \varphi_j \,.
$$

Comparing (4.26) with (3.29), we see that although the system matrix in (4.26) is no longer constant, it has preserved the important structural properties it had in (3.29): its diagonal terms are negative and its off-diagonal terms are skew-symmetric. The additional term $\frac{\partial \alpha_1}{\partial \theta} \Gamma \phi_3$ in the $(2,3)$-element of the matrix is caused by the fact that we used $\Gamma \tau_2$ instead of the actual update law in the definition of α_2 (see (4.13)). However, the skew-symmetry of the matrix was preserved by the addition of the term $-\frac{\partial \alpha_1}{\partial \theta} \Gamma \phi_3$ in the $(3,2)$-element (see (4.22) and (4.23)).

Furthermore, we see that, as in (3.29), the matrix that multiplies the parameter error $\tilde{\theta}$ in the $\dot{z}$-equation is used (in its transposed form) in the tuning function τ_3 and thus in the update law for the parameter estimates. It is also interesting to compare the expressions for the parameter update laws in (4.25) and (3.28): the tuning function τ_3 appears as the sum of the update laws for ϑ_1, ϑ_2 and ϑ_3:

$$
\Gamma \tau_3 = \dot{\vartheta}_1 + \dot{\vartheta}_2 + \dot{\vartheta}_3 \,.
$$

However, one must be careful to note that this equality is not true: since the expressions (3.14) and (4.13) for α_2 are different, the partial derivatives $\frac{\partial \alpha_2}{\partial x_1}$ and $\frac{\partial \alpha_2}{\partial x_2}$ will have different analytical expressions in (3.26) and (4.20).

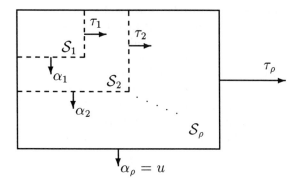

FIG. 4.1. *The design procedure with tuning functions.*

Due to the structure of the error system (4.26), its stability and convergence properties are derived in a manner almost identical to those of (3.29), and are therefore omitted here.

As is the case with the overparametrized design procedure of the previous section, the tuning functions scheme illustrated here can also be generalized to systems of arbitrary order which can be transformed into the strict-feedback (2.1) or the partially strict-feedback (2.2) canonical form. The number of design steps required is again equal to the relative degree ρ of the system. At each step, a new stabilizing function α_i and a new tuning function ϑ_i is generated. At the final step, the stabilizing function α_ρ becomes the control law u and the tuning function τ_ρ becomes the update law $\dot{\theta} = \Gamma\tau_\rho$. A schematic representation of this design procedure is given in Figure 4.1.

5. Output-feedback schemes. The design procedures illustrated so far required full-state feedback. As we mentioned in Section 2, however, these procedures can be modified to handle the case where only the output of a system is measured, provided that the system can be transformed into the output-feedback canonical form (2.4), in which the nonlinearities depend only on the measured output.

To illustrate the output-feedback schemes, let us first consider the nonlinear system in output-feedback canonical form

$$(5.1) \qquad \begin{aligned} \dot{x}_1 &= x_2 + \theta\varphi_1(y) \\ \dot{x}_2 &= x_3 + \theta\varphi_2(y) + u \\ \dot{x}_3 &= u \\ y &= x_1, \end{aligned}$$

where now we assume that only the output $y = x_1$ is measured, i.e., the states x_2 and x_3 are not available for feedback. We want to design an adaptive nonlinear output-feedback controller that guarantees asymptotic tracking of the reference signal $y_r(t)$ by the output y while keeping all the states of the closed-loop system bounded.

Since x_2 and x_3 are not measured and θ is an unknown parameter, we must attempt to reconstruct the full state of the system through the use of filters. In this case, we need only two filters: one for the known part of the system (i.e., the part that does not contain θ), and one for the unknown part (since we only have one unknown parameter). That is, we are trying to reconstruct the state through two filters ξ_0 and ξ_1 as:

$$(5.2) \qquad\qquad x = \xi_0 + \theta\xi_1 + \varepsilon\,.$$

It is clear that the estimate $\xi_0 + \theta\xi_1$ is only a "virtual estimate", since it depends on the unknown parameter θ. Nevertheless, its components ξ_0 and ξ_1 are known and this is good enough for our control purposes, provided that the error ε tends to zero asymptotically. To ensure this, the two filters are defined as follows:

$$(5.3) \quad \begin{aligned} \dot{\xi}_{01} &= -k_1(\xi_{01}-y)+\xi_{02} & \dot{\xi}_{11} &= -k_1\xi_{11}+\xi_{12}+\varphi_1(y) \\ \dot{\xi}_{02} &= -k_2(\xi_{01}-y)+\xi_{03}+u & \dot{\xi}_{12} &= -k_2\xi_{11}+\xi_{13}+\varphi_2(y) \\ \dot{\xi}_{03} &= -k_3(\xi_{01}-y)+u & \dot{\xi}_{13} &= -k_3\xi_{11}\,, \end{aligned}$$

where k_1,k_2,k_3 are chosen so that the matrix

$$(5.4) \qquad\qquad A_0 = \begin{bmatrix} -k_1 & 1 & 0 \\ -k_2 & 0 & 1 \\ -k_3 & 0 & 0 \end{bmatrix}$$

is Hurwitz. To justify these definitions and to show why they achieve our goal, let us rewrite the system in the x-coordinates in the following form:

$$\dot{x} = A_0 x + ky + bu + \theta\varphi(y)\,, \quad k = \begin{bmatrix} -k_1 \\ -k_2 \\ -k_3 \end{bmatrix},\ b = \begin{bmatrix} 0 \\ 1 \\ 1 \end{bmatrix},\ \varphi(y) = \begin{bmatrix} \varphi_1(y) \\ \varphi_2(y) \\ 0 \end{bmatrix}.$$
$$(5.5)$$

Now the filters ξ_0 and ξ_1 are rewritten in the more transparent form:

$$(5.6) \qquad\qquad \dot{\xi}_0 = A_0\xi_0 + ky + bu\,, \quad \dot{\xi}_1 = A_0\xi_1 + \varphi(y)\,.$$

Combining these equations, we obtain for the error $\varepsilon = x - \xi_0 - \theta\xi_1$:

$$\begin{aligned} \dot{\varepsilon} &= \dot{x} - \dot{\xi}_0 - \theta\dot{\xi}_1 \\ &= A_0 x + ky + bu + \theta\varphi(y) - A_0\xi_0 - ky - bu - A_0\theta\xi_1 - \theta\varphi(y) \\ &= A_0(x - \xi_0 - \theta\xi_1) \\ (5.7) \quad &= A_0\varepsilon\,. \end{aligned}$$

Since A_0 is a Hurwitz matrix, we can find a positive definite symmetric matrix P_0 which satisfies the Lyapunov matrix equation:

$$(5.8) \qquad\qquad P_0 A_0 + A_0^{\mathrm{T}} P_0 = -I\,.$$

This implies that

(5.9)
$$\frac{d}{dt}\left(\varepsilon^{\mathrm{T}} P_0 \varepsilon\right) = -\varepsilon^{\mathrm{T}} \varepsilon.$$

We are now ready to design our adaptive nonlinear output-feedback controller.

Step 1. The control objective is to track the reference signal $y_r(t)$ with the output y, so the first error variable is the tracking error:

(5.10)
$$z_1 = y - y_r.$$

The derivative of z_1 is:

(5.11)
$$\dot{z}_1 = \dot{y} - \dot{y}_r = x_2 + \theta\varphi_1(y) - \dot{y}_r.$$

If x_2 were measured, it would be our virtual control. Since it is not measured, we replace it with the sum of its "virtual estimate" and the corresponding error:

(5.12)
$$x_2 = \xi_{02} + \theta\xi_{12} + \varepsilon_2,$$

to obtain

(5.13)
$$\dot{z}_1 = \xi_{02} + \theta\underbrace{\left[\varphi_1(y) + \xi_{12}\right]}_{\omega} - \dot{y}_r + \varepsilon_2.$$

Now we must pick one of the known variables appearing in the above equation to play the role of the virtual control. Looking at the filter equations, we see that only ξ_{02} contains the control u in its first derivative, so it is the only candidate. Defining the second error variable as

(5.14)
$$z_2 = \xi_{02} - \alpha_1,$$

we get

(5.15)
$$\dot{z}_1 = z_2 + \alpha_1 + \theta\omega - \dot{y}_r + \varepsilon_2.$$

The first stabilizing function α_1 is chosen as

(5.16)
$$\alpha_1 = -c_1 z_1 - d_1 z_1 + \dot{y}_r - \hat{\theta}\omega.$$

With this choice, the $\dot{z}_1$-equation becomes

(5.17)
$$\dot{z}_1 = z_2 - c_1 z_1 - d_1 z_1 + \tilde{\theta}\omega + \varepsilon_2.$$

The first partial Lyapunov function is chosen as

(5.18)
$$V_1 = \frac{1}{2}z_1^2 + \frac{1}{2\gamma}\tilde{\theta}^2 + \frac{1}{d_1}\varepsilon^{\mathrm{T}} P_0 \varepsilon,$$

where $\gamma > 0$ is the adaptation gain. Its derivative satisfies:

$$
\begin{aligned}
\dot{V}_1 &= z_1 \dot{z}_1 - \frac{1}{\gamma} \tilde{\theta} \dot{\theta} + \frac{1}{d_1} \frac{d}{dt} \left(\varepsilon^{\mathrm{T}} P_0 \varepsilon \right) \\
&= z_1 z_2 - c_1 z_1^2 - d_1 z_1^2 + \tilde{\theta} \left(z_1 \omega - \frac{1}{\gamma} \dot{\theta} \right) + z_1 \varepsilon_2 - \frac{1}{d_1} \varepsilon^{\mathrm{T}} \varepsilon \\
&= z_1 z_2 - c_1 z_1^2 + \tilde{\theta} \left(z_1 \omega - \frac{1}{\gamma} \dot{\theta} \right) \\
&\quad - d_1 \left[z_1 - \frac{1}{2 d_1} \varepsilon_2 \right]^2 + \frac{1}{4 d_1} \varepsilon_2^2 - \frac{1}{d_1} \varepsilon^{\mathrm{T}} \varepsilon
\end{aligned}
$$

$$
(5.19) \qquad \leq z_1 z_2 - c_1 z_1^2 + \tilde{\theta} \left(z_1 \omega - \frac{1}{\gamma} \dot{\theta} \right) - \frac{3}{4 d_1} \varepsilon^{\mathrm{T}} \varepsilon .
$$

The reason for including the term $-d_1 z_1^2$ in α_1 should now be clear: that term was used to complete squares with the cross-term $z_1 c_2$. But the question that is left unanswered is why this completion of squares is necessary. After all, the term ε_2 is exponentially decaying, so why do we need to explicitly account for its presence? The answer is that in this first equation, where ε_2 is multiplied only with z_1, the presence of the d_1-term is not necessary. However, in subsequent steps we will see that this exponentially decaying error ε_2 is multiplied with nonlinear functions of y. In that case, we do indeed need to explicitly counteract the effect of ε_2, since its presence may even cause finite escape times. To prevent such phenomena, our design procedure incorporates *nonlinear damping* terms such as the term $-d_1 z_1^2$.

If this were the final step of the design procedure, we would eliminate the $\tilde{\theta}$-term from the derivative of V_1 by choosing the update law $\dot{\theta} = \gamma \tau_1$, where τ_1 is the first tuning function

$$
(5.20) \qquad\qquad \tau_1 = z_1 \omega .
$$

Since the control u has not appeared yet, the design procedure must continue. Therefore, the choice of update law is postponed. Using the tuning function τ_1, the V_1-inequality is rewritten as:

$$
(5.21) \qquad \dot{V}_1 \leq z_1 z_2 - c_1 z_1^2 + \tilde{\theta} \left(\tau_1 - \frac{1}{\gamma} \dot{\theta} \right) - \frac{3}{4 d_1} \varepsilon^{\mathrm{T}} \varepsilon .
$$

The first error subsystem $\mathcal{S}_1$ is:

$$
(5.22) \qquad \mathcal{S}_1 : \quad
\begin{aligned}
\dot{z}_1 &= -c_1 z_1 - d_1 z_1 + z_2 + \tilde{\theta} \omega + \varepsilon_2 \\
\dot{\varepsilon} &= A_0 \varepsilon \\
\tau_1 &= \omega z_1 .
\end{aligned}
$$

Step 2. The control u appears in the derivative of z_2, and hence this is the last design step:

$$
\begin{aligned}
\dot{z}_2 &= \dot{\xi}_{02} - \dot{\alpha}_1 \\
&= u - k_2(\xi_{01} - y) + \xi_{03} - \frac{\partial \alpha_1}{\partial y}\dot{y} \\
&\quad - \frac{\partial \alpha_1}{\partial \xi_{12}}\underbrace{(-k_2\xi_{11} + \xi_{13} + \varphi_2(y))}_{\dot{\xi}_{12}} - \frac{\partial \alpha_1}{\partial y_r}\dot{y}_r - \underbrace{\frac{\partial \alpha_1}{\partial \dot{y}_r}\ddot{y}_r}_{1} - \frac{\partial \alpha_1}{\partial \hat{\theta}}\dot{\hat{\theta}} \\
&= u - k_2(\xi_{01} - y) + \xi_{03} - \frac{\partial \alpha_1}{\partial y}\underbrace{(\xi_{02} + \theta\omega + \varepsilon_2)}_{\dot{y}} \\
&\quad - \frac{\partial \alpha_1}{\partial \xi_{12}}(-k_2\xi_{11} + \xi_{13} + \varphi_2(y)) - \frac{\partial \alpha_1}{\partial y_r}\dot{y}_r - \ddot{y}_r - \frac{\partial \alpha_1}{\partial \hat{\theta}}\dot{\hat{\theta}} \\
(5.23) \quad &= u + \beta_2 - \frac{\partial \alpha_1}{\partial y}\tilde{\theta}\omega - \frac{\partial \alpha_1}{\partial y}\varepsilon_2 - \ddot{y}_r - \frac{\partial \alpha_1}{\partial \hat{\theta}}\dot{\hat{\theta}},
\end{aligned}
$$

where β_2 encompasses all the known terms except u and $\ddot{y}_r$:

$$
\beta_2 = -k_2(\xi_{01}-y) + \xi_{03} - \frac{\partial \alpha_1}{\partial y}(\xi_{02}+\hat{\theta}\omega) - \frac{\partial \alpha_1}{\partial \xi_{12}}(-k_2\xi_{11}+\xi_{13}+\varphi_2(y)) - \frac{\partial \alpha_1}{\partial y_r}\dot{y}_r.
$$
(5.24)

To design the update law $\dot{\hat{\theta}}$ and the control u, we consider the augmented partial Lyapunov function

$$
(5.25) \qquad V_2 = V_1 + \frac{1}{2}z_2^2 + \frac{1}{d_2}\varepsilon^{\mathrm{T}}P_0\varepsilon,
$$

whose derivative is

$$
\begin{aligned}
\dot{V}_2 &= \dot{V}_1 + z_2\dot{z}_2 - \frac{1}{d_2}\varepsilon^{\mathrm{T}}\varepsilon \\
&\leq z_1 z_2 - c_1 z_1^2 + \tilde{\theta}\left(\tau_1 - \frac{1}{\gamma}\dot{\hat{\theta}}\right) - \frac{3}{4d_1}\varepsilon^{\mathrm{T}}\varepsilon \\
&\quad + z_2\left[u + \beta_2 - \tilde{\theta}\frac{\partial \alpha_1}{\partial y}\omega - \frac{\partial \alpha_1}{\partial y}\varepsilon_2 - \ddot{y}_r - \frac{\partial \alpha_1}{\partial \hat{\theta}}\dot{\hat{\theta}}\right] - \frac{1}{d_2}\varepsilon^{\mathrm{T}}\varepsilon \\
&= -c_1 z_1^2 + \tilde{\theta}\left(\tau_1 - z_2\frac{\partial \alpha_1}{\partial y}\omega - \frac{1}{\gamma}\dot{\hat{\theta}}\right) - \frac{3}{4d_1}\varepsilon^{\mathrm{T}}\varepsilon \\
&\quad + z_2\left[u + z_1 + \beta_2 - \ddot{y}_r - \frac{\partial \alpha_1}{\partial \hat{\theta}}\dot{\hat{\theta}}\right] - z_2\frac{\partial \alpha_1}{\partial y}\varepsilon_2 - \frac{1}{d_2}\varepsilon^{\mathrm{T}}\varepsilon \\
&= -c_1 z_1^2 + \tilde{\theta}\left(\tau_1 - z_2\frac{\partial \alpha_1}{\partial y}\omega - \frac{1}{\gamma}\dot{\hat{\theta}}\right) - \frac{3}{4d_1}\varepsilon^{\mathrm{T}}\varepsilon \\
&\quad + z_2\left[u + z_1 + \beta_2 - \ddot{y}_r - \frac{\partial \alpha_1}{\partial \hat{\theta}}\dot{\hat{\theta}}\right] \\
&\quad + d_2 z_2^2\left(\frac{\partial \alpha_1}{\partial y}\right)^2 - d_2\left[z_2\frac{\partial \alpha_1}{\partial y} - \frac{1}{2d_2}\varepsilon_2\right]^2 + \frac{1}{4d_2}\varepsilon_2^2 - \frac{1}{d_2}\varepsilon^{\mathrm{T}}\varepsilon
\end{aligned}
$$

$$\leq -c_1 z_1^2 + \tilde{\theta}\left(\tau_1 - z_2\frac{\partial\alpha_1}{\partial y}\omega - \frac{1}{\gamma}\dot{\hat{\theta}}\right) - \frac{3}{4}\left(\frac{1}{d_1} + \frac{1}{d_2}\right)\varepsilon^T\varepsilon$$

$$(5.26)\qquad +z_2\left[u + z_1 + d_2 z_2\left(\frac{\partial\alpha_1}{\partial y}\right)^2 + \beta_2 - \ddot{y}_r - \frac{\partial\alpha_1}{\partial\hat{\theta}}\dot{\hat{\theta}}\right].$$

The $\tilde{\theta}$-term is eliminated with the update law:

$$(5.27)\qquad \dot{\hat{\theta}} = \gamma\tau_2 = \gamma\left(\tau_1 - z_2\frac{\partial\alpha_1}{\partial y}\omega\right),$$

and the last term in (5.26) is rendered equal to $-c_2 z_2^2$ with the control law:

$$(5.28)\qquad u = -c_2 z_2 - z_1 - d_2\left(\frac{\partial\alpha_1}{\partial y}\right)^2 z_2 - \beta_2 + \ddot{y}_r + \frac{\partial\alpha_1}{\partial\hat{\theta}}\gamma\tau_2.$$

With these choices, the derivative of V_2 satisfies the inequality:

$$(5.29)\qquad \dot{V}_2 \leq -c_1 z_1^2 - c_2 z_2^2 - \frac{3}{4}\left(\frac{1}{d_1} + \frac{1}{d_2}\right)\varepsilon^T\varepsilon,$$

which guarantees the boundedness of $z_1, z_2, \tilde{\theta}, \varepsilon$ and the convergence of z_1, z_2, ε to zero. In particular, this implies that the tracking error $y - y_r$ converges to zero and that ξ_0, ξ_1, x_1, x_2 are bounded. The only thing that remains is to show the boundedness of x_3. To this end, we define the variable $\zeta = x_3 - x_2$. If we can show that ζ is bounded, then $x_3 = \zeta + x_2$ will also be bounded. We have:

$$
\begin{aligned}
\dot{\zeta} &= \dot{x}_3 - \dot{x}_2 \\
&= u - x_3 - \theta\varphi_2(y) - u \\
(5.30)\qquad &= -\zeta - x_2 - \theta\varphi_2(y).
\end{aligned}
$$

This shows that ζ is the output of a strictly proper linear system which has an asymptotically stable pole at $s = -1$ and whose input $(x_2 + \theta\varphi_2(y))$ is bounded. Thus, the output ζ of the system is also bounded.

To demonstrate the effect of the nonlinear damping terms, let us write the error system in its matrix form:

$$
\begin{bmatrix} \dot{z}_1 \\ \dot{z}_2 \end{bmatrix} = \begin{bmatrix} -c_1 - d_1 & 1 \\ -1 & -c_2 - d_2\left(\frac{\partial\alpha_1}{\partial y}\right)^2 \end{bmatrix}\begin{bmatrix} z_1 \\ z_2 \end{bmatrix} + \begin{bmatrix} 1 \\ -\frac{\partial\alpha_1}{\partial y} \end{bmatrix}\left(\omega\tilde{\theta} + \varepsilon_2\right)
$$

$$(5.31)\ \dot{\varepsilon} = A_0\varepsilon$$

$$
\tau_2 = \begin{bmatrix} 1 & -\frac{\partial\alpha_1}{\partial y} \end{bmatrix}\omega\begin{bmatrix} z_1 \\ z_2 \end{bmatrix},
$$

We see that the nonlinear damping terms strengthen the negativity of the diagonal entries by including the squares of the terms which multiply the state estimation error ε_2.

6. Application to linear systems. The output-feedback design procedure illustrated in the previous section is applicable to nonlinear systems that can be transformed into the output-feedback canonical form, which is repeated here for convenience:

$$
\begin{aligned}
\dot{x}_1 &= x_2 + \varphi_{0,1}(y) + \theta^{\mathrm{T}}\varphi_1(y) \\
\dot{x}_2 &= x_3 + \varphi_{0,2}(y) + \theta^{\mathrm{T}}\varphi_2(y)
\end{aligned}
$$

$$\vdots$$

(6.1)
$$
\begin{aligned}
\dot{x}_{\rho-1} &= x_\rho + \varphi_{0,\rho-1}(y) + \theta^{\mathrm{T}}\varphi_{\rho-1}(y) \\
\dot{x}_\rho &= x_{\rho+1} + \varphi_{0,\rho}(y) + \theta^{\mathrm{T}}\varphi_\rho(y) + b_{n-\rho}\sigma(y)u
\end{aligned}
$$

$$\vdots$$

$$
\begin{aligned}
\dot{x}_n &= \varphi_{0,n}(y) + \theta^{\mathrm{T}}\varphi_n(y) + b_0\sigma(y)u \\
y &= x_1,
\end{aligned}
$$

where θ and $b_0,\ldots,b_{n-\rho}$ are unknown, $\sigma(y) \neq 0$ for all $y \in \mathbb{R}$, the sign of $b_{n-\rho}$ is known, and the polynomial $B(s) = b_{n-\rho}s^{n-\rho} + \cdots + b_1 s + b_0$ is Hurwitz.

Clearly, any minimum-phase linear system can be transformed into this form, which in that case becomes the well-known observer canonical form. Indeed, any linear system

(6.2)
$$
y(s) = \frac{b_m s^m + \cdots + b_1 s + b_0}{s^n + a_{n-1}s^{n-1} + \cdots + a_1 s + a_0}u(s)
$$

can be expressed in the state-space representation

$$
\dot{x}_1 = x_2 - a_{n-1}y
$$

$$\vdots$$

(6.3)
$$
\begin{aligned}
\dot{x}_{\rho-1} &= x_\rho - a_{m+1}y \\
\dot{x}_\rho &= x_{\rho+1} - a_m y + b_m u
\end{aligned}
$$

$$\vdots$$

$$
\begin{aligned}
\dot{x}_n &= -a_0 y + b_0 u \\
y &= x_1,
\end{aligned}
$$

which is identical to (6.1) with $\theta^{\mathrm{T}} = [-a_{n-1},\ldots,-a_0]$, $\rho = n - m$, and $\sigma(y) = 1$.

To briefly illustrate the design procedure for linear systems, let us consider the unstable relative-degree-three plant

(6.4)
$$
y(s) = \frac{1}{s^2(s - \theta)}u(s) ,
$$

where $\theta > 0$ is considered to be unknown. The control objective is to asymptotically track the output of the reference model

$$(6.5) \qquad y_r(s) = \frac{1}{(s+1)^3} r(s).$$

To derive the adaptive controller resulting from our nonlinear design, the plant (6.4) is first rewritten in the state-space form (6.3):

$$(6.6) \qquad \begin{aligned} \dot{x}_1 &= x_2 + \theta x_1 \\ \dot{x}_2 &= x_3 \\ \dot{x}_3 &= u \\ y &= x_1 \,. \end{aligned}$$

The filters required for the "virtual estimates" of x_2 and x_3 are

$$(6.7) \qquad \dot{\eta} = A_0 \eta + e_3 y \,, \quad \xi_2 - A_0^2 \eta \,, \quad \xi_3 = -A_0^3 \eta$$

$$(6.8) \qquad \dot{\lambda} = A_0 \lambda + e_3 u \,, \quad v = \lambda \,, \quad A_0 = \begin{bmatrix} -k_1 & 1 & 0 \\ -k_2 & 0 & 1 \\ -k_3 & 0 & 0 \end{bmatrix} .$$

The signals $y_r, \dot{y}_r, \ddot{y}_r, y_r^{(3)}$ are implemented from the reference model (6.5) as follows:

$$(6.9) \qquad y_r = r_1 \,, \quad \dot{y}_r = r_2 \,, \quad \ddot{y}_r = r_3 \,, \quad y_r^{(3)} = -3r_3 - 3r_2 - r_1 + r \,,$$

where $\dot{r}_1 = r_2, \dot{r}_2 = r_3, \dot{r}_3 = -3r_3 - 3r_2 - r_1 + r$.

The virtual estimate of x is $\xi_3 + \theta\xi_2 + v$, and by defining $\omega = \xi_{2,2} + y$ the results of the three steps of our design procedure are:

Step 1.

$$(6.10) \qquad z_1 = y - y_r$$

$$(6.11) \qquad \tau_1 = \omega z_1$$

$$(6.12) \qquad \alpha_1 = -c_1 z_1 - d_1 z_1 - \xi_{3,2} + \dot{y}_r - \omega\hat{\theta} \,.$$

Step 2.

$$(6.13) \quad z_2 = v_2 - \alpha_1$$

$$(6.14) \quad \tau_2 = \tau_1 - \frac{\partial\alpha_1}{\partial y}\omega z_2$$

$$(6.15) \quad \alpha_2 = -c_2 z_2 - d_2\left(\frac{\partial\alpha_1}{\partial y}\right)^2 z_2 - z_1 + k_2 v_1 + \frac{\partial\alpha_1}{\partial y}(v_2 + \xi_{3,2})$$

$$+ \frac{\partial\alpha_1}{\partial y_r}\dot{y}_r + \frac{\partial\alpha_1}{\partial \dot{y}_r}\ddot{y}_r + \frac{\partial\alpha_1}{\partial\xi_3}(A_0\xi_3 + ky) + \frac{\partial\alpha_1}{\partial\xi_2}(A_0\xi_2 + e_1 y)$$

$$+ \frac{\partial\alpha_1}{\partial y}\omega\hat{\theta} + \frac{\partial\alpha_1}{\partial\hat{\theta}}\gamma\tau_2 \,.$$

Step 3.

(6.16)　　$z_3 = v_3 - \alpha_2$

(6.17)　　$\tau_3 = \tau_2 - \dfrac{\partial \alpha_2}{\partial y} \omega z_3$

(6.18)　　$u = -c_3 z_3 - d_3 \left(\dfrac{\partial \alpha_2}{\partial y} \right)^2 z_3 - z_2 + k_3 v_1 + \dfrac{\partial \alpha_2}{\partial y}(v_2 + \xi_{3,2})$

$$+ \frac{\partial \alpha_2}{\partial y_r} \dot{y}_r + \frac{\partial \alpha_2}{\partial \dot{y}_r} \ddot{y}_r + \frac{\partial \alpha_2}{\partial \ddot{y}_r} y_r^{(3)} + \frac{\partial \alpha_2}{\partial \xi_3}(A_0 \xi_3 + ky)$$

$$+ \frac{\partial \alpha_2}{\partial \xi_2}(A_0 \xi_2 + e_1 y) + \frac{\partial \alpha_2}{\partial v_1}(v_2 - k_1 v_1) + \frac{\partial \alpha_2}{\partial v_2}(v_3 - k_2 v_1)$$

$$+ \frac{\partial \alpha_2}{\partial y} \omega \hat{\theta} + \frac{\partial \alpha_2}{\partial \hat{\theta}} \gamma \tau_3 - \gamma z_2 \frac{\partial \alpha_1}{\partial \hat{\theta}} \frac{\partial \alpha_2}{\partial y} \omega .$$

The matrix form of the error system is:

$$
\begin{bmatrix} \dot{z}_1 \\ \dot{z}_2 \\ \dot{z}_3 \end{bmatrix}
=
\begin{bmatrix}
-c_1 - d_1 & 1 & 0 \\
-1 & -c_2 - d_2 \left(\frac{\partial \alpha_1}{\partial y} \right)^2 & 1 + \gamma \sigma \omega \\
0 & -1 - \gamma \sigma \omega & -c_3 - d_3 \left(\frac{\partial \alpha_2}{\partial y} \right)^2
\end{bmatrix}
\begin{bmatrix} z_1 \\ z_2 \\ z_3 \end{bmatrix}
$$

(6.19)
$$
+ \begin{bmatrix} 1 \\ -\frac{\partial \alpha_1}{\partial y} \\ -\frac{\partial \alpha_2}{\partial y} \end{bmatrix} \omega \tilde{\theta}
+ \begin{bmatrix} 1 \\ -\frac{\partial \alpha_1}{\partial y} \\ -\frac{\partial \alpha_2}{\partial y} \end{bmatrix} \varepsilon_2
$$

$$\dot{\varepsilon} = A_0 \varepsilon$$

$$
\tau_3 = \begin{bmatrix} 1 & -\frac{\partial \alpha_1}{\partial y} & -\frac{\partial \alpha_2}{\partial y} \end{bmatrix} \omega
\begin{bmatrix} z_1 \\ z_2 \\ z_3 \end{bmatrix} ,
$$

where $\sigma = \dfrac{\partial \alpha_1}{\partial \hat{\theta}} \dfrac{\partial \alpha_2}{\partial y}$. Note again the skew-symmetry of the off-diagonal entries and the stabilizing role of the diagonal entries, whose negativity has been strengthened by the nonlinear damping terms.

This adaptive scheme, developed in [13], was compared in the same paper with a standard indirect certainty-equivalence scheme [2,18] on the basis of transient performance and control effort. In the indirect scheme, the plant equation $s^2(s - \theta)y(s) = u(s)$ is filtered by a Hurwitz observer polynomial $s^3 + k_1 s^2 + k_2 s + k_3$ to obtain the estimation equation:

$$\phi = \psi \theta$$

(6.20)
$$\phi = \frac{s^3}{s^3 + k_1 s^2 + k_2 s + k_3} y(s) - \frac{1}{s^3 + k_1 s^2 + k_2 s + k_3} u(s)$$

$$\psi = \frac{s^2}{s^3 + k_1 s^2 + k_2 s + k_3} y(s) ,$$

and the parameter update law is a normalized gradient (the simulation results with a least-squares update law were virtually identical):

$$(6.21) \qquad \dot{\theta} = \gamma \frac{\psi e}{1 + \psi^2} , \quad e = \phi - \psi \hat{\theta} .$$

The control law is

$$(6.22) \qquad u = r + \left[\frac{\delta_0 s^2 + \delta_1 s + \delta_2}{s^2 + m_1 s + m_2} \right] y + \left[\frac{\delta_3 s + \delta_4}{s^2 + m_1 s + m_2} \right] u ,$$

where $s^2 + m_1 s + m_2$ is a Hurwitz polynomial and the controller parameters $\delta_0, \ldots, \delta_4$ are computed from the Bezout identity

$$s^5 + s^4[m_1 - \delta_3 - \theta] + s^3[m_2 - \delta_4 - a(m_1 - \delta_3)] - s^2[\delta_0 + (m_2 - \delta_4)\theta] =$$
$$(6.23) \qquad\qquad = (s + 1)^3(s^2 + m_1 s + m_2) + \delta_1 s + \delta_2 ,$$

which gives $\delta_3 = -(3 + \hat{\theta})$, $\delta_4 = -[3 + 3m_1 + \hat{\theta}(m_1 - \delta_3)]$, $\delta_0 = -[1 + 3m_1 + 3m_2 + (m_2 - \delta_4)\hat{\theta}]$, $\delta_1 = -m_1 - 3m_2$, $\delta_2 = -m_2$.

The above indirect adaptive linear scheme and the adaptive nonlinear scheme were applied to the plant (6.3) with the true parameter $\theta = 3$. In all tests the initial parameter estimate was $\hat{\theta}(0) = 0$, so that, with the adaptation switched off, both closed-loop systems were unstable. The reference input was $r(t) = \sin t$. As is evident from the simulation results in Figure 6.1, the nonlinear scheme consistently demonstrated much better transient performance without an increase in control effort.[3] This improvement can be attributed to the different construction of the control law in the two schemes, whose corresponding block diagrams are given in Figures 6.2 and 6.3. Figure 6.2 shows the standard structure of the input and output filters feeding into an estimator/controller block which consists of a parameter estimator and a certainty-equivalence "linear" controller. As seen in Figure 6.3, the adaptive nonlinear controller retains much of the familiar filter structure. The fundamental difference is that the estimator/controller block now becomes a nonlinear controller designed via the three-step procedure outlined above: in the control law (6.18) both parameter estimates and filter signals enter nonlinearly.

7. κ-terms. The design procedures we have presented so far guarantee global stability and asymptotic tracking. However, they provide no guarantees in terms of transient performance and, furthermore, adaptation is required for boundedness and stability. In this section we discuss a recently proposed [4,5] modification of the design procedures which allows the systematic improvement of transient performance and guarantees boundedness even without adaptation. Hence, with the addition of these "κ-terms", adaptation is required only for asymptotic tracking. The κ-terms

---

[3] For the control plots as well as for further details, the reader is referred to the original paper [13].

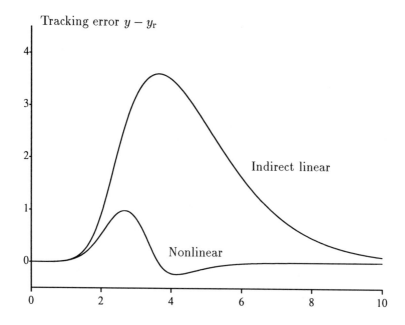

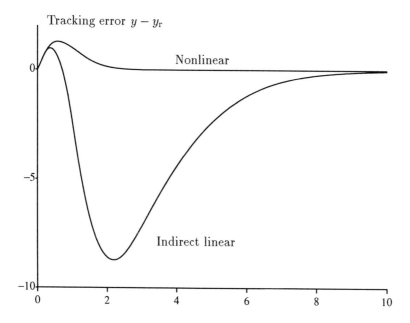

FIG. 6.1. *Transient performance comparison of indirect linear scheme with new nonlinear scheme when $y(0) = 0$ (top) and when $y(0) = 1$ (bottom). In both cases, the nonlinear scheme achieves a dramatic performance improvement without any increase in control effort.*

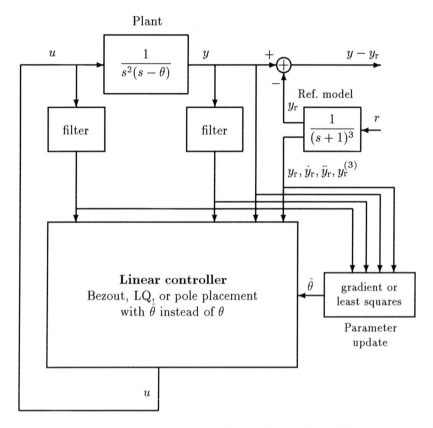

FIG. 6.2. *The indirect scheme is designed according to the traditional structure of adaptive "linear" control which is based on the certainty-equivalence principle.*

were instrumental in the recent development of estimation-based schemes for adaptive control of nonlinear systems presented in a companion paper in this volume [14].

Disturbance-induced instability. Consider the scalar nonlinear system depicted in Figure 7.1:

$$\dot{x} = u + \Delta(t)\varphi(x), \tag{7.1}$$

where $\varphi(x)$ is a known smooth nonlinearity, and $\Delta(t)$ is an exponentially decaying disturbance:

$$\Delta(t) = \Delta(0)e^{-kt}. \tag{7.2}$$

Because of its exponentially decaying nature, one might be tempted to ignore the disturbance $\Delta(t)$ and use the simple control $u = -cx$, which results in the closed-loop system

$$\dot{x} = -cx + \Delta(0)e^{-kt}\varphi(x). \tag{7.3}$$

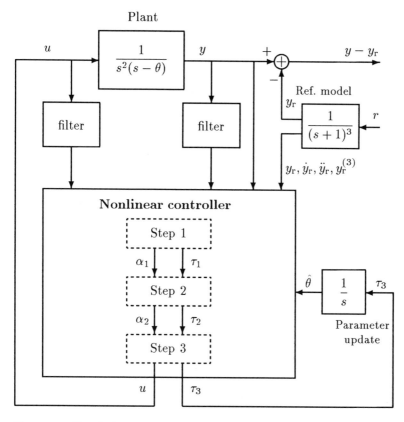

FIG. 6.3. *The distinguishing feature of the new adaptive system is the "Nonlinear controller" block. In contrast to the certainty-equivalence design of Fig. 6.2, the three-step nonlinear procedure produces a control law in which both parameter estimates and filter signals enter nonlinearly.*

While this design may be satisfactory in the case where $\varphi(x)$ is bounded by a constant or even a linear function of x, it is inadequate if $\varphi(x)$ is allowed to be any smooth nonlinear function. For example, consider the case $\varphi(x) = x^2$, which yields

$$(7.4) \qquad \dot{x} = -cx + \Delta(0)e^{-kt}x^2 .$$

The solution $x(t)$ of this system can be calculated explicitly using the change of variable $w = 1/x$:

$$(7.5) \qquad \dot{w} = -\frac{1}{x^2}\dot{x} = c\frac{1}{x} - \Delta(0)e^{-kt} = cw - \Delta(0)e^{-kt} ,$$

which yields

$$(7.6) \qquad w(t) = \left[w(0) - \frac{\Delta(0)}{c+k}\right]e^{ct} + \frac{\Delta(0)}{c+k}e^{-kt} .$$

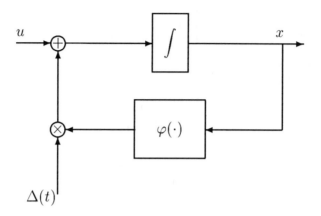

FIG. 7.1. *The perturbed system (7.1).*

The substitution $w = 1/x$ gives

$$(7.7) \qquad x(t) = \frac{x(0)(c+k)}{[c+k-\Delta(0)x(0)]e^{ct} + \Delta(0)x(0)e^{-kt}} .$$

From (7.7) we see that the behavior of the closed-loop system (7.4) depends critically on the initial conditions $\Delta(0), x(0)$:

(i) If $\Delta(0)x(0) < c+k$, the solutions $x(t)$ are bounded and converge asymptotically to zero.

(ii) The situation changes dramatically when $\Delta(0)x(0) > c+k > 0$. The solutions $x(t)$ which start from such initial conditions not only diverge to infinity, but do so in *finite time:*

$$(7.8) \qquad x(t) \to \infty \ \text{ as } t \to t_f = \frac{1}{c+k} \ln \left\{ \frac{\Delta(0)x(0)}{\Delta(0)x(0) - (c+k)} \right\} .$$

Note that this finite escape can not be eliminated by making c larger: for any values of c and k and for any nonzero value of $\Delta(0)$ there exist initial conditions $x(0)$ which satisfy the inequality $\Delta(0)x(0) > c+k$. This example shows that in a nonlinear system, neglecting the effects of exponentially decaying disturbances or nonzero initial conditions can be catastrophic: assuming that $x(0) = 0$ or $\Delta(0) = 0$ artificially eliminates the finite escape time phenomenon.

Nonlinear damping with κ-terms. To overcome this problem and guarantee that the solutions $x(t)$ starting from any initial condition $x(0)$ will remain bounded, we augment the control law $u = -cx$ with a *nonlinear damping term* $s(x)$ which is to be designed:

$$(7.9) \qquad u = -cx + s(x) .$$

Returning to the system (7.1), we use the quadratic Lyapunov function $V(x) = \frac{1}{2}x^2$ whose derivative is:

$$
\begin{aligned}
\dot{V} &= x\,u + x\,\varphi(x)\Delta(t) \\
&= -cx^2 + xs(x) + x\,\varphi(x)\Delta(t)\,.
\end{aligned}
$$
(7.10)

The objective of guaranteeing global boundedness of solutions can be equivalently expressed as rendering the derivative of the Lyapunov function negative outside a compact region. This is achieved with the choice

$$
s(x) = -\kappa\,x\,\varphi^2(x)\,, \quad \kappa > 0\,,
$$
(7.11)

which yields the control

$$
u = -cx - \kappa\,x\,\varphi^2(x)
$$
(7.12)

and the Lyapunov function derivative

$$
\begin{aligned}
\dot{V} &= -cx^2 - \kappa\,x^2\varphi^2(x) + x\,\varphi(x)\Delta(t) \\
&= -cx^2 - \kappa\left[x\,\varphi(x) - \frac{\Delta(t)}{2\kappa}\right]^2 + \frac{\Delta^2(t)}{4\kappa} \\
&\leq -cx^2 + \frac{\Delta^2(t)}{4\kappa}\,.
\end{aligned}
$$
(7.13)

It is clear that the choice of the nonlinear damping term in (7.11) allows the completion of squares in (7.13). This simple completion of squares can be viewed as a special case of *Young's Inequality*, which, in a simplified form, states that if the constants $p > 1$ and $q > 1$ are such that $(p-1)(q-1) = 1$, then for all $\varepsilon > 0$ and all $(x, y) \in \mathbb{R}^2$ we have

$$
xy \leq \frac{\varepsilon^p}{p}\,|x|^p + \frac{1}{q\varepsilon^q}\,|y|^q\,.
$$
(7.14)

Choosing $p = q = 2$ and $\varepsilon^2 = 2\kappa$, (7.14) becomes

$$
xy \leq \kappa x^2 + \frac{1}{4\kappa}\,y^2\,,
$$
(7.15)

which is the form used in (7.13):

$$
x\,\varphi(x)\Delta(t) \leq \kappa x^2\varphi^2(x) + \frac{\Delta^2(t)}{4\kappa}\,.
$$
(7.16)

Global boundedness and convergence. Returning to (7.13), we see that the Lyapunov function derivative is negative whenever $|x(t)| \geq \frac{\Delta(t)}{2\sqrt{\kappa c}}$. Since $\Delta(t)$ is a bounded disturbance, we conclude that $\dot{V}$ is negative outside the compact residual set

$$
\mathcal{R} = \left\{x : |x| \leq \frac{\|\Delta\|_\infty}{2\sqrt{\kappa c}}\right\}\,.
$$
(7.17)

Recalling that $V(x) = \frac{1}{2}x^2$, we conclude that $|x(t)|$ decreases whenever $x(t)$ is outside the set $\mathcal{R}$, and hence $x(t)$ is bounded:

$$(7.18) \qquad \|x\|_\infty \leq \max\left\{|x(0)|, \frac{\|\Delta\|_\infty}{2\sqrt{\kappa c}}\right\}.$$

Moreover, we can draw some conclusions about the asymptotic behavior of $x(t)$. Let us rewrite (7.13) as:

$$(7.19) \qquad \frac{d}{dt}\left(x^2\right) \leq -2cx^2 + \frac{\Delta^2(t)}{2\kappa}.$$

To obtain explicit bounds on $x(t)$, we consider the signal $x(t)e^{ct}$. Using (7.19) we get

$$(7.20) \qquad \frac{d}{dt}\left(x^2 e^{2ct}\right) = \frac{d}{dt}\left(x^2\right)e^{2ct} + 2cx^2 e^{2ct} \leq \frac{\Delta^2(t)}{2\kappa}e^{2ct}.$$

Integrating both sides over the interval $[0, t]$ yields

$$
\begin{aligned}
x^2(t)e^{2ct} &\leq x^2(0) + \int_0^t \frac{1}{2\kappa}\Delta^2(\tau)e^{2c\tau}\,d\tau \\
(7.21) \qquad &\leq x^2(0) + \frac{1}{4\kappa c}\left[\sup_{0\leq\tau\leq t}\Delta^2(\tau)\right]\left(e^{2ct}-1\right).
\end{aligned}
$$

Multiplying both sides with e^{-2ct} and using the fact that $a^2 \leq b^2 + c^2 \Rightarrow |a| \leq |b| + |c|$, we obtain an explicit bound for $x(t)$:

$$(7.22) \qquad |x(t)| \leq |x(0)|e^{-ct} + \frac{1}{2\sqrt{\kappa c}}\left[\sup_{0\leq\tau\leq t}|\Delta(\tau)|\right]\sqrt{1-e^{-2ct}}.$$

Since $\sup_{0\leq\tau\leq t}|\Delta(\tau)| \leq \sup_{0\leq\tau<\infty}|\Delta(\tau)| \triangleq \|\Delta\|_\infty$, (7.22) leads to

$$(7.23) \qquad |x(t)| \leq |x(0)|e^{-ct} + \frac{\|\Delta\|_\infty}{2\sqrt{\kappa c}}\sqrt{1-e^{-2ct}},$$

which shows that $x(t)$ converges to the compact set $\mathcal{R}$ defined in (7.17):

$$(7.24) \qquad \lim_{t\to\infty} \text{dist}\{x(t),\mathcal{R}\} = 0.$$

It is important to note that these properties of boundedness (cf. (7.18)) and convergence (cf. (7.24)) are guaranteed for any bounded disturbance $\Delta(t)$ and for any smooth nonlinearity $\varphi(x)$ (including $\varphi(x) = x^2$). Furthermore, the nonlinear control law (7.12) does not assume knowledge of a bound on the disturbance, nor does it have to use large values for the gains κ and c. Indeed, the residual set $\mathcal{R}$ defined in (7.17) is compact for any finite value of $\|\Delta\|_\infty$ and for any positive value of κ and c. Hence, *global boundedness is guaranteed in the presence of bounded disturbances with possibly unknown bounds, regardless of how small the gains κ and c are chosen.* Of course the size of $\mathcal{R}$ can not be estimated *a priori* if no bound for $\|\Delta\|_\infty$ is given, but it can be reduced *a posteriori* by increasing the values of κ and c.

This property is achieved by the "κ-term" $-\kappa x \varphi^2(x)$ in (7.12), which renders the effective gain of (7.12) "selectively high": when κ and c are chosen to be small, the gain is low around the origin, but it becomes high when x is in a region where $\varphi(x)$ is large enough to make the term $\kappa \varphi^2(x)$ large. If we interpret the nonlinearity $\varphi(x)$ as the "disturbance gain", since it multiplies the disturbance $\Delta(t)$ in (7.1), we see that the term $-\kappa \varphi^2(x)$ in (7.12) guarantees that when the disturbance gain becomes large, the control gain becomes large enough to keep the state bounded.

Finally, we should note that if the disturbance $\Delta(t)$ converges to zero in addition to being bounded, then the control (7.12) guarantees convergence of $x(t)$ to zero in addition to global boundedness. To show this, let $\bar{\Delta}(t)$ be a continuous nonnegative *monotonically decreasing* function such that $|\Delta(t)| \leq \bar{\Delta}(t)$ and $\lim_{t\to\infty} \bar{\Delta}(t) = 0$. Then, starting with the first inequality from (7.21), we obtain

$$
\begin{aligned}
|x(t)|^2 \;\leq\; & |x(0)|^2 e^{-2ct} + \frac{1}{2\kappa} \int_0^t e^{-2c(t-\tau)} \bar{\Delta}^2(\tau) d\tau \\
\leq\; & |x(0)|^2 e^{-2ct} + \frac{1}{2\kappa} \int_0^{t/2} e^{-2c(t-\tau)} \sup_{0\leq\tau\leq t/2}\{\bar{\Delta}^2(\tau)\} d\tau \\
& + \frac{1}{2\kappa} \int_{t/2}^t e^{-2c(t-\tau)} \sup_{t/2\leq\tau\leq t}\{\bar{\Delta}^2(\tau)\} d\tau \\
=\; & |x(0)|^2 e^{-2ct} + \frac{1}{2\kappa} \bar{\Delta}^2(0) \int_0^{t/2} e^{-2c(t-\tau)} d\tau \\
& + \frac{1}{2\kappa} \bar{\Delta}^2(t/2) \int_{t/2}^t e^{-2c(t-\tau)} d\tau \\
=\; & |x(0)|^2 e^{-2ct} + \frac{1}{4\kappa c} \bar{\Delta}^2(0) e^{-ct} \left(1 - e^{-ct}\right) \\
& + \frac{1}{4\kappa c} \bar{\Delta}^2(t/2) \left(1 - e^{-ct}\right) ,
\end{aligned}
$$

(7.25)

which leads to

(7.26) $\qquad |x(t)| \leq |x(0)| e^{-ct} + \frac{1}{2\sqrt{\kappa c}} \left(\bar{\Delta}(0) e^{-\frac{c}{2}t} + \bar{\Delta}(t/2)\right)$.

Since $\lim_{t\to\infty} \bar{\Delta}(t/2) = 0$, we see that $\lim_{t\to\infty} x(t) = 0$.

Nonlinear operator interpretation. The effect of the term $-\kappa x \varphi^2(x)$ in (7.12) can also be interpreted from an operator point of view on the basis of (7.18), which is repeated here for convenience:

$$(7.27) \qquad \|x\|_\infty \leq \max\left\{|x(0)|, \frac{\|\Delta\|_\infty}{2\sqrt{\kappa c}}\right\}.$$

Assuming that the initial condition $|x(0)|$ is small enough, we obtain

$$(7.28) \qquad \|x\|_\infty \leq \frac{1}{2\sqrt{\kappa c}}\|\Delta\|_\infty,$$

which shows that the nonlinear operator K mapping the disturbance $\Delta(t)$ to the output $x(t)$, depicted in Figure 7.2, is bounded, and its $\mathcal{L}_\infty$-induced gain is

$$(7.29) \qquad \|K\|_{\infty\,\text{ind}} \leq \frac{1}{2\sqrt{\kappa c}}.$$

Note that the nonlinear damping term renders the operator K bounded for *any* positive values of c and κ. Note also that (7.27) provides a more complete description of this operator than (7.28), because it explicitly shows the effect of initial conditions. In contrast to the linear operator case, neglecting the effects of initial conditions can be quite dangerous for nonlinear systems, as the finite escape time example (7.3)–(7.8) demonstrates.

ISS interpretation. Perhaps the most appropriate interpretation of the effect of the nonlinear damping term $-\kappa x \varphi^2(x)$ in (7.12) is that it renders the closed-loop system ISS (input-to-state stable [23,24]) with respect to the disturbance input $\Delta(t)$. Let us recall from [23,24] that the system $\dot{x} = f(t, x, u)$ is ISS with respect to u if there exist a class $\mathcal{KL}$ function $\beta_{\mathcal{KL}}$ and a class $\mathcal{K}$ function $\gamma_{\mathcal{K}}$ such that, for any $x(0)$ and for any input $u(\cdot)$ which is continuous and bounded on $[0, \infty)$, the solution $x(t)$ exists for all $t \geq 0$ and satisfies

$$(7.30) \qquad |x(t)| \leq \beta_{\mathcal{KL}}(|x(t_0)|, t - t_0) + \gamma_{\mathcal{K}}\left(\sup_{t_0 \leq \tau \leq t} |u(\tau)|\right)$$

for all t_0 and t such that $0 \leq t_0 \leq t$. To show that this is true for our closed-loop system, we repeat the argument that led from (7.20) to (7.22), this time integrating over the interval $[t_0, t]$. The result is

$$(7.31) \qquad |x(t)| \leq |x(t_0)|e^{-c(t-t_0)} + \frac{1}{2\sqrt{\kappa c}}\left[\sup_{t_0 \leq \tau \leq t} |\Delta(\tau)|\right],$$

which is identical to (7.30) with $\beta_{\mathcal{KL}}(r, s) = re^{-cs}$, $\gamma_{\mathcal{K}}(r) = \frac{1}{2\sqrt{\kappa c}}r$ and $u(\tau)$ replaced by the disturbance $\Delta(\tau)$.

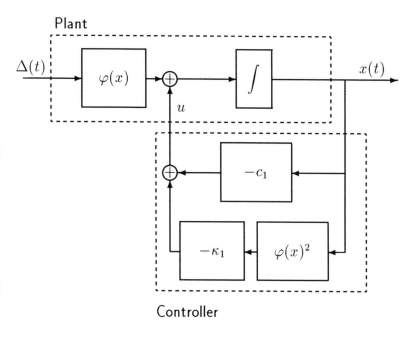

Fig. 7.2. *The bounded nonlinear operator* $K : \Delta(t) \rightarrow x(t)$.

8. Adaptive design with κ-terms. The κ-terms presented in the previous section can easily be incorporated into the design procedures we have discussed so far. The resulting adaptive controllers guarantee boundedness even when the adaptation is switched off, and their transient performance can be improved in a systematic way through *trajectory initialization* and the choice of design parameters.

Using available information about the initial state of the plant, it is always possible to design a reference trajectory which minimizes the initial values of the error variables in the adaptive system. In the case of model reference control, this is achieved by adjusting the initial conditions of the reference model. If, on the other hand, the reference trajectory is given as a precomputed function of time, then it can be initialized through the addition of exponentially decaying terms which define the *reference transients*. The full Lyapunov function that encompasses all the states of the adaptive system is our criterion for initialization. Since this function is comprised of quadratic functions of error variables with respect to the reference trajectory, its value can be used as a measure of the distance between the actual trajectory of the system and the reference trajectory. Our initialization procedure essentially *places the initial point of the reference trajectory as close as possible to the initial point of the system trajectory, thereby reduc-*

ing the initial value of this Lyapunov function. This in turn results in a significant reduction of the bounds on transient errors, since these bounds are directly related to the initial value of the Lyapunov function.

To illustrate the design with κ-terms and the process of trajectory initialization, we consider again the system (3.1) with the output $y = x_1$:

$$(8.1) \qquad \begin{aligned} \dot{x}_1 &= x_2 + \theta^{\mathrm{T}}\varphi_1(x_1) \\ \dot{x}_2 &= x_3 + \theta^{\mathrm{T}}\varphi_2(x_1, x_2) \\ \dot{x}_3 &= u + \theta^{\mathrm{T}}\varphi_3(x_1, x_2, x_3) \\ y &= x_1 . \end{aligned}$$

The control objective is to asymptotically track a reference output $y_r(t)$ with the output y of the system (8.1). We assume that not only y_r, but also its first three derivatives $\dot{y}_r$, $\ddot{y}_r$ and $y_r^{(3)}$ are known and uniformly bounded, and, in addition, $y_r^{(3)}$ is piecewise continuous.

Step 1. The first error variable is now the tracking error

$$(8.2) \qquad z_1 = y - y_r = x_1 - y_r ,$$

and its derivative is

$$(8.3) \qquad \dot{z}_1 = \dot{x}_1 - \dot{y}_r = x_2 + \theta^{\mathrm{T}}\varphi_1(x_1) - \dot{y}_r .$$

Viewing x_2 as the virtual control and defining the second error variable

$$(8.4) \qquad z_2 = x_2 - \alpha_1$$

and the stabilizing function

$$(8.5) \qquad \alpha_1 = -c_1 z_1 - \kappa_1 z_1 \|\varphi_1\|^2 - \hat{\theta}^{\mathrm{T}}\varphi_1 + \dot{y}_r ,$$

we rewrite (8.3) in the form

$$(8.6) \qquad \dot{z}_1 = -c_1 z_1 - \kappa_1 z_1 \|\varphi_1\|^2 + z_2 + \tilde{\theta}^{\mathrm{T}}\varphi_1 .$$

The derivative of the partial Lyapunov function $V_1 = \frac{1}{2}z_1^2 + \frac{1}{2}\tilde{\theta}^{\mathrm{T}}\Gamma^{-1}\tilde{\theta}$ becomes

$$(8.7) \qquad \dot{V}_1 = z_1 z_2 - c_1 z_1^2 - \kappa_1 z_1^2 \|\varphi_1\|^2 + \tilde{\theta}^{\mathrm{T}}\left(\tau_1 - \Gamma^{-1}\dot{\hat{\theta}}\right)$$

where the tuning function τ_1 is defined as

$$(8.8) \qquad \tau_1 = \varphi_1 z_1 .$$

The first error subsystem $\mathcal{S}_1$ becomes:

$$(8.9) \qquad \mathcal{S}_1 : \quad \begin{aligned} \dot{z}_1 &= -c_1 z_1 - \kappa_1 z_1^2 \|\varphi_1\|^2 + z_2 + \tilde{\theta}^{\mathrm{T}}\varphi_1 \\ \tau_1 &= \varphi_1 z_1 . \end{aligned}$$

Step 2. The derivative of z_2 is now expressed as

$$
\begin{aligned}
\dot{z}_2 &= \dot{x}_2 - \dot{\alpha}_1 \\
&= x_3 + \theta^{\mathrm{T}}\varphi_2 - \frac{\partial\alpha_1}{\partial z_1}\dot{z}_1 - \frac{\partial\alpha_1}{\partial y_{\mathrm{r}}}\dot{y}_{\mathrm{r}} - \ddot{y}_{\mathrm{r}} - \frac{\partial\alpha_1}{\partial\hat{\theta}}\dot{\hat{\theta}} \\
&= x_3 + \theta^{\mathrm{T}}\varphi_2 - \frac{\partial\alpha_1}{\partial z_1}\left(-c_1 z_1 - \kappa_1 z_1\|\varphi_1\|^2 + z_2 + \tilde{\theta}^{\mathrm{T}}\varphi_1\right) \\
&\quad - \frac{\partial\alpha_1}{\partial y_{\mathrm{r}}}\dot{y}_{\mathrm{r}} - \ddot{y}_{\mathrm{r}} - \frac{\partial\alpha_1}{\partial\hat{\theta}}\dot{\hat{\theta}} \\
&= x_3 + \hat{\theta}^{\mathrm{T}}\varphi_2 - \frac{\partial\alpha_1}{\partial z_1}\left(-c_1 z_1 - \kappa_1 z_1\|\varphi_1\|^2 + z_2\right) \\
&\quad - \frac{\partial\alpha_1}{\partial y_{\mathrm{r}}}\dot{y}_{\mathrm{r}} - \ddot{y}_{\mathrm{r}} + \tilde{\theta}^{\mathrm{T}}\left(\varphi_2 - \frac{\partial\alpha_1}{\partial z_1}\varphi_1\right) - \frac{\partial\alpha_1}{\partial\hat{\theta}}\dot{\hat{\theta}}
\end{aligned}
$$

$$
(8.10) \qquad = x_3 + \beta_2 - \ddot{y}_{\mathrm{r}} + \tilde{\theta}^{\mathrm{T}}\phi_2 - \frac{\partial\alpha_1}{\partial\hat{\theta}}\dot{\hat{\theta}}\,,
$$

where we have used (4.27) for $\phi_2 = \varphi_2 - \frac{\partial\alpha_1}{\partial z_1}\varphi_1$ and

$$
(8.11) \qquad \beta_2 = \hat{\theta}^{\mathrm{T}}\varphi_2 - \frac{\partial\alpha_1}{\partial z_1}\left(-c_1 z_1 - \kappa_1 z_1\|\varphi_1\|^2 + z_2\right) - \frac{\partial\alpha_1}{\partial y_{\mathrm{r}}}\dot{y}_{\mathrm{r}}\,.
$$

The function β_2 encompasses all the known terms on the right-hand side of (8.10), except $\ddot{y}_{\mathrm{r}}$ and x_3.

It is important to note that in (8.10) the analytical expression for the derivative $\dot{\alpha}_1$ is different than in (4.8). In particular, in (4.8) α_1 was treated as a function of x_1 and $\hat{\theta}$, because that simplified the resulting expressions for its derivative. From (8.5) we see that α_1 could now have been treated as a function of x_1, $\hat{\theta}$, y_{r} and $\dot{y}_{\mathrm{r}}$:

$$
\alpha_1 = -c_1(x_1 - y_{\mathrm{r}}) - \kappa_1(x_1 - y_{\mathrm{r}})\|\varphi_1(x_1)\|^2 - \hat{\theta}^{\mathrm{T}}\varphi_1(x_1) + \dot{y}_{\mathrm{r}}\,.
$$

However, since we are interested in trajectory initialization, it is more convenient to replace x_1 by $z_1 + y_{\mathrm{r}}$ in the representation of α_1:

$$
\alpha_1 = -c_1 z_1 - \kappa_1 z_1\|\varphi_1(z_1 + y_{\mathrm{r}})\|^2 - \hat{\theta}^{\mathrm{T}}\varphi_1(z_1 + y_{\mathrm{r}}) + \dot{y}_{\mathrm{r}}\,.
$$

The variable x_3 is now viewed as the virtual control in (8.10). Accordingly, the new error variable z_3 is defined as

$$
(8.12) \qquad\qquad\qquad z_3 = x_3 - \alpha_2\,.
$$

The design of α_2 is based on the derivative of the partial Lyapunov function

$$
(8.13) \qquad\qquad\qquad V_2 = V_1 + \frac{1}{2}z_2^2\,,
$$

which is computed using (8.7) and (8.10):

$$
\begin{aligned}
\dot{V}_2 &= \dot{V}_1 + z_2 \dot{z}_2 \\
&= z_1 z_2 - c_1 z_1^2 - \kappa_1 z_1^2 \|\varphi_1\|^2 + \tilde{\theta}^{\mathrm{T}} \left(\tau_1 - \Gamma^{-1} \dot{\hat{\theta}} \right) \\
&\quad + z_2 \left[z_3 + \alpha_2 + \beta_2 - \ddot{y}_r + \tilde{\theta}^{\mathrm{T}} \phi_2 - \frac{\partial \alpha_1}{\partial \hat{\theta}} \dot{\hat{\theta}} \right] \\
&= z_2 z_3 - c_1 z_1^2 - \kappa_1 z_1^2 \|\varphi_1\|^2 + \tilde{\theta}^{\mathrm{T}} \left(\tau_1 + \phi_2 z_2 - \Gamma^{-1} \dot{\hat{\theta}} \right) \\
&\quad + z_2 \left[\alpha_2 + z_1 + \beta_2 - \ddot{y}_r - \frac{\partial \alpha_1}{\partial \hat{\theta}} \dot{\hat{\theta}} \right] \\
&= z_2 z_3 - c_1 z_1^2 - \kappa_1 z_1^2 \|\varphi_1\|^2 + \tilde{\theta}^{\mathrm{T}} \left(\tau_2 - \Gamma^{-1} \dot{\hat{\theta}} \right) \\
&\quad + z_2 \left[\alpha_2 + z_1 + \beta_2 - \ddot{y}_r - \frac{\partial \alpha_1}{\partial \hat{\theta}} \dot{\hat{\theta}} \right] ,
\end{aligned}
$$
(8.14)

where the second tuning function is defined as

(8.15)
$$
\tau_2 = \tau_1 + \phi_2 z_2 .
$$

The second stabilizing function is defined as in (4.13), with the addition of the corresponding κ-term:

(8.16)
$$
\alpha_2 = -c_2 z_2 - \kappa_2 z_2 \|\phi_2\|^2 - z_1 - \beta_2 + \frac{\partial \alpha_1}{\partial \hat{\theta}} \Gamma \tau_2 + \ddot{y}_r .
$$

With the choices (8.15) and (8.16), the derivative of $\dot{V}_2$ satisfies:

(8.17)
$$
\begin{aligned}
\dot{V}_2 &= z_2 z_3 - c_1 z_1^2 - \kappa_1 z_1^2 \|\varphi_1\|^2 - c_2 z_2^2 - \kappa_2 z_2^2 \|\phi_2\|^2 \\
&\quad + \tilde{\theta}^{\mathrm{T}} \left[\tau_2 - \Gamma^{-1} \dot{\hat{\theta}} \right] - z_2 \frac{\partial \alpha_1}{\partial \hat{\theta}} \left(\dot{\hat{\theta}} - \Gamma \tau_2 \right) ,
\end{aligned}
$$

while the second error subsystem $\mathcal{S}_2$ becomes ($\phi_1 = \varphi_1$):

(8.18) $\mathcal{S}_2$:
$$
\begin{aligned}
\dot{z}_1 &= -c_1 z_1 - \kappa_1 z_1^2 \|\phi_1\|^2 + z_2 + \tilde{\theta}^{\mathrm{T}} \phi_1 \\
\dot{z}_2 &= -c_2 z_2 - \kappa_2 z_2^2 \|\phi_2\|^2 - z_1 + z_3 + \tilde{\theta}^{\mathrm{T}} \phi_2 - \frac{\partial \alpha_1}{\partial \hat{\theta}} \left(\dot{\hat{\theta}} - \Gamma \tau_2 \right) \\
\tau_2 &= \phi_1 z_1 + \phi_2 z_2 .
\end{aligned}
$$

Step 3. Using the fact that α_2, as defined in (8.16), can be expressed as a function of $z_1, z_2, y_r, \dot{y}_r, \ddot{y}_r$ and $\hat{\theta}$ only, we write

$$
\begin{aligned}
\dot{z}_3 &= \dot{x}_3 - \dot{\alpha}_2 \\
&= u + \theta^{\mathrm{T}} \varphi_3 - \frac{\partial \alpha_2}{\partial z_1} \dot{z}_1 - \frac{\partial \alpha_2}{\partial z_2} \dot{z}_2 - \frac{\partial \alpha_2}{\partial y_r} \dot{y}_r - \frac{\partial \alpha_2}{\partial \dot{y}_r} \ddot{y}_r - y_r^{(3)} - \frac{\partial \alpha_2}{\partial \hat{\theta}} \dot{\hat{\theta}} \\
(8.19) \quad &= x_3 + \beta_3 - y_r^{(3)} + \tilde{\theta}^{\mathrm{T}} \phi_3 - \frac{\vartheta \alpha_2}{\vartheta \hat{\theta}} \dot{\hat{\theta}} ,
\end{aligned}
$$

where $\phi_3 = \varphi_3 - \frac{\partial \alpha_2}{\partial z_1}\varphi_1 - \frac{\partial \alpha_2}{\partial z_2}\phi_2$ and

$$(8.20) \qquad \begin{aligned} \beta_3 &= \hat{\theta}^{\mathrm{T}}\varphi_3 - \frac{\partial \alpha_2}{\partial z_1}\left(-c_1 z_1 - \kappa_1 z_1 \|\phi_1\|^2 + z_2\right) \\ &\quad - \frac{\partial \alpha_2}{\partial z_2}\left(-c_2 z_2 - \kappa_2 z_2 \|\phi_2\|^2 - z_1 + z_3\right) \\ &\quad - \frac{\partial \alpha_2}{\partial y_{\mathrm{r}}}\dot{y}_{\mathrm{r}} - \frac{\partial \alpha_2}{\partial \dot{y}_{\mathrm{r}}}\ddot{y}_{\mathrm{r}} - \frac{\partial \alpha_2}{\partial z_2}\frac{\partial \alpha_1}{\partial \hat{\theta}}\Gamma \tau_2 \end{aligned}$$

$$(8.21) \qquad \frac{\vartheta \alpha_2}{\vartheta \hat{\theta}} = \frac{\partial \alpha_2}{\partial \hat{\theta}} - \frac{\partial \alpha_2}{\partial z_2}\frac{\partial \alpha_1}{\partial \hat{\theta}}.$$

Again, the function β_3 encompasses all the known terms on the right-hand side of (8.19), except $y_{\mathrm{r}}^{(3)}$ and u. Since this is the final step, we will design the actual control law and the actual update law based on the full Lyapunov function

$$(8.22) \qquad V_3 = V_2 + \frac{1}{2}z_3^2 = \frac{1}{2}\left(z_1^2 + z_2^2 + z_3^2\right) + \frac{1}{2}\tilde{\theta}^{\mathrm{T}}\Gamma^{-1}\tilde{\theta},$$

whose derivative is:

$$\begin{aligned} \dot{V}_3 &= \dot{V}_2 + z_3 \dot{z}_3 \\ &= z_2 z_3 - c_1 z_1^2 - \kappa_1 z_1^2 \|\phi_1\|^2 - c_2 z_2^2 - \kappa_2 z_2^2 \|\phi_2\|^2 \\ &\quad + \tilde{\theta}^{\mathrm{T}}\left[\tau_2 - \Gamma^{-1}\dot{\hat{\theta}}\right] - z_2 \frac{\partial \alpha_1}{\partial \hat{\theta}}\left(\dot{\hat{\theta}} - \Gamma \tau_2\right) \\ &\quad + z_3\left[u + \beta_3 - y_{\mathrm{r}}^{(3)} + \tilde{\theta}^{\mathrm{T}}\phi_3 - \frac{\vartheta \alpha_2}{\vartheta \hat{\theta}}\dot{\hat{\theta}}\right] \\ &= -c_1 z_1^2 - \kappa_1 z_1^2 \|\phi_1\|^2 - c_2 z_2^2 - \kappa_2 z_2^2 \|\phi_2\|^2 \\ &\quad + \tilde{\theta}^{\mathrm{T}}\left(\tau_2 + \phi_3 z_3 - \Gamma^{-1}\dot{\hat{\theta}}\right) - z_2 \frac{\partial \alpha_1}{\partial \hat{\theta}}\left(\dot{\hat{\theta}} - \gamma \tau_2\right) \end{aligned}$$

$$(8.23) \qquad + z_3\left[u + z_2 + \beta_3 - y_{\mathrm{r}}^{(3)} - \frac{\vartheta \alpha_2}{\vartheta \hat{\theta}}\dot{\hat{\theta}}\right].$$

The update law is now chosen to eliminate the $\tilde{\theta}$-term in (8.23):

$$(8.24) \qquad \dot{\hat{\theta}} = \Gamma \tau_3 = \Gamma(\tau_2 + \phi_3 z_3).$$

Using (8.24), equation (8.23) can be rewritten as

$$\dot{V}_3 = -c_1 z_1^2 - \kappa_1 z_1^2 \|\phi_1\|^2 - c_2 z_2^2 - \kappa_2 z_2^2 \|\phi_2\|^2$$

$$(8.25) \qquad + z_3\left[u + z_2 + \beta_3 - y_{\mathrm{r}}^{(3)} - \frac{\vartheta \alpha_2}{\vartheta \hat{\theta}}\Gamma \tau_3 - z_2 \frac{\partial \alpha_1}{\partial \hat{\theta}}\Gamma \phi_3\right].$$

Now the last term in (8.25) is set equal to $-c_3 z_3^2 - \kappa_3 z_3^2 \phi_3^2$ with the choice

$$(8.26) \qquad u = -c_3 z_3 - \kappa_3 z_3 \|\phi_3\|^2 - z_2 - \beta_3 + y_{\mathrm{r}}^{(3)} + \frac{\vartheta \alpha_2}{\vartheta \hat{\theta}}\Gamma \tau_3 + \frac{\partial \alpha_1}{\partial \hat{\theta}}\Gamma \phi_3 z_2.$$

With the choices (8.24) and (8.26), we have

$$(8.27) \quad \dot{V}_3 = -c_1 z_1^2 - \kappa_1 z_1^2 \|\phi_1\|^2 - c_2 z_2^2 - \kappa_2 z_2^2 \|\phi_2\|^2 - c_3 z_3^2 - \kappa_3 z_3^2 \|\phi_3\|^2 \,,$$

and the complete error system $\mathcal{S}_3$ becomes:

$$
\begin{aligned}
&\frac{d}{dt}
\begin{bmatrix} z_1 \\ z_2 \\ z_3 \end{bmatrix}
=
\begin{bmatrix}
-c_1 - \kappa_1 \|\phi_1\|^2 & 1 & 0 \\
-1 & -c_2 - \kappa_2 \|\phi_2\|^2 & 1 - \frac{\partial \alpha_1}{\partial \theta} \Gamma \phi_3 \\
0 & -1 + \frac{\partial \alpha_1}{\partial \theta} \Gamma \phi_3 & -c_3 - \kappa_3 \|\phi_3\|^2
\end{bmatrix}
\begin{bmatrix} z_1 \\ z_2 \\ z_3 \end{bmatrix} \\[2mm]
(8.28) \qquad &+
\begin{bmatrix} \phi_1^{\mathrm{T}} \\ \phi_2^{\mathrm{T}} \\ \phi_3^{\mathrm{T}} \end{bmatrix} \tilde{\theta} \\[2mm]
&\tau_3 = [\phi_1 \ \phi_2 \ \phi_3]
\begin{bmatrix} z_1 \\ z_2 \\ z_3 \end{bmatrix} .
\end{aligned}
$$

Comparing (8.28) to (4.26) we see that the κ-terms have strengthened the negativity of the diagonal terms in the system matrix, by including the squared norm of the terms which multiply the parameter error $\tilde{\theta}$.

Global stability and asymptotic tracking. Using (8.22) and (8.27) we conclude that the $(z, \tilde{\theta})$-system has a globally uniformly stable equilibrium at the origin. Furthermore, since $\dot{V}_3$ is integrable and $\ddot{V}_3$ is bounded, we obtain

$$(8.29) \qquad \lim_{t \to \infty} z(t) = 0 \,.$$

In particular, this implies that the state of the system (8.1) is globally uniformly bounded (since $y_r, \dot{y}_r, \ddot{y}_r, y_r^{(3)}$ are bounded), and that the tracking error $z_1 = y - y_r$ converges to zero asymptotically.

Boundedness without adaptation. It is also straightforward to see that the designed controller guarantees global uniform boundedness even when the adaptation is turned off, i.e., even with $\Gamma = 0$. In that case, the closed-loop system (8.28) becomes

$$
\begin{aligned}
&\frac{d}{dt}
\begin{bmatrix} z_1 \\ z_2 \\ z_3 \end{bmatrix}
=
\begin{bmatrix}
-c_1 - \kappa_1 \|\phi_1\|^2 & 1 & 0 \\
-1 & -c_2 - \kappa_2 \|\phi_2\|^2 & 1 \\
0 & -1 & -c_3 - \kappa_3 \|\phi_3\|^2
\end{bmatrix}
\begin{bmatrix} z_1 \\ z_2 \\ z_3 \end{bmatrix} \\[2mm]
(8.30) \qquad &+
\begin{bmatrix} \phi_1^{\mathrm{T}} \\ \phi_2^{\mathrm{T}} \\ \phi_3^{\mathrm{T}} \end{bmatrix} \tilde{\theta} \,.
\end{aligned}
$$

A candidate Lyapunov function for this system is given by

$$(8.31) \qquad V(z) = \frac{1}{2}\|z\|^2 = \frac{1}{2}\sum_{i=1}^{3} z_i^2 \,.$$

Its derivative along the solutions of (8.30) satisfies:

$$
\begin{aligned}
\dot{V}_{(8.30)} &= -\sum_{i=1}^{3}\left[c_i z_i^2 + \kappa_i z_i^2 \|\phi_i\|^2 - z_i \phi_i^{\mathrm{T}}\tilde{\theta}\right] \\
&\leq -\sum_{i=1}^{3}\left[c_i z_i^2 + \kappa_i z_i^2 \|\phi_i\|^2 - |z_i|\,\|\phi_i\|\,\|\tilde{\theta}\|\right] \\
&\leq -\sum_{i=1}^{3}\left[c_i z_i^2 + \kappa_i\left(|z_i|\,\|\phi_i\| - \frac{\|\tilde{\theta}\|}{2\kappa_i}\right)^2 - \frac{\|\tilde{\theta}\|^2}{4\kappa_i}\right] \\
&\leq -\sum_{i=1}^{3}\left[c_i z_i^2 - \frac{\|\tilde{\theta}\|^2}{4\kappa_i}\right]
\end{aligned}
$$

$$
(8.32) \qquad \leq -c_0\|z\|^2 + \frac{\|\tilde{\theta}\|^2}{4\kappa_0},
$$

where the constants c_0 and κ_0 are defined as

$$
(8.33) \qquad c_0 = \min\{c_1, c_2, c_3\}, \quad \kappa_0 = \left(\frac{1}{\kappa_1} + \frac{1}{\kappa_2} + \frac{1}{\kappa_3}\right)^{-1}.
$$

It is clear from (8.32) that, for any positive values of c_0 and κ_0, the state of the error system (and hence the state of the plant) is uniformly bounded, since $\dot{V} < 0$ whenever $\|z\|^2 > \|\tilde{\theta}\|^2/4\kappa_0 c_0$, where $\tilde{\theta} = \theta - \hat{\theta}(0)$ is constant since adaptation is turned off.

Transient performance improvement with trajectory initialization. Let us now investigate the transient performance of the adaptive closed-loop system (8.28). It is easy to see that the derivative of the nonnegative function $V(z)$ along the solutions of (8.28) satisfies the same inequality as in (8.32). From the definition (8.31), this implies that

$$
(8.34) \qquad \frac{d}{dt}\left(\frac{1}{2}\|z\|^2\right) \leq -c_0\|z\|^2 + \frac{\tilde{\theta}^2}{4\kappa_0}.
$$

Since the boundedness of $\tilde{\theta}$ has already been established from (8.22) and (8.27), we can strengthen the inequality in (8.34) by replacing $\|\tilde{\theta}\|^2$ with its bound $\|\tilde{\theta}\|_\infty^2$. This bound is estimated from (8.22) since V_3 is a nonincreasing function of time:

$$
\begin{aligned}
\frac{1}{\lambda_{\max}(\Gamma)}\|\tilde{\theta}(t)\|^2 &\leq \tilde{\theta}^{\mathrm{T}}(t)\Gamma^{-1}\tilde{\theta}(t) \leq \|z(t)\|^2 + \tilde{\theta}^{\mathrm{T}}(t)\Gamma^{-1}\tilde{\theta}(t) \\
&= 2V_3(t) \leq 2V_3(0) = \|z(0)\|^2 + \tilde{\theta}^{\mathrm{T}}(0)\Gamma^{-1}\tilde{\theta}(0) \\
(8.35) \qquad &\leq \|z(0)\|^2 + \frac{1}{\lambda_{\min}(\Gamma)}\|\tilde{\theta}(0)\|^2,
\end{aligned}
$$

which implies

$$(8.36) \qquad \|\tilde{\theta}\|_\infty^2 \leq \lambda_{\max}(\Gamma)\|z(0)\|^2 + \frac{\lambda_{\max}(\Gamma)}{\lambda_{\min}(\Gamma)}\|\tilde{\theta}(0)\|^2 .$$

Combining (8.34) and (8.36) we obtain

$$(8.37) \quad \frac{d}{dt}\left(\|z\|^2\right) \leq -2c_0\|z\|^2 + \frac{1}{2\kappa_0}\left[\lambda_{\max}(\Gamma)\|z(0)\|^2 + \frac{\lambda_{\max}(\Gamma)}{\lambda_{\min}(\Gamma)}\|\tilde{\theta}(0)\|^2\right].$$

Multiplying both sides of (8.37) by e^{2c_0t} and integrating over the interval $[0, t]$ results in

$$\|z(t)\|^2 \leq \|z(0)\|^2 e^{-2c_0t} + \frac{1}{4\kappa_0c_0}\left[\lambda_{\max}(\Gamma)\|z(0)\|^2 + \frac{\lambda_{\max}(\Gamma)}{\lambda_{\min}(\Gamma)}\|\tilde{\theta}(0)\|^2\right].$$
(8.38)

From (8.38) it is apparent that the transient behavior of the error system can be influenced through the choice of design constants c_0, κ_0 and Γ. What is not clear, however, is that an increase of $c_0\kappa_0$ alone will *not* necessarily produce a reduction of the $\mathcal{L}_\infty$-bound of z. In fact, it may even *increase* this bound by increasing the initial value $\|z(0)\|$. To clarify this point, let us recall the definitions of z_1 and z_2:

$$\begin{aligned} z_1 &= x_1 - y_r \\ z_2 &= x_2 - \alpha_1 = x_2 + c_1 z_1 + \kappa_1 z_1 \varphi_1^2 + \hat{\theta}^T \varphi_1 - \dot{y}_r . \end{aligned}$$

Suppose now that $z_1(0)$ is different than zero. In that case, an increase of c_1 and κ_1 may increase the value of $z_2(0)$ and thus also the value of $\|z(0)\|$. Moreover, this increase may more than offset the decreasing effect of the term $1/4c_0\kappa_0$ in (8.38), since $\|z(0)\|^2$ will increase in proportion to c_1^2 and κ_1^2. This phenomenon will become even more pronounced when $z_3(0)$ is taken into account.

It would seem that the dependence of $z(0)$ on the design constants c_i and κ_i eliminates any possibility of systematically improving the transient performance of the error system through the choice of c_0 and κ_0. Fortunately, it is not so. The remedy for this problem is to use *trajectory initialization* to render $z(0) = 0$ independently of the choice of c_i and κ_i. The initialization procedure is straightforward and is dictated by the definitions of the z-variables:

• Starting with z_1, set $z_1(0) = 0$ by choosing

$$(8.39) \qquad\qquad\qquad y_r(0) = x_1(0) .$$

• Since $z_1(0) = 0$, (8.5) shows that

$$(8.40) \qquad\qquad\qquad \alpha_1(0) = \dot{y}_r(0) - \hat{\theta}(0)^T\varphi_1(0) ,$$

where we use the notation $\varphi_i(0) = \varphi_i(x_1(0), \ldots, x_i(0))$. From (8.40) it is clear that we can set $z_2(0) = 0$ with the choice

$$(8.41) \qquad \dot{y}_r(0) = x_2(0) + \hat{\theta}(0)^T \varphi_1(0).$$

- Since $z_1(0) = z_2(0) = 0$, (8.5) and (8.15) result in

$$(8.42) \qquad \frac{\partial \alpha_1}{\partial y_r}(0) = -\hat{\theta}(0)^T \frac{\partial \varphi_1}{\partial x_1}(0), \quad \tau_2(0) = 0,$$

which, combined with (8.16) and (8.11), gives

$$(8.43) \qquad \alpha_2(0) = \ddot{y}_r(0) - \hat{\theta}(0)^T \left[\varphi_2(0) + \frac{\partial \varphi_1}{\partial x_1}(0) \dot{y}_r(0) \right].$$

Hence, $z_3(0) = 0$ is achieved with

$$(8.44) \qquad \ddot{y}_r(0) = x_3(0) + \hat{\theta}(0)^T \left[\varphi_2(0) + \frac{\partial \varphi_1}{\partial x_1}(0) \dot{y}_r(0) \right].$$

With the trajectory initialization defined by (8.39), (8.41), and (8.44), we have set $z(0) = 0$. Moreover, these equations are independent of the design constants c_i and κ_i. This means that different choices of c_0 and κ_0 will still result in $z(0) = 0$ with the same values of $y_r(0)$, $\dot{y}_r(0)$, and $\ddot{y}_r(0)$. Returning to (8.38), we substitute $z(0) = 0$ to obtain

$$(8.45) \qquad \|z(t)\|^2 \leq \frac{1}{4\kappa_0 c_0} \frac{\lambda_{\max}(\Gamma)}{\lambda_{\min}(\Gamma)} \|\tilde{\theta}(0)\|^2,$$

which implies

$$(8.46) \qquad \|z\|_\infty \leq \frac{1}{2\sqrt{\kappa_0 c_0}} \sqrt{\frac{\lambda_{\max}(\Gamma)}{\lambda_{\min}(\Gamma)}} \|\tilde{\theta}(0)\|.$$

Hence, the $\mathcal{L}_\infty$-bound on the transient performance of the error system is directly proportional to the initial parametric uncertainty and can be reduced by increasing the values of c_0 and κ_0. In particular, this implies that the transients of the tracking error $z_1 = y - y_r$ are directly influenced by the design constants c_i and κ_i.

To provide some further insight into the process of trajectory initialization, let us return to the Lyapunov function (8.22). When $z(0) = 0$, the initial value of this function is reduced to the initial value of the parametric uncertainty. If we interpret the value of this function as a distance between the actual system trajectory and the reference trajectory, we see that trajectory initialization places the initial point of the reference trajectory as close as possible to the initial point of the system trajectory. If the parametric uncertainty were zero, trajectory initialization would have placed the reference output and its derivatives at the true values of the plant output and its derivatives. This is easily seen if $\hat{\theta}(0)$ is replaced by θ in (8.39), (8.41), and (8.44). Since the parameter θ is unknown, however, trajectory initialization placed the reference output at the true value of the plant output and the derivatives of the reference output at the *estimated* values of the plant output derivatives.

It is important to note that the guidelines for this initialization are dictated by the design procedure itself, and correspond to the intuitive idea of matching the initial values of the reference and plant outputs and their derivatives as closely as possible.

Trajectory initialization is also possible in the output-feedback case, where only the initial value of the reference output can be chosen to match the initial value of the system output, while all other initial conditions of the reference trajectory and the filters are set to zero. This reduces the initial value of the Lyapunov function to the sum of the initial parameter and state estimation errors.

REFERENCES

[1] G. CAMPION AND G. BASTIN, *Indirect adaptive state-feedback control of linearly parametrized nonlinear systems*, International Journal of Adaptive Control and Signal Processing, vol. 4, pp. 345–358, 1990.

[2] G. C. GOODWIN AND D. Q. MAYNE, *A parameter estimation perspective of continuous time model reference adaptive control*, Automatica, vol. 23, pp. 57–70, 1987.

[3] Z. P. JIANG AND L. PRALY, *Iterative designs of adaptive controllers for systems with nonlinear integrators*, Proceedings of the 30th IEEE Conference on Decision and Control, Brighton, UK, pp. 2482–2487, 1991.

[4] I. KANELLAKOPOULOS, *Passive adaptive control of nonlinear systems*, International Journal of Adaptive Control and Signal Processing, vol. 7, pp. 339–352, 1993.

[5] I. KANELLAKOPOULOS, *'Low-gain' robust control of uncertain nonlinear systems*, submitted to IEEE Transactions on Automatic Control.

[6] I. KANELLAKOPOULOS, P. V. KOKOTOVIĆ, AND R. MARINO, *An extended direct scheme for robust adaptive nonlinear control*, Automatica, vol. 27, pp. 247–255, 1991.

[7] I. KANELLAKOPOULOS, P. V. KOKOTOVIĆ, AND A. S. MORSE, *Systematic design of adaptive controllers for feedback linearizable systems*, IEEE Transactions on Automatic Control, vol. 36, pp. 1241–1253, 1991.

[8] I. KANELLAKOPOULOS, P. V. KOKOTOVIĆ, AND A. S. MORSE, *Adaptive output-feedback control of a class of nonlinear systems*, Proceedings of the 30th IEEE Conference on Decision and Control, Brighton, UK, pp. 1082–1087, 1991.

[9] I. KANELLAKOPOULOS, P. V. KOKOTOVIĆ, AND A. S. MORSE, *A toolkit for nonlinear feedback design*, Systems & Control Letters, vol. 18, pp. 83–92, 1992.

[10] I. KANELLAKOPOULOS, P. V. KOKOTOVIĆ, AND A. S. MORSE, *Adaptive output-feedback control of systems with output nonlinearities*, IEEE Transactions on Automatic Control, vol. 37, pp. 1666–1682, 1992.

[11] I. KANELLAKOPOULOS, P. V. KOKOTOVIĆ, AND A. S. MORSE, *Adaptive nonlinear control with incomplete state information*, International Journal of Adaptive Control and Signal Processing, vol. 6, pp. 367–394, 1992.

[12] M. KRSTIĆ, I. KANELLAKOPOULOS, AND P. V. KOKOTOVIĆ, *Adaptive nonlinear control without overparametrization*, Systems & Control Letters, vol. 19, pp. 177–185, 1992.

[13] M. KRSTIĆ, I. KANELLAKOPOULOS, AND P. V. KOKOTOVIĆ, Nonlinear design of adaptive controllers for linear systems, IEEE Transactions on Automatic Control, vol. 39, pp. 738–752, 1994.

[14] M. KRSTIĆ AND P. V. KOKOTOVIĆ, Estimation-based schemes for adaptive nonlinear state-feedback control, this volume, pp. 165–198.

[15] R. MARINO AND P. TOMEI, Global adaptive observers for nonlinear systems via filtered transformations, IEEE Transactions on Automatic Control, vol. 37, pp. 1239–1245, 1992.

[16] R. MARINO AND P. TOMEI, Global adaptive output-feedback control of nonlinear systems, Part I: linear parameterization, IEEE Transactions on Automatic Control, vol. 38, pp. 17–32, 1993.

[17] R. MARINO AND P. TOMEI, Global adaptive output-feedback control of nonlinear systems, Part II: nonlinear parameterization, IEEE Transactions on Automatic Control, vol. 38, pp. 33–49, 1993.

[18] R. H. MIDDLETON, Indirect continuous time adaptive control, Automatica, vol. 23, pp. 793–795, 1987.

[19] K. NAM AND A. ARAPOSTATHIS, A model-reference adaptive control scheme for pure-feedback nonlinear systems, IEEE Transactions on Automatic Control, vol. 33, pp. 803–811, 1988.

[20] J. B. POMET AND L. PRALY, Adaptive nonlinear regulation: estimation from the Lyapunov equation, IEEE Transactions on Automatic Control, vol. 37, pp. 729–740, 1992.

[21] S. S. SASTRY AND A. ISIDORI, Adaptive control of linearizable systems, IEEE Transactions on Automatic Control, vol. 34, pp. 1123–1131, 1989.

[22] D. SETO, A. M. ANNASWAMY, AND J. BAILLIEUL, Adaptive control of a class of nonlinear systems with a triangular structure, Proceedings of the 31st IEEE Conference on Decision and Control, Tucson, AZ, pp. 278–283, 1992.

[23] E. D. SONTAG, Smooth stabilization implies coprime factorization, IEEE Transactions on Automatic Control, vol. 34, pp. 435–443, 1989.

[24] E. D. SONTAG, Input/output and state-space stability, in New Trends in System Theory, G. Conte et al., Eds., Boston: Birkhäuser, 1991.

[25] D. TAYLOR, P. V. KOKOTOVIĆ, R. MARINO, AND I. KANELLAKOPOULOS, Adaptive regulation of nonlinear systems with unmodeled dynamics, IEEE Transactions on Automatic Control, vol. 34, pp. 405–412, 1989.

[26] A. R. TEEL, R. R. KADIYALA, P. V. KOKOTOVIĆ AND S. S. SASTRY, Indirect techniques for adaptive input-output linearization of non-linear systems, International Journal of Control, vol. 53, pp. 193–222, 1991.

[27] A. R. TEEL, Error-based adaptive non-linear control and regions of feasibility, International Journal of Adaptive Control and Signal Processing, vol. 6, pp. 319–327, 1992.

DESIGN GUIDELINES FOR ADAPTIVE CONTROL WITH APPLICATION TO SYSTEMS WITH STRUCTURAL FLEXIBILITY*

JOY H. KELLY[†] AND B. ERIK YDSTIE[‡]

Abstract. In this paper we combine least squares identification with H_∞-control to provide a methodical approach to adaptive robust control. We use periodic resetting of the covariance matrix and update the control design less frequently than the sampling rate for the controller. This approach gives a closed-loop system with bounded ℓ_∞ and ℓ_2 gain when the model mismatch is small in the frequency range where the control gain is large. We apply the method to a model of the *Martin Marietta* flexible beam. A frequency domain interpretation of the estimator cost function is used to design the prefilter for the identifier. A post-projection scheme using a priori knowledge of the antiresonant damping is used to overcome poor identification of the antiresonances. Excitation ensures that the parameter estimator is stable and an adaptive stopping technique turns the estimator off once the parameter estimates have converged. These features, although not needed for global stability and boundedness, give improved performance of the algorithm. One of the main contributions of the paper is to show that adaptive control theory, in a natural way, leads to the application of H_∞ design methods for robust control.

1. Introduction. Recent efforts have provided substantial progress toward the goal of developing an adaptive robust controller. One of the first discussions of this perspective was Goodwin, Hill, & Palaniswami [2], in which the *adaptive robust control* problem was defined to be the design of a robust control law and a robust estimator. Applications of these ideas to flexible systems have been described by Sidman [12] and Rovner [9].

In this paper we develop design guidelines and give stability theory for adaptive robust control. The algorithm consists of a constrained least squares estimator and the H_∞ design method. The control design is executed at a lower rate than the sampling rate of the controller. This approach is sometimes referred to as iterative design. The iterative approach has two advantages. First, we can apply computationally expensive estimation, control design and signal processing methods. Second, the use of constant controls over finite windows allows us to develop frequency domain specifications for tolerated unmodeled dynamics. Links can then be made with robust control design methods. A theory for boundedness of signals in such systems is given in this paper. In fact, it is shown that application

---

* This work was carried out while one of the authors (J.H.K.) was employed by *Martin Marietta* in Denver, Colorado. Their support is greatly appreciated. We are especially thankful for the help Mike Stoughton gave in setting up the robust controller. The other author (B.E.Y.) acknowledges support from the *National Science Foundation* (# CTS-8903160).

† Sverdrup Technology, Bldg 260, P.O. Box 1935, Eglin Air Force Base, FL 32542. email: kellyj@teas.eglin.af.mil

‡ Department of Chemical Engineering, Carnegie Mellon University, Pittsburgh, PA 15213. email: ydstie@andrew.cmu.edu

of stability theory leads naturally to the application of H_∞ based control system design tools.

To illustrate the practical use of the H_∞ based adaptive control theory we describe an application to the *Martin Marietta* flexible beam under different loading conditions. The experimental setup has been used in a number of related studies. The system identification algorithm has been tested on the hardware and the results are presented in [3]. Closed-loop system identification experiments have been conducted using a real-time adaptive gain scheduled robust control design and results are summarized in Kelly, Stoughton, Ramey, & Schmitz [4]. H_∞ experiments are discussed in Stoughton [13].

The adaptive robust algorithm developed in the current paper was simulated using a high fidelity model of the flexible beam. In the simulation we use a frequency domain interpretation of the prediction error to design a prefilter which includes crude information about the frequencies of the antiresonnances. A post-projection scheme using *a priori* knowledge of the antiresonnant damping is developed to overcome the problem of poor identification of the antiresonnances. An adaptive deadzone is used to turn the estimator off once the parameters have converged to acceptable values. The results and algorithms are presented in this paper and discussed more extensively in [3].

Two application areas that fit into the category of problem discussed in this paper are single link flexible manipulators and launch vehicles. In the case of flexible manipulators, the purpose is to adapt the controller each time the payload changes. For the case of launch vehicles, system identification will be performed early in the flight (while a low performance controller is operating) to provide an accurate model for designing the high performance controller. As such systems become more complicated and performance requirements more stringent, the use of real-time adaptation becomes more appealing.

2. Testbed and performance requirements. The experimental hardware is the outer flexible beam of the mechanical system described in [3]. The manipulator is a six-foot planar arm with two lightweight aluminum links of rectangular cross-sections. The outer flexible link and the tip are supported by airpads floating on a flat table so that the link bending is in the horizontal plane. There are three sets of sensors: position and rate sensors at each joint, strain-gauges mounted at several locations along the links, and two accelerometers mounted at the tip along two orthogonal directions.

The inner flexible link is grounded for the experiments by shutting off the air supply to the outer link's airpad. This setup isolates the outer flexible beam motion and eliminates cross-coupled from the inner link. The performance requirements and design constraints for the flexible beam testbed are:

- The system bandwidth should be around 0.5 Hz.
- High frequency modes (> 20 Hz) must not be excited.
- Sampling rate is limited to 100 Hz.

In order to accommodate plant uncertainty we include the following in the synthesis model:

- ± 5% uncertainty in structural mode frequency
- ± 20% uncertainty in the structural mode damping
- Modes above 20 Hz not included in model

For control designs and system identification a finite-dimensional representation of the system dynamics is needed. Therefore, in practice, the infinite-dimensional transfer functions are truncated to the first n flexible modes:

$$
(2.1) \qquad
\begin{aligned}
\frac{y(s)}{u(s)} &= \frac{G_h}{s(I_T s + b)} \prod_{i=1}^{n} \frac{(s^2 + 2\zeta_i^z \Omega_i s + \Omega_i^2)}{(s^2 + 2\zeta_i^p \omega_i s + \omega_i^2)} \\
&\quad + \underbrace{\sum_{i=n+1}^{\infty} \frac{\sigma_i}{(s^2 + 2\zeta_i^p \omega_i s + \omega_i^2)}}_{\substack{\text{unmodeled} \\ \text{dynamics}}}
\end{aligned}
$$

where $y(s)$ is the Laplace transform of the hub velocity, $u(s)$ is the Laplace transform of the motor current and σ_i is the modal slope of the ith mode at the sensor. The accuracy of the resulting model then depends on the number of modes in the truncated model. The number of modes needed for the model can be determined analytically from the beam characteristics or experimentally using a dynamic signal analyzer on the hardware to determine the number of modes present within the frequency range of interest.

Fig. 1 is the frequency response (from the motor current-to-hub velocity) of the elastic link of the testbed with no payload. The two curves are the actual hardware frequency response and the corresponding response for a truncated transfer function with two flexible modes. This figure illustrates the accuracy of the reduced order model within the frequency range of interest.

For the plant simulation we used a model with four modes for the true plant. The magnitude response for the full-order model and the reduced order model is given in Fig. 2. Figs. 3 and 4 give the current-to-hub velocity function models for two different payloads on the flexible beam. These serve as examples of discontinuous plant variations and demonstrate the change in dynamics as a function of payload variation.

3. Parameter centering using adaptive control. Consider a discrete system with two inputs and one output

$$(3.1) \qquad z(t) = P_1(q^{-1})u(t) + P_2(q^{-1})w(t).$$

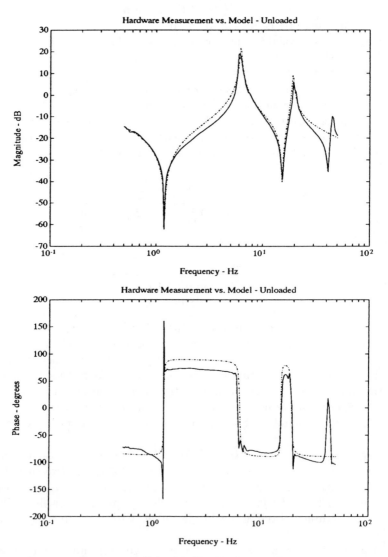

FIG. 1. *Frequency Response of Unloaded Flexible Beam - Hardware vs. Continuous Time Model from Motor Current to Hub Velocity (Hardware: solid, Reduced Order Model: dotted)*

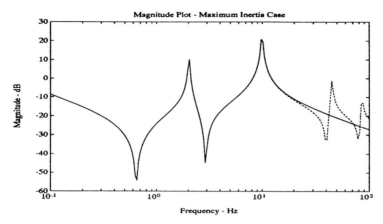

FIG. 2. *Magnitude Response of Flexible Beam Full-Order Model vs. Reduced-Order Model (Full-Order Model: dashed, Reduced-Order Model: solid)*

Here

$$P_1(q^{-1}) = \begin{bmatrix} P_{1,1}(q^{-1}) \\ P_{2,1}(q^{-1}) \end{bmatrix}, P_2(q^{-1}) = \begin{bmatrix} P_{1,2}(q^{-1}) \\ P_{2,2}(q^{-1}) \end{bmatrix},$$

are linear shift invariant operators in q^{-1}. $w' = (w_1, w_2)$ is the vector of external disturbances and $z' = (z_1, z_2)$ is the vector of outputs. w_1 is measured, w_2 is unmeasured, z_1 is the error signal we want to keep small and z_2 is a signal vector used for feedback.

The two step design approach, referred to as certainty equivalence control, consists of choosing a parameter dependent feedback controller and then tuning the parameter on line in order to optimize the system performance. The combination of tuning and linear control gives nonlinear feedback and thus linear stability analysis cannot be directly applied. Thus, one objective of the paper is to study the dynamics of the nonlinear feedback and show that stability can be achieved. In doing this we also show that H_∞ control is the rational choice for feedback design and that constrained least squares can be used for parameter adaptation. As in robust control, we assume that frequency domain bounds are available for the unmodeled dynamics. But, unlike robust control, we do not have an accurate nominal model. Instead, the nominal depends on a tunable parameter θ.

The result of the H_∞ calculation is a linear feedback law of the type

(3.2) $$u(t) = K_\theta(q^{-1})z_2(t)$$

Here $K_\theta(q^{-1})$ is a rational transfer function indexed by the parameter θ, viz.

(3.3) $$K_\theta(q^{-1}) = \frac{1}{R_\theta(q^{-1})} \left[T_\theta(q^{-1}), -S_\theta(q^{-1}) \right] = [K_{\theta,1}(q^{-1}), K_{\theta,2}(q^{-1})]$$

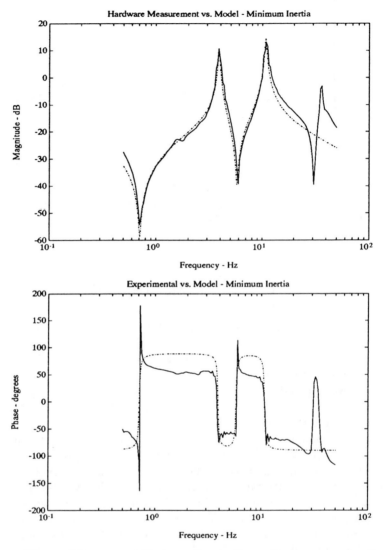

FIG. 3. *Frequency Response of Minimum Inertia Case - Hardware vs. Continuous Time Model from Motor Current to Hub Velocity (Hardware: solid, Reduced Order Model: dotted)*

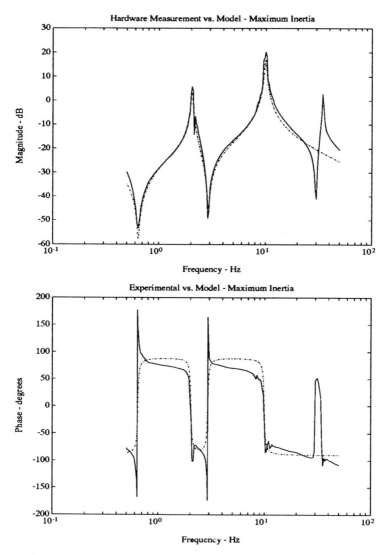

FIG. 4. *Frequency Response of Maximum Inertia Case - Hardware vs. Continuous Time Model from Motor Current to Hub Velocity (Hardware: solid, Reduced Order Model: dotted)*

where $R_\theta(q^{-1}), T_\theta(q^{-1}), S_\theta(q^{-1})$ are polynomials in q^{-1}. Typically, θ is a tunable parameter in a reduced order model describing the dynamical behavior of the process (3.1) .

To describe the closed-loop dynamics and define the optimal θ with respect to system (3.1) and the feedback law (3.2), we eliminate $u(t)$ and $z_2(t)$ to get

$$(3.4) \qquad\qquad z_1(t) = H_\theta(q^{-1})w(t).$$

The two input one output transfer vector

$$H_\theta(q^{-1}) = \left[H_{\theta,0}(q^{-1}), H_{\theta,1}(q^{-1}) \right]$$

is infinite dimensional and defined so that

$$H_\theta(q^{-1}) = P_{1,2} + P_{1,1}K_\theta(I - P_{2,1}K_\theta)^{-1}P_{2,2}.$$

The *centered parameter* can now be defined. Using the ℓ_2 performance objective we define

$$(3.5) \qquad \theta^* = \arg\min_{\theta \in \Theta} \lim_{N \to \infty} \frac{1}{N} \sum_{i=0}^{N} E\left\{ \|H_\theta(q^{-1})w(i)\|^2 \right\}.$$

Θ is a compact set so that K_θ is continuous and stabilizing for all $\theta \in \Theta$. Internal stability must also be maintained and Θ must be non-empty to have a well-defined optimization problem.

To be implementable both K and θ must be finite dimensional. $P(q^{-1})$ is in general infinite dimensional and not precisely known so any realistic optimization over θ must represent an approximation.

For the flexible beam we have

$$\begin{aligned} z_1(t) &= y(t) - y^*(t) \\ z_2(t) &= (y^*(t), y(t))' \\ w(t) &= (y^*(t), v(t))'. \end{aligned}$$

Here $y^*(t)$ is the reference and $v(t)$ is a bounded disturbance which is not measured. With these definitions we can write the system (3.1) so that

$$\begin{bmatrix} y(t) - y^*(t) \\ y^*(t) \\ y(t) \end{bmatrix} = \begin{bmatrix} P(q^{-1}) \\ 0 \\ P(q^{-1}) \end{bmatrix} u(t) + \begin{bmatrix} -1 & 1 \\ 1 & 0 \\ 0 & 1 \end{bmatrix} \begin{bmatrix} y^*(t) \\ v(t) \end{bmatrix}$$

where $P(q^{-1})$ is the discretized plant transfer function, derived from equation (2.1) using a zero-order hold and $v(t)$ results from unmodeled disturbances. The closed-loop from the disturbances and the setpoints to the error (equation 3.4) can now be written as

$$(3.6) \qquad y(t) - y^*(t) = H_{\theta,0}(q^{-1})y^*(t) + H_{\theta,1}(q^{-1})v(t).$$

Using this equation and equation (3.5) we apply the Parseval identity to define the centered parameter for the beam problem so that

$$(3.7) \quad \theta^* = \arg\min_{\theta \in \Theta} \int_{-\pi}^{\pi} |H_{\theta,0}(e^{j\omega})|\Phi_{y^*}(\omega) + |H_{\theta,1}(e^{j\omega})|\Phi_b(\omega)d\omega.$$

This expression shows that the feedback design, $K_\theta(q^{-1})$, should be chosen so that the effect of inputs is minimimized. This leads to sensitivity minimization and a classical trade-off between stability and performance. The parameter θ can be centered once the feedback design has been chosen. It is clear that the centered parameter depends on choice of feedback design and the spectrum of the input signals y^* and b. In other words, the best nominal model depends on control design as well as input signals.

Now we show how the centering problem can be approximated using adaptive control. We periodically match a finite dimensional regression model, called the a *nominal model*, to the plant data in closed-loop and then to use the estimated model for control system design. We use constrained least squares for estimation and an H_∞ method for control system design.

The nominal model is

$$(3.8) \qquad\qquad e(t) = A(q^{-1})y(t) - B(q^{-1})u(t)$$

where $e(t)$ is the model error and

$$A(q^{-1}) = 1 + a_1 q^{-1} + ... + a_n q^{-n}$$

and

$$B(q^{-1}) = b_0 + b_1 q^{-1} + ... + b_n q^{-n}$$

are polynomials in q^{-1} of order n. These definitions identify the parameter vector used in the control design procedure as

$$\theta = (a_1, ..., a_n, b_0, ..., b_n)'.$$

The centered parameter θ^* is defined as outlined above.

In order to define a finite dimensional adaptive control algorithm we re-write the closed-loop system defined by equation (3.6) in terms of the nominal model

$$(3.9) \qquad\qquad \begin{bmatrix} y^*(t) - y(t) \\ u(t) \end{bmatrix} = G_\theta(q^{-1}) \begin{bmatrix} y^*(t) \\ e(t) \end{bmatrix}$$

where

$$G_\theta(q^{-1}) = \frac{\begin{bmatrix} RA + BS - BT & -R \\ AT & -S \end{bmatrix}}{RA + BS} = \begin{bmatrix} G_{\theta,11}(q^{-1}) & G_{\theta,12}(q^{-1}) \\ G_{\theta,21}(q^{-1}) & G_{\theta,22}(q^{-1}) \end{bmatrix}.$$

This formulation defines the additive model mismatch

(3.10) $$\Delta_\theta(q^{-1}) = A(q^{-1})\left(P(q^{-1}) - \hat{P}_\theta(q^{-1})\right)$$

where

$$\hat{P}_\theta(q^{-1}) = \frac{B(q^{-1})}{A(q^{-1})}$$

For stability it is sufficient that the transfer functions $\Delta_\theta(q^{-1})$ and $G_\theta(q^{-1})$ are stable and that the small gain condition

$$|\Delta_\theta(q^{-1})G_{22,\theta}(q^{-1})|_{\sigma,\infty} \le K_v < 1$$

is satisfied.

We now define *admissible set*. The set consists of model parameters that give a well-defined solution to the H_∞ problem. For example, if there is no model mismatch then the admissible set contains all linear systems that do not have pole-zero cancellations. In our case the natural definition of the admissible set is the following

$$\begin{aligned}
\Theta^* \;=\; &\{\theta : |\Delta_{\theta^*}(q^{-1})G_{22,\theta}(q^{-1})|_{\sigma\infty} \le K_v, \\
&G_\theta(q^{-1}) \text{ stable and } K_\theta(q^{-1}) \text{ (Lipschitz) continuous}\}
\end{aligned}$$

Clearly $\Theta^* \supseteq \Theta$ and $\theta^* \in \Theta^*$. Also, note that any $\theta \in \Theta^*$ does not give stabilization of $P(q^{-1})$ since choosing $\theta \ne \theta^*$ gives parametric uncertainty. This property motivates the use of parameter adaptation to reduce parametric uncertainty.

Least Squares iterative adaptive control:

Step 0: Choose design law $K(\theta, q^{-1})$, $\theta(0) \in \Theta^*$, $N > 0$ and set $k = 0$.
Step 1: Compute $K(\theta(kN))$ and collect samples $t = kN+1, kN+2, \cdots, (k+1)N$.
Step 2: Estimate the parameters $\hat{\theta}((k+1)N) \in \Theta^*$ by constrained least squares.
Step 3: Set $k = k + 1$ and go to Step 1.

The constrained least squares algorithm minimizes the objective

$$J \;=\; \frac{1}{2}\sum_{i=kN}^{(k+1)N} e_f(i)^2 + (\theta - \theta(kN))'P(kN)^{-1}(\theta - \theta(kN)),$$
$$\text{Subject to:} \quad \theta \in \Theta^*$$

where $P(kN)$ is a positive definite matrix satisfying

$$0 < P_{\min}I \le P(kN) \le P_{\max}I.$$

We may reset $P(kN)$ so that $P(kN) = P(0) > 0$. The filtered error is defined so that

$$(3.11) \qquad\qquad e_f(t) = L_{\theta(kN)}(q^{-1})e(t).$$

$L_{\theta(kN)}(q^{-1})$ is a stable and stable invertible transfer function which may depend on estimated parameters. The set Θ^* consists of admissible models and physical constraints. For example, we have found it helpful to include *a priori* knowledge about the antiresonances. Guidelines for choice of $L(q^{-1})$ are given in Appendix B.

Result 1 *(Stability of indirect adaptive control):* Suppose that $\Theta^* \neq \emptyset$ for K_v small enough. Then there exist $N < \infty$ and constants $\mu_i, i = 1, 2, .., 6$ so that

$$\|y - y\|_2 \le \mu_1 + \mu_2\|v\|_2 + \mu_3\|y^*\|_2$$

and

$$\|y - y\|_\infty \le \mu_4 + \mu_5\|v\|_\infty + \mu_6\|y^*\|_\infty$$

Proof: (See Appendix A)

   Remarks:
   1. The result establishes sufficient conditions for stability and robustness of a broad class of indirect adaptive control algorithms. We use the H_∞ based method described below. Other methods such as LQR or pole-assignment can also be used. One significant novelty here is that we do not use any kind of normalization signal or relative deadzone. In this way the result represents a generalization of the result due to Ydstie [16].
   2. We use constrained parameter estimation to ensure that $\theta(t) \in \Theta^*$. This idea was put forward by Middleton et. al. [7]. Some progress has been made in developing optimization algorithms that solve problems of this type [10]. Experiments show that we obtain well-conditioned models and a feasible H_∞ problem when we excite the signals and that there is no need to apply as complicated an estimation scheme as indicated here. However, we have found it useful to constrain the parameters associated with the antiresonances.
   3. The stability result guarantees boundedness of all signals if the gain

   $$(3.12) \qquad\qquad |\Delta_{\theta^*}(q^{-1})G_{2,2}(\theta, q^{-1})|_{\sigma,\infty}$$

   is small enough. It is this expression which gives the motivation for the application of robust control theory. In the algorithm we use the *a priori* given frequency domain representation of unmodeled dynamics and performance specifications and then we use the estimated model together with the H_∞ method to ensure that the constraint is satisfied. Previous results in the literature on robust adaptive control put conditions on $|\Delta_{\theta^*}(q^{-1})|_\infty$, which is considerably more restrictive since the mismatch is now required to be small over the entire frequency range. The motivation for the use of robust control in the context of these results is less clear.

4. The constants, μ_i, can be computed by using the techniques described in Appendix A and in Ydstie [17]. It will be found that all signals converge exponentially fast to a residual set which may contain chaotic attractions.

4. H_∞ control design and adaptive estimation. The H_∞ design process is based on a *synthesis model* of the system; an augmented system description of the plant that includes the nominal plant model, the performance requirements, and a representation of the uncertainties in the model (both structured and unstructured). A complete description of the approach we use is given by Stoughton [14]. Using the H_∞ design algorithm on the synthesis model guarantees, provided a stabilizing controller exists, that the closed-loop system will meet the performance requirements over the entire range of plant variations accounted for in the representation of the plant uncertainties. In this way the controller guarantees performance robustness in the non-adaptive case.

The weighting functions used to generate the H_∞ controller do not account for any parametric variations in the plant. Without modification the "standard" H_∞ design approach allows stable plant inversion by cancelling left half plant poles and zeros to satisfy the design. We incorporate *parametric* uncertainties (albeit indirectly) into the synthesis model by including a disturbance input, d, with two scaling gains, k_{rgd} and k_{flx}, to the plant model as shown in Fig. 5.

The H_∞ design then yields a closed-loop system which is stable for all feedback connections between z and w and the system is stable for all uncertainties modeled in the Δ-block. A "feedback path" from z to w for a parametric uncertainty can be incorporated through an input disturbance to the plant, $G(s)$. In our application we include a disturbance input with scaling gains, k_{rgd} and k_{flx}, to account for parametric uncertainties in the rigid body and flexible dynamics.

The low frequency disturbance rejection is determined by the gain, k_{rgd} and the weighting function $W_S(s)$. At low frequencies the plant is dominated by the rigid body dynamics. Therefore, $k_{rgd} d$ is the primary input affecting the sensitivity function, $S(s)$. The low frequency disturbance rejection is bounded by $(k_{rgd} W_S(s))^{-1}$.

The final synthesis model for the flexible beam testbed is shown in Fig. 6. The model includes the weighting functions, a synthetic disturbance

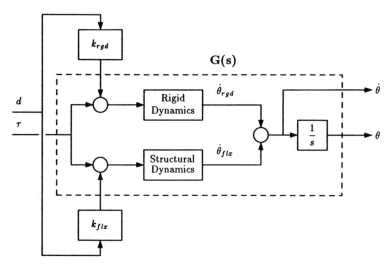

FIG. 5. *Nominal Plant, G(s) with Additional Disturbance Input*

signal, and sensor signals used for feedback to the controller.

For system identification we use least squares with prefiltering. The parameter estimates are constrained so that the model has an a priori specified value for the damping. The experiment design consists of the selection of the prefilter and the persistent excitation signal. The prefilter has the form

$$L(q^{-1}) = \frac{1}{\hat{B}(q^{-1})} \frac{1}{(1 + lq^{-1})}$$

where $\hat{B}(q^{-1})$ is a fixed estimate of the plant numerator and l is a filter pole. In our experiments we chose numerator dynamics $\hat{B}$ to have complex roots at 2.0 Hz and 12.0 Hz, with a damping factor $\zeta = 0.3$. The simple pole was placed at 0.1 Hz. This design is discussed in Appendix B. The goal is to produce a reasonably flat frequency response (of the relative weighting on the prediction error) over the bandwidth of interest. We have found that this prefilter design is effective and that the a priori uncertainty in the antiresonances can be large, (as much as 450% on the first antiresonance, and 100% on the second antiresonance), without causing significant deterioration in performance. Complete omission of the filter gives appreciable deterioration in performance.

We use an additive excitation signal, $d(t)$, to guarantee persistent excitation during the system identification process. The excitation signal was chosen to be a swept sine signal of magnitude 1.4 Nm, duration 2.5 seconds, and frequency sweep between 0.1 and 30 Hz.

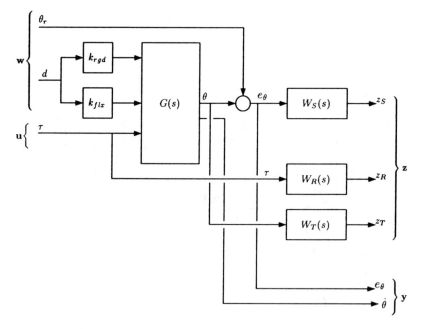

FIG. 6. *Final Synthesis Model for Flexible Beam Testbed*

An estimator deadzone is used to turn off the identifier.[1] The deadzone criterion (developed as a result of empirical studies) is given as follows:

$$s(t) = \sum_{t-N}^{t} e(i)^2$$

$$m(t) = \max(s(j), j = 0, \cdots, t)$$

$$\text{if } s(t) < \frac{m(t)}{c}, \text{ then } A(t) = 0$$

(4.1) $$\text{else } A(t) = 1.$$

N is the window length, c is a positive constant, $2 \le c \le 10$
$A(t)$ is a switching function $(0,1)$ which turns the algo-
    rithm on or off;
    if $A(t) = 0$, $\hat{\theta}(t) = \hat{\theta}(t-1)$ and $P(t) = P(t-1)$.
Once system identification has been triggered and the parameters have converged (i.e. the deadzone criterion has been satisfied), the H_∞ redesign is
performed. The redesign involves the following steps. First, the discrete
time identified model is converted to a continuous time plant model via the

---

[1] We assume that information for activating the estimator is available. For the case
of adaptive control of manipulators, knowledge of a payload change is typically available
to the controller from a higher level in the hierarchical structure.

inverse Tustin approximation. Next, the synthesis model is formed, using the new plant model and the a priori selected weighting functions and scaling gains (see Fig. 6). The H_∞ optimization is performed on the synthesis model to compute the continuous time controller. The new controller is then discretized using the Tustin approximation.

5. Experimental results. The plant we use for the simulation consists of one of three load configurations (unloaded, maximum inertia case, and minimum inertia case), combined with two higher frequency unmodeled structural modes. Two cases are addressed: 1) initial tuning of the controller (for use in applications such as launch vehicles), and 2) adaptation following a payload change (to achieve a high performance design for different payloads).

The sensor signals used for control design are hub position and velocity. The velocity signal is used since the tachometer and resolver on the hardware testbed are both colocated. To simulate the effects of the tachometer on the hardware testbed, low level white noise was added to the velocity signal. The level of noise was selected so that the power spectral density functions of the velocity signals from the hardware testbed and the simulation closely matched. The "true" plant is tenth order (ninth order for the velocity signal). The plant dynamics are numerically integrated using the Matlab function, ODE45, which uses a fourth and fifth order Runge-Kutta formula. The reduced order transfer function between the motor torque and hub velocity used for system identification and control design is 5th-order.

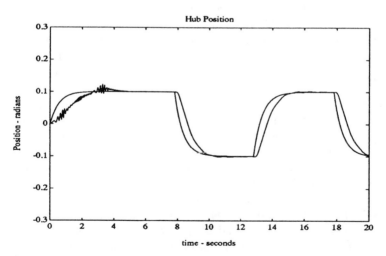

FIG. 7. *Hub Position (Actual vs. Reference) for Unloaded Case (0 ≤ t ≤ 3.88 sec: identification, 3.88 ≤ t ≤ 7.88 sec: T_{trans} for LPC, 7.88 sec: HPC)*

For the case of initial tuning, system identification is performed from time $t = 0$, to simulate the situation of tuning a controller on-line at startup. During the system identification phase we use a low performance (robustly stable) controller (LPC) and inject an additive excitation signal for persistent excitation.

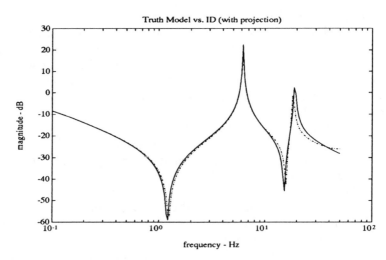

FIG. 8. *Frequency Response Magnitude of Reduced Order Model vs. Identified Model for Unloaded Case (Truth: solid, Identified: dashed)*

Once the identification phase is complete, i.e. the deadzone criterion is met, the LPC is left in operation for T_{trans} seconds, where T_{trans} is the settling time of the LPC. This allows the LPC states to settle to zero and the plant dynamics to reach steady state. During this transition phase, the reference command is held constant. Once this phase is completed, the newly designed high performance controller (HPC) is activated for the duration of the simulation. Figs. 7 through 9 are plots for the unloaded configuration. Fig. 7 shows the hub position versus the commanded hub position. The low-level additive excitation signal is evident in the output during the identification phase. The identifier is turned off at 3.88 seconds by the deadzone criterion. The frequency response magnitude plot given in Fig. 8 demonstrates that the H_∞ algorithm allows a margin of parametric uncertainty in the identified reduced order model. Fig. 9 shows the numerator and denominator parameter estimates.

A payload change at 5 seconds from the maximum payload to the minimum configuration is simulated to demonstrate the performance of the adaptive H_∞ control approach following a change in plant dynamics. The simulation is set up as follows. The HPC designed for the max inertia case is running at time $t = 0$. At $t = 5$ seconds, the plant dynamics are changed to the minimum inertia case. At this time, (i.e. when the payload change is detected by a higher level path planner, for example), the LPC

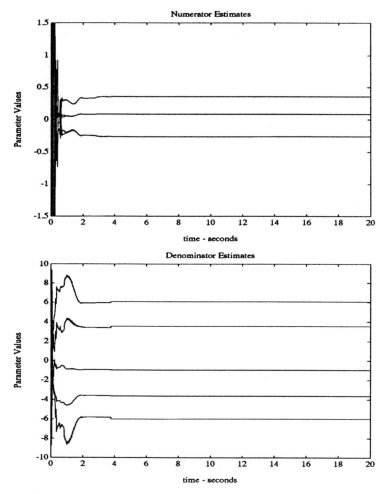

FIG. 9. *Numerator and Denominator Estimates for Unloaded Case*

is activated to guarantee system stability. A transition phase is entered
to allow the controller states to settle due to mismatch between the HPC
and LPC controller. The duration of the transition stage is equal to the
settling time of the LPC, T_{trans}. This phase is important in setting up the
system identification experiment properly. Otherwise, the controller state
mismatch (i.e. erroneous initial conditions for the LPC controller) will
look like a disturbance to the plant, and thus will result in a corresponding error in the identified model. When this transition phase is complete,
system identification (with an additive excitation signal) is activated. The
deadzone criterion for this case is satisfied at $t = 13.76$ seconds. Once the
deadzone criterion is met, the system identification process is terminated
and the H_∞ redesign is performed. A second transition phase is entered

to allow the LPC states to settle to zero and the plant dynamics to reach steady state. During this transition phase, the reference command is held constant. Once this phase is completed, the newly designed high performance controller is activated for the duration of the simulation. Plots for the maximum to minimum inertia configuration are shown in Figs. 10 and 11.

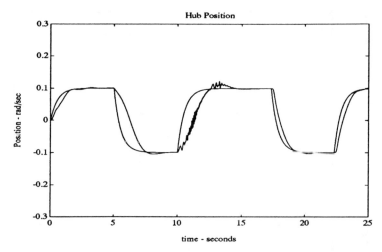

FIG. 10. *Hub Position (Actual vs. Reference) for Payload Change at 5 Seconds from Maximum Inertia to Minimum Inertia Case (*$0 \leq t \leq 5$* seconds: HPC for max inertia,* $5 < t \leq 10$ *seconds: LPC transition,* $10 < t \leq 13.76$ *sec: identification,* $13.76 < t \leq 17.76$ *sec:* T_{trans} *for LPC,* $t > 17.76$ *sec: HPC for minimum inertia)*

We have found through the simulations that the adaptive H_∞ control approach is robust to many types of disturbances. There are four main situations that can occur. The scenarios and effects are discussed below:

1. There can be an input or output disturbance to the plant during the system identification process. This will cause an erroneous (or poor) model to be identified, and will result in the failure of the H_∞ design. This event can be used to repeat the identification process to obtain a valid system model.

2. There can be a low frequency disturbance that occurs during the operation of the HPC. Robustness to disturbances of this nature are determined by the weighting function $W_S(s)$ in the synthesis model.

3. There can be higher order unmodeled dynamics or high frequency disturbances. Robustness to these effects is determined by the shape of the closed-loop control response, and is affected by the weighting function, $W_R(s)$.

4. Disturbances that occur during the operation of the HPC that are close to the bandwidth of the closed-loop system will affect the performance, as in any control design. Performance cannot be

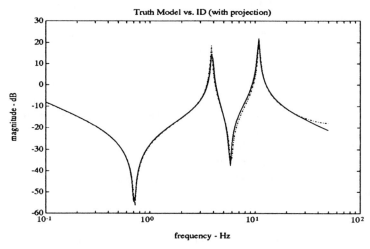

FIG. 11. *Frequency Response Magnitude of Reduced Order Model vs. Identified Model (Truth: solid, Identified: dashed)*

guaranteed in the presence of these types of disturbances, unless the disturbance is known and feedforward cancelation can be used.

6. Practical operation of the adaptive robust controller. This section discusses the practical operation of the adaptive H_∞ algorithm. Three topics are addressed: the computational requirements of the H_∞ redesign, the on-line selection of the appropriate identification prefilter,[2] and a summary of the design "variables".

Computational Requirements: The most intensive computational burden for the adaptive robust algorithm is the H_∞ redesign. The H_∞ redesign consists of the following steps:

1. Transformation of identified model to w-plane: 1.9 k flops,
2. Balanced Model Reduction: 36.7 k flops,
3. Generation of synthesis model: 15 k flops,
4. Balancing of synthesis model: 96.5 k flops,
5. H_∞ solution: 53.9 k flops,
6. Model reduction of controller: 59.4 k flops, and
7. Discretization of controller: 14.2 k flops.

The total number of flops is roughly 278 k flops. Although this is computationally intensive, these numbers are from the Matlab code, with no optimization for real-time implementation. Using these worst case numbers, and assuming the use of an Intel 80486-25 dedicated processor, the

---

[2] in situations where the dynamics vary over a sufficiently large range such that a single prefilter design is not adequate

time required to perform the redesign is 304 milliseconds.[3] There is a transition time (of duration equal to the settling time of the LPC) during which the LPC is operational following completion of the system identification procedure. It is therefore feasible to use this time to perform the H_∞ redesign. Furthermore, the problem can be set up so that the balanced model reduction of the identified plant (36.7 k flops) and the balancing of the synthesis model (96.5 k flops) are not required. Both of these operations are performed for numerical robustness, and can be omitted if the model structure is set up properly.

On-Line Selection of the Estimator Prefilter: There is a limit to the amount of initial uncertainty in the antiresonances that can be tolerated in order to design a prefilter that yields the required system identification accuracy. This becomes an issue in applications where the dynamics vary over a wide range, for example, with large payload variations. This requires that a *set* of prefilters be designed, and the appropriate one selected on-line, depending on the current payload range. There are a couple of ways this can be done. One approach is to partition the frequency range into a set of frequency bands, and design prefilters for each band. This approach was used in Sidman [12] and McGraw [6], although neither discussed the *selection* of different prefilters. A simpler approach than frequency partitioning is to perform a preliminary identification experiment to determine the frequency of the first resonance, and use this information to select the appropriate prefilter.

Figure 12 shows that even with a simple low pass filter, the resonances are correctly identified. It is observed from the frequency response plots of the different payload configurations that the first resonance is the most affected by a mass change. Furthermore, experiments have shown that convergence of the denominator coefficients is more rapid than the numerator coefficients. Thus a very brief identification phase can be used for the prefilter selection. This approach provides a simple, time-efficient method of selecting the appropriate prefilter on-line.

Design Variables: The following design variables must be chosen for the adaptive algorithm:

- system identification variables:
  - the design of prefilter(s), based on a priori "crude" knowledge of the antiresonance frequencies,
  - choice of the initial parameter estimates and $P(0)$ for the RLS algorithm,
  - an estimate of the damping for the antiresonances, to be used in the post-projection algorithm,

---

[3] The estimate on computational time was obtained using the Matlab benchmark program. The program uses Jack Dongarra's LINPACK benchmark to obtain a value for k flops/sec for a given processor.

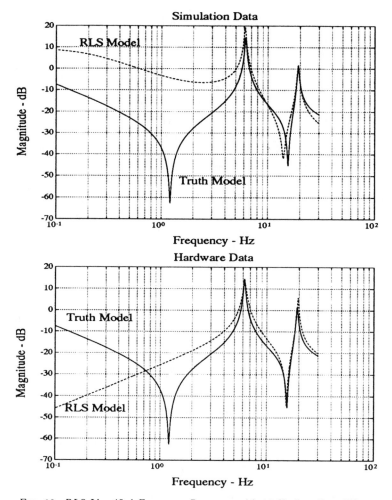

FIG. 12. *RLS Identified Frequency Response with 25 Hz Low Pass Filter*

- the value of the window length, N, for the mean squared error, $s(t)$, used in the deadzone criterion,
- the value of the constant, c, (the relative size of $s(t)$ compared to the peak value of $s(t)$), used to turn the identifier off.
- H_∞ redesign variables:
    - the weighting functions, $W_S(s)$, $W_T(s)$, and $W_R(s)$ to represent unmodeled dynamics.
    - the synthetic disturbances needed to incorporate parametric uncertainty into synthesis model for the H_∞ redesign.
- transition time variables:
    - the transition time, T_{trans}, (the settling time for the LPC).

7. Summary and conclusions. In this paper we have described an algorithm the adaptive robust control of systems with structural flexibility. The stability theory is developed and it is shown that closed-loop system is finite gain stable provided that the unmodeled dynamics have small gain in the frequency range where the control gain is high. This property can be achieved by application of the H_∞ control theory. Thus, the design philosophy provides a methodical approach for adaptive control.

A simulation study demonstrates the viability of the approach. We use a high fidelity model of the *Martin Marietta* flexible beam. This model closely matches the real plant data and noise has been added to make the simulation realistic. Two adaptive control scenarios are discussed: 1) the initial system identification phase to tune the controller parameters, and 2) the system identification and controller redesign following an abrupt change in plant dynamics (as in a payload change). In both cases the simulation shows that the adaptively updated controller has better performance than the fixed gain controller used to initialize the procedure. It is an important and remaining problem to show that the adaptive approach can converge to a performance similar to optimal.

8. Appendix A: Proof of Lemma 3.1. Define the exponentially weighted norm of a sequence $\{x(i), i = 0, 1, 2, ..., t\}$

$$\| x(t) \|_\sigma = \left(\sum_{i=0}^{t} \sigma^{-2i} \|x(i)\|^2 \right)^{1/2}$$

where $0 < \sigma \le 1$ is a real number. The case $\sigma = 1$ gives the truncated ell_2-norm of a function.

Let $F(\theta)$ be a matrix and let μ_0 and K_1, K_2 be numbers so that

$$\|F(\theta)^k\| \le K_1 \mu_0^k \qquad \text{and} \qquad \|F(\theta_1) - F(\theta_2)\| \le K_2 \|\theta_1 - \theta_2\|$$

We now define the vector norm $||| \cdot |||_\theta$ adapted to $F(\theta)$ [11]:

$$|||x|||_\theta = \sum_{i=0}^{\infty} \mu^{-i} \|F(\theta)^i x\|, \qquad \text{where } \mu_0 < \mu < 1$$

From this definition it follows that we have the triangle inequality and

Lemma A1: *(Properties of the adapted norm)*
(i) $\|x\| \le |||x|||_\theta \le \alpha\|x\|, \alpha = K_1 \frac{\mu}{\mu - \mu_0}.$
(ii) $|||F(\theta)x|||_\theta \le \sigma_0 |||x|||_\theta, \quad \sigma_0 = \mu - \frac{\mu - \mu_0}{K_1}.$
(iii) $|||x|||_{\theta(t)} \le (1 + \beta\|\theta(t) - \theta(t-1)\|)|||x|||_{\theta(t-1)}, \qquad \beta = 1 + \frac{K_1 K_2}{\mu - \mu_0}.$

Proof: See [8].

The centered parameters and equation (3.8) define the model error corresponding to the optimized system

$$(8.1) \qquad \gamma(t) = A^*(q^{-1})y(t) - B^*(q^{-1})u(t) = y(t) - \phi(t-1)'\theta^*$$

where

$$\phi(t) = (y(t), \cdots, y(t - n + 1), u(t), \cdots, u(t - n + 1))$$

is the regression vector and θ^* is the centered parameter.

Lemma A2: Suppose $\Theta^* \neq \emptyset$, that $|\Delta_{\theta^*}(q^{-1})G_{22}(q^{-1})|_{\sigma,\infty} \leq K_v$ with K_v sufficiently small. If N is sufficiently large; then there exist constants so that for all $t \geq 0$

(8.2) $\|\phi(t)\| \leq \sigma^t K_\phi (\|e_f(t)\|_{\sigma,t} + \|y_f^*(t)\|_{\sigma,t} + m(0))$

(8.3) $|\gamma_f(t)| \leq \sigma^t (K_v \|e_f(t)\|_{\sigma,t} + K_{y^*} \|y_f^*(t)\|_{\sigma,t} + m(0))$

(8.4) $+ ((t - kN)K_0 \sigma^{N(t-kN)} + \epsilon_N K_1)(\|e_f(t)\|_{\sigma,t} + \|y_f^*(t)\|_{\sigma,t} + m(0))$

where K_0 and K_1 are constants, $\lim_{N\to\infty} \epsilon_N = 0$, $m(0)$ is a constant chosen to overbound initial conditions and $\mu < \sigma < 1$.

Proof: First we note that equation (3.9) and (3.11) can be combined to define a state representation

(8.5) $\begin{aligned} x(t + 1) &= F(\theta(t))x(t) + g_1 e_f(t) + g_2 y_f^*(t), & x(0) = x_0 \\ \phi(t) &= h_1 x(t) \\ u_f(t) &= h_2 x(t), & L(q^{-1})u(t) \end{aligned}$

Since K_θ stabilizes the nominal model for all $\theta \in \Theta^*$ and $L(q^{-1})$ is stable and stably invertible, it follows that there exist two positive numbers K_1 and μ with $\mu < 1$ so that [1]:

$$\|F(\theta(t))^k\| \leq K_1 \mu^k \quad \text{for all } t \geq 0$$

Using continuity of the H_∞ design, it follows that there also exists a constant K_2 so that

$$\|F(\theta(t)) - F(\theta(t - 1))\| \leq K_2 \|\theta(t) - \theta(t - 1)\| \quad \text{for all } t \geq 0$$

From equation (8.5), the adapted metric and the triangle inequality

$$\begin{aligned} \||x(t + 1)\||_{\theta(t)} &\leq \||F(\theta(t))x(t)\||_{\theta(t)} + \alpha(\|g_1 e_f(t)\| + \|g_2 y_f^*(t)\|) \\ &\leq \sigma_0 \||x(t)\||_{\theta(t)} + \alpha_1 (|e_f(t)| + |y_f^*(t)|), \\ & \qquad \alpha_1 = \alpha \max\{|g_1|, |g_2|\} \\ &\leq \sigma_0 (1 + \beta(\|\theta(t) - \theta(t - 1)\|))\||x(t)\||_{\theta(t-1)} \\ & \quad + \alpha_1 (|e_f(t)| + |y_f^*(t)|) \end{aligned}$$

From compactness of Θ^* we have

$$\|\theta(t) - \theta(t - 1)\| \leq M \quad \text{for all } t \geq 0$$

where M is a constant. It follows from the above that we have

$$(8.6) \quad |||x(t+1)|||_{\theta(t)} \leq \sigma_0(1+m(t)\beta M)|||x(t)|||_{\theta(t-1)} \\ +\alpha_1(|e_f(t)|+|y_f^*(t)|)$$

where

$$m(t) = I_{\{||\theta(t)-\theta(t-1)||\neq 0\}}$$

We also have $||\theta(t) - \theta(t-1)|| = 0$ for $t \neq kN$, $k = 1, 2, \ldots$. Hence

$$\sum_{i=1}^{N} m(t+i) = 1 \qquad \text{for all } t \geq 0$$

and it follows that there exists $N < \infty$ and an associated number σ that

$$\prod_{i=1}^{N} \sigma_0(1+\beta m(t+i)M) = \sigma^N \qquad \text{with } \sigma_0 < \sigma < 1 \text{ and } \lim_{N \to \infty} \sigma = \sigma_0$$

It then follows from equation (8.6) and the definition of the exponentially weighted norm that we have a constant K_x so that

$$(8.7) \quad ||x(t+1)|| \leq \sigma^t K_x(||e_f(t)||_{2,\sigma_0} + ||y_f^*(t)||_{2,\sigma_0} + m(0))$$

$m(0)$ is a constant chosen to take care of the effects of non-zero initial conditions and $\sigma < 1$ provided N is large enough. Inequality (8.2) then follows by application of equation (8.5).

From definition (8.1) we have

$$\gamma(t) = A(q^{-1})(P(q^{-1}) - \hat{P}_{\theta*}(q^{-1}))u(t) + A(q^{-1})v(t)$$

By using equation (3.10) and (8.1) we then get

$$(8.8) \quad \gamma_f(t) = \Delta_{\theta*}(q^{-1})u_f(t) + v_f(t), \qquad v_f(t) = A(q^{-1})L(q^{-1})v(t)$$

Now, from equation (8.5) it follows that for every $t \in [kN, (k+1)N)$ we have (since $\theta(t)$ is constant over the etimation interval)

$$u_f(t) = \bar{G}_{t,kN}(q^{-1})e_f(t) + \bar{H}_{t,kN}(q^{-1})y_f^*(t) + h_2F(\theta(kN))^{t-kN}x(kN)$$
$$\text{for } t \in [kN, (1+k)N)$$

where

$$\bar{G}_{t,kN}(q^{-1}) = \sum_{i=0}^{\infty} \bar{g}_i q^{-i} \qquad \text{where } \bar{g}_i = \begin{cases} h_2F(\theta(kN))^i g_1 \text{ for } i \leq t - kN \\ \bar{g}_i = 0 \text{ otherwise} \end{cases}$$

$\bar{H}_{\theta(t)}(q^{-1})$ is defined likewise. Hence, from equation (8.8)

$$|\gamma_f(t)| = |\Delta_{\theta*}(q^{-1})\bar{G}_{t,kN}(q^{-1})e_f(t)| + |\Delta_{\theta*}(q^{-1})\bar{H}_{t,kN}(q^{-1})y_f^*(t)| \\ +|v_f(t)| + \xi(t)$$
$$(8.9)$$

where the last term is defined so that

$$(8.10) \quad \xi(t) = |\Delta_{\theta\bullet}(q^{-1})h_2 F(\theta(kN))^{t-kN}x(kN)|$$

$$(8.11) \quad \leq |\sum_{i=0}^{t-kN} \delta_i h_2 F(\theta(kN))^{t-kN-i}x(kN)|$$

$$(8.12) \quad + |\sum_{j=0}^{k}\sum_{i=0}^{N-1} \delta_{t-kN-i-1}h_2 F(\theta((k-j)N))^{N-i}x((k-j)N)|$$

$$+\mu^t m(0)$$

$$(8.13) \quad = \xi_0(t) + \xi_1(t) + \mu^t m(0)$$

$\{\delta_i\}$ is the impulse response of $\Delta_{\theta\bullet}(q^{-1})$ and $m(0)$ is a constant chosen to take care of non-zero initial conditions.

From the defintion of the weighted norm we have

$$|\Delta_{\theta\bullet}(q^{-1})\bar{G}_{t,kN}(q^{-1})|_{\sigma,\infty} \leq \sup_{k\geq 0} |\Delta_{\theta\bullet}(q^{-1})G_{22,\theta(kN)}(q^{-1})|_{\sigma,\infty} \leq K_v$$

$$|\Delta_{\theta\bullet}(q^{-1})\bar{H}_{\theta(kN)}(q^{-1})|_{\sigma,\infty} \leq \sup_{k\geq 0} |\Delta_{\theta\bullet}(q^{-1})\bar{H}_{\theta(kN)}(q^{-1})|_{\sigma,\infty} \leq K_{y^*}$$

where K_{y^*} is a constant and K_v is assumed to be small. So we can write

$$(8.14) \quad |\Delta_{\theta\bullet}(q^{-1})\bar{G}_{t,kN}(q^{-1})e_f(t)| \leq \sigma^t(K_v\|e_f(t)\|_\sigma + m(0))$$

$$(8.15) \quad |\Delta_{\theta\bullet}(q^{-1})\bar{G}_{t,kN}(q^{-1})y_f^*(t)| \leq \sigma^t(K_{y^*}\|y_f^*(t)\|_\sigma + m(0))$$

From the definitions in equation (8.13) we have

$$\xi_0(t) = |\sum_{i=0}^{t-kN} \delta_i h_2 F(\theta(kN))^{t-kN-i}x(kN)|$$

By defining $\delta = \max\{\mu, \bar{\delta}\}$ where $\bar{\delta}$ corresponds to the magnitude of the slowest pole of $\Delta_{\theta\bullet}(q^{-1})$ it follows that we have a constant $\bar{K}_0$ so that

$$\xi_0(t) \leq \bar{K}_0 \sum_{i=0}^{t-kN} \delta^i \delta^{t-kN-i}\|x(kN)\|$$

$$= \bar{K}_0(t - kN)\delta^{t-kN}\|x(kN)\|^2$$

It then follows from equation (8.7) that

$$(8.16) \quad \xi_0(t) \leq \sigma^t K_0(t - kN)\delta^{t-kN}(\|e_f(t)\|_\sigma + \|y_f^*(t)\|_\sigma + m(0))$$

Similar manipulations yield

$$\xi_1(t) \leq \bar{K}_1 \sum_{j=0}^{k}\sum_{i=0}^{N-1} \delta^{i+jN}\delta^{N-i}\|x((k-j)N)\|$$

$$= \bar{K}_1 \sum_{j=0}^{k} N\delta^{(j+1)N}\|x((k-j)N)\|$$

Now, $\lim_{N\to\infty} N\delta^N = 0$, so we have a number ϵ_N so that

$$(8.17) \quad \xi_1(t) \leq \epsilon_N \sum_{j=0}^{k} \delta^{j+1}\|x((k-j)N)\| \qquad \text{with } \lim_{N\to\infty} \epsilon_N = 0$$

Using equation (8.7) again we then get

$$(8.18) \quad \xi_1(t) \leq \epsilon_N \sum_{j=0}^{k} \delta^{j+1}\sigma^{jN} K_x(\|e_f(jN)\|_\sigma + \|y_f^*(jN)\|_\sigma + m(0))$$

Now, $\|e_f(jN)\|_\sigma \leq \sigma^{(k-j)N}\|e_f(kN)\|_\sigma$ for $k \geq j$ so that from equations (8.17) and (8.18) we get

$$
\begin{aligned}
\xi_1(t) &\leq \epsilon_N K_1 \sigma^{kN}((\|e_f(kN)\|_\sigma + \|y_f^*(kN)\|_\sigma + m(0)) \\
(8.19) \quad &\leq \epsilon_N K_1 \sigma^t((\|e_f(t)\|_\sigma + \|y_f^*(t)\|_\sigma + m(0))
\end{aligned}
$$

where K_1 is a constant which is independent of the inital conditions. The result then follows by using equations (8.14), (8.15), (8.16) and (8.19) in (8.13).   **q.e.d.**

Lemma A2 can be used to show that the normalized modeling errors are small in the mean provided that the filtered prediction errors are small in the mean. A number of techniques can then be used to show stability of the adaptive system provided that a suitable normalization signal is used. If we use the simple static normalizer of recursive least squares wih resetting then we can for example follow Ydstie (1991) and define the comparison variable:

$$(8.20) \qquad a(t+1) = \sigma^2 a(t) + e_f(t)^2 \qquad \text{with } a(0) = m(0)$$

and the zero-one indicator function

$$A(t) = I_{\{\sigma^2 a(t) + e_f(t)^2 \geq \sigma a(t)^2\}}$$

It follows immediately that

$$\sigma^t(\|e(t)\|_\sigma^2 + m(0)) \leq a(t+1)$$

Moreover, since $\theta(t) \in \Theta^*$, it follows from Lemma A1 that we have bounded rate of growth. Thus

$$(8.21) \qquad a(t+1) \leq g_1 a(t) + k_1, \qquad \text{for all } t \geq 0,$$

where g_1 and k_1 are positive constants. Lemma A2 now corresponds to equation (12) and (13) in Ydstie (1992) for direct adaptive control and the same method can be applied here to show finite gain stability.

9. **Appendix B: Frequency domain analysis of system identification performance and prefilter selection.** The distribution of the bias can be examined by obtaining a frequency domain expression for the limiting estimate, θ^* as described by Wahlberg & Ljung [5,15]. We review this theory here and describe how the estimation and control objectives can be better matched.

The asymptotic value of the parameter estimates in least squares estimation is defined as

$$(9.1) \qquad \theta^* = \operatorname*{arg\,min}_{\theta \in \Theta^*} \lim_{N \to \infty} \frac{1}{N} \sum_{t=1}^{N} E\left[e_f^2(t)\right],$$

where $e_f(t)$ is the filtered prediction error, and Θ^* is the model set to which $\hat{P}(q^{-1})$ belongs. In this expression the filtered prediction error including an additive excitation signal, $d(t)$, is

$$e_f(t) = \hat{L}(q^{-1})\hat{H}^{-1}(q^{-1})\left[\left(P(q^{-1}) - \hat{P}(q^{-1})\right)(u(t) + d(t)) + v(t)\right].$$
(9.2)

where the system model is expressed as $y(t) = P(q^{-1})(u(t) + d(t)) + H(q^{-1})v(t)$, the noise model is $H(q^{-1}) = A^{-1}(q^{-1})$ for the equation error formulation of RLS [3]. The control signal is expressed as $u(t) = K_1(q^{-1})y^*(t) - K_2(q^{-1})y(t)$.

The frequency domain formulation for the cost function is obtained by substituting the expression for $u(t)$ into Eq. (9.2), replacing $y(t)$ by the closed-loop expression for $y(t)$, and then transforming the equation via Parseval's theorem. The resulting cost function is

$$\theta^* = \operatorname*{arg\,min}_{\theta \in \Theta} \int_{-\pi}^{\pi} \left[|P - \hat{P}|^2 \left(\frac{|K_1|^2}{|1 + K_2 P|^2} \Phi_{y^*}(\omega) + \frac{1}{|1 + K_2 P|^2} \Phi_d(\omega) \right) \right.$$

$$(9.3) \qquad \left. + \frac{|1 + K_2 \hat{P}|^2}{|1 + K_2 P|^2} \Phi_v(\omega) \right] \frac{|\hat{L}|^2}{|\hat{H}|^2} d\omega .$$

Neglecting the effects of noise (i.e. $\Phi_v(\omega)$ is white or is high frequency and can be eliminated via low pass filtering), the resulting cost function is

$$\theta^* = \operatorname*{arg\,min}_{\theta \in \Theta} \int_{-\pi}^{\pi} |P - \hat{P}|^2 \left[\frac{\dfrac{|K_1|^2}{|1 + K_2 P|^2} \Phi_{y^*}(\omega) +}{|W_1(e^{j\omega})|^2} \right.$$

(9.4)
$$\underbrace{\frac{1}{|1 + K_2 P|^2}}_{|W_2(e^{j\omega})|^2} \Phi_d(\omega) \Bigg] \frac{\left|\hat{L}\right|^2}{\left|\hat{H}\right|^2} d\omega.$$

Eq. (9.4) indicates that the model fit, $| P - \hat{P}) |^2$, is weighted by the sum of the closed-loop transfer function ($W_1(q^{-1})$) between $y^*(t)$ and $u(t)$ times the reference signal spectrum, $\Phi_{y^*}(\omega)$, and the sensitivity function ($W_2(q^{-1})$) times the spectrum of the additive excitation signal, $\Phi_d(\omega)$. Therefore, even if $\Phi_{y^*}(\omega)$ is not persistently exciting, persistent excitation over the range of the closed-loop bandwidth of the system is achieved through the additive excitation signal, $d(t)$.

The typical method of assessing the transfer function match is via Bode plots. The Bode plot has an important physical interpretation in that *relative* errors are displayed rather than absolute errors (Ljung [5]). Furthermore, since the integral is a function of linear frequency, $d\omega$, higher frequencies are weighted more than lower frequencies. To incorporate these effects into the frequency domain cost function yields

$$\theta^* = \underset{\theta \in \Theta^*}{\arg\min} \int_{-\pi}^{\pi} \left|\frac{P - \hat{P}}{P}\right|^2 |P|^2 \left(|W_1|^2 \Phi_{y^*}(\omega) + |W_2|^2 \Phi_d(\omega)\right) \frac{|L|^2}{\left|\hat{H}\right|^2} d\omega.$$

(9.5)

where $d\omega$ can be interpreted as a weighting proportional to frequency; $d\omega \propto c\omega$, (c is a positive scalar). The selection of the system identification prefilter, $L(q^{-1})$ can be done by examining the effective weighting of the *relative* transfer function error of Eq. (9.5):

(9.6)
$$\left|\frac{P - \hat{P}}{P}\right|^2 |P|^2 |W_2|^2 |L|^2 \left|\hat{A}\right|^2 \Phi_d(\omega) d\omega$$

where $\left|\hat{H}\right|$ is given by $\frac{1}{|\hat{A}|}$ for the equation error formulation of RLS.

Note that only the term associated with the additive excitation signal is considered; it is assumed that the signal will provide the appropriate frequency content for good identification and thus the effects of the term associated with $\Phi_{y^*}(\omega)$ are ignored. The reference signal can be set to zero during system identification to ensure no interaction in the weighting function. Neglecting the contribution due to $W_2(q^{-1})$, the effective weighting on the relative transfer function error is

$$\begin{aligned} W_{eff} &= |P|^2 \left|\hat{A}\right|^2 |L|^2 \, d\omega \\ &= \left|\frac{B}{A}\right|^2 \left|\hat{A}\right|^2 |L|^2 \\ &\approx |B|^2 |L|^2 \, d\omega. \end{aligned}$$

(9.7)

Notice that this expression indicates that the weighting on the transfer function is determined by the numerator dynamics, $B(q^{-1})$, (McGraw [6]) rather than by the denominator dynamics, $A(q^{-1})$, as indicated by the expression in Eq. (9.4). Furthermore, there is an additional weighting, $d\omega$ (Ljung [5]). The effect of the frequency weighting, $d\omega$, is that lower frequencies have a smaller effect on the cost function than higher frequencies.

REFERENCES

[1] J.J. FUCHS, On the good use of the spectral radius of a matrix, IEEE Transactions on Automatic Control, **AC-27** (1982).

[2] G.C. GOODWIN, D.J. HILL, AND M. PALANISWAMI, Towards an adaptive robust controller, Proceedings of IFAC Identification and System Parameter Estimation, York, UK (1985), pp. 997–1002.

[3] J.H. KELLY, Adaptive robust control of systems with structural flexibility, PhD dissertation, Colorado State University, Ft. Collins, CO, Spring 1992.

[4] J.H. KELLY, R.M. STOUGHTON, M.F. RAMEY, E. SCHMITZ, Adaptive gain-scheduled robust control for flexible beam systems, Proceedings of the 13th Annual AAS Guidance and Control Conference, Keystone, CO, February 1990.

[5] L. LJUNG, System Identification—Theory for the User, Prentice-Hall, Englewood Cliffs, NJ, 1987.

[6] G.A. MCGRAW, Robust adaptive control design techniques, Aerospace Report No. TQR-0090(5025-05)-1, December 1989.

[7] R.H. MIDDLETON, G.C. GOODWIN, D.J. HILL, AND D.A. MAYNE, Design issues in adaptive control, IEEE Transactions on Automatic Control, **33** (1) (1988).

[8] L. PRALY, Adaptive linear control: bounded solutions and their properties, PhD thesis, Universite de Paris IX, 1988.

[9] D.M. ROVNER, Experiments in adaptive control of a very flexible one link manipulator, Ph.D. thesis, Stanford University, Department of Aeronautics and Astronautics, August 1987.

[10] H.D. SHERALI AND A. ALAMEDDINE, A new formulation-linearization technque for bilinear programing problems, J. of Global Opt. **2** (1992), pp. 379–410.

[11] M. SHUB, Stability of Dynamical Systems, Springer-Verlag, NY 1987.

[12] M.D. SIDMAN, Adaptive control of a flexible structure, Ph.D. dissertation, SU-DAAR 556, Stanford University, Stanford, CA, June 1986.

[13] R.M. STOUGHTON, An improved set of weighting functions for H_∞ control of flexible beam-like systems, Proceedings of the American Control Conference, May 1990.

[14] R.M. STOUGHTON AND C.T. VOTH, Vibration suppression for a large space structure using H_∞ control, (to be published in) Journal of Guidance, Navigation, and Control.

[15] B. WAHLBERG AND L. LJUNG, Design variables for bias distribution in transfer function estimation, IEEE Transactions on Automatic Control, **AC-31** (2) February (1986).

[16] B.E. YDSTIE, Stability of discrete model reference control—revisited, Syst. & Control Lett. **13** (1989), pp. 429–438.

[17] B.E. YDSTIE, Transient performance and robustness of direct adaptive control, IEEE Transactions on Automatic Control, **37** August (1992), pp. 1091–1105.

ESTIMATION-BASED SCHEMES FOR ADAPTIVE NONLINEAR STATE-FEEDBACK CONTROL*

MIROSLAV KRSTIĆ† AND PETAR V. KOKOTOVIĆ†

Abstract. We present a new approach to adaptive nonlinear control based on a complete controller-identifier separation which has long been a goal in adaptive system design. Our controllers guarantee certain input/state stability properties with respect to the parameter error $\tilde{\theta}$ and its derivative $\dot{\tilde{\theta}}$ as inputs. The parameter identifiers, in turn, guarantee $\tilde{\theta} \in \mathcal{L}_\infty$, and either $\dot{\tilde{\theta}} \in \mathcal{L}_\infty$ or $\dot{\tilde{\theta}} \in \mathcal{L}_2$ or both. This estimation-based approach encompases two families of schemes: swapping-based and observer-based. Swapping-based schemes allow the use of a wide variety of update laws — gradient and least-squares, normalized and unnormalized. Observer-based schemes use parameter identifiers of lower dynamic order. All these schemes achieve systematic improvement of transient performance.

1. Introduction. Adaptive nonlinear control is a new and increasingly active area of research. Following its first results [38,34,28,39,12], the 1990 Grainger Lectures [19,30,24,14] laid the foundations for a rapid development of backstepping and tuning functions designs [13,15,9,20,40], filtered transformations [23]– [25], and stability enhancement techniques [32]. Thanks to these efforts, it is now possible to design Lyapunov-based *direct adaptive schemes* for a broad class of nonlinear systems without conic or global Lipschitz restrictions. However, there are no similar results for estimation-based *indirect schemes* and other update laws, such as normalized gradient and least squares. Early attempts in this direction [2,10,11,39] introduced additional restrictive assumptions under which "certainty equivalence" controllers can be employed. Major advances in the development of adaptive nonlinear estimation-based schemes within a unified framework were presented in [30]. Since in the absence of matching conditions, all the schemes in [30] involved some growth restrictions, it soon became clear that, combined with normalized update laws, "certainty equivalence" controllers were unable to achieve boundedness without restrictions on nonlinearities.

The indirect estimation-based approach [6] has been very successful in adaptive linear control where it has unified a wide variety of adaptive schemes. This unification is due to the modularity feature: the identifier module achieves boundedness of the parameter error $\tilde{\theta}$ independently of the controller module.

The goal of this paper is to achieve modularity in adaptive nonlinear control. It turns out that in the nonlinear case we need to go a step further

---

* This work was supported in part by the National Science Foundation under Grant ECS-9203491 and in part by the Air Force Office of Scientific Research under Grant F-49620-92-J-0495.

† Department of Electrical and Computer Engineering, University of California, Santa Barbara, CA 93106.
miroslav@tesla.ece.ucsb.edu, (805) 893-4691 and petar@ece.ucsb.edu, (805) 893-7011.

and achieve a complete separation of the controller and identifier modules. The complete separation is made possible by the controller module's ability to guarantee boundedness of all closed-loop states whenever $\tilde{\theta} \in \mathcal{L}_\infty$, and either $\dot{\tilde{\theta}} \in \mathcal{L}_\infty$ or $\dot{\tilde{\theta}} \in \mathcal{L}_2$ (or both). The main tool for controller design is a form of nonlinear damping, including the "kappa-terms" introduced in [17]. In contrast to an exact compensation of $\dot{\tilde{\theta}}$ in the Lyapunov-based tuning functions scheme [20], the new controllers tolerate $\dot{\tilde{\theta}}$ thanks to the presence of nonlinear damping.

Our presentation encompasses two families of estimation-based schemes: swapping-based and observer-based.

Design and analysis of the swapping-based schemes rely on our nonlinear time-varying generalization of the ubiquitous swapping lemma [27]. We show that these schemes allow the use of a wide variety of update laws — gradient and least-squares, normalized and unnormalized.

Rather than in state estimation for output-feedback, the role of the observer in observer-based schemes is in parameter estimation. An advantage of using observer-based identifiers over those based on tuning functions is a less involved derivation of the control law, but this comes at the expense of an increase in the dynamic order with an observer. On the other hand, the dynamic order of observer-based identifiers is significantly lower than that of the swapping-based identifiers.

For linear systems, the issue of transient performance improvement has started receiving attention in the adaptive control literature [3,22]. We prove that with the presented estimation-based schemes we can systematically improve the $\mathcal{L}_\infty$, mean-square, and $\mathcal{L}_2$ performance.

The paper is organized as follows. We introduce the class of nonlinear uncertain systems and state the control objective in Section 2. In Section 3 we present our controller design and, considering $\tilde{\theta}$ and $\dot{\tilde{\theta}}$ as inputs, establish input/state stability properties. Section 4 is devoted to our nonlinear swapping lemma. Swapping-based and observer-based schemes are developed, and the stability proofs for the resulting adaptive systems are given in Sections 5 and 6, respectively. In Section 7 we analyze performance properties of the new adaptive systems and illustrate them by an example in Section 8.

2. Problem statement. The problem is to adaptively control nonlinear systems transformable into the *parametric-strict-feedback* form

$$\dot{x}_i = x_{i+1} + \theta^T \varphi_i(x_1, \ldots, x_i), \quad 1 \le i \le n-1$$

(2.1) $$\dot{x}_n = \beta_0(x)u + \theta^T \varphi_n(x)$$

$$y = x_1,$$

where $\theta \in \mathbb{R}^p$ is the vector of unknown constant parameters, β_0, and the components of $\varphi_i, 1 \le i \le n$, are smooth nonlinear functions in $\mathbb{R}^n$, $\varphi_1(0) =$

$\cdots = \varphi_n(0) = 0$, and $\beta_0(x) \neq 0$, $\forall x \in \mathbb{R}^n$.

The control objective is to force the output y of the system (2.1) to asymptotically track the output y_r of a known linear reference model while keeping all the closed-loop signals bounded. The reference model has the form

$$
(2.2) \quad \dot{x}_m = \begin{bmatrix} 0 & & & \\ \vdots & & I_{n-1} & \\ 0 & & & \\ -m_0 & -m_1 & \cdots & -m_{n-1} \end{bmatrix} x_m + \begin{bmatrix} 0 \\ \vdots \\ 0 \\ k_m \end{bmatrix} r
$$

$$
y_r = x_{m,1}
$$

where $M(s) = s^n + m_{n-1}s^{n-1} + \cdots + m_1 s + m_0$ is Hurwitz, $k_m > 0$, and $r(t)$ is bounded and piecewise continuous. This is feasible for all known fixed θ because (2.1) is feedback-linearizable. An alternative objective, as in [13], is to asymptotically track a given reference signal $y_r(t)$ with the assumption that its first n derivatives are known, bounded and piecewise continuous.

For strict-feedback nonlinear systems, the problem of global adaptive stabilization has been solved in [13] using for p unknown parameters, np estimates, a number subsequently reduced in half in [9]. The overparametrization was completely removed in [20] by the use of "tuning functions". An indirect, error-based scheme for this class of systems was designed in [40], with the same overparametrization as in [13].

Notation: For vectors we use $|x|_P \triangleq \left(x^{\mathrm{T}} P x\right)^{1/2}$ to denote the weighted Euclidean norm of x. For matrices, $|X|_{\mathcal{F}} \triangleq \left(\mathrm{tr}\{X^{\mathrm{T}}X\}\right)^{1/2} = \left(\mathrm{tr}\{XX^{\mathrm{T}}\}\right)^{1/2}$ denotes the Frobenius, and $|X|_2$ the induced 2-norm of X. The $\mathcal{L}_\infty$, $\mathcal{L}_2$ and $\mathcal{L}_1$ norms for signals are denoted by $\|\cdot\|_\infty$, $\|\cdot\|_2$ and $\|\cdot\|_1$ respectively. By referring to a matrix $A(t)$ as exponentially stable we mean that the corresponding LTV system $\dot{x} = A(t)x$ is exponentially stable. The spaces of all signals which are globally bounded, locally bounded and square-integrable on $[0, t_f)$, $t_f > 0$, are denoted by $\mathcal{L}_\infty[0, t_f)$, $\mathcal{L}_{\infty e}[0, t_f)$ and $\mathcal{L}_2[0, t_f)$, respectively.

3. Controller design and properties. Our controller design is based on integrator backstepping and an input/state version of the nonlinear damping lemma presented in the Appendix.

3.1. Controller design. The adaptive nonlinear controller is recursively defined by:

$$z_i = x_i - x_{m,i} - \alpha_{i-1}$$

$$\alpha_i(x_1, \ldots, x_i, \hat{\theta}, x_m) = -z_{i-1} - c_i z_i - \hat{\theta}^T w_i$$

$$+ \sum_{k=1}^{i-1} \left(\frac{\partial \alpha_{i-1}}{\partial x_k} x_{k+1} + \frac{\partial \alpha_{i-1}}{\partial x_{m,k}} x_{m,k+1} \right)$$

(3.1)
$$- s_i(x_1, \ldots, x_i, \hat{\theta}, x_m) z_i$$

$$w_i(x_1, \ldots, x_i, \hat{\theta}, x_m) = \varphi_i - \sum_{k=1}^{i-1} \frac{\partial \alpha_{i-1}}{\partial x_k} \varphi_k, \qquad i = 1, \ldots, n$$

$$u = \frac{1}{\beta_0(x)} \Big[\alpha_n(x, \hat{\theta}, x_m)$$

$$- m_0 x_{m,1} - \cdots - m_{n-1} x_{m,n} + k_m r \Big]$$

where $c_i > 0$, $i = 1, \ldots, n$, and, for notational convenience $z_0 \overset{\triangle}{=} 0$, $\alpha_0 \overset{\triangle}{=} 0$. The remaining design freedom is in the choice of the nonlinear damping functions $s_i(x_1, \ldots, x_i, \hat{\theta}, x_m)$. The resulting system, called the *error system*, is:

(3.2) $$\dot{z} = A_z(z, \hat{\theta}, t)z + W(z, \hat{\theta}, t)^T \tilde{\theta} + D(z, \hat{\theta}, t)\dot{\hat{\theta}}, \qquad z \in \mathbb{R}^n$$

where $z_1 = x_1 - x_{m,1} = y - y_r$ represents the tracking error, and

$$A_z(z, \hat{\theta}, t) = \begin{bmatrix} -c_1 - s_1 & 1 & 0 & \cdots & 0 \\ -1 & -c_2 - s_2 & 1 & \cdots & 0 \\ 0 & -1 & \ddots & \cdots & \vdots \\ \vdots & \vdots & \vdots & \ddots & 1 \\ 0 & 0 & \cdots & -1 & -c_n - s_n \end{bmatrix}$$

(3.3)

$$W(z, \hat{\theta}, t)^T = \begin{bmatrix} w_1^T \\ w_2^T \\ \vdots \\ w_n^T \end{bmatrix} \in \mathbb{R}^{n \times p}, \quad D(z, \hat{\theta}, t) = \begin{bmatrix} 0 \\ -\frac{\partial \alpha_1}{\partial \hat{\theta}} \\ \vdots \\ -\frac{\partial \alpha_{n-1}}{\partial \hat{\theta}} \end{bmatrix} \in \mathbb{R}^{n \times p}.$$

Except for the term $D(z, \hat{\theta}, t)\dot{\hat{\theta}}$, the form of this system is the same as in the recent backstepping designs [13,20,40]. While in [20] we eliminated the term $D(z, \hat{\theta}, t)\dot{\hat{\theta}}$ using *tuning functions*, here we let both $\tilde{\theta}$ and $\dot{\hat{\theta}}$ appear as disturbance inputs. Their boundedness properties ($\mathcal{L}_\infty$, $\mathcal{L}_2$) will later be guaranteed by parameter identifiers. Now we design the nonlinear damping functions $s_i(x_1, \ldots, x_i, \hat{\theta}, x_m)$ to achieve certain input/state stability properties (similar in spirit to the ISS of [36,37]) of the error system (3.2). Applying Lemma A.2, we choose

$$
(3.4) \qquad s_i(x_1, \ldots, x_i, \hat{\theta}, x_m) = \kappa_i |w_i|^2 + g_i \left| \frac{\partial \alpha_{i-1}}{\partial \hat{\theta}}^{\mathrm{T}} \right|^2
$$

where $\kappa_i, g_i, \ i = 1, \ldots, n$ are positive scalar constants. It is easy to verify that this choice makes the nonlinear time-varying system $\dot{z} = A_z(z, \hat{\theta}, t)z$ exponentially stable for all bounded $\hat{\theta}(t)$. The "kappa–terms" $\kappa_i |w_i|^2$ have recently been introduced by Kanellakopoulos [17].

3.2. Input/state properties of the error system. To prove input/state stability properties of the error system (3.2)–(3.4), we make use of the following constants: $c_0 = \min_{1 \le i \le n} c_i$, $\dfrac{1}{\kappa_0} = \sum_{i=1}^n \dfrac{1}{\kappa_i}$ and $\dfrac{1}{g_0} = \sum_{i=1}^n \dfrac{1}{g_i}$.

LEMMA 3.1. *In the error system (3.2)–(3.4), the following input/state properties hold:*

(i) If $\tilde{\theta}, \dot{\hat{\theta}} \in \mathcal{L}_\infty$ then $z, x \in \mathcal{L}_\infty$, and

$$
(3.5) \qquad |z(t)| \le \frac{1}{2\sqrt{c_0}} \left(\frac{1}{\kappa_0} \|\tilde{\theta}\|_\infty^2 + \frac{1}{g_0} \|\dot{\hat{\theta}}\|_\infty^2 \right)^{\frac{1}{2}} + |z(0)| e^{-c_0 t}.
$$

(ii) If $\tilde{\theta} \in \mathcal{L}_\infty$ and $\dot{\hat{\theta}} \in \mathcal{L}_2$ then $z, x \in \mathcal{L}_\infty$, and

$$
(3.6) \qquad |z(t)| \le \left(\frac{1}{4c_0\kappa_0} \|\tilde{\theta}\|_\infty^2 + \frac{1}{2g_0} \|\dot{\hat{\theta}}\|_2^2 \right)^{\frac{1}{2}} + |z(0)| e^{-c_0 t}.
$$

Proof. Differentiating $\frac{1}{2}|z|^2$ along the solutions of (3.2) we compute

$$
\begin{aligned}
\frac{d}{dt}\left(\frac{1}{2}|z|^2\right) =\ & -\sum_{i=1}^n c_i z_i^2 - \sum_{i=1}^n \left(\kappa_i |w_i|^2 + g_i \left| \frac{\partial \alpha_{i-1}}{\partial \hat{\theta}}^{\mathrm{T}} \right|^2 \right) z_i^2 \\
& + \sum_{i=1}^n z_i \left(w_i^{\mathrm{T}} \tilde{\theta} - \frac{\partial \alpha_{i-1}}{\partial \hat{\theta}} \dot{\hat{\theta}} \right) \\
(3.7) \qquad \le\ & -c_0 |z|^2 - \sum_{i=1}^n \kappa_i \left| w_i z_i - \frac{1}{2\kappa_i} \tilde{\theta} \right|^2 \\
& - \sum_{i=1}^n g_i \left| \frac{\partial \alpha_{i-1}}{\partial \hat{\theta}}^{\mathrm{T}} z_i + \frac{1}{2g_i} \dot{\hat{\theta}} \right|^2 \\
& + \left(\sum_{i=1}^n \frac{1}{4\kappa_i} \right) |\tilde{\theta}|^2 + \left(\sum_{i=1}^n \frac{1}{4g_i} \right) |\dot{\hat{\theta}}|^2
\end{aligned}
$$

and arrive at

$$(3.8) \qquad \frac{d}{dt}\left(\frac{1}{2}|z|^2\right) \leq -c_0|z|^2 + \frac{1}{4}\left(\frac{1}{\kappa_0}|\tilde{\theta}|^2 + \frac{1}{g_0}|\dot{\tilde{\theta}}|^2\right).$$

From Lemma A.1*(i)*, it follows that

$$|z(t)|^2 \leq |z(0)|^2 e^{-2c_0 t} + \frac{1}{2}\int_0^t e^{-2c_0(t-\tau)}\left(\frac{1}{\kappa_0}|\tilde{\theta}(\tau)|^2 + \frac{1}{g_0}|\dot{\tilde{\theta}}(\tau)|^2\right)d\tau.$$
(3.9)

This yields

$$(3.10) \qquad |z(t)|^2 \leq |z(0)|^2 e^{-2c_0 t} + \frac{1}{4c_0}\left(\frac{1}{\kappa_0}\|\tilde{\theta}\|_\infty^2 + \frac{1}{g_0}\|\dot{\tilde{\theta}}\|_\infty^2\right)$$

which proves $z \in \mathcal{L}_\infty$ and (3.5), and by (3.1), $x \in \mathcal{L}_\infty$. On the other hand, (3.9) implies

$$(3.11) \qquad |z(t)|^2 \leq |z(0)|^2 e^{-2c_0 t} + \frac{1}{4c_0\kappa_0}\|\tilde{\theta}\|_\infty^2 + \frac{1}{2g_0}\|\dot{\tilde{\theta}}\|_2^2$$

which proves $z \in \mathcal{L}_\infty$ and (3.6), and by (3.1), $x \in \mathcal{L}_\infty$. □

A consequence of this lemma is that, even when the adaptation is switched off, the state of the error system (3.2)–(3.4) remains bounded and converges exponentially to a positively invariant compact set. In this case the terms $-g_i\left|\frac{\partial \alpha_{i-1}}{\partial \hat{\theta}}^{\mathrm{T}}\right|^2 z_i$ are not needed. Moreover, when the adaptation is switched off, this boundedness result holds even when the unknown parameter is time varying.

COROLLARY 3.2. (Boundedness without adaptation.) *If* $\theta : \mathbb{R}_+ \to \mathbb{R}^p$ *is piecewise continuous and bounded, and* $\dot{\hat{\theta}}(t) \equiv 0$, *then* $z, x \in \mathcal{L}_\infty$, *and*

$$(3.12) \qquad |z(t)| \leq \frac{1}{2\sqrt{c_0\kappa_0}}\sup_{\tau\geq 0}|\theta(\tau) - \hat{\theta}| + |z(0)|e^{-c_0 t}.$$

Proof. In this case $\hat{\theta}$ is constant and (3.8) holds with $\tilde{\theta}(t) = \theta(t) - \hat{\theta}$.i □

4. Nonlinear swapping. The well known Swapping Lemma [27] is ubiquitous in adaptive linear control. Thus far this powerful tool has been lacking its nonlinear counterpart which we now provide.

LEMMA 4.1. (Nonlinear Swapping Lemma.) *Consider the nonlinear time-varying system*

$$(4.1) \quad \Sigma_1: \quad \begin{aligned} \dot{z} &= A(z,t)z + g(z,t)\left[W(z,t)^{\mathrm{T}}\tilde{\theta} - D(z,t)\dot{\tilde{\theta}}\right] \\ y_1 &= h(z,t)z + l(z,t)\left[W(z,t)^{\mathrm{T}}\tilde{\theta} - D(z,t)\dot{\tilde{\theta}}\right] \end{aligned}$$

where $\tilde{\theta} : \mathbb{R}_+ \to \mathbb{R}^p$ *is differentiable,* $A : \mathbb{R}^n \times \mathbb{R}_+ \to \mathbb{R}^{n\times n}$, $g : \mathbb{R}^n \times \mathbb{R}_+ \to \mathbb{R}^{m\times n}$, $W : \mathbb{R}^n \times \mathbb{R}_+ \to \mathbb{R}^{p\times m}$, $D : \mathbb{R}^n \times \mathbb{R}_+ \to \mathbb{R}^{m\times p}$ *are locally*

Lipschitz in z and continuous and bounded in t, and $h : \mathbb{R}^n \times \mathbb{R}_+ \to \mathbb{R}^{r \times n}$,
$l : \mathbb{R}^n \times \mathbb{R}_+ \to \mathbb{R}^{r \times m}$ are continuous in z and bounded in t. Along with
(4.1) consider the linear time-varying systems

$$(4.2) \qquad \Sigma_2 : \qquad \begin{aligned} \dot\chi &= A(z,t)\chi + g(z,t)W(z,t)^{\mathrm{T}} \\ y_2 &= h(z,t)\chi + l(z,t)W(z,t)^{\mathrm{T}} \end{aligned}$$

$$(4.3) \qquad \Sigma_3 : \qquad \begin{aligned} \dot\psi &= A(z,t)\psi + \chi\dot{\tilde\theta} + g(z,t)D(z,t)\dot{\tilde\theta} \\ y_3 &= -h(z,t)\psi - l(z,t)D(z,t)\dot{\tilde\theta} . \end{aligned}$$

Assume that the system $\dot\zeta = A(z,t)\zeta$ has the strong exponential stability
property that there exists a continuously differentiable function $V : \mathbb{R}^n \times$
$\mathbb{R}_+ \to \mathbb{R}_+$ such that

$$(4.4) \qquad \alpha_1|\zeta|^2 \leq V(\zeta,t) \leq \alpha_2|\zeta|^2 ,$$

and for each $z \in \mathcal{L}_\infty$,

$$(4.5) \qquad \frac{\partial V}{\partial \zeta}A(z,t)\zeta + \frac{\partial V}{\partial t} \leq -\alpha_3|\zeta|^2$$

$\forall t \geq 0$, $\forall \zeta \in \mathbb{R}^n$, $\alpha_1, \alpha_2, \alpha_3 > 0$. (In particular V can be quadratic in ζ.)
Then for $\forall z(0), \psi(0) \in \mathbb{R}^n$, $\forall \chi(0) \in \mathbb{R}^{n \times p}$, $\forall t \geq 0$ the outputs of systems
(4.1)–(4.3) are related by

$$(4.6) \qquad y_1 = y_2\tilde\theta + y_3 + y_\epsilon$$

where $y_\epsilon(t)$ decays exponentially.

   Proof. From (4.4)–(4.5), uniform boundedness of $g(z(t),t)$, $W(z(t),t)$
and $D(z(t),t)$, and differentiability of $\tilde\theta(t)$, it follows that $\chi \in \mathcal{L}_\infty$ and
$\psi \in \mathcal{L}_{\infty e}$. Differentiating $\tilde\epsilon = z + \psi - \chi\tilde\theta$, we readily obtain

$$(4.7) \qquad \dot{\tilde\epsilon} = \dot z + \dot\psi - \dot\chi\tilde\theta - \chi\dot{\tilde\theta} = A(z,t)\tilde\epsilon$$

which together with (4.4), (4.5) yields

$$(4.8) \qquad \dot V(\tilde\epsilon) = \frac{\partial V}{\partial \tilde\epsilon}A(z,t)\tilde\epsilon + \frac{\partial V}{\partial t} \leq -\alpha_3|\tilde\epsilon|^2 \leq -\frac{\alpha_3}{\alpha_2}V .$$

Therefore $V(t) \leq V(0)e^{-\frac{\alpha_3}{\alpha_2}t}$, and consequently

$$(4.9) \qquad |\tilde\epsilon(t)| \leq \sqrt{\frac{\alpha_2}{\alpha_1}}|\tilde\epsilon(0)|e^{-\frac{\alpha_3}{2\alpha_2}t} .$$

Now, (4.1)–(4.3) imply that $y_\epsilon = y_1 - y_2\tilde\theta - y_3 = h(z,t)\tilde\epsilon$, and therefore

$$(4.10) \quad |y_\epsilon(t)| \leq \sup_{\tau \geq 0}|h(z(\tau),\tau)|_2\sqrt{\frac{\alpha_2}{\alpha_1}}|z(0) - \chi(0)\tilde\theta(0) + \psi(0)|e^{-\frac{\alpha_3}{2\alpha_2}t}$$

$\forall t \geq 0$, $\forall z(0), \psi(0) \in \mathbb{R}^n$, $\forall \chi(0) \in \mathbb{R}^{n \times p}$. $\square$

REMARK 4.1. If instead of $z \in \mathcal{L}_\infty$, the maximal interval of existence of $z(t)$ is $[0, t_f)$, then the lemma holds on this interval.

REMARK 4.2. When $D(z,t) \equiv 0$, the result of Lemma 4.1 is reminiscent of Morse's linear Swapping Lemma

$$(4.11) \qquad T_z[W^T \tilde{\theta}] = T[W^T]\tilde{\theta} + T_h \left[T_g[W^T]\dot{\tilde{\theta}} \right] + y_\epsilon \,.$$

In this notation $T_z : W^T \tilde{\theta} \mapsto y_1$ is the nonlinear operator defined by (4.1) with $D(z,t) \equiv 0$, while the system

$$(4.12) \qquad \begin{aligned} \dot{\xi} &= A(z(t), t)\xi + g(z(t), t)u \\ y &= h(z(t), t)\xi + l(z(t), t)u \end{aligned}$$

is used to define the linear time-varying operators: $T : u \mapsto y$, $T_g : u \mapsto y$ for $h = I$ and $l = 0$, $T_h : u \mapsto y$ for $g = I$ and $l = 0$. When A, g, h and l are constant, then the operator $T_z(s) = T(s) = h(sI - A)^{-1}g + l$ is a proper stable rational transfer function, $T_g(s) = (sI - A)^{-1}g$, $T_h(s) = -h(sI - A)^{-1}$, and Lemma 4.1 reduces to Lemma 3.6.5 from [33].

In some texts on adaptive linear control, an extended result which guarantees that $\dot{\tilde{\theta}} \in \mathcal{L}_2 \cap \mathcal{L}_\infty \Rightarrow T_z[W^T \tilde{\theta}] - T[W^T]\tilde{\theta} \in \mathcal{L}_2$ is also referred to as Swapping Lemma. Our next lemma is a nonlinear time-varying generalization of this result.

LEMMA 4.2. Consider systems (4.1)–(4.3) with the same set of assumptions as in Lemma 4.1. If $\dot{\tilde{\theta}} \in \mathcal{L}_2$, then

$$(4.13) \qquad y_1 - y_2 \tilde{\theta} \in \mathcal{L}_2 \,.$$

If $\dot{\tilde{\theta}} \in \mathcal{L}_2 \cap \mathcal{L}_\infty$ and $l(z,t) \equiv 0$, then

$$(4.14) \qquad \lim_{t \to \infty} \left[y_1(t) - y_2(t)\tilde{\theta}(t) \right] = 0 \,.$$

Proof. By Lemma 4.1, $y_\epsilon \in \mathcal{L}_2$. We need to prove that $y_3 \in \mathcal{L}_2$. The solution of (4.3) is:

$$(4.15) \quad \psi(t) = \Phi_z(t,0)\psi(0) + \int_0^t \Phi_z(t,\tau)[\chi(\tau) + g(z(\tau), \tau)D(z(\tau), \tau)]\dot{\tilde{\theta}}(\tau)d\tau$$

where (4.4)–(4.5) guarantee that the state transition matrix $\Phi_z : \mathbb{R}_+ \times \mathbb{R}_+ \to \mathbb{R}^{n \times n}$ is such that $|\Phi_z(t,\tau)|_2 \leq ke^{-\alpha(t-\tau)}$, $k, \alpha > 0$. Since χ, g and

D are bounded then

$$
\begin{aligned}
|\psi(t)| \quad &\leq k e^{-\alpha t}|\psi(0)| + k\|\chi + gD\|_\infty \int_0^t e^{-\alpha(t-\tau)}|\dot{\tilde{\theta}}(\tau)| d\tau \\
&\leq k e^{-\alpha t}|\psi(0)| + k\|\chi + gD\|_\infty \left(\int_0^t e^{-\alpha(t-\tau)} d\tau \right)^{\frac{1}{2}} \\
&\qquad \left(\int_0^t e^{-\alpha(t-\tau)}|\dot{\tilde{\theta}}(\tau)|^2 d\tau \right)^{\frac{1}{2}} \\
&\leq k e^{-\alpha t}|\psi(0)| + k\|\chi + gD\|_\infty \frac{1}{\sqrt{\alpha}} \left(\int_0^t e^{-\alpha(t-\tau)}|\dot{\tilde{\theta}}(\tau)|^2 d\tau \right)^{\frac{1}{2}}
\end{aligned}
$$
(4.16)

where the second inequality is obtained using the Schwartz inequality. By squaring (4.16) and integrating over $[0,t]$ we obtain

$$
\int_0^t |\psi(\tau)|^2 d\tau \leq \frac{k^2}{2\alpha}|\psi(0)|^2 + \frac{k^2}{\alpha}\|\chi + gD\|_\infty^2 \int_0^t \left[\int_0^\tau e^{-\alpha(\tau-s)}|\dot{\tilde{\theta}}(s)|^2 ds \right] d\tau .
$$
(4.17)

Changing the sequence of integration, (4.17) becomes

$$
\begin{aligned}
\int_0^t |\psi(\tau)|^2 d\tau \quad &\leq \frac{k^2}{2\alpha}|\psi(0)|^2 + \frac{k^2}{\alpha}\|\chi + gD\|_\infty^2 \int_0^t e^{\alpha s}|\dot{\tilde{\theta}}(s)|^2 \\
&\qquad \left(\int_s^t e^{-\alpha\tau} d\tau \right) ds \\
&\leq \frac{k^2}{2\alpha}|\psi(0)|^2 + \frac{k^2}{\alpha}\|\chi + gD\|_\infty^2 \int_0^t e^{\alpha s}|\dot{\tilde{\theta}}(s)|^2 \frac{1}{\alpha} e^{-\alpha s} ds
\end{aligned}
$$
(4.18)

because $\int_s^t e^{-\alpha\tau} d\tau = \frac{1}{\alpha}(e^{-\alpha s} - e^{-\alpha t}) \leq \frac{1}{\alpha} e^{-\alpha s}$. Now, the cancellation $e^{\alpha s} e^{-\alpha s} = 1$ in (4.18) yields $\|\psi\|_2 \leq \frac{k}{\sqrt{2\alpha}}|\psi(0)| + \frac{k}{\alpha}\|\chi + gD\|_\infty \|\dot{\tilde{\theta}}\|_2 < \infty$ which proves $\psi \in \mathcal{L}_2$. Due to the uniform boundedness of W, D, h, l, it follows that $y_3 \in \mathcal{L}_2$. This proves (4.13). When $\dot{\tilde{\theta}} \in \mathcal{L}_2 \cap \mathcal{L}_\infty$ then $\psi \in \mathcal{L}_\infty \cap \mathcal{L}_2$ and $\dot{\psi} \in \mathcal{L}_\infty$. Thus, by Barbalat's lemma, $\psi(t) \to 0$ as $t \to \infty$. Since $l(z,t) \equiv 0$, then $y_3(t) \to 0$ as $t \to \infty$. This proves (4.14) because $y_\epsilon(t) \to 0$ as $t \to \infty$. $\square$

REMARK 4.3. When $D(z,t) \equiv 0$, we rewrite (4.13) as

$$
T_z[W^{\mathrm{T}}\tilde{\theta}] - T[W^{\mathrm{T}}]\tilde{\theta} \in \mathcal{L}_2
$$
(4.19)

and (4.14) as

$$
\lim_{t \to \infty} \left\{ T_z[W^{\mathrm{T}}\tilde{\theta}](t) - \left(T[W^{\mathrm{T}}]\tilde{\theta} \right)(t) \right\} = 0
$$
(4.20)

with T_z and T as in Remark 4.2. For constant A, g, h and l, the operator $T_z = T$ is a proper stable rational transfer function, and Lemma 4.2 reduces to Lemma 2.11 from [29].

5. Swapping-based schemes. We present two swapping-based schemes: the z-swapping scheme and the x-swapping scheme. In the z-swapping scheme, the parameter identifier is based on the error system (z), whereas in the x-swapping scheme, the parameter identifier is based on the original plant (x). For each of the two we use two different update laws: gradient and least-squares, either normalized or unnormalized. The parameter identifiers of this section are variants of the regressor filtering identifiers in [30].

5.1. z-swapping scheme. We define the *augmented error* vector

$$(5.1) \qquad \epsilon = z + \bar{\chi} - \chi\hat{\theta}$$

where $\bar{\chi}$ and χ are the states of the filters

$$(5.2) \quad \dot{\bar{\chi}} = A_s(z,\hat{\theta},t)\bar{\chi} + W(z,\hat{\theta},t)^{\mathrm{T}}\hat{\theta} - D(z,\hat{\theta},t)\dot{\hat{\theta}}, \qquad \bar{\chi} \in \mathbb{R}^n$$

$$(5.3) \quad \dot{\chi} = A_z(z,\hat{\theta},t)\chi + W(z,\hat{\theta},t)^{\mathrm{T}}, \qquad\qquad \chi \in \mathbb{R}^{n\times p}.$$

By substituting (3.2), (5.2) and (5.3) into (5.1) we obtain

$$(5.4) \qquad \epsilon = \chi\tilde{\theta} + \tilde{\epsilon}$$

where $\tilde{\epsilon}$ is governed by

$$(5.5) \qquad \dot{\tilde{\epsilon}} = A_z(z,\hat{\theta},t)\tilde{\epsilon}, \qquad \tilde{\epsilon} \in \mathbb{R}^n.$$

The update law for $\hat{\theta}$ is either the gradient:

$$(5.6) \qquad \dot{\hat{\theta}} = \Gamma\frac{\chi^{\mathrm{T}}\epsilon}{1+\nu|\chi|_{\mathcal{F}}^2}, \qquad \Gamma = \Gamma_0 = \Gamma_0^{\mathrm{T}} > 0,\ \nu \geq 0$$

or the least squares:

$$(5.7) \qquad \begin{aligned} \dot{\hat{\theta}} &= \Gamma\frac{\chi^{\mathrm{T}}\epsilon}{1+\nu|\chi|_{\mathcal{F}}^2} \\ \dot{\Gamma} &= -\Gamma\frac{\chi^{\mathrm{T}}\chi}{1+\nu|\chi|_{\mathcal{F}}^2}\Gamma \qquad \Gamma(0) = \Gamma_0 = \Gamma_0^{\mathrm{T}} > 0,\ \nu \geq 0, \end{aligned}$$

where by allowing $\nu = 0$, we encompass unnormalized update laws.

With the regressor χ^{T} being a matrix, our use of the Frobenius norm $|\chi|_{\mathcal{F}}$ avoids unnecessary algebraic complications in the stability arguments that arise from applying the normalized gradient update law in the form $\dot{\hat{\theta}} = \Gamma\chi^{\mathrm{T}}\left(I_n + \nu\chi\Gamma\chi^{\mathrm{T}}\right)^{-1}\epsilon$ or the normalized least-squares with $\dot{\Gamma} = -\Gamma\chi^{\mathrm{T}}\left(I_n + \nu\chi\Gamma\chi^{\mathrm{T}}\right)^{-1}\chi\Gamma$. It also eliminates the need for on-line matrix inversion.

LEMMA 5.1. *Suppose $\chi : \mathbb{R}_+ \to \mathbb{R}^{n\times p}$ is piecewise continuous, and x is bounded on $[0,t_f)$. The update laws (5.6) and (5.7) guarantee that*

(i) if $\nu = 0$ then $\tilde{\theta} \in \mathcal{L}_\infty[0, t_f)$ and $\epsilon \in \mathcal{L}_2[0, t_f)$,

(ii) if $\nu > 0$ then $\tilde{\theta} \in \mathcal{L}_\infty[0, t_f)$ and $\dot{\hat{\theta}}, \dfrac{\epsilon}{\sqrt{1 + \nu |\chi|_{\mathcal{F}}^2}} \in \mathcal{L}_2[0, t_f) \cap \mathcal{L}_\infty[0, t_f)$.

Proof. (Sketch) Noting from (5.5) that

$$(5.8) \qquad \frac{d}{dt}\left(\frac{1}{2}|\tilde{\epsilon}|^2\right) = -\sum_{i=1}^{n} c_i \tilde{\epsilon}_i^2 \le -c_0 |\tilde{\epsilon}|^2$$

which implies that $|\tilde{\epsilon}(t)| \le |\tilde{\epsilon}(0)|e^{-c_0 t}$, it is clear that the positive definite function $V_{\tilde{\theta}} = \dfrac{1}{2}|\tilde{\theta}|_{\Gamma^{-1}}^2 + \dfrac{1}{2}|\tilde{\epsilon}|^2$ can be used as in [6,33,7] to prove the lemma. □

As explained in [6], various modifications of the least-squares algorithm — covariance resetting, exponential data weighting, etc., do not affect the properties established by Lemma 5.1. A priori knowledge of parameter bounds can also be included in the form of projection.

THEOREM 5.2. (**z-swapping scheme**) *All the signals in the adaptive system consisting of the plant (2.1), controller (3.1), filters (5.2),(5.3), and either the gradient (5.6) or the least-squares (5.7) update law, are globally uniformly bounded for all $t \ge 0$, and $\lim\limits_{t \to \infty} z(t) = 0$. This means, in particular, that global asymptotic tracking is achieved:*

$$(5.9) \qquad \lim_{t \to \infty}[y(t) - y_r(t)] = 0.$$

Furthermore, if $\lim\limits_{t \to \infty} r(t) = 0$ then $\lim\limits_{t \to \infty} x(t) = 0$.

Proof. Due to the continuity of x_m and the smoothness of the nonlinear terms appearing in (2.1), (3.1), (5.2), (5.3), (5.6), (5.7), the solution of the closed-loop adaptive system exists and is unique. Let its maximum interval of existence be $[0, t_f)$.

For the normalized update laws, from Lemma 5.1 we obtain $\tilde{\theta}, \hat{\theta}, \dot{\hat{\theta}}, \dfrac{\epsilon}{\sqrt{1 + \nu|\chi|_{\mathcal{F}}^2}} \in \mathcal{L}_\infty[0, t_f)$. When the update laws are unnormalized, Lemma 5.1 gives only $\tilde{\theta}, \hat{\theta} \in \mathcal{L}_\infty[0, t_f)$ and we have to establish boundedness of $\dot{\hat{\theta}}$. To this end, we treat (5.3) in a fashion similar to (3.7) and obtain:

$$(5.10) \qquad \frac{d}{dt}\left(\frac{1}{2}|\chi|_{\mathcal{F}}^2\right) \le -c_0 |\chi|_{\mathcal{F}}^2 + \frac{p}{4\kappa_0}.$$

This proves that $\chi \in \mathcal{L}_\infty[0, t_f)$. Therefore, by (5.4) and because of the boundedness of $\tilde{\epsilon}$ we conclude that $\epsilon \in \mathcal{L}_\infty[0, t_f)$. Now by (5.6) or (5.7), $\dot{\hat{\theta}} \in \mathcal{L}_\infty[0, t_f)$. Therefore, by Lemma 3.1, $z, x \in \mathcal{L}_\infty[0, t_f)$. Finally, by (5.1), $\bar{\chi} \in \mathcal{L}_\infty[0, t_f)$.

We have thus shown that all of the signals of the closed-loop adaptive system are bounded on $[0, t_f)$ by constants depending only on the initial conditions, design gains, the external signals x_m and r, and not depending on t_f. The independence of the bounds of t_f proves that $t_f = \infty$.

Now we set out to prove that $z \in \mathcal{L}_2$, and eventually that $z(t) \to 0$ as $t \to \infty$. For the normalized update laws, from Lemma 5.1 we obtain $\dot{\theta}, \dfrac{\epsilon}{\sqrt{1 + \nu |\chi|_{\mathcal{F}}^2}} \in \mathcal{L}_2$. Since $\chi \in \mathcal{L}_\infty$ then $\epsilon \in \mathcal{L}_2$. When the update laws are unnormalized Lemma 5.1 gives $\epsilon \in \mathcal{L}_2$, and since $\chi \in \mathcal{L}_\infty$ then by (5.6) or (5.7), $\dot{\tilde{\theta}}, \dot{\hat{\theta}} \in \mathcal{L}_2$. Consequently in both the normalized and the unnormalized cases $\chi \tilde{\theta} \in \mathcal{L}_2$ because $\tilde{\epsilon} \in \mathcal{L}_2$. With $V = \frac{1}{2}|\zeta|^2$, all the conditions of Lemmas 4.1 and 4.2 are satisfied. Thus, by Lemma 4.2, $z - \chi\tilde{\theta} \in \mathcal{L}_2$. Hence $z \in \mathcal{L}_2$. To prove the convergence of z to zero, we note that (3.2), (3.3) implies that $\dot{z} \in \mathcal{L}_\infty$. Therefore, by Barbalat's lemma $z(t) \to 0$ as $t \to \infty$. When $r(t) \to 0$ then $x_m(t) \to 0$ as $t \to \infty$, and from the definitions in (3.1) we conclude that $x(t) \to 0$ as $t \to \infty$. □

5.2. x-swapping scheme. For the plant (2.1) rewritten in the form

$$(5.11) \qquad \dot{x} = Ex + e_n u + \phi(x)^{\mathrm{T}}\theta$$

where $E = \begin{bmatrix} 0 & \\ \vdots & I_{n-1} \\ 0 & \cdots\ 0 \end{bmatrix}$ and $\phi^{\mathrm{T}} = \begin{bmatrix} \varphi_1^{\mathrm{T}} \\ \vdots \\ \varphi_n^{\mathrm{T}} \end{bmatrix}$, we define the *equation error* vector

$$(5.12) \qquad \epsilon_x = x - \Omega_0 - \Omega\hat{\theta}$$

where Ω_0 and Ω are governed by

$$(5.13) \qquad \dot{\Omega}_0 = \bar{A}(t)(\Omega_0 - x) + Ex + e_n u, \qquad \Omega_0 \in \mathbb{R}^n$$
$$(5.14) \qquad \dot{\Omega} = \bar{A}(t)\Omega + \phi(x)^{\mathrm{T}}, \qquad\qquad \Omega \in \mathbb{R}^{n \times p}$$

and $\bar{A}(t)$ is an exponentially stable matrix. A similar identifier structure was proposed in [31] and [1], and used in [39]. By substituting (5.11), (5.13) and (5.14) into (5.12) we obtain

$$(5.15) \qquad \epsilon_x = \Omega\tilde{\theta} + \tilde{\epsilon}_x$$

where $\tilde{\epsilon}_x$ satisfies

$$(5.16) \qquad \dot{\tilde{\epsilon}}_x = \bar{A}(t)\tilde{\epsilon}_x, \qquad \tilde{\epsilon}_x \in \mathbb{R}^n.$$

The update law for $\hat{\theta}$ is either the gradient:

$$(5.17) \qquad \dot{\hat{\theta}} = \Gamma \frac{\Omega^{\mathrm{T}}\epsilon_x}{1 + \nu|\Omega|_{\mathcal{F}}^2}, \qquad\qquad \Gamma = \Gamma_0 = \Gamma_0^{\mathrm{T}} > 0,\ \nu \geq 0$$

or the least squares:

$$
\begin{aligned}
\dot{\theta} &= \Gamma \frac{\Omega^{\mathrm{T}} \epsilon_x}{1 + \nu |\Omega|_{\mathcal{F}}^2} \\
\dot{\Gamma} &= -\Gamma \frac{\Omega^{\mathrm{T}} \Omega}{1 + \nu |\Omega|_{\mathcal{F}}^2} \Gamma \qquad \Gamma(0) = \Gamma_0 = \Gamma_0^{\mathrm{T}} > 0, \ \nu \geq 0.
\end{aligned}
$$

(5.18)

Again, by allowing $\nu = 0$, we encompass unnormalized gradient and least-squares. Concerning the update law modifications, the same comments from the preceding subsection are also in order here.

   LEMMA 5.3. *Suppose $\Omega : \mathbb{R}_+ \to \mathbb{R}^{n \times p}$ is piecewise continuous, x is bounded, and $\bar{A}(t)$ is continuous and bounded on $[0, t_f)$, and (5.16) is exponentially stable. The update laws (5.17) and (5.18) guarantee that*
   (i) if $\nu = 0$ then $\tilde{\theta} \in \mathcal{L}_\infty[0, t_f)$ and $\epsilon_x \in \mathcal{L}_2[0, t_f)$,
   (ii) if $\nu > 0$ then $\tilde{\theta} \in \mathcal{L}_\infty[0, t_f)$ and $\dot{\theta}, \dfrac{\epsilon_x}{\sqrt{1 + \nu |\Omega|_{\mathcal{F}}^2}} \in \mathcal{L}_2[0, t_f) \cap \mathcal{L}_\infty[0, t_f)$.

   Proof. (Sketch) There exists a continuously differentiable, bounded, positive definite, symmetric $P : \mathbb{R}_+ \to \mathbb{R}^{n \times n}$ such that $\dot{P} + P\bar{A} + \bar{A}^{\mathrm{T}} P = -I$, $\forall t \in [0, t_f)$, and the positive definite function

$$
V_{\tilde{\theta}} = \frac{1}{2} \tilde{\theta}^{\mathrm{T}} \Gamma^{-1} \tilde{\theta} + \frac{1}{4} |\bar{\epsilon}_x|_P^2 \tag{5.19}
$$

can be used as in [6,33,7] to prove the lemma. $\square$

   Now we proceed to prove stability of the x-swapping scheme. With normalized gradient (5.17) and least-squares (5.18) update laws, the proof is similar to the proof of Theorem 5.1. With the unnormalized update laws, it is not clear how to prove boundedness of all signals for an arbitrary exponentially stable $\bar{A}(t)$. We avoid this difficulty by designing

$$
\bar{A}(t) = \bar{A}_0 - \lambda \phi^{\mathrm{T}}(x) \phi(x) P_0 \tag{5.20}
$$

where $\lambda > 0$ and A_0 can be an arbitrary constant matrix that satisfies $P_0 \bar{A}_0 + \bar{A}_0^{\mathrm{T}} P_0 = -I, P_0 = P_0^{\mathrm{T}} > 0$. With this design the matrix $\bar{A}(t)$ is exponentially stable because

$$
P_0 \bar{A}(t) + \bar{A}^{\mathrm{T}}(t) P_0 = -I - 2\lambda P_0 \phi^{\mathrm{T}} \phi P_0 \leq -I. \tag{5.21}
$$

   THEOREM 5.4. (*x-swapping scheme*) *All the signals in the adaptive system consisting of the plant (2.1), controller (3.1), filters (5.13),(5.14), and either the gradient (5.17) or the least-squares (5.18) update law are globally uniformly bounded for all $t \geq 0$, and $\lim_{t \to \infty} z(t) = 0$. This means, in particular, that global asymptotic tracking is achieved:*

$$
\lim_{t \to \infty} [y(t) - y_r(t)] = 0. \tag{5.22}
$$

Furthermore, if $\lim_{t \to \infty} r(t) = 0$ then $\lim_{t \to \infty} x(t) = 0$.

Proof. We first consider the normalized update laws. As in the proof of Theorem 5.1, we show that $\hat{\theta}, \dot{\hat{\theta}}, z, x \in \mathcal{L}_\infty[0, t_f)$ and hence $u \in \mathcal{L}_\infty[0, t_f)$. From (5.13) and (5.14) it follows that Ω_0, Ω, and therefore ϵ_x are in $\mathcal{L}_\infty[0, t_f)$. Now, by the same argument as in the proof of Theorem 5.1 we conclude that $t_f = \infty$.

Second, we consider the unnormalized update laws (5.17) and (5.18) with $\bar{A}(t)$ given by (5.20). Along the solutions of (5.14) we have

$$
(5.23) \quad \begin{aligned}
\frac{d}{dt}\left(\Omega^{\mathrm{T}} P_0 \Omega\right) &= -\Omega^{\mathrm{T}}\Omega - \lambda \Omega^{\mathrm{T}} P_0 \phi^{\mathrm{T}}\phi P_0 \Omega + \Omega^{\mathrm{T}} P_0 \phi^{\mathrm{T}} \\
&= -\Omega^{\mathrm{T}}\Omega - \lambda\left|\phi P_0 \Omega - \frac{1}{\lambda}I_p\right|^2 + \frac{1}{4\lambda}I_p
\end{aligned}
$$

which implies

$$
(5.24) \quad \frac{d}{dt}\left(\mathrm{tr}\left\{\Omega^{\mathrm{T}} P_0 \Omega\right\}\right) \leq -\mathrm{tr}\left\{\Omega^{\mathrm{T}}\Omega\right\} + \frac{p}{4\lambda}.
$$

Hence $\Omega \in \mathcal{L}_\infty[0, t_f)$. Lemma 5.2 gives $\tilde{\theta}, \dot{\hat{\theta}} \in \mathcal{L}_\infty[0, t_f)$, and from (5.15) and (5.16) we conclude that $\epsilon_x \in \mathcal{L}_\infty[0, t_f)$. Now by (5.17) or (5.18), $\dot{\hat{\theta}} \in \mathcal{L}_\infty[0, t_f)$. Therefore, by Lemma 3.1, $z, x \in \mathcal{L}_\infty[0, t_f)$. Finally, by (5.1), $\Omega_0 \in \mathcal{L}_\infty[0, t_f)$. As before, $t_f = \infty$.

Now we set out to prove that $z \in \mathcal{L}_2$. For normalized update laws, from Lemma 5.2, we have that $\dot{\hat{\theta}}, \dfrac{\epsilon_x}{\sqrt{1 + \nu|\Omega|_\mathcal{F}^2}} \in \mathcal{L}_2$. Since $\Omega \in \mathcal{L}_\infty$ then $\epsilon_x \in \mathcal{L}_2$. When the update laws are unnormalized, Lemma 5.2 gives $\epsilon_x \in \mathcal{L}_2$, and since $\Omega \in \mathcal{L}_\infty$ then by (5.17) or (5.18), $\dot{\hat{\theta}} \in \mathcal{L}_2$. Consequently for both the normalized and the unnormalized cases, $\Omega\tilde{\theta} \in \mathcal{L}_2$ because $\tilde{\epsilon}_x \in \mathcal{L}_2$. Now, as in Theorem 5.1, we invoke Lemma 4.2 to deduce that $z - \chi\tilde{\theta} \in \mathcal{L}_2$. In order to show that $z \in \mathcal{L}_2$, we need to prove that $\Omega\tilde{\theta} \in \mathcal{L}_2$ implies $\chi\tilde{\theta} \in \mathcal{L}_2$, or, in the notation of Lemma A.3 from the Appendix, that $T_{\bar{A}}[\phi^{\mathrm{T}}]\tilde{\theta} \in \mathcal{L}_2$ implies $T_{A_z}[W^{\mathrm{T}}]\tilde{\theta} \in \mathcal{L}_2$. To apply this lemma to our adaptive system we note from (3.3) and (3.1) that

$$
(5.25) \quad W^{\mathrm{T}}(z, \hat{\theta}, t) = \begin{bmatrix} 1 & 0 & \cdots & 0 \\ -\frac{\partial \alpha_1}{\partial x_1} & 1 & \cdots & 0 \\ \vdots & \vdots & \ddots & \vdots \\ -\frac{\partial \alpha_{n-1}}{\partial x_1} & -\frac{\partial \alpha_{n-1}}{\partial x_2} & \cdots & 1 \end{bmatrix} \phi^{\mathrm{T}}(x) \stackrel{\triangle}{=} M(z, \hat{\theta}, t)\phi^{\mathrm{T}}(x).
$$

Since $M(z(t), \hat{\theta}(t), t)$ satisfies the conditions of Lemma A.3 then $\chi\tilde{\theta} \in \mathcal{L}_2$ and hence $z \in \mathcal{L}_2$. The rest of the proof is the same as for Theorem 5.1. □

6. Observer-based schemes. We present two observer-based schemes: the z-observer scheme and the x-observer scheme. In the z-observer scheme, the parameter identifier is based on the error system (z), whereas in the x-observer scheme, the parameter identifier is based on the original plant (x). Similar observer-based identifiers have earlier been used in [2,30,39,40,5] and are also known as equation error filtering identifiers.

6.1. z-observer scheme. We implement an "observer" for the error state z of (3.2) by dropping the $\tilde{\theta}$-term, that is,

$$(6.1) \qquad \dot{\hat{z}} = A_z(z,\hat{\theta},t)\hat{z} + D(z,\hat{\theta},t)\dot{\hat{\theta}}.$$

With (6.1) the *observer error*

$$(6.2) \qquad \epsilon = z - \hat{z}$$

is governed by the equation in which the $\tilde{\theta}$-term reappears:

$$(6.3) \qquad \dot{\epsilon} = A_z(z,\hat{\theta},t)\epsilon + W(z,\hat{\theta},t)^{\mathrm{T}}\tilde{\theta}.$$

As the parameter update law we employ

$$(6.4) \qquad \dot{\hat{\theta}} = \Gamma W \epsilon, \qquad \Gamma = \Gamma^{\mathrm{T}} > 0.$$

It is important to note that our closed-loop adaptive system with the controller (3.1) has two equivalent state representations (2.1), (6.1), (6.4) – and – (3.2), (6.3), (6.4).

Now we prove stability of the z-observer scheme. Although the proof can be carried out using input/state arguments, we use a direct Lyapunov analysis. This is because we can prove global uniform stability in the sense of Lyapunov, as opposed to only global uniform boundedness for other schemes in this paper.

THEOREM 6.1. (*z-observer*) *The closed-loop adaptive system consisting of the plant (2.1), controller (3.1), observer (6.1), and update law (6.4), has a globally uniformly stable equilibrium at the origin $z = 0, \epsilon = 0, \tilde{\theta} = 0$, and* $\lim_{t\to\infty} z(t) = \lim_{t\to\infty} \epsilon(t) = 0$. *This means, in particular, that global asymptotic tracking is achieved:*

$$(6.5) \qquad \lim_{t\to\infty} [y(t) - y_{\mathrm{r}}(t)] = 0.$$

Furthermore, if $\lim_{t\to\infty} r(t) = 0$ *then* $\lim_{t\to\infty} x(t) = 0$.

Proof. Starting from the update law (6.4) we obtain the following inequalities:

$$(6.6) \quad |\dot{\hat{\theta}}|^2 \leq \bar{\lambda}(\Gamma)^2 |W\epsilon|^2 = \bar{\lambda}(\Gamma)^2 \left| \sum_{i=1}^{n} w_i \epsilon_i \right|^2 \leq \bar{\lambda}(\Gamma)^2 n \sum_{i=1}^{n} |w_i|^2 \epsilon_i^2 . \cdot$$

We make use of the following constants: $\kappa_m = \min_{1 \le i \le n} \kappa_i$, and $\mu > 0$ to be chosen later. Along the solutions of (6.1), (6.3), (6.4), we have

$$\frac{d}{dt}\left(\frac{\mu}{2}|\hat{z}|^2 + \frac{1}{2}|\epsilon|^2 + \frac{1}{2}|\tilde{\theta}|^2_{\Gamma^{-1}}\right) \le -\mu \sum_{i=1}^{n}\left(c_i + \kappa_i|w_i|^2 + g_i\left|\frac{\partial\alpha_{i-1}}{\partial\hat{\theta}}^{\mathrm{T}}\right|^2\right)$$

$$\hat{z}_i^2 - \mu \sum_{i=1}^{n}\hat{z}_i\frac{\partial\alpha_{i-1}}{\partial\hat{\theta}}\dot{\hat{\theta}}$$

$$-\sum_{i=1}^{n}\left(c_i + \kappa_i|w_i|^2 + g_i\left|\frac{\partial\alpha_{i-1}}{\partial\hat{\theta}}^{\mathrm{T}}\right|^2\right)$$

$$\epsilon_i^2 + \epsilon^{\mathrm{T}}W^{\mathrm{T}}\tilde{\theta} - \tilde{\theta}^{\mathrm{T}}\Gamma^{-1}\dot{\hat{\theta}}$$

$$\le -\mu c_0|\hat{z}|^2 - \mu \sum_{i=1}^{n}g_i\left|\frac{\partial\alpha_{i-1}}{\partial\hat{\theta}}^{\mathrm{T}}\hat{z}_i + \frac{1}{2g_i}\dot{\hat{\theta}}\right|^2 + \frac{\mu}{4g_0}|\dot{\hat{\theta}}|^2$$

$$-c_0|\epsilon|^2 - \kappa_m \sum_{i=1}^{n}|w_i|^2\epsilon_i^2$$

$$\le -\mu c_0|\hat{z}|^2 - c_0|\epsilon|^2 - \left(\kappa_m - \mu\frac{\bar{\lambda}(\Gamma)^2 n}{4g_0}\right)\sum_{i=1}^{n}|w_i|^2\epsilon_i^2.$$

(6.7)

Choosing $\mu < \dfrac{4g_0\kappa_m}{n\bar{\lambda}(\Gamma)^2}$ we get

(6.8) $$\frac{d}{dt}\left(\frac{\mu}{2}|\hat{z}|^2 + \frac{1}{2}|\epsilon|^2 + \frac{1}{2}|\tilde{\theta}|^2_{\Gamma^{-1}}\right) \le -\mu c_0|\hat{z}|^2 - c_0|\epsilon|^2$$

which proves that $z = 0, \epsilon = 0, \tilde{\theta} = 0$ is g.u.s. From LaSalle's invariance theorem, it further follows that $z(t), \epsilon(t) \to 0$ as $t \to \infty$. □

Note that the stability enhancing terms $\kappa_i|w_i|^2$ in the matrix A_z of the observer error equation (6.3) are crucial for counteracting the destabilizing effects of $\dot{\hat{\theta}}$.

6.2. x-observer scheme. For the plant (2.1) rewritten in the form

(6.9) $$\dot{x} = Ex + e_n u + \phi(x)^{\mathrm{T}}\theta$$

where $E = \begin{bmatrix} 0 & & \\ \vdots & I_{n-1} & \\ 0 & \cdots & 0 \end{bmatrix}$ and $\phi^{\mathrm{T}} = \begin{bmatrix} \varphi_1^{\mathrm{T}} \\ \vdots \\ \varphi_n^{\mathrm{T}} \end{bmatrix}$, we implement an "observer"

(6.10) $$\dot{\hat{x}} = \left(\bar{A} - \lambda\phi(x)^{\mathrm{T}}\phi(x)\bar{P}\right)(\hat{x} - x) + Ex + e_n u + \phi(x)^{\mathrm{T}}\hat{\theta}$$

where $\bar{A}$ satisfies $\bar{P}\bar{A} + \bar{A}^{\mathrm{T}}\bar{P} = -qI$, $\bar{P} = \bar{P}^{\mathrm{T}} > 0$, and $\lambda, q > 0$. The *observer error*

(6.11) $$\epsilon_x = x - \hat{x}$$

is governed by

(6.12)
$$\dot{\epsilon}_x = \left(\bar{A} - \lambda \phi(x)^{\mathrm{T}} \phi(x) \bar{P}\right) \epsilon_x + \phi(x)^{\mathrm{T}} \tilde{\theta} \,.$$

The stability enhancing matrix $-\lambda \phi(x)^{\mathrm{T}} \phi(x) \bar{P}$ plays a crucial role in counteracting the destabilizing effect of $\hat{\theta}$. The update law is

(6.13)
$$\dot{\hat{\theta}} = \Gamma \phi \bar{P} \epsilon_x \,, \qquad \Gamma = \Gamma^{\mathrm{T}} > 0 \,.$$

LEMMA 6.2. *If $x \in \mathcal{L}_{\infty e}[0, t_f)$, then the update law (6.13) guarantees that*
(i) $\tilde{\theta} \in \mathcal{L}_{\infty}[0, t_f)$,
(ii) $\epsilon_x \in \mathcal{L}_{\infty}[0, t_f) \cap \mathcal{L}_2[0, t_f)$,
(iii) $\dot{\hat{\theta}} \in \mathcal{L}_2[0, t_f)$.

Proof. (Sketch) Parts (i) and (ii) are standard. (iii) Using (6.12)–(6.13), we have

$$\frac{d}{dt} \left(|\epsilon_x|_{\bar{P}}^2 + |\tilde{\theta}|_{\Gamma^{-1}}^2 \right) \leq -q |\epsilon_x|^2 - \lambda |\phi \bar{P} \epsilon_x|^2 \leq -q |\epsilon_x|^2 - \frac{\lambda}{\bar{\lambda}(\Gamma)^2} |\dot{\hat{\theta}}|^2$$

which implies

$$\|\dot{\hat{\theta}}\|_2 \leq \frac{\bar{\lambda}(\Gamma)}{\sqrt{\lambda}} \left(|\epsilon_x(0)|_{\bar{P}}^2 + |\tilde{\theta}(0)|_{\Gamma^{-1}}^2 \right)^{1/2} \,.$$

$\square$

THEOREM 6.3. (**x-observer**) *All the signals in the closed-loop adaptive system consisting of the plant (2.1), controller (3.1), observer (6.10), and the update law (6.13), are globally uniformly bounded, and $\lim\limits_{t \to \infty} z(t) = \lim\limits_{t \to \infty} \epsilon_x(t) = 0$. This means, in particular, that global asymptotic tracking is achieved:*

(6.14)
$$\lim_{t \to \infty} [y(t) - y_r(t)] = 0 \,.$$

Furthermore, if $\lim\limits_{t \to \infty} r(t) = 0$ then $\lim\limits_{t \to \infty} x(t) = 0$.

Proof. Due to the continuity of $x_m(t)$ and the smoothness of the nonlinearities in (2.1), the solution of the closed-loop adaptive system exists and is unique on a maximum interval of existence $[0, t_f)$. From Lemma 6.1 we have $\tilde{\theta} \in \mathcal{L}_{\infty}[0, t_f)$ and $\dot{\hat{\theta}} \in \mathcal{L}_2[0, t_f)$, which in view of Lemma 3.1 implies that $z \in \mathcal{L}_{\infty}[0, t_f)$. Since all the signals of the closed-loop adaptive system are bounded on $[0, t_f)$ by constants depending only on the initial conditions, then $t_f = \infty$.

To prove convergence of z to zero, we recall first that from Lemma 6.1 that $\epsilon_x, \dot{\hat{\theta}} \in \mathcal{L}_2$. Factoring the regressor matrix W as in (5.25) — $W^{\mathrm{T}}(z, \hat{\theta}, t)$

$= M(z, \hat{\theta}, t)\phi^{\mathrm{T}}(x)$, we consider $\zeta \overset{\triangle}{=} z - M\epsilon_x$ and obtain

$$\dot{\zeta} = A_z(z, \hat{\theta}, t)\zeta + \left[\dot{M} + A_z(z, \hat{\theta}, t)M - M\left(\bar{A} - \lambda\phi^{\mathrm{T}}\phi\bar{P}\right)\right]\epsilon_x + D(z, \hat{\theta}, t)\dot{\hat{\theta}}$$

(6.15)

where $\dot{M} + A_z(z, \hat{\theta}, t)M - M\left(\bar{A} - \lambda\phi^{\mathrm{T}}\phi\bar{P}\right)$ is bounded. It is now straightforward to derive

(6.16)
$$\frac{d}{dt}\left(\tfrac{1}{2}|\zeta|^2\right) \leq -\tfrac{c_0}{2}|\zeta|^2$$
$$+\frac{1}{2c_0}\left|\dot{M} + A_z(z, \hat{\theta}, t)M - M\left(\bar{A} - \lambda\phi^{\mathrm{T}}\phi\bar{P}\right)\right|_2^2 |\epsilon_x|^2 + \frac{1}{4g_0}|\dot{\hat{\theta}}|^2,$$

and since $\epsilon_x, \dot{\hat{\theta}} \in \mathcal{L}_2$, it follows by Lemma A.1 that $\zeta \in \mathcal{L}_2$. Therefore $z \in \mathcal{L}_2$. We recall that $z, \epsilon_x \in \mathcal{L}_\infty$ and note that (3.2) implies $\dot{z} \in \mathcal{L}_\infty$ and (6.12) implies $\dot{\epsilon}_x \in \mathcal{L}_\infty$. Therefore, by Barbalat's lemma $z(t), \epsilon_x(t) \to 0$ as $t \to \infty$. □

REMARK 6.1. The results of the last two sections are presented for nonlinear systems transformable into the parametric-strict-feedback form (2.1) without zero dynamics. As in [13], they can be readily modified for the strict-feedback systems with zero-dynamics

(6.17)
$$\dot{x}_i = x_{i+1} + \theta^{\mathrm{T}}\varphi_i(x_1, \ldots, x_i), \quad 1 \leq i \leq n - 1$$
$$\dot{x}_n = \beta_0(x)u + \theta^{\mathrm{T}}\varphi_n(x)$$
$$\dot{x}^{\mathrm{r}} = \Phi_0(y, x^{\mathrm{r}}) + \Phi(y, x^{\mathrm{r}})\theta$$
$$y = x_1,$$

where the x^{r}-subsystem has a bounded-input bounded-state (BIBS) property with respect to y as its input. The procedure can also be modified, as in [13], to obtain a local result for the systems transformable into the parametric-pure-feedback form in which φ_i also depends on x_{i+1}. A broadening of the class of systems that can be adaptively stabilized using the approach of [13] was suggested in [35].

7. $\mathcal{L}_\infty$, mean-square, and $\mathcal{L}_2$ performance. In this section we will derive $\mathcal{L}_\infty$, mean-square, and $\mathcal{L}_2$ norm bounds for the error state z, which incorporate the bounds for the tracking error $y - y_{\mathrm{r}}$.

7.1. Swapping-based schemes. First we give performance bounds for parameter identifiers and use them to establish $\mathcal{L}_\infty$ and mean-square bounds for z that are valid for both the z-swapping and the x-swapping schemes. Then we derive an $\mathcal{L}_2$ norm bound on z for the z-swapping scheme. For the x-swapping scheme a similar $\mathcal{L}_2$ bound is not yet available.

We analyze the performance of adaptive nonlinear systems which employ identifiers with the normalized gradient update law. We will comment on how to modify the derivations to obtain similar results with the unnormalized gradient, the normalized and the unnormalized least-squares update laws.

Without loss of generality we assume in our analysis, and recommend for implementation, that $\tilde{\epsilon}(0)$, $\chi(0)$, in the z-swapping scheme, and $\tilde{\epsilon}_x(0)$, $\Omega(0)$, in the x-swapping scheme, be set to zero. This can be achieved by initializing $\bar{\chi}(0) = -z(0)$, $\chi(0) = 0$, in the z-swapping scheme, and $\Omega_0(0) = x(0)$, $\Omega(0) = 0$, in the x-swapping scheme. We also let $\Gamma_0 = \gamma I$.

LEMMA 7.1. *For both the z-swapping (5.6) and the x-swapping (5.17) normalized ($\nu > 0$) gradient update laws, the following bounds hold:*

(7.1)   (i) $\|\tilde{\theta}\|_\infty = |\tilde{\theta}(0)|$

(7.2)   (ii) $\|\dot{\hat{\theta}}\|_\infty \leq \dfrac{\gamma}{\nu}|\tilde{\theta}(0)|$

(7.3)   (iii) $\|\dot{\hat{\theta}}\|_2 \leq \sqrt{\dfrac{\gamma}{2\nu}}|\tilde{\theta}(0)|$.

Proof. The proof is given for the z-swapping identifier (5.2),(5.3), (5.6). The proof for the x-swapping identifier (5.13), (5.14),(5.17) is identical.

Consider the positive definite function $V_{\tilde{\theta}} = \dfrac{1}{2\gamma}|\tilde{\theta}|^2$. Its derivative along the solutions of (5.6),(5.4) is

(7.4) $$\dot{V}_{\tilde{\theta}} = -\dfrac{|\epsilon|^2}{1+\nu|\chi|_{\mathcal{F}}^2} \leq 0 .$$

(i) Due to the nonnegativity of $\dot{V}_{\tilde{\theta}}$ we have $V_{\tilde{\theta}}(t) \leq V_{\tilde{\theta}}(0)$ which implies (7.1).

(ii) From (5.6) we can write

(7.5)
$$|\dot{\hat{\theta}}|^2 \leq \gamma^2 \dfrac{\epsilon^{\mathrm{T}}\chi\chi^{\mathrm{T}}\epsilon}{(1+\nu|\chi|_{\mathcal{F}}^2)^2} \leq \gamma^2 \dfrac{|\epsilon|^2\lambda_{\max}(\chi\chi^{\mathrm{T}})}{(1+\nu|\chi|_{\mathcal{F}}^2)^2} \leq \gamma^2 \dfrac{|\epsilon|^2|\chi|_{\mathcal{F}}^2}{(1+\nu|\chi|_{\mathcal{F}}^2)^2} \leq \dfrac{\gamma^2}{\nu} \dfrac{|\epsilon|^2}{1+\nu|\chi|_{\mathcal{F}}^2} .$$

By using (5.4) we get

(7.6) $$|\dot{\hat{\theta}}|^2 \leq \dfrac{\gamma^2}{\nu} \dfrac{\tilde{\theta}^{\mathrm{T}}\chi^{\mathrm{T}}\chi\tilde{\theta}}{1+\nu|\chi|_{\mathcal{F}}^2} \leq \dfrac{\gamma^2}{\nu} \dfrac{|\tilde{\theta}|^2|\chi|_{\mathcal{F}}^2}{1+\nu|\chi|_{\mathcal{F}}^2} \leq \left(\dfrac{\gamma}{\nu}\right)^2 |\tilde{\theta}|^2$$

which, in view of (7.1), proves (7.2).

(iii) By integrating (7.4) over $[0,\infty)$ we obtain

(7.7) $$\left\|\dfrac{\epsilon}{\sqrt{1+\nu|\chi|_{\mathcal{F}}^2}}\right\|_2 \leq \sqrt{V_{\tilde{\theta}}(0)} = \dfrac{1}{\sqrt{2\gamma}}|\tilde{\theta}(0)| .$$

Integration of (7.5) over $[0,\infty)$ and substitution of (7.7) yields (7.3). $\square$

REMARK 7.1. The only difference in the case of the normalized least-squares is that (7.3) becomes $\|\dot{\hat{\theta}}\|_2 \leq \sqrt{\dfrac{\gamma}{\nu}}|\tilde{\theta}(0)|$.

THEOREM 7.2. *In the adaptive system (2.1),(3.1) using either the parameter identifier (5.2), (5.3), (5.6) or (5.13), (5.14), (5.17) with normalized update laws, the following inequalities hold:*

$$(7.8) \quad (i) \quad |z(t)| \le \frac{|\tilde{\theta}(0)|}{2\sqrt{c_0}} \left(\frac{1}{\kappa_0} + \frac{\gamma^2}{g_0\nu^2} \right)^{\frac{1}{2}} + |z(0)|e^{-c_0 t},$$

$$(7.9) \quad (ii) \quad \left(\frac{1}{t} \int_0^t |z(\tau)|^2 d\tau \right)^{\frac{1}{2}} \le \frac{|\tilde{\theta}(0)|}{2\sqrt{c_0}} \left(\frac{1}{\kappa_0} + \frac{1}{t} \frac{\gamma^2}{2g_0\nu^2} \right)^{\frac{1}{2}} + \frac{1}{\sqrt{2c_0}}|z(0)|.$$

Proof. (i) This bound follows by substituting (7.1) and (7.2) into (3.5). *(ii)* By integrating (3.9) we get

$$(7.10) \quad \begin{aligned} \int_0^t |z(\tau)|^2 d\tau \quad &\le \frac{1}{2c_0}|z(0)|^2 + \frac{1}{4c_0\kappa_0}\|\tilde{\theta}\|_\infty^2 t \\ &+ \frac{1}{2g_0} \int_0^t \left(\int_0^\tau e^{-2c_0(\tau-s)}|\dot{\hat{\theta}}(s)|^2 ds \right) d\tau. \end{aligned}$$

Now, to arrive at (7.9), the sequence of integration in (7.10) is interchanged as in the proof of Lemma A.1.*(ii)*. □

REMARK 7.2. Although the initial states $z_2(0), \ldots, z_\rho(0)$ may depend on c_i, κ_i, g_i, this dependence can be removed by setting $z(0) = 0$ with the following initialization of the reference model:

$$x_{m,i}(0) = x_i(0) - \alpha_{i-1}(x_1(0), \ldots, x_{i-1}(0), \hat{\theta}(0), x_{m,1}(0), \ldots, x_{m,i-1}(0)).$$
(7.11)
It can also be proven that in this initialization $x_m(0)$ does not depend on c_i, κ_i, g_i. Therefore, the bounds (7.8),(7.9) can be made as small as desired by a choice of c_0, and/or κ_0, g_0.

LEMMA 7.3. *For the adaptive system (2.1), (3.1), (5.2), (5.3), (5.6), the following inequalities hold:*

$$(7.12) \qquad (i) \qquad \||\chi|_{\mathcal{F}}\|_\infty \le \sqrt{\frac{p}{4c_0\kappa_0}}$$

$$(7.13) \qquad (ii) \qquad \|\epsilon\|_\infty \le \sqrt{\frac{p}{4c_0\kappa_0}}|\tilde{\theta}(0)|$$

$$(7.14) \qquad (iii) \qquad \|\epsilon\|_2 \le \sqrt{\frac{1}{2\gamma} + \frac{\nu}{2\gamma}\frac{p}{4c_0\kappa_0}}|\tilde{\theta}(0)|.$$

Proof. (i) By Lemma A.1*(i)*, and since $\chi(0) = 0$, inequality (5.10) is rewritten as

$$(7.15) \quad |\chi(t)|_{\mathcal{F}}^2 \le |\chi(0)|_{\mathcal{F}}^2 e^{-2c_0 t} + \int_0^t e^{-2c_0(t-\tau)}\frac{p}{2\kappa_0}d\tau \le \frac{p}{4c_0\kappa_0}$$

and (7.12) follows.

(ii) Now (5.4) implies $\|\epsilon\|_\infty \leq \||\chi|_{\mathcal{F}}\|_\infty \|\tilde{\theta}\|_\infty \leq \sqrt{\dfrac{p}{4c_0\kappa_0}}|\tilde{\theta}(0)|$ which proves (7.13).

(iii) The bound on the $\mathcal{L}_2$ norm of ϵ is obtained using

(7.16)
$$\int_0^\infty |\epsilon(\tau)|^2 d\tau \leq \int_0^\infty \frac{|\epsilon(\tau)|^2}{1+\nu|\chi|_{\mathcal{F}}^2}(1+\nu\||\chi|_{\mathcal{F}}\|_\infty^2)d\tau$$
$$\leq (1+\nu\||\chi|_{\mathcal{F}}\|_\infty^2)\left\|\frac{\epsilon}{\sqrt{1+\nu|\chi|_{\mathcal{F}}^2}}\right\|_2^2 .$$

By substituting (7.7) and (7.12) into (7.16) we prove (7.14). $\square$

REMARK 7.3. With the bounds (7.12)–(7.14) for the z-swapping scheme we can tighten the bounds on $\|\dot{\hat{\theta}}\|_2$ and $\|\dot{\hat{\theta}}\|_\infty$ in Lemma 7.1 and make them valid for the unnormalized update laws with $\nu = 0$. It is straightforward to show that $\dfrac{1}{\nu}$ in (7.2)–(7.3) can be replaced by $\min\left\{\dfrac{1}{\nu}, \dfrac{p}{4c_0\kappa_0}\right\}$. The same is true for (7.8)–(7.9). We can also show that for the x-swapping scheme $\dfrac{1}{\nu}$ can be replaced by $\min\left\{\dfrac{1}{\nu}, \dfrac{p\lambda_{\max}(P_0)}{4\lambda\lambda_{\min}(P_0)}\right\}$.

THEOREM 7.4. *In the adaptive system (2.1), (3.1) with the z-swapping identification scheme (5.2), (5.3), (5.6) the $\mathcal{L}_2$ norm of z is bounded by*

$$\|z\|_2 \leq \frac{|\tilde{\theta}(0)|}{2\sqrt{c_0}}\left[\sqrt{\frac{p}{2\kappa_0}}\left(\sqrt{\frac{\gamma}{c_0^2\nu}}+\sqrt{\frac{\nu}{\gamma}}\right)+\sqrt{\frac{\gamma}{g_0\nu}}\right]+\frac{1}{\sqrt{c_0}}|z(0)|+\frac{|\tilde{\theta}(0)|}{\sqrt{2\gamma}} .$$
(7.17)

Proof. We will calculate the $\mathcal{L}_2$ norm bound for z as $\|z\|_2 \leq \|\epsilon\|_2 + \|\psi\|_2$, where

(7.18)
$$\psi \overset{\triangle}{=} \bar{\chi} - \chi\hat{\theta} .$$

A bound on $\|\epsilon\|_2$ is given by (7.14). To obtain a bound on $\|\psi\|_2$, we examine

(7.19)
$$\dot{\psi} = A_z(z,\hat{\theta},t)\psi - D(z,\hat{\theta},t)\dot{\hat{\theta}} - \chi\dot{\hat{\theta}} .$$

By using (3.3) and repeating the sequence of inequalities (3.7), we derive

(7.20)
$$\frac{d}{dt}\left(\frac{1}{2}|\psi|^2\right) \leq -c_0|\psi|^2 + \frac{1}{4g_0}|\dot{\hat{\theta}}|^2 - \psi^T\chi\dot{\hat{\theta}}$$
$$\leq -\frac{c_0}{2}|\psi|^2 - \frac{c_0}{2}\left|\psi - \frac{1}{c_0}\chi\dot{\hat{\theta}}\right|^2 + \frac{1}{4g_0}|\dot{\hat{\theta}}|^2 + \frac{1}{2c_0}|\chi\dot{\hat{\theta}}|^2$$

which gives

(7.21)
$$\frac{d}{dt}\left(\frac{1}{2}|\psi|^2\right) \leq -\frac{c_0}{2}|\psi|^2 + \frac{1}{2}\left(\frac{1}{2g_0}|\dot{\hat{\theta}}|^2 + \frac{1}{c_0}|\chi\dot{\hat{\theta}}|^2\right) .$$

By applying Lemma A.1.*(ii)* to (7.21), we arrive at

$$(7.22) \qquad \|\psi\|_2 \leq \frac{1}{\sqrt{c_0}} \left(|\psi(0)| + \frac{1}{\sqrt{2g_0}} \|\dot{\hat{\theta}}\|_2 + \frac{1}{\sqrt{c_0}} \|\chi\dot{\hat{\theta}}\|_2 \right).$$

By substituting (7.3) and (7.12) into (7.22) we get

$$(7.23) \qquad \begin{aligned} \|\psi\|_2 &\leq \frac{1}{\sqrt{c_0}} \left(|\psi(0)| + \frac{1}{\sqrt{2g_0}} \|\dot{\hat{\theta}}\|_2 + \frac{1}{\sqrt{c_0}} \||\chi|_{\mathcal{F}}\|_\infty \|\dot{\hat{\theta}}\|_2 \right) \\ &\leq \frac{|\tilde{\theta}(0)|}{2\sqrt{c_0}} \sqrt{\frac{\gamma}{\nu}} \left(\frac{1}{\sqrt{g_0}} + \frac{1}{c_0} \sqrt{\frac{p}{2\kappa_0}} \right) + \frac{1}{\sqrt{c_0}} |z(0)| \end{aligned}$$

where we have assumed that $\bar{\chi}(0) = z(0)$. Combining this and (7.14), and rearranging the terms, we obtain (7.17). □

The form of the bound (7.17) is favorable because it is linear in $|\tilde{\theta}(0)|$. It may not be possible to make the $\mathcal{L}_2$ norm of z as small as desired by c_0 alone because of the term $\dfrac{|\tilde{\theta}(0)|}{\sqrt{2\gamma}}$. However, with the standard initialization $z(0) = 0$, a possibility to improve the $\mathcal{L}_2$ performance is by simultaneously increasing c_0, g_0 and γ.

7.2. Observer-based schemes. We give $\mathcal{L}_2$ and $\mathcal{L}_\infty$ performance bounds for the z-observer scheme and comment later on the x-observer scheme. Without loss of generality, we assume that $\hat{z}(0) = z(0)$ and $\Gamma = \gamma I$.

THEOREM 7.5. *In the adaptive system (2.1), (3.1), (6.1), (6.4), the following inequalities hold:*

$$(7.24) \qquad (i) \qquad \|z\|_2 \leq \frac{|\tilde{\theta}(0)|}{\sqrt{2c_0\gamma}} \left(1 + \sqrt{\frac{n\gamma^2}{2g_0\kappa_m}} \right) + \frac{1}{\sqrt{2c_0}} |z(0)|,$$

$$(7.25) \qquad (ii) \qquad |z(t)| \leq \frac{|\tilde{\theta}(0)|}{2\sqrt{c_0\kappa_0}} \left(1 + \sqrt{\frac{2n\gamma^2}{g_0\kappa_m}} \right) + |z(0)|e^{-c_0 t}.$$

Proof. (i) Along the solutions of (6.3)–(6.4), we have

$$(7.26) \qquad \frac{d}{dt} \left(\frac{1}{2}|\epsilon|^2 + \frac{1}{2\gamma}|\tilde{\theta}|^2 \right) \leq -c_0|\epsilon|^2.$$

Since $\epsilon(0) = z(0) - \hat{z}(0) = 0$, this implies that $\|\tilde{\theta}\|_\infty = |\tilde{\theta}(0)|$ and

$$(7.27) \qquad \|\epsilon\|_2 \leq \frac{1}{\sqrt{2c_0\gamma}} |\tilde{\theta}(0)|.$$

Now from (6.8), for $\mu < \dfrac{4g_0\kappa_m}{n\gamma^2}$, we get

$$(7.28) \qquad \mu\|\hat{z}\|_2^2 + \|\epsilon\|_2^2 \leq \frac{1}{c_0} \left(\frac{\mu|\hat{z}(0)|^2 + |\epsilon(0)|^2}{2} + \frac{1}{2\gamma}|\tilde{\theta}(0)|^2 \right),$$

and, since $\hat{z}(0) = z(0)$, then

(7.29) $$\|\hat{z}\|_2 \leq \frac{1}{\sqrt{2c_0\gamma\mu}}|\tilde{\theta}(0)| + \frac{1}{\sqrt{2c_0}}|z(0)| .$$

Letting $\mu = \dfrac{2g_0\kappa_m}{n\gamma^2}$ and adding (7.27) and (7.29) in $\|z\|_2 \leq \|\epsilon\|_2 + \|\hat{z}\|_2$, we arrive at (7.24).

(ii) In a fashion similar to (6.7), we compute

(7.30)
$$
\begin{aligned}
\frac{d}{dt}\left(\frac{\mu|z|^2 + |\epsilon|^2}{2}\right) \leq{}& -c_0(\mu|z|^2 + |\epsilon|^2) - \mu\sum_{i=1}^{n}\kappa_i|w_i|^2 z_i^2 \\
& -\sum_{i=1}^{n}\kappa_i|w_i|^2\epsilon_i^2 \\
& +\mu\sum_{i=1}^{n}z_i w_i^{\mathrm{T}}\tilde{\theta} + \sum_{i=1}^{n}\epsilon_i w_i^{\mathrm{T}}\tilde{\theta} + \frac{\mu}{4g_0}|\dot{\hat{\theta}}|^2 \\
\leq{}& -c_0(\mu|z|^2 + |\epsilon|^2) + \frac{\mu}{4\kappa_0}|\tilde{\theta}|^2 + \frac{1}{2\kappa_0}|\tilde{\theta}|^2 \\
& -\left(\frac{\kappa_m}{2} - \mu\frac{n\gamma^2}{4g_0}\right)\sum_{i=1}^{n}|w_i|^2\epsilon_i^2 .
\end{aligned}
$$

Choosing $\mu = \dfrac{g_0\kappa_m}{n\gamma^2}$ we get

(7.31)
$$\mu|z(t)|^2 + |\epsilon(t)|^2 \leq \left(\mu|z(0)|^2 + |\epsilon(0)|^2\right)e^{-2c_0 t} + \frac{\mu+2}{2\kappa_0}\int_0^t e^{-2c_0(t-\tau)}|\tilde{\theta}(\tau)|^2 d\tau$$

which implies

(7.32) $$|z(t)| \leq \frac{1}{2\sqrt{c_0\kappa_0}}\left(1 + \sqrt{\frac{2}{\mu}}\right)\|\tilde{\theta}\|_\infty + |z(0)|e^{-c_0 t} .$$

The last inequality proves (7.25) and also establishes an ISS property from $\tilde{\theta}$ to z. It is easy to see that $|\epsilon(t)| \leq \frac{1}{2\sqrt{c_0\kappa_0}}\|\tilde{\theta}\|_\infty$ describes the ISS property from $\tilde{\theta}$ to ϵ. $\square$

The bounds (7.24), (7.25) can be made as small as desired by initializing $z(0) = 0$ and increasing c_0.

REMARK 7.4. For the x-observer scheme, an $\mathcal{L}_\infty$ performance bound equal to the one from Theorem 7.3 can be derived as follows. For $\mu > 0$ we readily obtain

(7.33)
$$
\begin{aligned}
\frac{d}{dt}\left(\frac{\mu}{2}|z|^2 + |\epsilon_x|_{\bar{P}}^2\right) \leq{}& -c_0\mu|z|^2 + \frac{\mu}{4\kappa_0}|\tilde{\theta}|^2 \\
& +\frac{\mu}{4g_0}|\dot{\hat{\theta}}|^2 - q|\epsilon_x|^2 - 2\lambda|\phi\bar{P}\epsilon_x|^2 + 2\tilde{\theta}^{\mathrm{T}}\phi\bar{P}\epsilon_x \\
\leq{}& -c_0\mu|z|^2 - q|\epsilon_x|^2 + \frac{\mu}{4\kappa_0}|\tilde{\theta}|^2 \\
& +\frac{1}{\lambda}|\tilde{\theta}|^2 - \left(\lambda - \mu\frac{\bar{\lambda}(\Gamma)^2}{4g_0}\right)|\phi\bar{P}\epsilon_x|^2 .
\end{aligned}
$$

Choosing $\mu < \dfrac{4g_0\lambda}{\lambda(\Gamma)^2}$ we get

$$(7.34) \quad \frac{d}{dt}\left(\frac{\mu}{2}|z|^2 + |\epsilon_x|_{\tilde{P}}^2\right) \leq -c_0\mu|z|^2 - q|\epsilon_x|^2 + \left(\frac{\mu}{4\kappa_0} + \frac{1}{\lambda}\right)|\tilde{\theta}|^2 .$$

In the case $\bar{A} = -c_0 I, \bar{P} = \frac{1}{2}I, q = c_0, \lambda = 2\kappa_0$, and (w.l.o.g.) $\hat{x}(0) = x(0), \Gamma = 2\gamma I$, by proceeding from (7.34), as in the proof of Theorem 7.3, we get

$$(7.35) \qquad |z(t)| \leq \frac{|\tilde{\theta}(0)|}{2\sqrt{c_0\kappa_0}}\left(1 + \sqrt{\frac{2n\gamma^2}{g_0\kappa_m}}\right) + |z(0)|e^{-c_0 t} .$$

It is not clear, however, how to derive a useful $\mathcal{L}_2$ tracking performance bound similar to (7.25).

8. Example. Let us illustrate the new estimation-based designs on the relative-degree two plant:

$$(8.1) \qquad \begin{aligned} \dot{x}_1 &= x_2 + \theta\varphi(x_1), \qquad \varphi(x_1) = x_1^2 \\ \dot{x}_2 &= u \end{aligned}$$

with the objective to regulate x to zero (without a reference model).

Design. We define the error variables:

$$(8.2) \qquad \begin{aligned} z_1 &= x_1 \\ z_2 &= x_2 - \alpha_1(x_1, \hat{\theta}) . \end{aligned}$$

The two-step controller design

$$\begin{aligned} \alpha_1 &= -c_1 z_1 - \kappa_1\varphi^2 z_1 - \hat{\theta}\varphi \\ u &= -z_1 - c_2 z_2 - \kappa_2\left(\frac{\partial\alpha_1}{\partial x_1}\right)^2\varphi^2 z_2 - g_2\left(\frac{\partial\alpha_1}{\partial\hat{\theta}}\right)^2 z_2 + \frac{\partial\alpha_1}{\partial x_1}(x_2 + \hat{\theta}\varphi) \end{aligned}$$

(8.3)
results in the error system

$$(8.4) \qquad \begin{aligned} \dot{z} &= \begin{bmatrix} -c_1 - \kappa_1\varphi^2 & 1 \\ -1 & -c_2 - \kappa_2\left(\dfrac{\partial\alpha_1}{\partial x_1}\right)^2\varphi^2 - g_2\left(\dfrac{\partial\alpha_1}{\partial\hat{\theta}}\right)^2 \end{bmatrix} z \\ &+ \begin{bmatrix} \varphi \\ -\dfrac{\partial\alpha_1}{\partial x_1}\varphi \end{bmatrix}\tilde{\theta} + \begin{bmatrix} 0 \\ -\dfrac{\partial\alpha_1}{\partial\hat{\theta}} \end{bmatrix}\dot{\hat{\theta}} . \end{aligned}$$

The z-swapping identifier is designed with the following filters:

$$(8.5) \quad \dot{\bar{\chi}} = \begin{bmatrix} -c_1 - \kappa_1\varphi^2 & 1 \\ -1 & -c_2 - \kappa_2\left(\dfrac{\partial\alpha_1}{\partial x_1}\right)^2\varphi^2 - g_2\left(\dfrac{\partial\alpha_1}{\partial\hat\theta}\right)^2 \end{bmatrix}$$
$$\bar{\chi} + \begin{bmatrix} \varphi \\ -\dfrac{\partial\alpha_1}{\partial x_1}\varphi \end{bmatrix}\hat\theta - \begin{bmatrix} 0 \\ -\dfrac{\partial\alpha_1}{\partial\hat\theta} \end{bmatrix}\dot{\hat\theta}$$

$$(8.6) \quad \dot{\chi} = \begin{bmatrix} -c_1 - \kappa_1\varphi^2 & 1 \\ -1 & -c_2 - \kappa_2\left(\dfrac{\partial\alpha_1}{\partial x_1}\right)^2\varphi^2 - g_2\left(\dfrac{\partial\alpha_1}{\partial\hat\theta}\right)^2 \end{bmatrix}$$
$$\chi + \begin{bmatrix} \varphi \\ -\dfrac{\partial\alpha_1}{\partial x_1}\varphi \end{bmatrix},$$

which are used to implement the augmented error:

$$(8.7) \qquad \epsilon = z + \bar\chi - \chi\hat\theta,$$

and the gradient update law:

$$(8.8) \qquad \dot{\hat\theta} = \gamma\frac{\chi^{\mathrm{T}}\epsilon}{1 + \nu\chi^{\mathrm{T}}\chi}.$$

The x-swapping identifier is designed with $\bar A(t) = -(\bar a + \lambda\varphi^2)I$, $\bar a, \lambda > 0$, and the following filters:

$$(8.9) \qquad \dot\Omega_0 = -(\bar a + \lambda\varphi^2)(\Omega_0 - x) + \begin{bmatrix} x_2 \\ u \end{bmatrix}, \qquad \Omega_0 \in \mathbb{R}^2$$

$$(8.10) \qquad \dot\Omega = -(\bar a + \lambda\varphi^2)\Omega + \varphi, \qquad \Omega \in \mathbb{R},$$

which are used to implement the equation error:

$$(8.11) \qquad \epsilon_x = x - \Omega_0 - \begin{bmatrix} \Omega \\ 0 \end{bmatrix}\hat\theta,$$

and the gradient update law:

$$(8.12) \qquad \dot{\hat\theta} = \gamma\frac{\Omega^{\mathrm{T}}\epsilon_{x,1}}{1 + \nu\Omega^2}.$$

The z-observer identifier is designed with the following observer:

$$\dot{\hat z} = \begin{bmatrix} -c_1 - \kappa_1\varphi^2 & 1 \\ -1 & -c_2 - \kappa_2\left(\dfrac{\partial\alpha_1}{\partial x_1}\right)^2\varphi^2 - g_2\left(\dfrac{\partial\alpha_1}{\partial\hat\theta}\right)^2 \end{bmatrix}\hat z + \begin{bmatrix} 0 \\ -\dfrac{\partial\alpha_1}{\partial\hat\theta} \end{bmatrix}\dot{\hat\theta}$$
$$(8.13)$$

which is used to implement the observer error:

$$(8.14) \qquad\qquad \epsilon = z - \hat{z},$$

and the unnormalized gradient update law:

$$(8.15) \qquad\qquad \dot{\hat{\theta}} = \gamma\varphi \left[1, \ -\frac{\partial \alpha_1}{\partial x_1} \right] \epsilon.$$

The x-observer identifier is designed with the following observer:

$$(8.16) \qquad\qquad \dot{\hat{x}}_1 = -(\bar{a} + \lambda\varphi^2)(\hat{x}_1 - x_1) + x_2 + \varphi\hat{\theta}$$

which is used to implement the equation error:

$$(8.17) \qquad\qquad \epsilon_{x,1} = x_1 - \hat{x}_1,$$

and the unnormalized gradient update law:

$$(8.18) \qquad\qquad \dot{\hat{\theta}} = \gamma\varphi\epsilon_{x,1}.$$

Equations (8.9)–(8.10) and (8.16) reveal that the x-schemes are uncertainty specific in the sense that only the terms φ_i multiplying the unknown parameter θ need to be filtered. This opens a possibility for a reduction in the dynamic order.

In simulations, the only difference between the schemes was in the value of γ needed to achieve the same speed of adaptation. Therefore we choose to show results only for the z-swapping scheme.

Simulations. Simulations were carried out with nominal values $c_1 = c_2 = c_0 = \kappa_1 = \kappa_2 = \kappa_0 = g_2 = g_0 = 1$, $\gamma = 10$, $\theta = 5$, $\hat{\theta}(0) = 0$ which were judged to give representative responses. All simulations are with following initial conditions: $x_1(0) = -\bar{\chi}_1(0) = 0$, $x_2(0) = -\bar{\chi}_2(0) = 10$ (to set $\tilde{\epsilon}(0) = 0$), $\chi(0) = 0$.

Figure 8.1.a) illustrates Theorems 7.1, 7.2 and 7.3. The design parameter c_0 can be used for systematically improving the transient performance. However, Fig. 8.1.a) also illustrates that the error transients and the control effort are decreasing simultaneously as c_0 increases, up to some point, beyond which the control effort starts increasing. The control u is given in an expanded time scale in order to clearly display the main qualitative differences among the three cases. Figure 8.1.b) illustrates Corollary 3.1. When adaptation is switched off, the states are uniformly bounded and converge to (or are confined inside) a compact residual set. Corollary 3.1 does not describe the behavior inside the residual set, which may contain multiple equilibria, limit cycles, etc. For this example, there is always an equilibrium at the origin. While, for small values of c_0, the equilibrium at

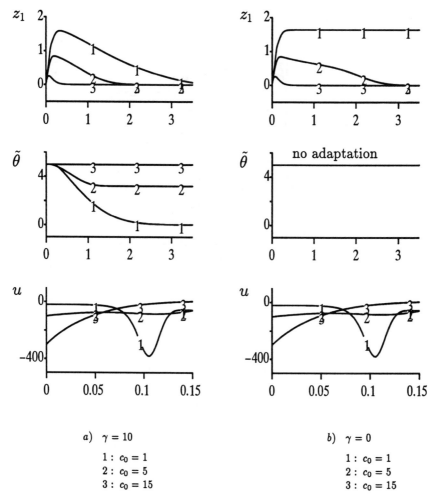

FIG. 8.1. *Dependence of the transients on c_0 with $\kappa_0 = g_0 = 1$. (Note an expanded time scale for control u.)*

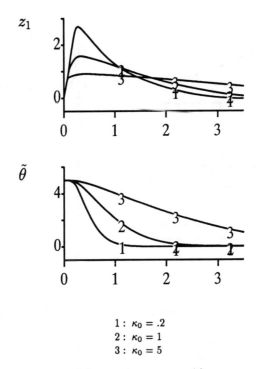

1 : $\kappa_0 = .2$
2 : $\kappa_0 = 1$
3 : $\kappa_0 = 5$

FIG. 8.2. *Dependence of the transients on* κ_0 *with* $c_0 = g_0 = 1$, $\gamma = 10$.

the origin is only locally stable (and the state z converges to another equilibrium), for higher values of c_0 it becomes globally asymptotically stable, which may be confirmed by analysis.

Figure 8.2 shows the influence of κ_0 on transients. According to Theorems 7.1(i) and 7.3(ii) and Remark 7.3, the peak values can be decreased by increasing κ_0, which is confirmed by the plot. However, as it may be expected from Theorems 7.2 and 7.3(i), the $\mathcal{L}_2$ performance may not be improved by increasing κ_0 for the following reason. By severely (through nonlinear gains) penalizing the peaks in z, we reduce $\dot{\hat{\theta}}$, so that the terms $-\kappa_1 \varphi^2 z_1$ and $-\kappa_2 \left(\frac{\partial \alpha_1}{\partial x_1}\right)^2 \varphi^2 z_2$ act as a sort of normalization of the update law. The effect of the term $-g_2 \left(\frac{\partial \alpha_1}{\partial \hat{\theta}}\right)^2 z_2$ was shown to be significant only for very small c_0 and κ_0 or for very large γ and therefore it is not discussed here.

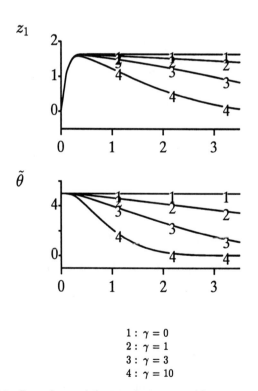

$$1 : \gamma = 0$$
$$2 : \gamma = 1$$
$$3 : \gamma = 3$$
$$4 : \gamma = 10$$

FIG. 8.3. *Dependence of the transients on γ with $c_0 = g_0 = \kappa_0 = 1$.*

Figure 8.3 demonstrates the influence of the adaptation gain γ on transients. Due to the slow initial adaptation, which should be attributed not only to the normalized gradient update law but also to the fact that the regressor is filtered, there is a clear separation of the action of the nonadaptive controller, which at the beginning brings the state z quickly to the residual set, and the adaptive controller which takes over to drive the state to the origin. The property that the $\mathcal{L}_\infty$ bounds are increasing functions of γ, to be expected from Theorems 7.1 and 7.3, was not exhibited in simulations except with extremely high values of γ. This indicates that some of the bounds derived may not be very tight over the entire range of design parameter values.

9. Conclusions. The proposed adaptive nonlinear control design is modular in the sense that the controller and identifier can be separately designed. In the nonlinear swapping-based approach, various adaptive schemes can be designed with a wide variety of update laws: gradient and least-squares, normalized and unnormalized. The observer-based designs offer a possibility to both reduce the dynamic order of the swapping-based schemes and the complexity of the tuning functions design.

We have shown analytically and illustrated by an example that the proposed adaptive schemes provide possibilities for systematic improvement of $\mathcal{L}_\infty$, mean-square and $\mathcal{L}_2$ performance.

A. Appendix

LEMMA A.1. *Let* $v, \rho : \mathbb{R}_+ \to \mathbb{R}$, $c, b > 0$. *If*

$$(A.1) \qquad \dot{v} \le -cv + b\rho^2, \qquad v(0) \ge 0$$

(i) then

$$(A.2) \qquad v(t) \le v(0)e^{-ct} + b\int_0^t e^{-c(t-\tau)}\rho(\tau)^2 d\tau .$$

(ii) If, in addition, $\rho \in \mathcal{L}_2$, *then* $v \in \mathcal{L}_\infty \cap \mathcal{L}_1$ *and*

$$(A.3) \qquad \|v\|_1 \le \frac{1}{c}\left(v(0) + b\|\rho\|_2^2\right) .$$

Proof. (i) Upon multiplication of (A.1) by e^{ct}, it becomes

$$(A.4) \qquad \frac{d}{dt}\left(v(t)e^{ct}\right) \le b\rho(t)^2 e^{ct} .$$

Integrating (A.4) over $[0, t]$, we arrive at (A.2).

(ii) Noting that (A.2) implies that

$$(A.5) \qquad v(t) \le v(0)e^{-ct} + b \sup_{\tau \in [0,t]}\left\{e^{-c(t-\tau)}\right\}\int_0^t \rho(\tau)^2 d\tau$$

we conclude that $v \in \mathcal{L}_\infty$. By integrating (A.2) over $[0,t]$, we get

(A.6)
$$
\begin{aligned}
\int_0^t v(\tau)d\tau &\leq \int_0^t v(0)e^{-c\tau}d\tau + b\int_0^t \left[\int_0^\tau e^{-c(\tau-s)}\rho(s)^2 ds\right]d\tau \\
&\leq \frac{1}{c}v(0) + b\int_0^t e^{cs}\rho(s)^2 \left(\int_s^t e^{-c\tau}d\tau\right)ds \\
&\leq \frac{1}{c}v(0) + b\int_0^t e^{cs}\rho(s)^2 \frac{1}{c}e^{-cs}ds \\
&\leq \frac{1}{c}\left[v(0) + b\int_0^t \rho(\tau)^2 d\tau\right]
\end{aligned}
$$

which proves (A.3). □

LEMMA A.2. (**Nonlinear Damping** – $\mathcal{L}_\infty$ **and** $\mathcal{L}_2$ **stabilization.**) *Assume that for the system*

(A.7) $\dot{x} = f(x,t) + g(x,t)\left[u + p(x,t)^{\mathrm{T}} d(t)\right]$, $x \in \mathbb{R}^n, u \in \mathbb{R}^m$

a feedback control $u = \mu(x,t)$ *guarantees*

(A.8) $\dfrac{\partial V}{\partial x}\left[f(x,t) + g(x,t)\mu(x,t)\right] + \dfrac{\partial V}{\partial t} \leq -U(x,t)$, $\forall x \in \mathbb{R}^n, \forall t \geq 0$

where $V, U : \mathbb{R}^n \times \mathbb{R}_+ \to \mathbb{R}_+$ *are positive definite and radially unbounded and* V *is decrescent,* $f : \mathbb{R}^n \times \mathbb{R}_+ \to \mathbb{R}^n$, $g : \mathbb{R}^n \times \mathbb{R}_+ \to \mathbb{R}^{n\times m}$, $p : \mathbb{R}^n \times \mathbb{R}_+ \to \mathbb{R}^{p\times m}$, $\mu : \mathbb{R}^n \times \mathbb{R}_+ \to \mathbb{R}^m$ *are continuously differentiable in* x *and piecewise continuous in* t, *and* $d : \mathbb{R}_+ \to \mathbb{R}^{p\times m}$ *is piecewise continuous. Then the feedback control*

(A.9) $u = \mu(x,t) - p(x,t)^{\mathrm{T}}\Lambda(t)p(x,t)\left[\dfrac{\partial V}{\partial x}(x,t)g(x,t)\right]^{\mathrm{T}}$,

for any continuous $\Lambda : \mathbb{R}_+ \to \mathbb{R}^{p\times p}$, $\Lambda(t) = \Lambda^{\mathrm{T}}(t) \geq \lambda I > 0$, *guarantees that:*

(i) If $d \in \mathcal{L}_\infty$ *then* $x \in \mathcal{L}_\infty$.

(ii) If $d \in \mathcal{L}_2$ *and* $\alpha_1|x|^2 \leq V(x,t) \leq \alpha_2|x|^2$, $U(x,t) \geq \alpha_3|x|^2$, $\forall x \in \mathbb{R}^n, \forall t \geq 0$, $\alpha_1, \alpha_2, \alpha_3 > 0$, *then* $x \in \mathcal{L}_2 \cap \mathcal{L}_\infty$. *If, in addition,* $d \in \mathcal{L}_\infty$ *then* $x(t) \to 0$ *as* $t \to \infty$.

Proof. (i) Due to (A.8), the derivative of V along (A.7)–(A.9) is

(A.10)
$$
\begin{aligned}
\dot{V} &= \frac{\partial V}{\partial x}\left\{f + g\mu + g\left[-p^{\mathrm{T}}\Lambda p\left(\frac{\partial V}{\partial x}g\right)^{\mathrm{T}} + p^{\mathrm{T}}d\right]\right\} + \frac{\partial V}{\partial t} \\
&\leq -U - \left|p\left(\frac{\partial V}{\partial x}g\right)^{\mathrm{T}} - \frac{1}{2}\Lambda^{-1}d\right|_\Lambda^2 + \frac{1}{4}|d|_{\Lambda^{-1}}^2 \\
&\leq -U + \frac{1}{4\lambda}|d|^2
\end{aligned}
$$

and, hence $x \in \mathcal{L}_\infty$.

(ii) Since $U(x,t) \geq \alpha_3|x|^2 \geq \frac{\alpha_3}{\alpha_2}V$, then from (A.10) we obtain

$$(A.11) \qquad \dot{V} \leq -\frac{\alpha_3}{\alpha_2}V + \frac{1}{4\lambda}|d|^2.$$

By applying Lemma A.1*(ii)*, it follows that $x \in \mathcal{L}_2 \cap \mathcal{L}_\infty$, and if $d \in \mathcal{L}_2 \cap \mathcal{L}_\infty$, then also $\dot{x} \in \mathcal{L}_\infty$. Thus, by Barbalat's lemma, $x(t) \to 0$ as $t \to \infty$. □

LEMMA A.3. *Let $T_i : u \mapsto \zeta_i$, $i = 1,2$ be linear time-varying operators defined by*

$$(A.12) \qquad \dot{\zeta}_i = A_i(t)\zeta_i + u$$

where $A_i : \mathbb{R}_+ \to \mathbb{R}^{n \times n}$ are continuous, bounded and exponentially stable. Suppose $\tilde{\theta} : \mathbb{R}_+ \to \mathbb{R}^p$ is differentiable, $\phi : \mathbb{R}_+ \to \mathbb{R}^{p \times m}$ is piecewise continuous and bounded, and $M : \mathbb{R}_+ \to \mathbb{R}^{n \times n}$ is bounded and has a bounded derivative on $\mathbb{R}_+$. If $\dot{\tilde{\theta}} \in \mathcal{L}_2$ then

$$(A.13) \qquad T_1[\phi^{\mathrm{T}}]\tilde{\theta} \in \mathcal{L}_2 \;\Rightarrow\; T_2[M\phi^{\mathrm{T}}]\tilde{\theta} \in \mathcal{L}_2.$$

If moreover, $M(t)$ is nonsingular $\forall t$, and M^{-1} is bounded and has a bounded derivative on $\mathbb{R}_+$ then (A.13) holds in both directions.

Proof. Suppose that $T_1[\phi^{\mathrm{T}}]\tilde{\theta} \in \mathcal{L}_2$. By Lemma 4.2, $T_1[\phi^{\mathrm{T}}\tilde{\theta}] - T_1[\phi^{\mathrm{T}}]\tilde{\theta} \in \mathcal{L}_2$ and therefore $\zeta_1 \overset{\triangle}{=} T_1[\phi^{\mathrm{T}}\tilde{\theta}] \in \mathcal{L}_2$. We will show first that $\zeta_2 \overset{\triangle}{=} T_2[M\phi^{\mathrm{T}}\tilde{\theta}] \in \mathcal{L}_2$. By substituting $\phi^{\mathrm{T}}\tilde{\theta} = \dot{\zeta}_1 - A_1(t)\zeta_1$ into the variation of constants formula and applying partial integration we calculate:

$$
\begin{aligned}
\zeta_2(t) &= \Phi_2(t,0)\zeta_2(0) + \int_0^t \Phi_2(t,\tau)M(\tau)\phi^{\mathrm{T}}(\tau)\tilde{\theta}(\tau)d\tau \\
(A.14) \quad &= \Phi_2(t,0)\zeta_2(0) + \int_0^t \Phi_2(t,\tau)M(\tau)\left[\dot{\zeta}_1(\tau) - A_1(\tau)\zeta_1(\tau)\right]d\tau \\
&= \Phi_2(t,0)\zeta_2(0) + M(t)\zeta_1(t) - \Phi_2(t,0)M(0)\zeta_1(0) \\
&\quad + \int_0^t \Phi_2(t,\tau)\left[\dot{M}(\tau) + A_2(\tau)M(\tau) - M(\tau)A_1(\tau)\right]\zeta_1(\tau)d\tau
\end{aligned}
$$

where $\Phi_2(t,\tau)$ is the state transition matrix of $A_2(t)$ that satisfies $|\Phi_2(t,\tau)|_2 \leq ke^{-\alpha(t-\tau)}$, $k,\alpha > 0$. It is clear that $\Phi_2(t,0)\zeta_2(0) + M(t)\zeta_1(t) - \Phi_2(t,0)M(0)\zeta_1(0) \in \mathcal{L}_2$ because $\Phi_2(t,0)$ is exponentially decaying, $M(t)$ is bounded and $\zeta_1 \in \mathcal{L}_2$. Since

$$
\begin{aligned}
(A.15) \quad &\left|\int_0^t \Phi_2(t,\tau)\left[\dot{M}(\tau) + A_2(\tau)M(\tau) - M(\tau)A_1(\tau)\right]\zeta_1(\tau)d\tau\right|^2 \\
&\leq \|\dot{M} + A_2M - MA_1\|_\infty^2 k^2 \int_0^t e^{-2\alpha(t-\tau)}|\zeta_1(\tau)|^2 d\tau
\end{aligned}
$$

then similarly to (4.17)–(4.18) from the proof of Lemma 4.2, we can show that the expression (A.15) is in $\mathcal{L}_2$. Thus $\zeta_2 = T_2[M\phi^{\mathrm{T}}\tilde{\theta}] \in \mathcal{L}_2$. By Lemma 4.2, $T_2[M\phi^{\mathrm{T}}\tilde{\theta}] - T_2[M\phi^{\mathrm{T}}]\tilde{\theta} \in \mathcal{L}_2$ and therefore $T_2[M\phi^{\mathrm{T}}\tilde{\theta}] \in \mathcal{L}_2$. The proof of the other direction of (A.13) when $M(t)$ is nonsingular $\forall t$, and M^{-1} is bounded and has a bounded derivative on $\mathbb{R}_+$ is identical. □

REFERENCES

[1] G. BASTIN AND G. CAMPION, "Indirect adaptive control of linearly parametrized nonlinear systems," *Proceedings of the 3rd IFAC Symposium on Adaptive Systems in Control, and Signal Processing*, Glasgow, UK, 1989.

[2] G. CAMPION AND G. BASTIN, "Indirect adaptive state-feedback control of linearly parametrized nonlinear systems," *International Journal of Adaptive Control and Signal Processing*, vol. 4, 1990, pp. 345–358.

[3] A. DATTA AND P. IOANNOU, "Performance improvement versus robust stability in model reference adaptive control," *Proceedings of the 30th IEEE Conference on Decision and Control*, Brighton, UK, 1991, pp. 748–753.

[4] C. A. DESOER, AND M. VIDYASAGAR, *Feedback Systems: Input–Output Properties*, New York: Academic Press, 1975.

[5] R. GHANADAN AND G. L. BLANKENSHIP, "Adaptive control of nonlinear systems via approximate linearization," Report ISR TR93–23, Institute for Systems Research, University of Maryland, presented at the IMA Period of Concentration on Nonlinear Feedback Design.

[6] G. C. GOODWIN AND D. Q. MAYNE, "A parameter estimation perspective of continuous time model reference adaptive control," *Automatica*, vol. 23, 1987, pp. 57-70.

[7] P. A. IOANNOU AND J. SUN, *Stable and Robust Adaptive Control*, in preparation.

[8] A. ISIDORI, *Nonlinear Control Systems*, Berlin: Springer-Verlag, 1989.

[9] Z. P. JIANG AND L. PRALY, "Iterative designs of adaptive controllers for systems with nonlinear integrators," *Proceedings of the 30th IEEE Conference on Decision and Control*, Brighton, UK, December 1991, pp. 2482–2487.

[10] I. KANELLAKOPOULOS, P. V. KOKOTOVIĆ, AND R. H. MIDDLETON, "Observer-based adaptive control of nonlinear systems under matching conditions," *Proceedings of the 1990 American Control Conference*, San Diego, CA, pp. 549–552.

[11] I. KANELLAKOPOULOS, P. V. KOKOTOVIĆ, AND R. H. MIDDLETON, "Indirect adaptive output-feedback control of a class of nonlinear systems," *Proceedings of the 29th IEEE Conference on Decision and Control*, Honolulu, HI, December 1990, pp. 2714–2719 .

[12] I. KANELLAKOPOULOS, P. V. KOKOTOVIĆ, AND R. MARINO, "An extended direct scheme for robust adaptive nonlinear control," *Automatica*, vol. 27, 1991, pp. 247–255.

[13] I. KANELLAKOPOULOS, P. V. KOKOTOVIĆ, AND A. S. MORSE, "Systematic design of adaptive controllers for feedback linearizable systems," *IEEE Transactions on Automatic Control*, vol. 36, 1991, pp. 1241–1253.

[14] I. KANELLAKOPOULOS, P. V. KOKOTOVIĆ, AND A. S. MORSE, "Adaptive output-feedback control of systems with output nonlinearities," pp. 495–525 in [18].

[15] I. KANELLAKOPOULOS, P. V. KOKOTOVIĆ, AND A. S. MORSE, "Adaptive output-feedback control of a class of nonlinear systems," *Proceedings of the 30th IEEE Conference on Decision and Control*, Brighton, UK, December 1991, pp. 1082–1087.

[16] I. KANELLAKOPOULOS, P. V. KOKOTOVIĆ, AND A. S. MORSE, "A toolkit for nonlinear feedback design," *Systems & Control Letters*, vol. 18, 1992, pp. 83–92.

[17] I. KANELLAKOPOULOS, "Passive adaptive control of nonlinear systems," *International Journal of Adaptive Control and Signal Processing*, to appear, 1993.

[18] P. V. KOKOTOVIĆ, Ed., *Foundations of Adaptive Control*, Berlin: Springer-Verlag, 1991.

[19] P. V. KOKOTOVIĆ, I. KANELLAKOPOULOS, AND A. S. MORSE, "Adaptive feedback linearization of nonlinear systems," pp. 311–346 in [18].

[20] M. KRSTIĆ, I. KANELLAKOPOULOS, AND P. V. KOKOTOVIĆ, "Adaptive nonlinear control without overparametrization," *Systems & Control Letters*, vol. 19, 1992, pp. 177–185.

[21] M. KRSTIĆ, I. KANELLAKOPOULOS AND P. V. KOKOTOVIĆ, "A new generation of

adaptive controllers for linear systems," *Proceedings of the 31st IEEE Conference on Decision and Control*, Tucson, AZ, December 1992, pp. 3644-3651.

[22] M. KRSTIĆ, P. V. KOKOTOVIĆ AND I. KANELLAKOPOULOS, "Transient performance improvement with a new class of adaptive controllers," *Systems & Control Letters*, vol. 21, 1993, pp. 451–461.

[23] R. MARINO AND P. TOMEI, "Global adaptive observers for nonlinear systems via filtered transformations," *IEEE Transactions on Automatic Control*, vol. 37, 1992, pp. 1239–1245.

[24] R. MARINO AND P. TOMEI, "Global adaptive observers and output-feedback stabilization for a class of nonlinear systems," in *Foundations of Adaptive Control*, pp. 455–493 in [18].

[25] R. MARINO AND P. TOMEI, "Global adaptive output-feedback control of nonlinear systems, Part I: linear parametrization," *IEEE Transactions on Automatic Control*, vol. 38, 1993, pp. 17–32.

[26] R. MARINO AND P. TOMEI, "Global adaptive output-feedback control of nonlinear systems, Part II: nonlinear parametrization," *IEEE Transactions on Automatic Control*, vol. 38, 1993, pp. 33–49.

[27] A. S. MORSE, "Global stability of parameter-adaptive control systems," *IEEE Transactions on Automatic Control*, vol. 25, 1980, pp. 433–439.

[28] K. NAM AND A. ARAPOSTRATHIS, "A model-reference adaptive control scheme for pure-feedback nonlinear systems," *IEEE Transactions on Automatic Control*, vol. 33, 1988, pp. 803–811.

[29] K. S. NARENDRA AND A. M. ANNASWAMY, *Stable Adaptive Systems*, Englewood Cliffs, NJ: Prentice-Hall, 1989.

[30] L. PRALY, G. BASTIN, J.-B. POMET AND Z. P. JIANG, "Adaptive stabilization of nonlinear systems," pp. 347–434 in [18].

[31] J. B. POMET AND L. PRALY, "Indirect adaptive nonlinear control," *Proceedings of the 27th IEEE Conference on Decision and Control*, Austin, TX, December 1988, pp. 2414–2415.

[32] J. B. POMET AND L. PRALY, "Adaptive nonlinear regulation: estimation from the Lyapunov equation," *IEEE Transactions on Automatic Control*, vol. 37, 1992, pp. 729–740.

[33] S. S. SASTRY AND M. BODSON, *Adaptive Control: Stability, Convergence and Robustness*, Englewood Cliffs, NJ: Prentice-Hall, 1989.

[34] S. S. SASTRY AND A. ISIDORI, "Adaptive control of linearizable systems," *IEEE Transactions on Automatic Control*, vol. 34, 1989, pp. 1123–1131.

[35] D. SETO, A. M. ANNASWAMY AND J. BAILLIEUL, "Adaptive control of a class of nonlinear systems with a triangular structure," *Proceedings of the 31st IEEE Conference on Decision and Control*, Tucson, AZ, December 1992, pp. 278–283.

[36] E. D. SONTAG, "Smooth stabilization implies coprime factorization," *IEEE Transactions on Automatic Control*, vol. 34, 1989, pp. 435–443.

[37] E. D. SONTAG, "Input/output and state-space stability," in *New Trends in System Theory*, G. Conte et al., Eds., Boston: Birkhäuser, 1991.

[38] D. TAYLOR, P. V. KOKOTOVIĆ, R. MARINO AND I. KANELLAKOPOULOS, "Adaptive regulation of nonlinear systems with unmodeled dynamics," *IEEE Transactions on Automatic Control*, vol. 34, 1991, pp. 405–412.

[39] A. R. TEEL, R. R. KADIYALA, P. V. KOKOTOVIĆ AND S. S. SASTRY, "Indirect techniques for adaptive input-output linearization of non-linear systems," *International Journal of Control*, vol. 53, 1991, pp. 193–222.

[40] A. R. TEEL, "Error-based adaptive non-linear control and regions of feasibility," *International Journal of Adaptive Control and Signal Processing*, vol. 6, 1992, pp. 319–327.

AN ADAPTIVE CONTROLLER INSPIRED BY RECENT RESULTS ON LEARNING FROM EXPERTS*

P.R. KUMAR[†]

Abstract. In computational learning theory there have been some interesting developments recently on the problem of "learning from experts."

In this paper, we "adapt" the learning problem to an adaptive control formulation. What results is an adaptive controller which is reminiscent of a certainty equivalence scheme using the "posterior mean" for the parameter estimator. We show that this scheme can be analyzed in a somewhat novel way, for ideal linear systems. The analysis techniques may be of some interest to researchers in the theory of adaptive control.

Key words. Adaptive Control, Learning Theory

1. Introduction. Recently, Littlestone and Warmuth [3] and Cesa-Bianchi, et al. [1] have addressed the interesting problem of learning from experts.

Here we show how these results may be "adapted" to an adaptive control framework. Briefly, we regard each parameter vector θ as giving an "expert prediction" of the next value of the output. Over time, we acquire more confidence in some "experts" and less in others. We adopt the learning scheme from [3] and [1] to fashion a parameter estimator. Inspired by the techniques there, we also provide a somewhat novel analysis of our adaptive controller, which may be of interest in its own right to those interested in the theory of adaptive control, e.g., techniques to establish stability and other asymptotic properties.

2. System description. Consider a standard "ideal" linear system,

$$y(t) = \phi^T(t-1)\theta^\circ$$

where

$$\phi(t-1) := (y(t-1), \ldots, y(t-p), u(t-1), \ldots, u(t-p))^T,$$

and

$$\theta^\circ := (a_1, \ldots, a_p, b_1, \ldots, b_p)^T.$$

---

* This research was conducted while the author was visiting the IMA in Winter/Spring 1993. The author wishes to express his appreciation for the extremely warm hospitality in an extremely cold climate. The research reported here was supported in part by the U.S. Army Research Office under Contract Nos. DAAL-03-91-G-0182 and DA/DAAH04-93-G-0197, by the National Science Foundation under Grant No. ECS-92-16487, and by the Joint Services Electronics Program under Contract No. N00014-90-J1270.

† University of Illinois, Department of Electrical and Computer Engineering and the Coordinated Science Lab, 1308 West Main Street, Urbana, IL 61801, USA.

Here $u(t)$ and $y(t)$ are, respectively, the input and output to the system.

We assume that θ° is in the interior of Θ, a closed sphere of unit volume, centered at the origin. (These assumptions can be generalized somewhat). We also suppose that the system is of strictly minimum phase.

Except for these assumptions, we assume that the parameter vector θ° is unknown. Our goal is to adaptively control the system in such a way that

$$\lim_{t \to \infty} y(t) = 0,$$

while $u(t)$ is kept bounded.

3. A New adaptive controller. Let $n(t-1) = 1 + 2\max\{1, \sup_\Theta \|\theta\|\}\|\phi(t-1)\|$ be a "normalization" signal. Note that

$$\frac{\|\phi(t-1)\|}{n(t-1)} \leq \frac{1}{2} \quad \text{and} \quad \frac{|y(t)|}{n(t-1)} \leq \frac{1}{2}.$$

Let $0 < \mu < 1$. Set

(3.1) $$q(0, \theta) \equiv 1, \qquad \text{for all } \theta \in \Theta,$$

and recursively define,

$$q(t, \theta) = F(t)q(t-1, \theta),$$

where $F(t)$ only needs to satisfy

(3.2) $$(1-\mu)^{\frac{|y(t) - \phi^T(t-1)\theta|}{n(t-1)}} \leq F(t) \leq 1 - \mu \frac{|y(t) - \phi^T(t-1)\theta|}{n(t-1)}.$$

(We note that $F(t)$ is allowed to depend on past measurements, and at each t should be chosen to satisfy the bounds given above). Intuitively, one can think of $q(t, \theta)$ as our "confidence," at time t, that the value of θ° is θ. Note, however, that $q(t, \cdot)$ is "unnormalized" since $\int_\Theta q(t, \theta) d\theta$ need not be 1; hence $q(t, \theta)$ can be regarded as an "unnormalized" density function.

Two examples of confidence updating schemes which satisfy (3.2) are given below.

Example 1. Consider the system,

$$y(t) = \phi^T(t-1)\theta^\circ + w(t),$$

where $\theta^\circ \sim U(\Theta)$, i.e., uniformly distributed over Θ. The term $w(t)$ represents an additive noise. Assume that $\{w(t)\}$ is a sequence of independent random variables with density,

$$
\begin{aligned}
p_{w(t)}(w) &= 1 - \mu\frac{|w|}{n(t-1)}, \quad \text{for } \frac{-1 + \sqrt{1 - \mu/n}}{\mu} \leq \frac{w}{n(t-1)} \\
&\leq \frac{1 - \sqrt{1 - \mu/n}}{\mu}, \\
&= 0, \qquad\qquad\qquad \text{otherwise.}
\end{aligned}
$$

Then the unnormalized posterior density for θ° is given by,

$$q(t, \theta) = q(t - 1, \theta) - \mu q(t - 1, \theta) \frac{|y(t) - \phi^T(t - 1)\theta|}{n(t - 1)}.$$

This is the expression for the upper bound in (3.2). Thus, the "confidence" $q(t, \theta)$ has a Bayesian interpretation. □

Example 2. Consider the same system as in Example 1, except that

$$p_{w(t)}(w) = \frac{n(t - 1)(1 - \mu)^{\frac{|w|}{n(t-1)}}}{-2\ln(1 - \mu)}.$$

Then

$$q(t, \theta) = q(t - 1, \theta)(1 - \mu)^{\frac{|y(t) - \phi^T(t-1)\theta|}{n(t-1)}}$$

is the recursion for the unnormalized posterior density. It corresponds to the lower bound in (3.2). □

Let us define

(3.3) $$\hat{\theta}(t) := \frac{\int_\Theta \theta q(t, \theta) d\theta}{\int_\Theta q(t, \theta) d\theta}.$$

It can be regarded as the "mean value" of our confidence distribution.

We will adopt a "certainty equivalent" approach, and apply a control $u(t)$ which results in

(3.4) $$\phi^T(t)\hat{\theta}(t) = 0.$$

(For simplicity, we suppose that this is always feasible, i.e., $\hat{b}_1(t) \neq 0$). This corresponds to a deadbeat control.

4. The analysis. We will first develop some properties of the parameter estimator, without invoking the form of the control applied.

Define

$$s(t) := \int_\Theta q(t, \theta) d\theta.$$

This is the normalizing factor in (3.3). Note that

$$s(0) \equiv 1$$

from (3.1), since Θ has unit volume.

Lemma 1

$$\sum_{t=1}^{T} \frac{|y(t) - \phi^T(t - 1)\hat{\theta}(t - 1)|}{n(t - 1)} \leq -\frac{1}{\mu}\ln s(T).$$

Proof. From (3.2),

$$q(t, \theta) \le q(t - 1, \theta) \left[1 - \mu \frac{|y(t) - \phi^T(t - 1)\theta|}{n(t - 1)} \right].$$

Hence

$$\begin{aligned}
s(t) &= \int_{\Theta} q(t, \theta) \mathrm{d}\theta \\
&\le \int_{\Theta} \left[1 - \mu \frac{|y(t) - \phi^T(t - 1)\theta|}{n(t - 1)} \right] q(t - 1, \theta) \mathrm{d}\theta \\
&= s(t - 1) - \int_{\Theta} \mu \frac{|y(t) - \phi^T(t - 1)\theta|}{n(t - 1)} q(t - 1, \theta) \mathrm{d}\theta \\
&\le s(t - 1) - \left| \int_{\Theta} \frac{\mu(y(t) - \phi^T(t - 1))}{n(t - 1)} q(t - 1, \theta) \mathrm{d}\theta \right| \\
&= s(t-1) - \left| \mu \frac{y(t)}{n(t-1)} s(t-1) - \mu \frac{\phi^T(t-1)\hat{\theta}(t-1)}{n(t - 1)} s(t-1) \right| \quad \text{(from (3.3))} \\
&= s(t - 1) - \mu \frac{s(t - 1)}{n(t - 1)} \left| y(t) - \phi^T(t - 1)\hat{\theta}(t - 1) \right| \\
&= \left[1 - \mu \frac{|y(t) - \phi^T(t - 1)\hat{\theta}(t - 1)|}{n(t - 1)} \right] s(t - 1).
\end{aligned}$$

So,

$$s(T) \le s(0) \prod_{t=1}^{T} \left[1 - \mu \frac{|y(t) - \phi^T(t - 1)\hat{\theta}(t - 1)|}{n(t - 1)} \right].$$

Taking logarithms, we obtain

$$\ln \frac{s(0)}{s(T)} \ge \sum_{t=1}^{T} -\ln \left[1 - \mu \frac{|y(t) - \phi^T(t - 1)\hat{\theta}(t - 1)|}{n(t - 1)} \right].$$

Noting $s(0) = 1$, and using $\ln(1 - x) \le -x$, yields the desired result. $\quad \square$

Lemma 2. *Let $S(r, \theta^*)$ denote a small closed sphere of radius r centered at θ^*. Assume $S(r, \theta^*) \subseteq \Theta$. Then, for some constant c,*

$$\sum_{t=1}^{T} \frac{|y(t) - \phi^T(t - 1)\hat{\theta}(t - 1)|}{n(t - 1)}$$

$$\le \frac{1}{\mu} \left[-\ln(cr^{2p}) + r \ln \left(\frac{1}{1 - \mu} \right) \sum_{t=1}^{T} \frac{\|\phi(t - 1)\|}{n(t - 1)} \right.$$

$$\left. + \ln \left(\frac{1}{1 - \mu} \right) \sum_{t=1}^{T} \frac{|y(t) - \phi^T(t - 1)\theta^*|}{n(t - 1)} \right].$$

Proof. Clearly,

$$\max_{\theta \in S(r,\theta^*)} \sum_{t=1}^{T} \frac{|y(t) - \phi^T(t-1)\theta|}{n(t-1)} \le \sum_{t=1}^{T} \frac{|y(t) - \phi^T(t-1)\theta^*|}{n(t-1)} + r \sum_{t=1}^{T} \frac{\|\phi(t-1)\|}{n(t-1)}.$$

Now,

$$
\begin{aligned}
s(T) &= \int_{\Theta} q(T,\theta)\mathrm{d}\theta \\
&\ge \int_{S(r,\theta^*)} q(T,\theta)\mathrm{d}\theta \\
&\ge \int_{S(r,\theta^*)} q(0,\theta)(1-\mu)^{\sum_{t=1}^{T} \frac{|y(t)-\phi^T(t-1)\theta|}{n(t-1)}}\mathrm{d}\theta \quad \text{(from (3.2))} \\
&\ge cr^{2p}(1-\mu)^{[\sum_{t=1}^{T} \frac{|y(t)-\phi^T(t-1)\theta^*|}{n(t-1)} + r\sum_{t=1}^{T} \frac{\|\phi(t-1)\|}{n(t-1)}]}.
\end{aligned}
$$

Above, cr^{2p} is the volume of $S(r,\theta^*)$. Taking logarithms and using Lemma 1 yields the result. $\qquad\square$

Clearly, $\theta^\circ \in \arg\min_\Theta \sum_{t=1}^{T} \frac{|y(t)-\phi^T(t-1)\theta|}{n(t-1)}$; in fact $\sum_{t=1}^{T} \frac{|y(t)-\phi^T(t-1)\theta^\circ|}{n(t-1)} = 0$. Hence, by choosing $\theta^* = \theta^\circ$, for small enough r,

$$
\begin{aligned}
(4.1) \quad \sum_{t=1}^{T} \frac{|y(t)-\phi^T(t-1)\hat\theta(t-1)|}{n(t-1)} &\le \frac{1}{\mu}\left[-\ln(cr^{2p}) + r\ln\left(\frac{1}{1-\mu}\right)\sum_{t=1}^{T}\frac{\|\phi(t-1)\|}{n(t-1)}\right] \\
&= \frac{1}{\mu}\left[-\ln c - 2p\ln r + r\ln\left(\frac{1}{1-\mu}\right)\sum_{t=1}^{T}\frac{\|\phi(t-1)\|}{n(t-1)}\right].
\end{aligned}
$$

Now note that for $r = \frac{2p}{x}$,

$$-2p\ln r + rx = 2p + 2p\ln\left(\frac{x}{2p}\right).$$

Hence, with $x = \ln\left(\frac{1}{1-\mu}\right)\sum_{t=1}^{T}\frac{\|\phi(t-1)\|}{n(t-1)}$, we obtain from (4.1),

$$(4.2) \quad \sum_{t=1}^{T} \frac{|y(t)-\phi^T(t-1)\hat\theta(t-1)|}{n(t-1)} \le c_1 + c_2 \ln\sum_{t=1}^{T}\frac{\|\phi(t-1)\|}{n(t-1)}.$$

(Above, we are only treating the case $\sum_{t=1}^{\infty}\frac{\|\phi(t-1)\|}{n(t-1)} = +\infty$, for otherwise $\phi(t) \to 0$, and we are done).

From (4.2), by using the strict minimum phase property of the system, and the Key Technical Lemma from Goodwin and Sin [2], it is easy to conclude that the adaptive controller gives $y(t) \to 0$, while keeping signals bounded.

5. Concluding remarks. We have provided a somewhat novel method of analysis, for an adaptive controller that is not too different from traditional adaptive controllers. This method of analysis may be of interest to others.

REFERENCES

[1] NICCOLO CESA-BIANCHI, YOAV FREUD, DAVID P. HELMBOLD, DAVID HAUSSLER, ROBERT E. SCHAPIRE AND MANFRED K. WARMUTH, *How to use expert advice*, Technical report, Universita di Milano, UC Santa Cruz and AT&T Bell Labs, 1992.
[2] G. C. GOODWIN AND K. S. SIN, *Adaptive Filtering, Prediction and Control.* Prentice-Hall, Englewood Cliffs, NJ, 1984.
[3] N. LITTLESTONE AND MANFRED K. WARMUTH, *Weighted majority learning*, Technical Report UCSC-CRL-91-28, University of California, Santa Cruz, Santa Cruz, CA 95064, October 1992. Baskin Center for Computer Engineering and Information Sciences.

STOCHASTIC APPROXIMATION WITH AVERAGING AND FEEDBACK: FASTER CONVERGENCE

HAROLD J. KUSHNER[‡][*] AND JICHUAN YANG[‡][†]

Abstract. Consider the stochastic approximation algorithm

(∗) $$X_{n+1} = X_n + a_n g(X_n, \xi_n).$$

The problem of selecting the gain or step size sequences a_n has been a serious handicap in applications. In a fundamental paper, Polyak and Juditsky [17] showed that (loosely speaking) if the coefficients a_n go to zero slower than $O(1/n)$, then the averaged sequence $\sum_{i=1}^{n} X_i/n$ converged to its limit at an optimum rate, for any coefficient sequence. This result implies that we should use "larger" than usual" gains, and let the off line averaging take care of the increased noise effects, with substantial overall improvement. Here we give a simpler proof under weaker conditions. Basically, it is shown that the averaging works whenever there is a "classical" rate of convergence theorem. I.e., results of this type are generic to stochastic approximation. Intuitive insight is provided by relating the behavior to that of a two time scale discrete algorithm. The value of the method has been supported by simulations. Since the averaged estimate is "off line," it is not the actual value used in the SA iteration (∗) itself. We show how the averaged value can be partially fed back into the actual operating algorithm for improved performance. Numerical data are presented to support the theoretical conclusions. An error in the tightness part of the proof in [14] is corrected.

1. Introduction. We will discuss improved methods for selecting the step sizes or gain sequence for the stochastic approximation (SA) process

(1.1) $$X_{n+1} = X_n + a_n g(X_n, \xi_n),$$

following the basic idea in [17]. As usual, $0 < a_n \to 0, \sum_n a_n = \infty$. The $\{\xi_n\}$ is a "driving noise" sequence, either exogeneous or state dependent. The procedure (1.1) has been of interest since the early 1950's, but the selection of good gain sequences has never been satisfactorally resolved. The usual idea is to select the gains so that some appropriate measure of rate of convergence is maximized.

Suppose that there is a vector θ such that $X_n \to \theta$ either with probability one or in probability. Then, under appropriate conditions $(X_n - \theta)/\sqrt{a_n}$ converges in distribution to a normally distributed random variable with mean zero and some positive definite covariance matrix V_0. The matrix V_0 is often considered to be a measure of the "rate of convergence," taken together with the scale factors or gains $\{a_n\}$.

Suppose that $a_n \to 0$ "slower" than $O(1/n)$. In particular, suppose that

(1.2) $$a_n/a_{n+1} = 1 + o(a_n).$$

---

[*] Supported by AFOSR Contract F49620-92-0081 and NSF grant ECS-8913351.
[†] Supported by AFOSR Contract F49620-92-0081.
[‡] Division of Applied Mathematics, Brown University, Providence, R.I. 02912.

Define

$$(1.3) \qquad \overline{X}_n = \frac{1}{n} \sum_1^n X_i.$$

Then, in one of the more interesting developments in SA in many years, [17] showed that $\sqrt{n}(\overline{X}_n - \theta)$ converged in distribution to a normally distributed random variable with mean zero and covariance V, where V was the smallest possible in some sense (to be defined in Section 2). Under (1.2), the value of V did not depend on the particular sequence $\{a_n\}$. This weakening of the requirements on $\{a_n\}$ has important implications for applications. It says that the gains should be "relatively large." (Recall that a sequence of the form $a_n = O(1/n)$ was the "classical" one recommended for the "tail" of the procedure, although it was well known that such a sequence led to poor "finite time" behavior.) Indeed, in an asymptotic sense, one cannot do better than (1.3) even if the gains a_n in (1.1) were matrix valued. Simulations by many people have now supported the theoretical conclusions Keep in mind that this advantage would not hold if a_n decreased as $O(1/n)$.

These developments have simplified the problem of choosing a_n. The results of Ruppert [18] for a one dimensional case contain conclusions similar to those in [16], [17]. Further work on this problem occurs in [14,13,19,20,4]. This paper presents somewhat simpler proofs under weaker conditions. We attempt to outline the essential points. The use of and advantages to feedback of the averages into the original algorithm is also discussed.

The proofs in [17] did not use previously known results in SA. They essentially started from 'the beginning', and much detail was required. The conditions were stronger than necessary, and the basic underlying reasons for success were not really evident.

It was shown in [14] that a straightforward application of known results in SA yield results of the above type under fairly general conditions, and we will discuss those results here. First, in Section 2 it will be shown that a useful averaging result can be readily obtained directly from "classical" rate of convergence results. This result uses a "minimal window" of averaging, smaller than in (1.3). The "window of averaging" is extended to the maximal one in Section 4. There was an incomplete point in the tightness part of the proof in [14]. In particular, the assertion below (5) in the Appendix is not always correct for the range of indices indicated there. This is corrected here by a related proof and under slightly weaker conditions.

It is noted in [14] that the success of the averaging idea is due to the fact that the "time scales" of the original sequence X_n and the averaged sequence $\bar{X}_n$ are "separated." The time scale of the former sequence being the "faster" one. Of course, time scale separation arguments have been used in the analysis of SA's from the very beginning. But, the separation of the scales of the X_n and the noise ξ_n sequences was the issue. Now there is an additional time scale to be considered, that of the $\bar{X}_n$-sequence. This idea is discussed in Section 3.

The averaging method is an "off line" procedure in the sense that the iteration (1.1) is not influenced by the $\bar{X}_n$. In many cases, it is X_n which is of prime interest since that is the "operating" or physical parameter. We cannot simply substitute $\bar{X}_n$ for X_n either on the right side of (1.1) or in the dynamical term $g(\cdot)$ function alone without effectively ruining the procedure. Thus, there is still the question of how we might be able to exploit the averaging to improve the behavior of the $\{X_n\}$. This issue is dealt with in [13] and is also discussed in Section 5. Due to problems in the stability analysis, we deal with "linear" algorithms of the type appearing in parameter estimators and adaptive noise cancellers.

Data is presented in Section 6, and shows the advantages of both averaging and feedback.

2. A rate of convergence theorem for the averaged iterates: minimal window of averaging. Define the "interpolated time scale "
$$t_n = \sum_0^{n-1} a_i.$$
Without loss of generality, we set $\theta = 0$, unless noted otherwise. For each $n \geq 0$, define the interpolations $X^n(\cdot)$ and $U^n(\cdot)$ by

$$\left. \begin{array}{l} X^n(t) = X_{n+i} \\[2mm] U^n(t) = X_{n+i}/\sqrt{a_{n+i}} \end{array} \right\} \text{ for } t \in [t_{n+i} - t_n, t_{n+i+1} - t_n), \ i \geq 0.$$

The above is referred to as the a_n scale. We will also work with a $1/n$ scale. Since $a_n n \to \infty$, the interpolations in the $1/n$ scale are "squeezed" versions of those in the a_n scale. The symbol $\Rightarrow$ is used to denote weak convergence in the Skorohod topology on the path space $D^r[0, \infty)$ of functions which are $\mathbb{R}^r$-valued, right continuous, and have left hand limits [2,5]. The aim of Theorem 2.1 is to show that the basic averaging idea works (for the minimal window of averaging) under essentially any conditions which guarantee the classical asymptotic normality of the (suitably normalized) SA. There is a vast literature on the subject. In order to reduce the problem to the classical one, we use the following condition.

A2.1. *There is a matrix G whose eigenvalues lie in the open left half plane and a positive definite symmetric matrix R_0 such that $X^n(\cdot) \Rightarrow$ zero process and $U^n(\cdot) \Rightarrow U(\cdot)$, where $U(\cdot)$ is the stationary solution to*

$$(2.1) \qquad\qquad dU = GU\,dt + R_0^{\frac{1}{2}}\,dw.$$

Comment. It is useful to state the condition in the given form since there is large literature which gives various sets of conditions guaranteeing (A2.1). See, e.g., [10,1,6,8,15].

For $T > 0$, define the normalized average $Z^n(\cdot)$ by

(2.2)
$$Z^n(T) = \frac{1}{\sqrt{T/a_n}} \sum_{i=n}^{n+T/a_n} X_i.$$

In sums of the type $\sum_\alpha^\beta$ for real α, β, we always use the integer parts of α, β. In (2.2), the window of averaging is T/a_n for arbitrary real T. This is less than $O(n)$, and indeed will define the minimal window for which an averaging result can be obtained. The value of T can be made as large as desired in (2.2), and can go to infinity slowly with n. Two sided averages can also be used instead of the one sided average in (2.2).

Constant gain coefficients. Consider the constant gain algorithm

(2.3) $X_{n+1}^\varepsilon = X_n^\varepsilon + \varepsilon g(X_n^\varepsilon, \xi_n), \quad \varepsilon > 0.$

Define $X^\varepsilon(\cdot)$ and $U^\varepsilon(\cdot)$ by

$$X^\varepsilon(t) = X_n^\varepsilon, \ U^\varepsilon(t) = X_n^\varepsilon / \sqrt{\varepsilon}$$

on $[n\varepsilon, n\varepsilon + \varepsilon)$. Again, we wish to get a result for an averaged sequence by exploiting known results in SA. Therefore, we suppose that there are $t_\varepsilon \xrightarrow{\varepsilon} \infty$ such that $U^\varepsilon(t_\varepsilon + \cdot) \Rightarrow U(\cdot)$, where $U(\cdot)$ is a stationary process which satisfies (2.1). See [6,8]. The $t_\varepsilon \to \infty$ accounts for the transient period. Let $t_i \geq 0, T = t_1 + t_2$ and define the normalized sequence of averages

$$Z^\varepsilon(t) = \sqrt{\frac{\varepsilon}{T}} \sum_{(t_\varepsilon - t_2)/\varepsilon}^{(t_\varepsilon + t_1)/\varepsilon} X_i.$$

A proof very close to that of Theorem 2.1 gives the same conclusion; namely, that for each T, $Z^\varepsilon(T)$ converges in distribution to a normally distributed random variable with mean zero and covariance $V + O(1/T)$.

In (2.2), the "window" of the averaging is $O(1/a_n)$. By, Theorem 2.1 we see that $O(1/a_n)$ is the smallest window which can be used. Consider the case where $a_n = 1/n^\gamma, \gamma \in (0, 1)$. Then as $\gamma \to 0$, the minimal window size decreases. Roughly, for smaller rates of decrease of $\{a_n\}$, the iterates $\{X_n\}$ jump about the limit point much more, and thus less averaging is needed.

The "minimal window" convergence theorem for the averages

Theorem 2.1. *Assume (1.2) and (A2.1) and define $V = G^{-1} R_0 (G')^{-1}$. For each t, $Z^n(t)$ converges in distribution to a random variable with mean zero and covariance $V_t = V + O(1/t)$.*

Proof. Only the essential details will be given. See [14]. Define the processes

$$\tilde{Z}^n(t) = \frac{1}{\sqrt{t}} \int_0^t U^n(s)ds, \quad \tilde{Z}(t) = \frac{1}{\sqrt{t}} \int_0^t U(s)ds.$$

By the weak convergence in (A2.1), $\tilde{Z}^n(\cdot) \Rightarrow \tilde{Z}(\cdot)$. Define the covariance matrix $R(s) = EU(t)U'(t+s)$, where $U(\cdot)$ is the stationary solution to (2.1). Since $R(s) \to 0$ exponentially as $s \to \infty$, we can write

(2.4)
$$\text{Cov } \tilde{Z}(t) = \frac{1}{t} \int_0^t \int_0^t R(s-\tau)dsd\tau$$
$$= \int_{-\infty}^{\infty} R(s)ds + O(1/t).$$

But $\int_{-\infty}^{\infty} R(s)ds = G^{-1}R_0(G^{-1})'$.

The basic result on the averaged iterates is obtained by relating $Z^n(t)$ to $\tilde{Z}^n(t)$. (1.2) implies that for any $t < \infty$,

(2.5)
$$\max\{i - n : 0 \le t_i - t_n \le t\} \cdot a_n/t \xrightarrow{n} 1.$$

The relation (2.5) would not hold if $a_n = O(1/n)$. By the definition of $U^n(\cdot)$, for $i \ge n$

$$\sqrt{t}\tilde{Z}^n(T) = \sum_{i:t_i - t_n \le T} (X_i a_i^{-1/2})a_i$$
$$= \sum_{i:t_i - t_n \le T} X_i(a_i^{1/2} - a_n^{1/2}) + a_n^{1/2} \sum_{i:t_i - t_n \le T} X_i.$$

By the weak convergence of $U^n(\cdot)$ in (A2.1) and the use of (2.5), the first sum goes to zero in probability as $n \to \infty$. Similarly, the second sum converges in distibution to $\tilde{Z}(T)$, which yields the desired conclusion. $\square$

The optimality of the "rate of convergence" of the sequence of averages. Let us suppose that the normalized error defined by $U_n = X_n/\sqrt{a_n}$ converges in distribution to a normally distributed random variable $\hat{U}$ with mean zero, as essentially asserted by (A2.1). The covariance of $\sqrt{a_n}\hat{U}$, taken together with the gain sequence a_n, is a traditional measure of the rate of convergence of X_n to the limit zero. In this sense, the best (asymptotic) value of a_n is $O(1/n)$. Pursuing this value, let $a_n = A/n$, where A is a positive definite matrix. To get the best asymptotic rate, one now needs to get the best A. Under appropriate conditions, It is a classical result [10,1] that $U^n(\cdot) \Rightarrow \hat{U}(\cdot)$, where $\hat{U}(\cdot)$ is the stationary solution to

(2.6)
$$d\tilde{U} = \left(\frac{I}{2} + AG\right)\tilde{U}dt + AR_0^{\frac{1}{2}}dw.$$

Here, both G and R_0 are as defined in (A2.1). In addition, it is obviously required that the matrix $(\frac{I}{2} + AG)$ is stable, a condition that is not needed for the averaging result to hold. By minimizing the trace of the covariance matrix of (2.6) over A, as in [17] we get the best value of the matrix A to be

$$A = -G^{-1}.$$

With this value of A, the covariance of $\tilde{U}(0)$ is just the V defined in Theorem 2.1. In this sense, the rate of convergence of the sequence of averages in [16,17] and of Theorem 2.1 is optimal. Note that we need not use a matrix valued gain in (1.1).

3. A two time scale point of view. A key to understanding why the averaging works when (1.2) holds, but not when $a_n = O(1/n)$, can be seen by rewriting the two recursions (1.1), (1.3) in the same time scale. The discussion will be motivational only. Thus, in order not to encumber the notation, we work with a linear one dimensional model. Let $a_n = A/n^\gamma$, where $A > 0$ and $\gamma \in (0,1)$. For $G < 0$, let (1.1) take the special form

$$X_{n+1} = \left(1 + \frac{AG}{n^\gamma}\right) X_n + \frac{A\xi_n}{n^\gamma}.$$

Define the normalized (multiplying the average $\bar{X}_n$ by $\sqrt{n}$) average $\overline{U}_n = \sum_1^n X_i/\sqrt{n}$. The algorithm for the pair $(X_n, \overline{U}_n)$ can be written in the form

$$(3.1) \qquad \frac{1}{n^{1-\gamma}}(X_{n+1} - X_n) = \frac{AGX_n}{n} + \frac{A\xi_n}{n},$$

$$(3.2) \qquad \overline{U}_{n+1} - \overline{U}_n = -\frac{\overline{U}_n}{2n}(1 + O(\frac{1}{n})) + \frac{X_{n+1}}{\sqrt{n+1}}.$$

Due to the presence of the $1/n^{1-\gamma}$ factor, it is seen that the algorithm (3.1), (3.2) can be viewed as a two time scale or singularly perturbed SA, although the time scale of the first component is "time varying." If $\gamma = 1$, then the scales of the $\{X_n\}$ and $\{U_n\}$ are the same and the driving term X_n in the U–equation has a correlation which is of the order of that of U_n itself. If $\gamma < 1$, then the correlation of the driving X_n process gets shorter and shorter (as seen from the point of view of the $1/n$ scale used in (3.1),(3.2)) as $n \to \infty$, and the X_n behaves more and more like a "white noise." Then, the form of (3.2) leads us to expect that the averaging will yield the optimal result.

The scheme of (3.1), (3.2) loosely resembles the continuous time two time scale system

$$(3.3) \qquad \begin{aligned} \varepsilon dz^\varepsilon &= A_{11} z^\varepsilon \, dt + dw_1 \\ dx^\varepsilon &= A_{22} x^\varepsilon \, dt + A_{12} z^\varepsilon \, dt + dw_2, \end{aligned}$$

for a small parameter ε. It is shown in [9], that (under suitable stability conditions) $\int_0^t z^\varepsilon(s)ds$ converges in the weak sense to a Wiener process. This result and the resemblance of (3.3) to the form (3.1), (3.2) suggest that the function defined by

$$\frac{1}{\sqrt{n}} \sum_{i=0}^{nt} X_i$$

might converge weakly to a Wiener process with covariance matrix V. This would be the extension of Theorem 2.2 to the maximal window case. This situation is dealt with in the next section.

4. Maximal window of averaging in (2.2). In this section, we will let window of averaging go to infinity faster than allowed in (2.2). Let $q_n \leq n$ be a sequence of integers that goes to infinity as $n \to \infty$. Define the normalized average $M^n(t)$ by

$$(4.1) \qquad M^n(t) = \frac{1}{\sqrt{q_n}} \sum_{i=n}^{n+q_n t} X_i.$$

In the sum (4.1) one could let the indices be symmetric about n or be below n with the same results. In order to extend the "window of averaging" beyond the range t/a_n, we need $q_n a_n \to \infty$.

Theorem 2.1 used $q_n = O(1/a_n)$. In this section we suppose that

$$(\mathbf{A4.1}) \qquad q_n a_n^2 \to 0, \; q_n a_n \to \infty, q_n \leq k_0 n \text{ for some positive } k_0.$$

Thus if $q_n = n$ and $a_n = 1/n^\gamma$, then $\gamma \in (1/2, 1)$ is needed. It is not entirely clear why γ is limited to $(1/2, 1)$, but the same restriction appears in the works of the other authors. We will show that $M^n(\cdot) \Rightarrow W(\cdot)$, a Wiener process with covariance matrix Vt, which is the assertion made at the end of the last section.

Comments concerning the noise processes. The conditions stated below are intended to cover many types of the noise processes which have been used in the SA literature. The processes can be roughly divided into two classes. In the first class the sequence $\{\xi_n\}$ is an "exogenous" noise processes. Here, the statistical evolution of $\{X_n\}$ does not affect $\{\xi_n\}$. The second class is that of "state dependent" noise processes. Here, the pair (X_n, ξ_n) is jointly Markov and is usually defined as follows. The value of the state X_{n+1} in terms of the last value X_n and the noise ξ_n is given by (1.1). In addition, the noise evolves by postulating a transition function $p(\xi, \cdot | x)$ such that $P\{\xi_{n+1} \in \cdot | \xi_n = \xi, X_n = x\} = p(\xi, \cdot | x)$. For each n and x, define the Markov process $\{\xi_j(x), j \geq n\}$ which starts at time n with initial condition $\xi_n(x) = \xi_n$ and transition function $p(\xi, \cdot | x)$. Such models were introduced in [6,11] and were also used in [1,8].

The following conditions will be used. The symbol E_n denotes the expectation conditioned on $\{X_i, i \le n, \xi_i, i < n\}$.

A4.2. *There is a continuously differentiable "centering" function $\overline{g}(\cdot)$ such that with the definition $\psi_j(x) = g(x, \xi_j) - \overline{g}(x)$ (for the exogenous noise) and $\psi_j(x) = g(x, \xi_j(x)) - \overline{g}(x), j \ge n$ (for the state dependent noise) we have for each n and x,*

$$(4.2) \qquad \sum_{j=n}^{\infty} a_j E_n \psi_j(x) = O(a_n),$$

where $O(a_n)$ is uniform in n, ω, x (where ω is the canonical point of the sample space), and $\psi_j(x)$ is bounded. In (4.2), for the state dependent noise case, the initial condition of $\{\xi_j(x), j \ge n\}$ is $\xi_n(x) = \xi_n$, following the usage in [8,11].

A4.3. *$\overline{g}(x) = Gx + \delta g(x)$, where G has its eigenvalues in the open left half plane and $|\delta g(x)| = O(|x|^2)$.*

A4.4.

$$\sum_{j=n}^{\infty} a_j E_n[\psi_j(x) - \psi_j(y)] = O(a_n)|y - x|,$$

$$|g(x, \xi)| = O(1)[|x| + 1].$$

Comments on (A4.2), (A4.4). Conditions of this type appear to have been initially introduced in [6,7] and have been used in many of the other references; e.g., [19,20,1,8,15]. These were generalizations of the perturbations used by [3] for multiscale Markovian diffusion processes. They are conditions on the rate of mixing of the processes. Many specific examples are in the books [1,8]. For the state dependent noise case, the transition kernel $p(\xi, \cdot|x)$ is often weak sense continuous in x. This is used in various examples in [8] to show that $\int g(x, \xi')p(\xi, d\xi'|x)$ is smooth enough so that (A4.4) holds.

The conditions can be extended in many directions, including letting the estimates in (4.2) and (A4.4) grow with x in a way that can be dominated by the "good" terms which appear in the analysis, but we prefer to concentrate on the main ideas and keep the development simple.

Stability of (1.1). In any asymptotic analysis of SA algorithms, one needs to prove certain "tightness" results. Here, a central issue in extending the window of averaging from that used in Theorem 2.1 concerns the tightness of $\{X_n/\sqrt{a_n}\}$. Indeed the proofs of (A2.1) require that such a tightness be shown. With tightness given, it is often the case that standard averaging methods then lead to (A2.1). Theorem 4.2 requires the bounds on the moments which are proved in Theorem 4.1. The following stability condition will be used. Recall that we set $\theta = 0$ without loss of generality.

A4.5. *There is a non-negative continuous function $V(\cdot)$ which goes to infinity as $x \to \infty$ and whose first and second mixed partial derivatives exist and are continuous. For some positive definite symmetric matrix P, and some $\gamma > 0, K < \infty$,*

$$V(x) = x' P x + o(|x|^2),$$

$$V_x'(x)\bar{g}(x) \leq -\gamma V(x), \quad |V_x(x)|^2 \leq KV(x),$$

$$|g(x,\xi)|^2 \leq K(V(x) + 1).$$

and $V_{xx}(x)$ is uniformly bounded.

For each integer p, let $\mathcal{F}_p$ denote the minimal σ-algebra which measures $\{X_i, i \leq p; \xi_i, i < p\}$.

Theorem 4.1. [Stability.] *Assume the conditions (A4.2)–(A4.5). Then for each integer k, $\limsup_n E|X_n|^{2k}/a_n^k < \infty$.*

Proof. Set $k = 1$. A perturbed Liapunov function method will be used. Define the perturbation to the Liapunov function:

$$V_1(x, n) = \sum_{j=n}^{\infty} a_j V_x'(x) E_n \psi_j(x) = O(a_n)|V_x(x)|.$$

The right hand estimate follows from (A4.2). Now, expanding and using the definition of $V_1(x, n)$ yields the equations

(4.3)
$$E_n V(X_{n+1}) - V(X_n) = a_n V_x'(X_n)\bar{g}(X_n)$$
$$+ a_n V_x'(X_n)\psi_n(X_n) + a_n^2 O(1)|g(X_n, \xi_n)|^2$$

(4.4)
$$E_n V_1(X_{n+1}, n+1) - V_1(X_n, n) = -a_n V_x'(X_n)\psi_n(X_n)$$
$$+ \sum_{j=n+1}^{\infty} E_n a_j [V_x'(X_{n+1})\psi_j(X_{n+1}) - V_x'(X_n)\psi_j(X_n)].$$

We next estimate the sum on the right side of (4.4) in terms of the bounds in (4.5a), (4.5b):

(4.5a)
$$|\sum_{j=n+1}^{\infty} E_n a_j V_x'(X_n)(\psi_j(X_{n+1}) - \psi_j(X_n))|$$
$$= |V_x(X_n)|O(a_n^2)[V^{1/2}(X_n) + 1] = O(a_n^2)[V(X_n) + 1],$$

$$(4.5b) \quad |\sum_{j=n+1}^{\infty} a_j E_n[V_x'(X_{n+1}) - V_x'(X_n)]\psi_j(X_{n+1})|$$

$$= O(a_n^2)|g(X_n, \xi_n)| = O(a_n^2)[V(X_n) + 1].$$

The right sides follow from (A4.2)–(A4.5).

We next define the perturbed Liapunov function $\tilde{V}_n = V(X_n) + V_1(X_n, n)$. By (4.4) and (4.5) we can write

$$E_n \tilde{V}_{n+1} - \tilde{V}_n \leq -a_n \gamma V(X_n) + O(a_n^2)$$

$$+ O(a_n^2)[1 + V(X_n)],$$

from which follows

$$(4.6) \quad E_n \tilde{V}_{n+1} - \tilde{V}_n \leq -\frac{a_n}{2}\gamma \tilde{V}_n + O(a_n^2).$$

The boundedness from above of $\{E\tilde{V}_n/a_n, n < \infty\}$ is implied by (4.6). This boundedness and the estimate $V_1(x, n) = O(a_n)[V(x) + 1]$ yields that

$$\{EV(X_n)/a_n, n < \infty\}$$

is bounded. This fact and the first equation in (A4.5) yields the boundedness of $\{E|X_n|^2/a_n, n < \infty\}$. The proof for general k follows the same procedure and is in [14] for $k = 2$. □

The following gives us the largest window of averaging. If $q_n = O(n)$, then the window size is $O(n)$. The mutual independence of the increments of the Wiener process which is the limit in the theorem supports the fundamental role played by the fact that the system is "two time scale." It is hard to use a weak convergence argument directly to get the theorem, since the time intervals of interest are too long: Recall that the time interval is characterized by $q_n a_n \to \infty$, which goes to infinity here (as opposed to the simpler case in Section 2). To get the desired result, we need to estimate the correlation between X_j, X_k for j, k very far apart, so that weak convergence arguments cannot be used directly. Part 1 of the following theorem gives the desired estimate.

Theorem 4.2. *Assume (1.2), (A2.1) and (A4.1)-(A4.5), and that for each $k > 0$,*

$$\sup_n \sup_{n+kq_n \geq i \geq n} a_n/a_i < \infty.$$

Then $M^n(\cdot) \Rightarrow W(\cdot)$, a Wiener process with covariance Vt.

Part 1. We can write (1.1) in the form

(4.7)
$$X_{n+1} = X_n + a_n \overline{g}(X_n) + a_n \psi_n(X_n)$$
$$= X_n + a_n G X_n + a_n \delta g(X_n) + a_n \psi_n(X_n).$$

It is shown in [14, Theorem 4.1, Part 1] that the asymptotic contribution of the δg terms to the limit is zero, and that in proving the tightness and asymptotic characterization of the sequence $M^n(\cdot)$, we can work as if each process $M^n(\cdot)$ were defined with $X_n = 0$.

Thus, in order to prove the theorem, we can replace $\{X_m, m \geq n\}$ by the $\{Y_m^n, m \geq n\}$ process, which is defined by

(4.8)
$$Y_{m+1}^n = (I + a_m G) Y_m^n + a_m \psi_m(X_m), \quad m \geq n,$$

where we define $Y_n^n = 0$. Note that the stability of G in (A4.3) and the boundedness of $\{\psi_m(X_m)\}$ imply that $(Y_m^n, m \geq n)$ is bounded, uniformly in n.

A perturbed test function method will be used to prove the tightness. We rewrite (4.8) to put it into a more convenient form. For $k > 0$ and $n + k q_n \geq j \geq n$, define the "perturbations"

(4.9)
$$\delta Y_j^n = \sum_{i=j}^{\infty} a_i E_j \psi_i(X_j) = O(a_j),$$

$$\tilde{Y}_j^n = Y_j^n + \delta Y_j^n,$$

where the $O(a_j)$ value in (4.9) is due to (4.2). Note that the argument of the $\psi_i(\cdot)$ in (4.9) is X_j, the state at the lower index of summation.

We can write the recursive equation for $\{\tilde{Y}_m^n, m \geq n\}$ as

(4.10)
$$\tilde{Y}_{m+1}^n = (I + a_m G)\tilde{Y}_m^n + a_m \left[\tilde{\psi}(X_m) + \tilde{S}_m(X_m) + \hat{S}_m(X_m) \right] + O(a_m^2),$$

where we define

$$\tilde{S}_m(X_m) = \frac{1}{a_m} \sum_{i=m+1}^{\infty} a_i \left[E_{m+1} \psi_i(X_m) - E_m \psi_i(X_m) \right],$$

$$\tilde{\psi}_m(X_m) = \psi_m(X_m) - E_m \psi_m(X_m),$$

$$\begin{aligned} \hat{S}_m(X_m) &= \frac{1}{a_m} \sum_{i=m+1}^{\infty} a_i \left[E_{m+1} \psi_i(X_{m+1}) - E_{m+1} \psi_i(X_m) \right] \\ &= O(a_m) \left[|X_m| + 1 \right]. \end{aligned}$$

Note that both terms $\tilde{S}_m(X_m), \tilde{\psi}_m(X_m)$ are $\mathcal{F}_p$−martingale differences, and are uniformly bounded.

Using the facts that $a_n q_n^{1/2} \to 0$ and $E|X_m| = O(a_m^{1/2})$, it can be shown by a direct calculation that the contributions of the $\tilde{S}_m(X_m)$ and the $O(a_m^2)$ terms to $M^n()$ are asymptotically negligible, and we ignore them henceforth. Also, a proof similar to that of Theorem 4.1 yields that, for each integer c, $E|\tilde{Y}_m^n|^{2c} = O(a_n^c)$ uniformly in $n, m : m \geq n$. Thus it is sufficient to work with the $\{\tilde{Y}_m^n, m \geq n\}$ sequence which is redefined by $\tilde{Y}_n^n = 0$, and

(4.11) $$\tilde{Y}_{m+1}^n = (I + a_m G)\tilde{Y}_m^n + a_m \rho_m^n,$$

where the $\rho_m^n = \tilde{S}_m(X_m) + \tilde{\psi}_m(X_m)$ are $\mathcal{F}_p$−martingale differences and are bounded.

Part 2. Using the new definition (4.11), define

$$F^n(t) = \frac{1}{\sqrt{q_n}} \sum_{i=n}^{n+q_n t} \tilde{Y}_i^n.$$

By the results in Part 1, it is sufficient to prove the theorem for $F^n(\cdot)$ replacing $M^n(\cdot)$.

Tightness of $\{F^n(\cdot)\}$. Recall the definition $t_n = \sum_0^{n-1} a_i$. Let $k > 0$. Let $\tau \in [0, k]$ be such that $n + q_n \tau \equiv r(n)$ is an $\mathcal{F}_p$-stopping time, with values in $[n, n + kq_n]$. To prove tightness, it is sufficient that ([2, Chapter3, Theorem 8.6], [5, Theorem 3.3]) $\sup_n E|F^n(t)| < \infty$ for each $t > 0$ and that

(4.12) $$\lim_{\delta \to 0} \overline{\lim_n} \sup_{\tau \leq k} \sup_{s \leq \delta} E|F^n(\tau + s) - F^n(\tau)|^2 = 0.$$

For notational simplicity, let the X_n and $\tilde{Y}_j^n$ be real valued henceforth in the proof. The proof for the general case follows the same lines. We can write

(4.13) $$E|F^n(\tau + s) - F^n(\tau)|^2 = \frac{1}{q_n} E \sum_{i,j=r(n)}^{r(n)+q_n s} \tilde{Y}_i^n \tilde{Y}_j^n.$$

The fact that for each $T < \infty$,

$$\sup_{n, t \leq T} E|F^n(t)|^2 < \infty$$

is shown by a calculation which is very similar to that used to get (4.12), and we concentrate on showing (4.12). Let us introduce some simplifying notation. For $j > i$ define

$$\hat{\Pi}(i, j) = \prod_{m=r(n)+i}^{r(n)+j} (I + a_m G), \quad \hat{\Pi}(i, i-1) = 1.$$

Define $\hat{a}_i = a_{r(n)+i}$, $\hat{Y}_i^n = \tilde{Y}_{r(n)+i}^n$, and, in general, let the hat $\hat{\ }$ over a variable indicate that the index of the variable is increased by $r(n)$ from whatever it was for the variable without the hat.

To get the tightness, it is sufficient to prove, for $r(n)$ of the above type and $\delta > 0$,

$$(4.14) \qquad \lim_{\delta}\overline{\lim_n}\sup_{r(n)}\sup_{s\leq\delta}\left[\frac{1}{q_n}E\sum_{q_n s\geq j\geq i\geq 0}\hat{Y}_i^n\hat{Y}_j^n\right] = 0.$$

For $j \geq i$, the martingale difference property implies that

$$\hat{E}_0\hat{Y}_j^n = \hat{\Pi}(i, j-1)\hat{Y}_i^n,$$

$$\hat{E}_i\left[\hat{Y}_i^n\right]^2 = \hat{\Pi}^2(0, i-1)\left(\hat{Y}_0^n\right)^2 + \hat{E}_0\left[\sum_{l=0}^{i-1}\hat{\Pi}(l+1, i-1)\hat{a}_l\hat{\rho}_l^n\right]^2.$$

Again owing to the $\mathcal{F}_p$—martingale difference and boundedness properties of the $\{\rho_m^n\}$, the term on the right equals $O(\hat{a}_0) = \hat{O}(a_j) = O(a_n)$ on the range of indices of interest.

With the use of these conditional expectations and $O(\hat{a}_i) = O(\hat{a}_j)$ for the range of indices of interest, the bracketed term in (4.14) can be written as the sum of the two terms given by (4.15a) and (4.15b):

$$(4.15a) \qquad \frac{O(1)}{q_n}E\sum_{q_n s\geq j\geq i\geq 0}\hat{\Pi}(i, j-1)O(\hat{a}_j) = O(s),$$

$$(4.15b) \qquad \frac{O(1)}{a_n^2 q_n}E\sum_{q_n s\geq j\geq i\geq 0}\hat{\Pi}(0, i-1)\hat{a}_i\hat{\Pi}(i, j-1)\hat{a}_j|\hat{Y}_0^n|^2 = \frac{E|\tilde{Y}_{r(n)}^n|^2}{a_n^2 q_n}.$$

We now proceed to estimate the expectation in (4.15b) via a Liapunov function method. We continue to work with the scalar case, but the general proof is similar. Let $m \geq n$. Then

$$E_m|\tilde{Y}_{m+1}^n|^4 = (I + 4Ga_m)|\tilde{Y}_m^n|^4 + O(a_m^2)|\tilde{Y}_m^n|^2 + O(a_m^3)|\tilde{Y}_m^n|$$
$$+ O(a_m^2)|\tilde{Y}_m^n|^4 + O(a_m^4).$$

For $k > 0$, $k_1 > 0$, and $m \leq n + q_n k$, define

$$F_m^n = k_1\sum_{i=m}^{n+q_n k}E_m\left[a_i^2|\tilde{Y}_i^n|^2 + a_i^2|\tilde{Y}_i^n|^4 + a_i^3|\tilde{Y}_i^n| + a_i^4\right].$$

For large enough k_1 and $m < n + q_n k$, we have

$$E_m\left[|\tilde{Y}_{m+1}^n|^4 + F_{m+1}^n\right] \leq |\tilde{Y}_m^n|^4 + F_m^n.$$

Thus, the sequence $\{|\tilde{Y}_{m+1}^n|^4 + F_{m+1}^n\}$ is a nonnegative $\mathcal{F}_p$–supermartingale on the time interval $[n, n + kq_n]$, and since $\tilde{Y}_n^n = 0$ and $F_m^n \geq 0$, we can write

$$(4.16) \qquad E|\tilde{Y}_{r(n)}^n|^4 \leq EF_0^n = q_n O(a_n^3).$$

Putting this into (4.15b) yields that (4.15b) equals

$$O(1)\sqrt{q_n a_n^3}/q_n a_n^2 = O(1)/\sqrt{a_n q_n} \to 0,$$

as $n \to \infty$. The order of magnitude of the estimates of the terms in (4.15a,b) does not depend on $r(n)$, provided that it takes values in the set $[n, n+kq_n]$ for some positive k not depending on n. Thus, (4.12) holds and tightness is proved.

The rest of the proof is exactly as in [14] and the reader is referred to that reference for the rest of the details.   □

5. Averaging with feedback. The algorithm (1.1), (1.2) is "off line" in the sense that the "best" values $\bar{X}_n$ are not used in the primary algorthm (1.1). One can ask whether there is any way of introducing $\bar{X}_n$ into the primary algorithm with profit. It appears from the results below that the best feedback is of the order of a_n. Indeed, we would like to work with algorithms of the intuitively reasonable form

$$X_{n+1} = X_n + a_n g(X_n, \xi_n) + a_n A(\bar{X}_n - X_n), \ A > 0.$$

The nonlinear case is difficult to treat at present, and we will work with a "linear" form which occurs frequently in applications.

We consider a well known algorithm which is used for parameter identification and related applications. The conditions which are used below are certainly not the best possible, but we wish to keep the essential ideas in the forefront. Let the vector θ denote the unknown parameter of the system, ϕ_n a sequence of random "input vectors" with $E\phi_n\phi_n' = G > 0$ (positive definite) and ψ_n a sequence of real valued random variables (which we suppose to have mean zero, without loss of generality). The observed output is the real valued process $y_n \equiv \theta'\phi_n + \psi_n$. Let $\hat{\theta}_n$ be the sequence of estimates of θ.

Consider the algorithm

$$(5.1) \qquad \hat{\theta}_{n+1} = \hat{\theta}_n + \varepsilon\phi_n[y_n - \hat{\theta}_n'\phi_n].$$

with the average

$$(5.2) \qquad \bar{\theta}_n = \frac{1}{n}\sum_{i=1}^n \hat{\theta}_i.$$

One also might use the average (T is a parameter which gives the window of averaging)

$$(5.2') \qquad \bar{\theta}_n = \frac{1}{T/\varepsilon} \sum_{n-T/\varepsilon}^{n} \hat{\theta}_i.$$

The theory of the previous parts of the paper or of the references tells us that for large T, $\bar{\theta}_n$ has a much better asymptotic covariance than does the classical estimator $\hat{\theta}_n$.

A feedback form of (5.1), (5.2). Now, let us feedback the average of the estimates into the primary algorithm. For $A > 0$, redefine $\hat{\theta}_n$ and $\bar{\theta}_n$ by

$$\hat{\theta}_{n+1} = \hat{\theta}_n + \varepsilon \phi_n [y_n - \hat{\theta}'_n \phi_n] + \varepsilon A [\bar{\theta}_n - \hat{\theta}_n]$$

$$(5.3)$$

$$\bar{\theta}_n = \frac{1}{n} \sum_{i=1}^{n} \hat{\theta}_i.$$

Using the definitions $X_n = \hat{\theta}_n - \theta, \bar{X}_n = \bar{\theta}_n - \theta$, let us replace (5.3) with

$$(5.4a) \qquad X_{n+1} = [I - \varepsilon G] X_n + \varepsilon \xi_n + \varepsilon A [\bar{X}_n - X_n],$$

$$(5.4b) \qquad \bar{X}_n = \frac{1}{n} \sum_{i=1}^{n} X_i,$$

where the "noise" term ξ_n is defined by

$$\xi_n = [G - \phi_n \phi'_n] X_n + \phi_n \psi_n \equiv \Phi_n X_n + \rho_n,$$

where the definitions of Φ_n and ρ_n are obvious.

For convenience, we can closely approximate the average by the solution to the equation

$$\bar{X}_0 = \bar{X}_1 = X_0 \text{ and, for } n > 1,$$

$$(5.5)$$

$$\bar{X}_{n+1} = (1 - \frac{1}{n}) \bar{X}_n + \frac{X_n}{n}.$$

More generally, we will have more flexibility by working with the more general forms which are defined by (5.6) or (5.7):

$$(5.6) \qquad \bar{X}_{n+1} = (1 - b_n) \bar{X}_n + b_n X_n,$$

where

$$b_n \to 0, \ \sum_{n=1}^{\infty} b_n = \infty;$$

or

$$(5.7) \qquad \bar{X}_{n+1} = (1 - \alpha\varepsilon)\bar{X}_n + \alpha\varepsilon X_n,$$

where $\alpha > 0$ is small.

Asymptotic orders. We will state some basic estimates from the reference [13]. These give some indication of what might be expected. Informal calculations suggest that the estimates (5.8b) and (5.8c) below could be conservative. The estimates tell us that the feedback does not hurt, but yields an average which is still better than the original estimate. This is further elaboratedupon below. The development of the estimates used the following assumptions.

Assumptions. Define E_n as the conditional expectation given $\{X_0, \phi_i, \psi_i, i < n\}$. For real K_0, we require, uniformly in n,

$$(A5.1) \qquad \left| \sum_{m=n}^{\infty} E_n \Phi_m \right| \le K_0,$$

$$(A5.2) \qquad \left| \sum_{m=n}^{\infty} E_n \rho_m \right| \le K_0.$$

$$(A5.3) \qquad \{\Phi_i, \rho_i, i < \infty\} \text{ is bounded.}$$

Note. (A5.1–A5.3) are convenient and not too restrictive. They involve a minimal rate of decrease of the correlation; more particularly, a minimal rate of decrease of the expectation of distant future "noise values" given the past. Some examples are in [5,8]

Theorem 5.1. [See proof of Theorem 4.1 in [13].] *Assume* (A5.1)–(A5.3). *Then, for (5.4a), (5.6),*

$$(5.8a) \qquad \limsup_n E|X_n|^2 = O(\varepsilon).$$

$$(5.8b) \qquad \limsup_n E|\bar{X}_n|^2 = O(\varepsilon^2).$$

If $E_n \rho_i = 0, i \ge n$, *then*

$$(5.8c) \qquad \limsup_n E|\bar{X}_n|^2 = O(\varepsilon^3).$$

If (5.7) is used, then the right hand sides of (5.8b), (5.8c) are replaced by $O(\varepsilon)$.

Note. $E_n \rho_i = 0, i \geq n$, if the noises $\{\psi_n\}$ are mutually independent, have mean zero and are independent of the $\{\phi_n\}$.

Rates of convergence: Limits of normalized iterates. By Theorem 5.1, the use of feedback yields an error of $O(\varepsilon)$ for X_n, and of $O(\varepsilon^2)$ or $O(\varepsilon^2)$ for $\bar{X}_n$. This information can be much improved. The improvement in behavior due to feedback will become clearer. Define $U_n = X_n/\sqrt{\varepsilon}, \bar{U}_n = \bar{X}_n/\sqrt{\varepsilon}$. The natural continuous parameter interpolations are defined by $U^\varepsilon(\cdot)$ by: $U^\varepsilon(t) = U_n$ for $t \in [n\varepsilon, n\varepsilon + \varepsilon)$, and define $\bar{U}^\varepsilon(\cdot)$ analogously. By (5.4a),

$$(5.9) \qquad U_{n+1} = (I - \varepsilon Q)U_n + \sqrt{\varepsilon}\xi_n + \varepsilon A \bar{U}_n, \quad Q = AI + G.$$

In order to get asymptotic results for the normalized processes, the following is assumed in [13]. Recall that if $X_n = 0$ (the limit value), then $\xi_n = \phi_n \psi_n \equiv \rho_n$.

(A5.4). The sequence of processes $\sqrt{\varepsilon} \sum_{i=n}^{n+t/\varepsilon} \rho_i$, $(\rho_n = \phi_n \psi_n)$ converges weakly to a Wiener process with covariance defined by $R_0 = \sum_{-\infty}^{\infty} E\rho_n \rho_0'$, as $\varepsilon \to \infty$.

(A5.5). There are $n_\varepsilon \to \infty$ as $\varepsilon \to 0$ such that $\varepsilon n_\varepsilon \to 0$ and such that (in the sense of probability, and uniformly in n)

$$\frac{1}{n_\varepsilon} \sum_{i=n}^{n+n_\varepsilon} \phi_i \phi_i' \to G,$$

$$\frac{1}{n_\varepsilon} \sum_{i=n}^{n+n_\varepsilon} \phi_i \psi_i \to 0.$$

Note on the assumptions. In the convergence methods and applications in [8], the limit of the second sum in (A5.5) can be any constant vector. This simply creates a bias, and can be used here too. Condition (A5.4) is also not very restrictive. It seems preferable to word the condition as it is due to the sizeable literature on convergence of such sums to a Wiener process [8,2,5], and the convergence occurs under many different sets of assumptions.

Theorem 5.2. [See proof of Theorem 5.1 in [13].] *Assume (A5.1-A5.5), and let $b_n \to 0$. Then there are $N_\varepsilon \to \infty$ such that $U^\varepsilon(\varepsilon N_\varepsilon + \cdot)$ converges weakly to the stationary solution of*

$$(5.10) \qquad dU = -QU\,dt + R_0^{1/2}dW$$

where $W(\cdot)$ is a standard Wiener process.

If $b_n = \alpha \varepsilon$ for small ε, α then by Theorem 5.1. $\limsup_n E|\bar{X}_n|^2 = O(\varepsilon)$. The actual situation is much better, and an analysis of this case gives additional insight into the case where $b_n \to 0$. The constant $b_n = \alpha \varepsilon$ case is of practical importance, since there is slow "exponential forgetting," and seems to be similar to an average with a window of size $O(1/\varepsilon \alpha)$. Note that in the simulations averages of the type $(5.2')$ performed better than did (5.7).

Theorem 5.3. [See proof of Theorem 5.2 in [13].] *Assume (A5.1-A5.5), and let $b_n = \alpha \varepsilon$, for small α. Then there are $N_\varepsilon \to \infty$ such that $(U^\varepsilon(\varepsilon N_\varepsilon + \cdot), \bar{U}^\varepsilon(\varepsilon N_\varepsilon + \cdot))$ converges weakly to the stationary solution of*

$$(5.11) \qquad \begin{aligned} dU &= -QU\,dt + A\bar{U}(t)\,dt + R_0^{1/2}dW \\ d\bar{U} &= \alpha U\,dt - \alpha \bar{U}\,dt, \end{aligned}$$

where $W(\cdot)$ is a standard vector valued Wiener process.

Note on Theorem 5.3. The stationarity of the solution to (5.11) implies that $X_n \sim N(0, \varepsilon V_0(Q))$, where

$$V_0(Q) = \int_0^\infty e^{-Qs} R_0 e^{-Q's}\,ds$$

is the stationary covariance. Hence, the stationary covariance of $U(t)$ decreases as A increases, and is $O(1/A)$ for large A. I.e., for small fixed ε, as A increases the asymptotic covariance of X_n decreases. We cannot let A grow indefinitely if ε is fixed, without reaching instability. In practice, it seems that the asymptotic results improve until near the point where stability is questionable.

The order of the errors in Theorem 5.3 for small α and large A. It is shown in [13] that (for the stationary solution)

$$(5.12a) \qquad E|U(t)|^2 = O(1/A) \times \text{ trace } R_0,$$

$$(5.12b) \qquad E|\bar{U}(t)|^2 = O(\alpha)E|U(t)|^2.$$

We now see that the asymptotic variance of $\bar{U}(t)$ is inversely proportional to the effective window of averaging, as for the case without feedback.

6. Simulation data. The following data are typical of the many runs taken. The simulations used a 5 dimensional system, and the vector ϕ_n was defined by: $\phi_n = (\tilde{\phi}_n, \tilde{\phi}_{n-1}, \tilde{\phi}_{n-2}, \tilde{\phi}_{n-3}, \tilde{\phi}_{n-4})$. The random variables $\tilde{\phi}_n$ were defined by the equation $\tilde{\phi}_{n+1} = \tilde{\phi}_n/2 + \zeta_n$, and here the ζ_n are mutually independent Gaussian random variables with mean zero and variance

1.0. Also, $\theta = (4.0, -4.2, 3.0, 2.7, -3.0)$. The standard deviation of the observation noise ψ_n was 6.0. The noise level was chosen to be large so that we could easily distinguish among the algorithms. For small noise levels all algorithms worked well. We noted that at very low observation noise levels the non feedback algorithm might be preferred. Asymptotically, the best values of ε are small, and decrease to zero as the run length increases to infinity. We have attempted to compare the performance of the various algorithms using step size parameters at which each works well. The averaged value that was actually used was

$$\bar{X}_n = \frac{1}{\text{window}} \sum_{n-\text{window}}^{n} X_i.$$

The averaging was started after a suitable transient period. The tables list the sample mean square errors in the estimates of the five components of θ at 5,000 iterations. The values were essentially constant after 1,000 iterations. Generally, the algorithm with feedback is not as sensitive to the value of ε and can work well at somewhat larger values (assuming stability) of ε.

In all cases, both with and without feedback, it can be seen that averaging can yield much improvement. This was true even when the original iterates X_n (without feedback) were poor. The performance improves as the window of averaging increases.

If the observation noise level is not too small, then the averaged iterate $\bar{X}_n$ for the feedback case is generally better than that for the non feedback case, and is often much better. Similarly for the errors in the X_n. This latter observation is important for good "on line" algorithms. Indeed, in some cases the errors X_n with feedback are as good as the errors $\bar{X}_n$ without feedback. We note that the gap between using feedback and not narrows as $\varepsilon \to 0$. But one might not want to work with very small ε. While more testing and comparison needs to be done, it is clear that the use of feedback is quite promising and has many advantages.

We also note that the use of feedback makes the behavior of the X_n less sensitive to the value of G, as seen from the forms of the stochastic differential equations in (5.14).

Table 1. $\varepsilon = .05$, $A = 5$, Window=150

case		1	2	3	4	5
no FB	X_n	1.133	1.212	1.173	1.251	1.014
no FB	$\bar{X}_n$	.230	.297	.257	.370	.207
FB	X_n	.275	.238	.216	.278	.255
FB	$\bar{X}_n$	.065	.037	.019	.086	.060

Table 2. $\varepsilon = .05$, $A = 22$, Window=150

case		1	2	3	4	5
no FB	X_n	1.133	1.212	1.173	1.251	1.014
no FB	$\overline{X}_n$	.230	.297	.257	.370	.207
FB	X_n	.206	.191	.217	.299	.325
FB	$\overline{X}_n$	.045	.014	.020	.083	.095

Table 3. $\varepsilon = .02$, $A = 9.5$, Window=300

case		1	2	3	4	5
no FB	X_n	0.350	0.441	0.385	0.489	0.350
no FB	$\overline{X}_n$	.086	.125	.085	.194	.088
FB	X_n	.061	.105	.131	.067	.096
FB	$\overline{X}_n$	.013	.064	.090	.027	.059

Table 4. $\varepsilon = .05$, $A = 5$, Window=300

case		1	2	3	4	5
no FB	X_n	1.133	1.212	1.173	1.251	1.014
no FB	$\overline{X}_n$	.124	.131	.096	.225	.110
FB	X_n	.274	.229	.245	.239	.242
FB	$\overline{X}_n$	.049	.016	.037	.038	.037

An application of feedback and averaging to an optimization problem. We now present some results of the use of averaging with and without feedback to a problem of parametric optimization of a nonlinear system. As yet there are no proofs of convergence for the feedback case, but the method worked well. The data are not conclusive, since we did not do an exhaustive study, but they support the view that the use of feedback in SA can be quite useful.

The problem is the design of a controller for tracking and capture in the presence of observation noise. The model of the tracking system is

$$dx_1 = x_2 dt, \; dx_2 = -ux_4 dt + \sigma_1 dw_1,$$

$$dx_3 = x_4 dt, \; dx_4 = ux_2 dt + \sigma_2 dw_2,$$

where u is the control. The model of the target is

$$dz_1 = z_2 dt, \; dz_2 = \sigma_3 dw_3,$$

$$dz_3 = z_4 dt, \; dz_4 = \sigma_4 dw_4.$$

The observations are

$$dy_1 = z_1 dt + \sigma_5 dw_5,$$

$$dy_2 = z_3 dt + \sigma_6 dw_6.$$

All the Wiener processes are standard and they are mutually independent. Define the error

$$e^2(t) = |x_1(t) - z_1(t)|^2 + |x_3(t) - z_3(t)|^2.$$

The objective is to maximize

$$P\{\sup_{t \leq T} |e(t)| \leq \Delta\}.$$

A Kalman filter is used to get the conditional mean of the target state.

The control is of the form $u = kb$, where b is the angle in Figure 1; i.e., control is over the angle of the thrust. It is linear in the angle between the line of sight from the tracker to the estimated position of the object being tracked and the current velocity vector of the tracker. (We have also used controls depending linearly on other states, with the same results.) The angle used here remains the most important variable. The parameter k is to be determined.

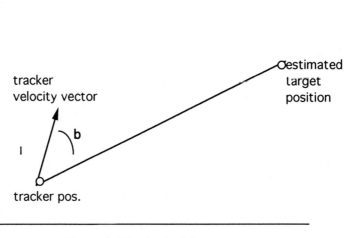

Figure 1. The control form

We used $\sigma_i = 1$, and the algorithm

$$k_{n+1} = k_n + \varepsilon Z_n + \varepsilon A(\bar{k}_n - k_n)$$

$$\bar{k}_n = \frac{1}{\text{window}} \sum_{n-\text{window}}^{n} k_i.$$

Table 5. The tracking problem. $\varepsilon = .02., A = 10$

iter.	no FB iter.	no FB av.	FB iter.	FB av.
500	121	140	154	140
1000	122	122	155	155
1500	152	147	160	160
2000	184	181	163	163
2500	200	186	166	165
3000	186	181	165	167
3500	165	180	174	171
4000	186	180	173	173
4500	210	215	176	176
5000	206	209	179	180
5500	184	184	186	185
6000	170	166	194	188
6500	152	153	193	193
7000	144	150	198	198
7500	210	202	202	201
8000	185	183	202	202
8500	201	186	201	206
9000	219	214	208	209
9500	204	208	212	212
10000	220	215	215	215

Of course, without feedback $A = 0$. The Z_n is an estimate of the gradient of the performance value with respect to the parameter, obtained using the method in [12], and was quite noisy for the problem at hand. The tracker started at the origin, and the target at position (12,11), and with quite different initial velocities. The optimal parameter value was about 218. The averaging window size was 200. Table 5 gives the results of one selected run. But bear in mind that the results are not claimed to be definitive, only suggestive. In the chosen run, all methods ended up near the best point. Generally there was much variation. But the data indicate the general features; the smoothing effect of the averagings, and the particularly smooth behavior of the feedback-average method. Generally, the feedback-averaging worked best. There was a particular risk in using the feedback algorithm here, since (unlike the identification algorithm studied earlier) the derivatives go to zero as the parameter increases beyond its optimal value. This points out a liability of the feedback algorithm; the feedback can slow down corrections of poor estimates if the current average is on the "wrong side" of the current iterate and the gradients are small. The tabulated run was one of the best for the non feedback algorithm, and it was better than average for the feedback algorithm.

For the most part, the iterates were not in the "asymptotic" region. Thus the various asymptotic estimates as well as the given logic for the advantages of averaging and/or feedback do not apply. Again, we note that the point here is simply to encourage the further study of the method, not to reach definitive conclusions.

REFERENCES

[1] A. BENVENISTE, M. METIVIER, AND P. PRIORET, Adaptive Algorithms and Stochastic Approximation, Springer-Verlag, New York, Berlin 1990.

[2] P. BILLINGSLEY, Convergence of Probability Measures, John Wiley, New York 1968.

[3] G. BLANKENSHIP AND G.C. PAPANICOLAOU, Stability and control of systems with wide band noise disturbances, SIAM J. Appl. Math., 34 (1978), pp. 437–476.

[4] B. DEYLON AND A. JUDITSKY, Stochastic optimization with averaging of trajectories, Stochastics, 39 (1992), pp. 107–118.

[5] S.N. ETHIER AND T.G. KURTZ, Markov Processes: Characterization and Convergence, Wiley, New York 1986.

[6] H.J. KUSHNER AND HAI HUANG, Averaging methods for the asymptotic analysis of learning and adaptive systems, SIAM J. on Control and Optimization, 19 (1981), pp. 635–650.

[7] H.J. KUSHNER, Stochastic approximation with discontinuous dynamics and state dependent noise, J. Math. Analysis and Applications, 82 (1981), pp. 527–542.

[8] H.J. KUSHNER, Approximation and Weak Convergence Methods for Random Processes with Applications to Stochastic System Theory, MIT Press, Cambridge, MA 1984.

[9] H.J. KUSHNER, Weak Convergence Methods and Singularly Perturbed Stochastic Control and Filtering Problems, Volume 3 of Systems and Control, Birkhauser, Boston 1990.

[10] H.J. KUSHNER AND D.S. CLARK, Stochastic Approximation for Constrained and Unconstrained Syetems, Springer-Verlag, Berlin and New York 1978.

[11] H.J. KUSHNER AND A. SHWARTZ, An invariant measure approach to the convergence of stochastic approximations with state dependent noise, SIAM J. on Optimization and Control, 22 (1984), pp. 13–27.

[12] H.J. KUSHNER AND J. YANG, A monte carlo method for the sensitivity analysis and parametric optimization of nonlinear stochastic systems, SIAM J. on Control and Optimization, 29 (1992), pp. 1216–1249.

[13] H.J. KUSHNER AND J. YANG, Stochastic approximation with averaging and feedback: rapidly convergent "on line" algorithms, and applications to adaptive systems, (Technical report) Brown University, Lefschetz Center for Dynamical Systems 1992.

[14] H.J. KUSHNER AND J. YANG, Stochastic approximation with averaging: optimal asymptotic rates of convergence for general processes, SIAM J. on Control and Optimization 31 (1993), pp. 1045–1062.

[15] H.J. KUSHNER AND G. YIN, Asymptotic properties of distributed and communicating stochastic approximation algorithms, SIAM J. on Control and Optimization, 25 (1987), pp. 1266–1290.

[16] B.T. POLYAK, *New stochastic approximation type procedures*, Autom. i Telemekh, **7** (1990), pp. 98–107.

[17] B.T. POLYAK AND A.B. JUDITSKY, *Acceleration of stochastic approximation by averaging*, SIAM J. on Control and Optimization, **30** (1992), pp. 838–855.

[18] D. RUPPERT, *Efficient estimators from a slowly convergent robbins-munro process*, Technical Report 787, School of Operations Research and Industrial Engineering, Cornell University 1988.

[19] G. YIN, *On extensions of polyak's averaging approach to stochastic approximation*, Stochastics, **36** (1992), pp. 245–264.

[20] G. YIN, *Stochastic approximation via averaging: Polyak's approach revisited*, in *Lecture Notes in Economics and Mathematical Systems* **374** (G. PFLUG AND U. DIETER, editors), Springer-Verlag, Berlin 1992, pp. 119–134.

BUILDING MODELS FROM FREQUENCY DOMAIN DATA

LENNART LJUNG*

Abstract. The interplay between the time and frequency domains for linear systems is well known and most useful. In the case of identifying linear models from observed data, this interplay manifests itself in two ways.

The observed data is of course primarily recorded in the time domain (even though the recording equipment may deliver them in the frequency domain). To build linear models from the data we can either do the fitting in the time domain and evaluate the resulting modle's properties in the frequency domain. We can also transfer the data themselves to the frequency domain and fit models directly there. In this contribution we shall consider some aspects of these two possibilities.

1. Introduction. For linear systems the connections and interplay between time-domain and frequency domain aspects have proved to be most fruitful in all applications. We shall in this contribution discuss some aspects in applications to linear system identification.

There are two sides of this interplay. One is to consider the primary observation to be in the time-domain, and then to interpret corresponding identification criteria, algorithms and properties in the frequency domain. There are many early results of this character, e.g. [12], [4], [1], [5]. More recently such results have been exploited and developed in [7].

The other side of the interplay is to consider the primary observations to be in the frequency domain. That is, the Fourier transforms of the measured signals (or certain ratios of them) are treated as the actual measurements. This view has been less common in the traditional system identification literature, but has been of great importance in the Mechanical Engineering community, vibrational analysis and so on. An early reference is [6]. An excellent recent account, with many references, of this view is given in the book [9].

This contribution will deal with a few questions of the latter view from a more traditional System Identification background.

2. Parameterized models. We shall throughout this paper consider linear models in discrete or continuous time, parameterized as follows:

$$(2.1) \qquad y(t) = G(q, \theta)u(t) + H(q, \theta)e(t)$$

(Discrete time)

$$(2.2) \qquad y(t) = G(p, \theta)u(t) + H(p, \theta)e(t)$$

(Continuous time)

---

* Department of Electrical Engineering, Linköping University, S-581 83 Linköping, Sweden.

Here y, u and e are the output, the input and the noise source, respectively. e is supposed to be white noise with variance (intensity) λ. q is the shift operator and p is the differentiation operator.

A typical parameterization, both in continuous and discrete time could be as a rational function

$$G(p, \theta) = \frac{b_1 p^{n-1} + \cdots + b_n}{p^n + f_1 p^{n-1} + \cdots + f_n} = \frac{B(p)}{F(p)}$$

(2.3)

$$\theta = (b_1, \ldots b_n, \quad f_1, \ldots f_n).$$

In discrete time we could, e.g. also use parameterizations that originate from an underlying continuous time state space model, discretized under the assumption that the input is piecewise constant over the sampling interval:

(2.4) $$G(q, \theta) = C(qI - e^{A(\theta)T})^{-1} \int_0^T e^{A(\theta)\tau} B(\theta) d\tau$$

See, e.g. [7] for many more examples of the parameterization (2.1).

Moreover, various basis function expressions could be used for black-box parameterizations, like

(2.5) $$G(\theta, q) = \sum_{k=1}^d \theta_k g_k(q)$$

where the functions $\{g_k\}$ could be, e.g. Laguerre or Kautz functions. See, e.g. [11].

In addition to these traditional time-domain parameterizations, one may also parameterize the transfer functions in a way that is more frequency domain oriented.

A simple case (taken from problem 7G.2 in [7]) is to let

(2.6) $$G(e^{i\omega}, \theta) = \sum_{k=1}^d (g_k^R + i g_k^I) W_\gamma(k, \omega - \omega_k)$$

(2.7) $$\theta = [g_1^R, g_1^I, \ldots, g_d^R, g_d^I]$$

(and correspondingly in continuous time)

One should typically think of the functions $W_k(\gamma, \omega)$ as bandpass filters, with a width that may be scaled by γ. The parameter g_k would then describe the frequency response around the frequency value ω_k. If the width of the passband increases with frequency we obtain parameterizations linked to wavelet transforms. See, e.g., the insightful discussion by [8].

3. Time domain data. Suppose input-output data in the time domain are given:

$$(3.1) \qquad z^N = \{y(t), u(t); t = T, 2T, \ldots, NT\}$$

Prediction error methods. A much used approach is to estimate θ in (2.1) is to form the corresponding predictions

$$(3.2) \quad \hat{y}(t|\theta) = G(q, \theta)u(t) + (I - H^{-1}(q, \theta))(y(t) - G(q, \theta)u(t))$$

and the associated prediction errors:

$$(3.3) \qquad \varepsilon(t, \theta) = y(t) - \hat{y}(t|\theta) = H^{-1}(q, \theta)(y(t) - G(q, \theta)u(t))$$

and then compute

$$(3.4) \qquad \hat{\theta}_N = \arg\min_\theta \sum_{t=1}^N \varepsilon^2(t, \theta)$$

Most frequency domain interpretations of this time domain method go back to the application of Parseval's relationship to the right hand side of (3.4):

$$(3.5) \qquad \hat{\theta}_N \sim \arg\min_\theta \int_{-\pi}^\pi |E_N(\omega, \theta)|^2 d\omega$$

where E is the Fourier transform of ε:

$$(3.6) \qquad E_N(\omega, \theta) = H^{-1}(e^{i\omega}, \theta)[Y(\omega) - G(e^{i\omega}, \theta)U_N(\omega)]$$

$$(3.7) \qquad Y_N(\omega) = \frac{1}{\sqrt{N}} \sum_{t=1}^N y(t)e^{-i\omega t}$$

and similarly for $U_N(\omega)$. If we introduce the "Empirical Transfer Function Estimate",

$$(3.8) \qquad \hat{G}_N(e^{i\omega}) = \frac{Y_N(\omega)}{U_N(\omega)}$$

then (3.4) - (3.7) can be rewritten

$$(3.9) \qquad \hat{\theta}_N \sim \arg\min_\theta \int_{-\pi}^\pi |\hat{G}_N(e^{i\omega}) - G(e^{i\omega}, \theta)|^2 \frac{|U_N(\omega)|^2}{|H(e^{i\omega}, \theta)|^2} d\omega$$

This shows the close relationship between time-domain prediction error methods and frequency domain expressions.

Spectral analysis. Another well known method to estimate frequency functions is spectral analysis (e.g. [2]). The well known Blackman-Tukey (windowing) approach to spectral analysis computes auto– and cross co-variance functions between input and output, windows them and applies the Fourier Transform in a straightforward way. The resulting estimate can also be written ([7], Chapter 6).

$$(3.10) \qquad \hat{G}_N(e^{i\omega_0}) = \frac{\int_{-\pi}^{\pi} W_\gamma(\omega_0, \zeta - \omega_0)|U_N(\zeta)|^2 \hat{G}_N(e^{i\zeta})d\zeta}{\int_{-\pi}^{\pi} W_\gamma(\omega_0, \zeta - \omega_0)|U_N(\zeta)|^2 d\zeta}$$

where $W_\gamma(\omega_0, \zeta)$ is (for fixed ω_0) the Fourier Transform of the (lag-) windowing function. In the traditional Blackman-Tukey approach $W_\gamma(\omega_0, \zeta)$ does not depend on ω_0 - which is a consequence of the fact that it normally is implemented in the time domain.

However, it is quite conceivable, and would be useful, to let the frequency smoothing function W_γ, which determines the frequency resolution, also depend on the frequency.

The following could be noted. We could apply the prediction error method (3.4) to the model structure (2.6). In the time domain that would imply that a bank of d bandpass filters with pass-bands given by W_γ would have to be applied to the data. Let us plug in this parameterization into (3.9) which gives a quadratic expression in θ. Let us also assume that the passbands are so narrow that we approximately have

$$H(e^{i\omega}, \eta) \cdot W_\gamma(k, \omega - \omega_k) \approx H(e^{i\omega_k}, \eta)W_\gamma(k, \omega - \omega_k)$$

$$G(e^{i\omega}, \theta) \cdot W_\gamma(k, \omega - \omega_k) \approx G(e^{i\omega_k}, \theta)W_\gamma(k, \omega - \omega_k)$$

(Here $H(q, \eta)$ is an arbitrarily parameterized noise mode). Then also (3.10) will hold approximately for the resulting model.

Consequently we can interpret spectral analysis as a prediction error method applied to the model structure (2.6).

Asymptotic properties. The asymptotic properties of the model that results from (3.4) can be calculated using time-domain methods and then be translated to the frequency domain. This gives (cf. [7], Chapters 8 and 9).

$$\hat{\theta}_N \to \theta^* = \arg\min_\theta \int_{-\pi}^{\pi} [|G_0(e^{i\omega}) - G(e^{i\omega}, \theta)|^2 \Phi_u(\omega) + \Phi_v(\omega)]/|H(e^{i\omega}, \theta)|^2 d\omega$$

where G_0 is the true frequency function and Φ_u and Φ_v are the input and noise spectra respectively.

If the noise model H is fixed, $H(q, \eta) = H^*(q)$ we obtain

$$(3.11) \qquad \theta^* = \arg\min_\theta \int_{-\pi}^{\pi} |G_0(e^{i\omega}) - G(e^{i\omega}, \theta)|^2 \frac{\Phi_u(\omega)}{|H^*(e^{i\omega})|} d\omega$$

The general expression for the covariance matrix of $\hat{\theta}_N$ can be given as

$$\text{cov}\hat{\theta}_N = \frac{\lambda}{N}[E\Psi(t,\theta_0)\Psi^T(t,\theta_0)]^{-1} =$$

(3.12)·
$$\frac{1}{N}[\int_{-\pi}^{\pi} G'_\theta(e^{i\omega},\theta_0)[G'_\theta(e^{i\omega},\theta_0)]^* \frac{\Phi_u(\omega)}{\Phi_v(\omega)}d\omega]^{-1}$$

Here Φ is the negative gradient of ε w.r.t θ, and the second equality holds when then noise model is fixed and equal to the true value: $\Phi_v(\omega) = \lambda|H_0(e^{i\omega})|^2$.

See Section 14.3 in [7].

In terms of the frequency functions we have asymptotically in the model order

$$\text{cov}\begin{bmatrix} \hat{G}_N(e^{i\omega}) \\ \hat{H}_N(e^{i\omega}) \end{bmatrix} = \text{cov}\begin{bmatrix} G(e^{i\omega},\hat{\theta}_N) \\ H(e^{i\omega},\hat{\theta}_N) \end{bmatrix} = \frac{n}{N}\Phi_v\begin{bmatrix} \Phi_u(\omega & \Phi_{eu}(\omega) \\ \Phi_{ue}(\omega) & \lambda \end{bmatrix}^{-1}$$

where $\Phi_{ue}(\omega)$ is the cross spectrum between noise source and input and n is the model order.

4. Frequency domain data. Suppose now that the original data are supposed to be

(4.1)
$$Z^N = \{Y(\omega_k), U(\omega_k), k = 1, \ldots N\}$$

where $Y(\omega_k)$ and $U(\omega_k)$ either are the discrete Fourier transforms of $y(t)$ and $u(t)$ as in (3.7) or are considered as Fourier transforms of the underlying continuous signals:

(4.2)
$$Y(\omega) = \int_{-\infty}^{\infty} y(t)e^{-i\omega t}dt$$

(or a normalized version). Which interpretation is more suitable depends of course on the signal character, sampling interval and so on.

How to estimate θ in (2.1) or (2.2) from (4.1)? In view of (3.9) it would be tempting to use

$$\hat{\theta}_N = \arg\min_\theta V(\theta)$$

(4.3)
$$V(\theta) = \sum_{k=1}^{N} |Y(\omega_k) - G(e^{i\omega_k T},\theta)U(\omega_k)|^2 \cdot \frac{1}{|H(e^{i\omega_k T},\theta)|^2}$$

(replacing $e^{i\omega_k T}$ by $i\omega_k$ for the continuous-time model (2.2).)

If H in fact does not depend on θ (*fixed* or *known noise model*) experience shows that (4.3) works well. Otherwise the estimate $\hat{\theta}_N$ may not be consistent.

To find a better estimator we turn to the maximum likelihood (ML) method for advice: (We give the expressions for the continuous time case; in the case of (2.1), just replace $i\omega_k$ by $e^{i\omega_k T}$)

If the data were generated by

$$y(t) = G(p, \theta)u(t) + H(p, \theta)e(t)$$

the Fourier transforms would be related by

(4.4) $$Y(\omega) = G(i\omega, \theta)U(\omega) + H(i\omega, \theta)E(\omega)$$

To be true, (4.4) should in many cases contain an error term that accounts for the fact that the measured data $Y(\omega_k)$ often are not exact realizations of (4.2). For periodic signals, observed over an integer number of periods, (4.4) may however hold exactly.

Now, if $e(t)$ is white noise, its Fourier transform (suitably normalized) will have a (complex) Normal distribution:

(4.5) $$E(\omega) \in N(0, \lambda I) \quad \text{complex}$$

This means that the real and imaginary parts are each normally distributed, with zero means and variances λ. The real and imaginary parts are independent and, moreover, $E(\omega_1)$ and $E(\omega_2)$ are independent for $\omega_1 \neq \omega_2$. This implies that

(4.6) $$Y(\omega_k) \in N(G(i\omega_k, \theta)U(\omega_k), \lambda|H(i\omega_k, \theta)|^2)$$

according to the model, so that the negative logarithm of the likelihood function becomes

$$V_N(\theta) = \sum_{k=1}^{N} \{2 \log |H(i\omega_k, \theta)| +$$

(4.7) $$\frac{1}{\lambda}|Y(\omega_k) - G(i\omega_k, \theta)U(\omega_k)|^2 \cdot \frac{1}{|H(i\omega_k, \theta)|^2}\} + N \log \lambda$$

The ML estimate is

(4.8) $$\hat{\theta}_N = \arg \min_{\theta} V_N(\theta)$$

If we perform analytical minimization w.r.t. λ, we obtain

(4.9) $$\hat{\theta}_N = \arg \min_{\theta} \left[N \cdot \log W_N(\theta) + 2 \sum_{k=1}^{N} \log |H(i\omega_k, \theta)| \right]$$

(4.10) $$W_N(\theta) = \frac{1}{N} \sum_{k=1}^{N} |Y(\omega_k) - G(i\omega_k)U(\omega_k)|^2 \cdot \frac{1}{|H(i\omega_k, \theta)|^2}$$

(4.11)
$$\hat{\lambda}_N = W_N(\hat{\theta}_N)$$

Compared to (4.3) we thus have an additional term

(4.12)
$$\sum_{k=1}^{N} \log |H(i\omega_k, \theta)|^2$$

We may note that for any monic, stable and inversely stable transfer function $H(q, \theta)$ we have

(4.13)
$$\int_{-\pi}^{\pi} \log |H(e^{i\omega}, \theta)|^2 d\omega \equiv 0$$

This is the reason why (4.12) is missing from criteria that use dense, equally spaced frequencies ω_k for discrete time models (like (3.9)).

[In fact (4.12) is the determinant from the change of variables from Y to E (outputs to innovations). In the discrete time domain this transformation is a triangular operator with 1's along the diagonal ($e(t) = y(t)$-past data). Hence this transformation has a determinant equal to 1, so it does not affect the ML criterion.]

It is apparently often assumed (as in [9]) that the noise model is given or known. Then of course the term (4.12) is again not essential.

5. Asymptotic properties. The asymptotic properties (as $N \to \infty$) of the estimate (4.7)-(4.8) can be developed in a rather straightforward fashion, using the standard techniques. We confine ourselves below to the case of a fixed noise model $H(i\omega, \theta) = H_*(i\omega)$ and a known λ. Suppose, as $N \to \infty$ the frequencies ω_k cover the frequency interval $[-\Omega, \Omega]$ with a density function $W(\omega)$. [That is, let $w_N(\Omega_1, \Omega_2)$ be the number of observed frequencies in the interval Ω_1 to Ω_2 when the total number of frequencies is N. Then

$$\lim_{N \to \infty} \frac{1}{N} w_N(\Omega_1, \Omega_2) = \int_{\Omega_1}^{\Omega_2} W(\omega) d\omega.]$$

Then the θ-dependent part of (4.7) converges, uniformly in θ and with probability 1 to

(5.1)
$$\bar{V}(\theta) = \int_{-\Omega}^{\Omega} |G_0(i\omega) - G(i\omega, \theta)|^2 \frac{\Phi_u(\omega) W(\omega)}{|H_*(i\omega)|^2} d\omega$$

where G_0 is the true transfer function, and $\Phi_u(\omega)$ is the input spectrum. Hence

(5.2)
$$\hat{\theta}_N \to \arg\min_{\theta} \bar{V}(\theta) \quad w.p.1 \text{ as } N \to \infty$$

If there exists a value θ_0 such that $G_0(i\omega) = G(i\omega, \theta_0)$ and $\Phi_u(\omega)W(\omega)$ is different from zero at sufficiently many frequencies it will follow that

$$\hat{\theta}_N \to \theta_0 \quad as \quad N \to \infty$$

In that case the covariance matrix of $\hat{\theta}_N$ will be, asymptotically,

$$(5.3) \qquad Cov\ \hat{\theta}_N \sim \lambda \cdot \left[\sum_{k=1}^{N} \frac{G'_\theta(i\omega_k, \theta_0)G'_\theta(i\omega_k, \theta_0)^* \Phi_u(\omega_k)}{|H_*(i\omega_k)|^2} \right]^{-1}$$

Here G'_θ is the gradient of $G(i\omega, \theta)$ with respect to θ and superscript $*$ denotes complex conjugation and matrix transpose.

6. Some practical aspects. There are several distinct features with the direct frequency domain approach that could be quite useful. We shall list a few (see also, e.g., [9])

- **Prefiltering** is known as quite useful in the time-domain approach. For frequency domain data it becomes very simple: It just corresponds to assigning different weights to different frequencies, which in turn is the same as using a frequency dependent $\lambda \equiv$ cheating on the assumed noise levels. It is of course particularly easy to implement perfect band-pass filtering effects in the frequency domain approach.

- **Condensing large data sets.** When dealing with systems with a fairly wide spread of time constants, large data sets have to be collected in the time domain. When converted to the frequency domain they can easily be condensed, so that, for example, logarithmically spaced frequencies are obtained. At higher frequencies one would thus decimate the data, which involves averaging over neighboring frequencies. Then the noise level (λ_k) is reduced accordingly.

- **Combining experiments.** Nothing in the approach of Section 4 says that the frequency response data at different frequencies have to come from the same experiment, or even that the frequencies involved ($\omega_k, k = 1...N$) all have to be different. It is thus very easy to combine data from different experiments.

- **Periodic inputs.** The main drawback with the frequency domain approach in that the underlying frequency domain model (4.4) is strictly correct only for a periodic input and assuming all transients have died out. On the other hand, typical use of the time domain method (3.2) - (3.4) assumes inputs and outputs prior to time 0 to be zero. Whichever assumption about past behavior is closer to the truth should thus affect the choice of approach.

- **Band-limited signals.** If the actual input signals are band-limited, (like no power above the Nyquist frequency) the continuous time Fourier transform (4.2) can be well computed from sampled data. It is then possible to directly build continuous-time models without any extra work.

- **Continuous-time models.** The comment above shows that direct continuous-time system identification from "continuous-time data" can be dealt with in a much more relaxed way than in the time-domain, with all its mathematical intricacies.
- **Trade-off noise/frequency resolution.** The approach also allows for a more direct and frequency dependent trade-off between frequency resolution and noise levels. That will be done as the original Fourier transform data are decimated to the selected range of frequencies $\omega_k, k = 1, ..., N$.

7. Some algorithmic questions. The criterion (4.9) to be minimized is non-quadratic in θ in most cases. This calls for iterative search procedures for the calculation of $\hat{\theta}_N$. This in turn raises two questions:

1. What method should be applied for the iterations?
2. At what parameter values should the search be initialized?

We shall deal with these questions in order.

Iterative minimization

If the noise model H is fixed (θ-independent), the remaining criterion to be minimized in $W_N(\theta)$, which is a sum of squared error terms. The classical method to deal with such a function minimization is the damped Gauss-Newton method [3]. This apparently is still the best approach around, and is the basic method used in System Identification. Indeed, the MATLAB Signal Processing Toolbox commands for solving (4.9) for a fixed noise model (`invfreqz` and `invfreqs`) implement this approach.

Unfortunately, it turns out that the additional term (4.12) may seriously deteriorate the performance of the damped Gauss-Newton procedure. This is, not unexpectedly, most pronounced for continuous time models and for very unequally spaced frequency samples. One probably then has to go to full Newton-methods, which however puts greater demands on the line search. Also, it is important to scale the parameterization, so that the criterion remains reasonably well conditioned.

Initial parameter estimates

Also in the time-domain approach it is very important to provide the Gauss-Newton iterative scheme with good initial conditions. In [7] (Section 10.5) several steps to achieve such initial estimations are described. They are based on the Instrumental Variable (IV) method and the so called repeated Least Squares (rLS) method (i.e. estimating a high order ARX-model, then compute the innovations from this and use them as measured inputs in the next step).

Fortunately these methods can be more or less directly carried over to direct frequency domain methods. *The IV method* (see also [10]) can be described as follows: The problem is to find an initial estimate

$$\hat{G}^{(0)}(e^{i\omega}) = \frac{B(e^{i\omega})}{A(e^{i\omega})}$$

Step i): Solve

$$\min_{a_i, b_i} \sum_k |A(e^{i\omega_k})Y(\omega_k) - B(e^{i\omega_k})U(\omega_k)|^2$$

for A^s, B^s. Let $\hat{G}^s = \frac{B^s}{A^s}$
Step ii): Solve

$$(7.1) \qquad 0 = \sum_k (A(e^{i\omega_k})Y(\omega_k) - B(e^{i\omega_k})U(\omega_k)) \cdot \zeta(\omega_k)$$

for A and B where

$$(7.2) \qquad \zeta(\omega_k) = \begin{bmatrix} \vdots \\ \hat{G}^s(e^{i\omega_k}) \cdot e^{i\ell\omega_k} U(\omega_k) \\ \vdots \\ e^{i\ell\omega_k} U(\omega_k) \\ \vdots \end{bmatrix}$$

where ℓ ranges from 1 to the orders of A and B respectively. The vector (7.2) is the vector of instruments.

The rLS method. is as follows in the frequency domain. The problem is to find $A(q)$ and $C(q)$ in an ARMA model

$$A(q)y(t) = C(q)e(t).$$

Step 1). Solve

$$\min_\alpha \sum_R |\alpha(e^{i\omega_k})Y(\omega_k)|^2$$

for $\hat{\alpha}(e^{i\omega})$ for a "high order" polynomial α.
Step 2). Treat

$$\hat{E}(\omega_k) = \hat{\alpha}(e^{i\omega_k})Y(\omega_k)$$

as measured input and solve

$$(7.3) \qquad \min_{A,C} \sum_k |A(e^{i\omega_k})Y(\omega_k) - (C(e^{i\omega_k}) - 1)\hat{E}(\omega_k)|^2$$

for $\hat{A}$, $\hat{C}$. It is my experience that these start-up procedures work well.

8. Conclusions. We have in this contribution discussed various aspects of frequency domain methods for linear system identification. Generally speaking, it could be said that the direct frequency domain approach has been underutilized in conventional system identification. The contribution has been partly of tutorial character, summarizing some main points. In addition the author's experiences with various implementations of the algorithms have been described.

REFERENCES

[1] H. AKAIKE, *Maximum likelihood identification of gaussian autoregregressive moving average models.*, Biometrika, 20 (1973), pp. 255–265.

[2] D. R. BRILLINGER, *Time Series: Data Analysis and Thoery*, Holden-Day, San Francisco, 1981.

[3] J. E. DENNIS AND R. B. SCHNABEL, *Numerical methods for unconstrained optimization and nonlinear equations*, Prentice-Hall, 1983.

[4] E. J. HANNAN, *Multiple Time Series*, Wiley, New York, 1970.

[5] P. V. KABAILA AND G. C. GOODWIN, *On the estimation of the parameters of an optimal interpolator when the class of interpolators is restricted*, SIAM J. Control and Optimization, 18 (1980), pp. 121–144.

[6] E. C. LEVI, *Complex-curve fitting*, IRE Trans. on Automatic Control, AC-4 (1959), pp. 37–44.

[7] L. LJUNG, *System Identification - Theory for the User*, Prentice-Hall, Englewood Cliffs, N.J., 1987.

[8] B. M. NINNESS, *Stochastic and Deterministic Modeling*, PhD thesis, Dept. of Electrical Engineering, University of Newcastle, NSW, Australia, August, 1993.

[9] J. SCHOUKENS AND R. PINTELON, *Identification of Linear Systems: A Practical Guideline to Accurate Modeling*, Pergamon Press, London (U.K.), 1991.

[10] A. VAN DEN BOS, *Identification of continuous-time systems using multiharmonic test signals*, Identification of Continuous-Time Systems, (1992). Edited by Sinha and Rao, Kluwer Academic, Dordrecht (The Netherlands).

[11] B. WAHLBERG, *System identification using Laguerre models*, IEEE Trans. Automatic Control, AC-36 (1991), pp. 551–562.

[12] P. WHITTLE, *Hypothesis Testing in Time Series Analysis*, PhD thesis, Uppsala University, Almqvist and Wiksell, Uppsala. Hafner, New York , 1951.

SUPERVISORY CONTROL*

A.S. MORSE†

Abstract. This paper describes a simple, 'high-level' controller called a 'supervisor' which is capable of switching into feedback with a siso process, a sequence of linear positioning or set-point controllers from a family $\mathcal{F}_C$ of candidate controllers, so as to cause the output of the process to approach and track a constant reference input. The process is assumed to be modeled by a siso linear system whose transfer function is in the union of a number of subclasses, each subclass being small enough so that one of the controllers in $\mathcal{F}_C$ would solve the positioning problem, were the process's transfer function to be one of the subclass's members. The supervisor decides which controller to put in feedback with the process, not by an exhaustive search - i.e., by experimentally evaluating each and every candidate controller's performance by briefly applying it to the process - but rather by continuously comparing in real time suitably defined normed values of 'output estimation errors' generated by the candidate controllers, whether or not they are in feedback with the process. It is shown that in the absence of unmodelled process dynamics and disturbances, the proposed supervisor can successfully perform its function provided $\mathcal{F}_C$ is finitely large. A sequel to the full length version of this paper will analyze the performance of the same supervisor when unmodelled dynamics and disturbances are present.

1. Introduction. Plant models are inherently inaccurate, and controllers regulating processes described by such models must be able to ensure satisfactory closed-loop performance in the presence of exogenous process disturbances which cannot be measured. Modern linear control theories {e.g., pole-placement/observer theory, linear quadratic theory, H^∞ theory, and the like} are by now very highly developed and can be used to design controllers with such capabilities for processes admitting linear models provided the model uncertainties are time-invariant and 'sufficiently small'. But for 'large' model uncertainties derived from real-time changes in plant dynamics, common sense suggests and simple examples prove that no single, fixed-parameter linear controller can possible regulate in a satisfactory way. Such large uncertainties might arise in real time, because of changes in operating environment, component aging or failure, or perhaps a sudden change in plant dynamics due to an external influence.

To cope with these types of uncertainties obviously requires a controller a good deal 'smarter' than that which linear feedback theory can provide. What's clearly needed is a controller which can *change* or be changed in response to perceived changes in plant dynamics. If plant changes can be

* This paper is a condensed version of [1]. The research was supported by the Institute for Mathematics and its Applications with funds provided by the National Science foundation, by the National Science Foundation under Grant No. ECS-9012551, by the U. S. Air Force Office of Scientific Research under Grant No. F49620-92-J-0077, and by the Dutch Systems and Control Theory Network.

† Part of this author's research was done while he was in the Netherlands visiting the Department of Applied Mathematics, University of Twente, Enschede and the Mathematics Institute, University of Groningen. Department of Electrical Engineering, Yale University, New Haven, Connecticut 06520.

predicted in advance or can be directly measured when they occur, then controller gain scheduling will often suffice. But if plant changes cannot be predicted or directly measured, on-line controller selection or 'tuning' must be carried out using whatever real-time operating data is available.

The aim of this paper is to describe a simply-structured, 'high-level' controller called a 'supervisor' which is capable of switching into feedback with a siso process, a sequence of linear positioning or set-point controllers from a family of candidate controllers so as to cause the output of the process to approach and track a constant reference input. What makes the proposed supervisor distinctly different from other algorithms which might be used for the same purpose [6,4,2,7], is the philosophy underlying the method it uses to carry out its task. In particular, the supervisor decides which controller to put in feedback with the process, not by going through an an exhaustive search - i.e., by experimentally evaluating each and every candidate controller's performance by briefly applying it to the process - but rather by continuously comparing in real time suitably defined normed values of 'output estimation errors' generated by the candidate controllers, whether or not they are in feedback with the process. Motivation for this idea is obvious: the controller which generates the 'smallest' output estimation error ought to have the best idea of what the process is and thus should be able to do the best job of controlling the process. The origin of this idea is of course the concept of certainty equivalence from parameter adaptive control [12].

The subject of supervisory control will be treated in two papers. This is a condensed version of the first [8]. The specific problem to which the papers are addressed is formulated in §2. The process to be controlled is assumed to modeled by a linear system whose transfer function is in the union of a number of given subclasses $C(p)$, p being a parameter vector taking values in some set $\mathcal{P}$ over which the union is taken. $\mathcal{P}$ may be either a finite set or a closed, bounded subset of a real finite dimensional linear space. Each subclass $C(p)$ contains a *nominal process model* transfer function ν_p about which the subclass is "centered." It is assumed that for each subclass $C(p)$, there is single, given, two degree of freedom controller transfer matrix κ_p which would solve the positioning control problem were the process model transfer function to be any member of $C(p)$. Each nominal process model transfer function, controller transfer matrix pair (ν_p, κ_p) determines an "identifier-based" controller Σ_p with two outputs - one a candidate feedback control signal for the process, and the other an 'output estimation error' which is used by the supervisor to assess the controller's potential closed-loop performance §3. The problem of having to realize as many identifier-based controllers Σ_p as there are points in $\mathcal{P}$ is dealt with by requiring all such controllers to share the same state x_C. Σ_p's two outputs are then generated as parameter-dependent outputs of a single, finite dimensional linear system Σ_C with state x_C §3. The supervisor is a specially structured hybrid dynamical system Σ_S whose output is a switching signal σ_S taking values in $\mathcal{P}$ and whose inputs are x_C and the process output y §4. The value σ_S at each instant of time, specifies which controller is in feedback with the process. The supervisor decides what the value of σ_S should be on specific time intervals, by comparing suitably defined, normed output estimation errors or "performance signals" π_p, one for each value of $p \in \mathcal{P}$. The supervisor's inputs provide all the time-dependent data needs to evaluate $\pi_p(t)$ for each $p \in \mathcal{P}$. The value of σ_S between two successive switching times t_1 and t_2 is determined by minimizing $\pi_p(t_1 - \tau_C)$ over $\mathcal{P}$, τ_C being a prespecified constant bounding from above the time it takes the supervisor to carry out the minimization. Chattering is avoided by requiring σ_S to dwell at each of its values for at least τ_D time units, τ_D being a prespecified positive number called a *dwell time*.

Closed-loop system behavior is analyzed in §5 for a system employing $\mathcal{L}^2$ performance signals. It is assumed that $\mathcal{P}$ is finite and that the transfer function of the process to be controlled is equal to one of the nominal process model transfer functions; i.e., unmodelled dynamics are not taken into account. It is also assumed that there is no measurement noise or disturbances. The effects of measurement noise, disturbances and unmodelled dynamics will be treated in the sequel to the full-length version of this paper.

It is shown in §5 that if τ_D is sufficiently large, then for all initializations and set-point values, the process input and output as well as x_C, are bounded on $[0, \infty)$ and the process output tends to its specified set-point value. It is also shown that the supervisor will always correctly classify the process model's "dc gain" in finite time, so long as the specified set-point value is nonzero. The effects of alternative choices of performance signals are briefly discussed in §6. It is shown that systems employing exponentially weighted $\mathcal{L}^2$ performance signals have globally bounded states as well as all of the properties of the systems analyzed in §5. Finally, in §7 it is shown that all of these findings are valid not just for τ_D sufficiently large, but for any value of τ_D greater than zero.

Notation: In the sequel prime denotes transpose. $\mathbb{R}^{n \times m}$ is the linear space of real $n \times m$ matrices. The norm of $M \in \mathbb{R}^{n \times m}$, written $|M|$ is the sum of the magnitudes of its entries. For $x \in \mathbb{R}^n$, $||x||$ denotes the Euclidean norm $||x|| = \sqrt{x'x}$. If $\mathcal{S}$ is a subspace of $\mathbb{R}^n$, $\mathcal{S}^\perp$ is its orthogonal complement with respect to the inner product $x'y$. If $f : [0, \infty) \to \mathbb{R}^n$ and $g : [0, \infty) \to \mathbb{R}^n$ are piecewise-continuous time functions we sometimes write $f \to g$ if normed difference $|f - g|$ goes to zero. $\mathbb{R}_S$ denotes the space of all real, rational, proper, stable transfer functions. As such $\mathbb{R}_S$ is both a principal ideal domain [10] and a linear space. For $\alpha \in \mathbb{R}_S$, $||\alpha||_\infty$ denotes the norm

$$||\alpha||_\infty = \sup_{\omega \in \mathbb{R}} |\alpha(j\omega)|$$

2. Problem formulation. In the sequel, we assume that the process to be controlled is a siso controllable, observable linear system Σ_P; i. e.

$$\dot{x}_P = A_P x_P + b_P u$$

(2.1)

$$y = c_P x_P$$

We further assume that Σ_P's transfer function is a member of a known class of admissible transfer functions $\mathcal{C}_P$. We take $\mathcal{C}_P$ to be of the form

$$\mathcal{C}_P = \bigcup_{p \in \mathcal{P}} C(p)$$

where $\mathcal{P}$ is $\mathcal{C}_P$'s *parameter space* which may be either a finite set or a closed, bounded subset of a real, finite-dimensional linear space. For each fixed $p \in \mathcal{P}$, $C(p)$ denotes the subclass

$$C(p) = \left\{ \frac{\alpha_p + \delta_p}{1 - \beta_p} : ||\delta_p||_\infty \leq \epsilon_p \right\}$$

where α_p, β_p and δ_p are strictly proper transfer functions in $\mathbb{R}_S$ and ϵ_p is a real, non-negative number. Here

$$\nu_p \overset{\Delta}{=} \frac{\alpha_p}{1 - \beta_p}$$

is a preselected, *nominal transfer function* and δ_p is a norm-bounded perturbation representing unmodelled dynamics of either the additive or multiplicative types [3]. We assume for each $p \in \mathcal{P}$ that α_p and $1 - \beta_p$ are coprime, and that the allowable values of $\delta_p \in \mathbb{R}_S$ exclude transfer functions for which $\alpha_p + \omega_p \delta$ and $1 - \beta_p$ are not coprime. All transfer functions in $\mathcal{C}_P$ are thus strictly proper, but not necessarily stable rational functions. For each $p \in \mathcal{P}$, $C(p)$ is required be at least small enough so that it can be robustly stabilized with a single, fixed-parameter, linear controller. Of course $\mathcal{C}_P$ need not have this property.

The overall design goal is to construct a positioning or set point control system capable of causing y to approach and track any constant reference input r. Towards this end we first introduce a standard integrating subsystem for the tracking error $r - y$; i.e.,

(2.2)

$$\dot{z} = (r - y)$$

Note that the transfer function from $r - y$ to z can be written as

(2.3)

$$\frac{\zeta}{1 - a\zeta}$$

where $\zeta = \frac{1}{s+a}$ and a is a positive number . Doing this expresses the transfer function as a ratio of $\mathbb{R}_S$ transfer functions.

As our concern is mainly with supervisory control, we take as given, a parameterized family of "two-degree of freedom" controller transfer matrices of the form

$$\kappa_p = \left[\frac{\gamma_p}{1-\rho_p} \quad \frac{\psi_p}{1-\rho_p} \right], \quad p \in \mathcal{P}$$

where for each $p \in \mathcal{P}$, γ_p, ψ_p, and ρ_p are in $\mathbb{R}_S$, ρ_p being strictly proper and coprime with the greatest common divisor of γ_p and ψ_p. Each of these transfer matrices is considered to be a candidate for the feedback connection from $[z \ y]'$ to u {cf., Figure 2.1}. For reasons to be made clear in §3, we assume the following.

ASSUMPTION 1. $\alpha_p, \beta_p, \gamma_p, \rho_p$, and ψ_p, all have the same parameter-independent, monic, stable, denominator polynomial $\omega(s)$.

Assumption 1 implies that the set of nominal process model transfer functions $\mathcal{N} \triangleq \{\nu_p : p \in \mathcal{P}$ and the set of controller transfer matrices $\mathcal{K} \triangleq \{\kappa_p : p \in \mathcal{P}\}$, are subsets of real linear spaces of dimensions $2n$ and $3n$ respectively, n being the degree of $\omega(s)$. We shall require the dependence of ν_p and κ_p on p to be smooth:

ASSUMPTION 2. The functions $p \longmapsto \nu_p$ and $p \longmapsto \kappa_p$ are continuous[1] on $\mathcal{P}$

We shall not demand that the nominal process models in $\mathcal{N}$ be distinct. We shall however require there to be exactly one controller transfer matrix in $\mathcal{K}$ for all nominal process model transfer functions which are the same:

ASSUMPTION 3. $\alpha_p, \beta_p, \gamma_p, \rho_p$, and ψ_p, have the property that $\gamma_p = \gamma_q$, $\rho_p = \rho_q$, and $\psi_p = \psi_q$ for each pair of points $p, q \in \mathcal{P}$ at which $\alpha_p = \alpha_q$ and $\beta_p = \beta_q$.

Assumption 3 implies that the assignment $\nu_p \longmapsto \kappa_p$, $p \in \mathcal{P}$ is a well-defined function from $\mathcal{N}$ to $\mathcal{K}$. We shall require this function to be smooth:

ASSUMPTION 4. The aforementioned function is continuous on $\mathcal{N}$.

For fixed $p, q \in \mathcal{P}$, the feedback interconnection of nominal process model transfer function ν_q with error integrating transfer function (2.3), and controller transfer matrix κ_p determines the *characteristic transfer function*

$$(2.4) \qquad \phi_{pq} \triangleq (1 - a\zeta)(1 - \beta_q)(1 - \rho_p) + \alpha_q\{\zeta\gamma_p - \psi_p(1 - a\zeta)\}$$

whose zeros, after cancellation of the stable common factor $(s + a)\omega^3(s)$, are the closed-loop poles of the interconnection.

We assume that for each $p \in \mathcal{P}$, controller transfer matrix κ_p stabilizes nominal process model transfer function ν_p with "stability margin" λ_S. More precisely we assume the following:

ASSUMPTION 5. There is a positive number λ_S with the property that for each $p \in \mathcal{P}$, the real parts of all zeros of ϕ_{pp}, after cancellation of the stable common factor $(s + a)\omega^3(s)$, are no greater than $-\lambda_S$.

---

[1] Here and elsewhere when we speak of continuity on $\mathcal{P}$ we implicitedly assume $\mathcal{P}$ is a subset of a finite dimensional linear space.

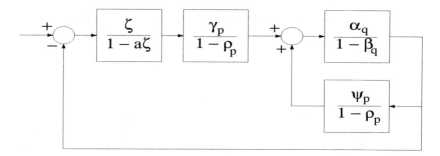

FIG. 2.1. *Feedback Interconnection*

We shall require the same of ω:

ASSUMPTION 6. The real parts of all zeros of ω are no greater than $-\lambda_S$.

In addition to the preceding we assume that for each $p \in \mathcal{P}$, controller transfer matrix κ_p at least robustly stabilizes its corresponding class of process model transfer functions $\mathcal{C}(p)$. By this we mean that for each $p \in \mathcal{P}$

$$[(1 - a\zeta)(1 - \beta_p)(1 - \rho_q) + (\alpha_p + \omega_p\delta_p)\{\zeta\gamma_q - \psi_q(1 - a\zeta)\}]^{-1} \in \mathbb{R}_S$$

for all admissible δ_p. As is well known, this will be true for $p \in \mathcal{P}$, just in case ϕ_{pp} has an inverse ϕ_{pp}^{-1} in $\mathbb{R}_S$ and, if $\epsilon_p > 0$,

$$||\phi_{pp}^{-1}\zeta\gamma_p||_\infty < \frac{1}{\epsilon_p}$$

It is also well known that for such controller transfer functions to exist, the numerator of each process model transfer function in $\mathcal{C}_P$ must be non-zero at $s = 0$. Using (2.4), together with the fact that $(1 - a\zeta)$ has a zero at $s = 0$, it is easy to see that ϕ_{pp} will have an inverse in $\mathbb{R}_S$ only if γ_p is nonzero at $s = 0$. For reasons to be made clear in the sequel, let us agree to call

$$(2.5) \qquad\qquad \Gamma_p \triangleq \frac{1 - \beta_p(0)}{\alpha_p(0)}$$

and

$$(2.6) \qquad\qquad \bar{\Gamma}_q \triangleq \left[\frac{1 - \rho_q(0)}{\gamma_q(0)} \quad \frac{\psi_q(0)}{\gamma_q(0)} \right]$$

the *dc gains* of nominal process model transfer function ν_p and controller transfer matrix κ_q respectively. In view of the preceding, for problem solvability we must make

ASSUMPTION 7. The norm of dc gain of each nominal process model transfer function and each controller transfer matrix is finite.

REMARK 2.1. In view of Assumption 2, $p \longmapsto \Gamma_p$ and $p \longmapsto \bar{\Gamma}_p$ are continuous functions on $\mathcal{P}$.♠

REMARK 2.2. The preceding equations, as well as those which follow, can easily be reduced to the equations relevant to the the somewhat simpler output regulation problem. This can be accomplished by by setting $r = 0$, $a = 0$, and $\zeta = 1$. For the regulation problem, Assumption 2 is unnecessary and γ_p may have a zero at zero. ♠

Roughly speaking, the problem of interest is to develop a supervisory logic for switching into the feedback loop between $[\,z \quad y\,]'$ and u a sequence of controllers with transfer matrices in the set $\mathcal{K}$ so as to achieve desired closed-loop 'performance'. By satisfactory performance we mean that for any constant input r, no signals grow without bound and $y \rightarrow r$. There are two important, practical requirements which make the problem especially challenging:

- Since we envision applications in which $\mathcal{P}$ might be infinitely large, we specifically rule out any supervisory logic which employs an exhaustive or 'round robin' search through the admissible controller set, along the lines of those switching algorithms considered previously in [6,4,2,7] and elsewhere.
- Since it is usually undesirable to intentional apply excitation to a system under operation, the supervisory controller is required to ensure satisfactory performance without the aid of a probing signal.

These two requirements have far reaching implications which are not widely appreciated. For example, in order to achieve acceptable closed-loop performance {e.g., stability and tracking} without an exhaustive search, the supervisor *must* be able to evaluate the potential closed-loop performance of each controller without actually trying it out in the feedback loop. Moreover, since process model classification {i.e., identification} cannot be accomplished without excitation, the supervisor must be capable of providing acceptable closed-loop performance {e.g., stability and tracking} *even under conditions which make correct process model classification impossible.*

3. State space systems. To deal with the assumption Σ_P's transfer function is in $\mathcal{C}_P$, but is otherwise unknown, we shall employ a parameterized family of "identifier-based" controllers $\{\Sigma_p \ : \ p \in \mathcal{P}\}$. Each Σ_p is a stable, linear system with inputs y, u, and z and outputs e_p and u_p. Here u_p is Σ_p's candidate control signal and e_p is Σ_p's "output estimation error;" e_p is used by the supervisor to assess Σ_p's potential closed-loop performance. The transfer matrix of Σ_p from $[\,y \quad u \quad z\,]'$ to $[\,e_p \quad u_p\,]'$ is specified to be

$$(3.1) \qquad \begin{bmatrix} \beta_p - 1 & \alpha_p & 0 \\ \psi_p & \rho_p & \gamma_p \end{bmatrix}$$

The preceding implies that no matter how Σ_p is realized, the following are true:

- The feedback law $u = u_p$ is well-defined since ρ_p, and Σ_P's transfer functions are strictly proper.
- If $u \triangleq u_p$, the transfer matrix from $[z \quad y]$ to u is the controller transfer matrix κ_p specified earlier.
- If the transfer function of Σ_P is the same as the nominal process model transfer function ν_p, then no matter how u is defined, $e_p \to 0$ as fast as $e^{-\lambda_s t}$.

We shall require all controllers to share the same state x_C. In other words, e_p and u_p will be generated as parameter-dependent outputs of a single, stable, finite dimensional linear system Σ_C with inputs y, u and z and state x_C. Σ_C is defined as follows: First pick any single-input controllable pair $(\tilde{A}, \tilde{b})$ in such a way that $\omega(s)$ is $\tilde{A}$'s characteristic polynomial {c.f., Assumption 1}. Second define $A_C = $ block diag.$\{\tilde{A}, \tilde{A}, \tilde{A}\}$, $d_C = $ block diag.$\{\tilde{b}, 0, 0\}$, $b_C = $ block diag.$\{0, \tilde{b}, 0\}$, and $h_C = $ block diag.$\{0, 0, \tilde{b}\}$. Third define c_p, f_p, g_p and h_p to be the unique solution to the equation

$$(3.2) \quad \begin{bmatrix} c_p \\ f_p \end{bmatrix} (sI - A_C)^{-1} [d_C \quad b_C \quad h_C] + \begin{bmatrix} 0 & 0 & 0 \\ h_p & 0 & g_p \end{bmatrix} = \begin{bmatrix} \beta_p & \alpha_p & 0 \\ \psi_p & \rho_p & \gamma_p \end{bmatrix}$$

The existence of the parameter-dependent vectors c_p, f_p, g_p and h_p is a direct consequence of Assumption 1.

Finally, define Σ_C to be the parameter dependent system

$$
\begin{aligned}
\dot{x}_C &= A_C x_C + d_C y + b_C u + h_C z \\
u_p &= f_p x_C + g_p z + h_p y \\
(3.3) \qquad e_p &= c_p x_C - y
\end{aligned}
$$

It is straightforward to verify that the transfer matrix of Σ_C from $[y \quad u \quad z]'$ to $[e_p \quad u_p]'$ is the transfer matrix of Σ_p specified in (3.1). It is also not difficult see that if Σ_P's transfer function were the same as nominal process model transfer function ν_p, then e_p would be of the form

$$(3.4) \qquad e_p = c_p e^{A_C t}(x_C(0) - M x_P(0))$$

where M is a time-independent matrix depending only on Σ_P and p. Finally we note that since the triple (c_p, A_C, h_C) has a zero transfer function,

$$(3.5) \qquad c_p A_C^{-1} h_C = 0, \quad p \in \mathcal{P}$$

The feedback control to the process can now be written as

$$(3.6) \qquad u = u_\sigma$$

where σ is a piecewise-constant, "switching signal" taking values in $\mathcal{P}$. Closing the "supervisory-loop" means setting

$$(3.7) \qquad \sigma = \sigma_S$$

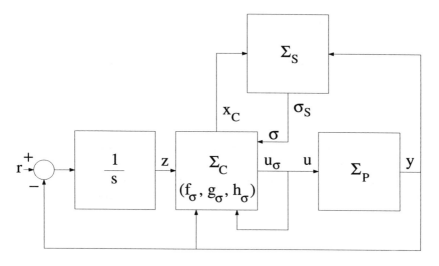

FIG. 3.1. *Closed-Loop Supervisory Control System*

where σ_S is the output of a suitably defined "supervisory control system" whose structure we discuss in the next section.

REMARK 3.1. The uniqueness of the parameter-dependent vectors c_p, f_p, g_p and h_p defined by (3.2) together with Assumption 2 ensures that $p \longmapsto c_p$, $p \longmapsto f_p$, $p \longmapsto g_p$, and $p \longmapsto h_p$ are continuous functions on $\mathcal{P}$. In addition, Assumptions 3 and 4 imply that the assignment $c_p \longmapsto (f_p, g_p, h_p)$ is a well-defined function, continuous function on $\{c_p : p \in \mathcal{P}\} \subset \mathbb{R}^{3n}$ ♠

REMARK 3.2. The preceding definitions together with Assumption 7 imply that $g_p - f_C A_C^{-1} h_C \neq 0$ and that

$$(3.8) \qquad\qquad c_p A_C^{-1} b_C \neq 0$$

These definitions also imply that

$$\Gamma_p = -\frac{1 + c_p A_C^{-1} d_C}{c_p A_C^{-1} b_C}$$

$$\bar{\Gamma}_p = \left[\frac{(1 + f_p A_C^{-1} b_C)}{g_p - f_C A_C^{-1} h_C} \quad \frac{h_p - f_p A_C^{-1} d_C}{g_p - f_C A_C^{-1} h_C} \right]$$

where Γ_p and $\bar{\Gamma}_p$ are the process model and controller dc gains defined previously by (2.5) and (2.6) respectively. ♠

4. Supervisor. By a supervisor is meant a specially structured causal dynamical system Σ_S whose output σ_S is a switching signal taking values in $\mathcal{P}$ and whose inputs are x_C and y. Internally a supervisor consists of two subsystems, one a a *performance weight generator* Σ_W and the other a *dwell-time switching logic* Σ_D. Σ_W is a causal dynamical system whose

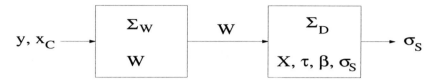

FIG. 4.1. *Supervisor* Σ_S

inputs are x_C and y and whose state and output W is a "weighting matrix" which takes values in a
linear space $\mathcal{W}$. W together with a suitably defined *performance function* $\Pi : \mathcal{W} \times \mathcal{P} \to \mathbb{R}$ determine, for each $p \in \mathcal{P}$, a scalar-valued *performance signal* of the form

$$(4.1) \qquad \pi_p = \Pi(W, p), \ p \in \mathcal{P}$$

which is viewed by the supervisor as a measure of the expected closed-loop performance of controller p. One possible pair of definitions for Σ_W and Π is

$$(4.2) \qquad \dot{W} = -\lambda W + \begin{bmatrix} x_C \\ -y \end{bmatrix} \begin{bmatrix} x_C \\ -y \end{bmatrix}'$$

and

$$(4.3) \qquad \Pi(W, p) = [\, c_p \quad 1\,]\, W\, [\, c_p \quad 1\,]'$$

respectively where λ is a positive number. In the light of (3.3) and (4.1) it is easy to see that these definitions imply that

$$(4.4) \qquad \dot{\pi}_p = -\lambda \pi_p + e_p^2, \quad p \in \mathcal{P}$$

Although we will be concerned primarily with performance signals of this form, it should be noted that it is possible to realize other types of performance signals by defining W and Π in other ways. For example, if $\mathcal{P}$ is a finite set {say $\mathcal{P} = \{1, 2, \ldots, m\}$} and if Σ_W is the dynamical system $\dot{w}_p = |e_p|$, $p \in \mathcal{P}$ with state $w \overset{\Delta}{=} [\, w_1 \quad w_2 \quad \cdots \quad w_m\,]'$, defining $\Pi(w, p) \overset{\Delta}{=} w_p$ would realize the $\mathcal{L}^1$ performance signal $\dot{\pi}_p = |e_p|$. Note however that if $\mathcal{P}$ were not finite, this particular performance signal could not be realized with $\mathcal{W}$ finite dimensional.

   The supervisor's other subsystem is a hybrid dynamical system Σ_D, whose input and output are W and σ_S respectively, and whose state is the ordered quadruple $\{X, \tau, \beta, \sigma_S\}$. Here X is a discrete-time matrix which takes on sampled values of W, τ is a continuous-time variable called a *timing signal*, and β is a logic variable taking values in $\{0, 1\}$. τ takes values in the closed interval $[0, \tau_D]$, where τ_D is a prespecified positive

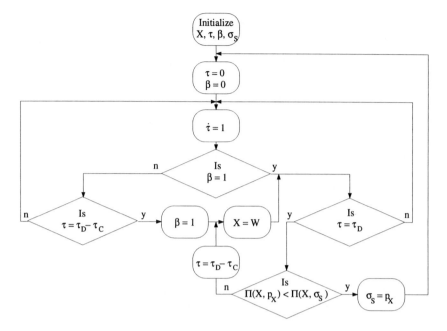

FIG. 4.2. *Dwell-Time Switching Logic* Σ_D

number called a *dwell time*. It is assumed that there is a second prespecified nonnegative number $\tau_C \le \tau_D$ called a *computation time*, which bounds from above for any $X \in \mathcal{W}$, the time it would take the supervisor to compute a value $p = p_X \in \mathcal{P}$ which minimizes $\Pi(X, p)$. Between "event times" τ is generated by a reset integrator according to the rule

$$\dot{\tau} = 1$$

Event times occur when the value of τ reaches either $\tau_D - \tau_C$ or τ_D; as such times τ is reset to either 0 or $\tau_D - \tau_C$ depending on the value of Σ_D's state. Σ_D's internal logic is defined by the following computer diagram in which p_X denotes a value of $p \in \mathcal{P}$ which minimizes $\Pi(X, p)$.

The functioning of Σ_D can be explained roughly as follows. Suppose that at some time t_0, Σ_S has just changed the value of σ_S to p. At this instant τ is reset to 0. After $\tau_D - \tau_C$ time units have elapsed, W is sampled and X is set equal to this value. During the next τ_C time units, a value $p = p_X$ is computed which minimizes $\Pi(X, p)$. At the end of this period, when $\tau = \tau_D$, if $\Pi(X, p_X)$ is smaller than $\Pi(X, \sigma_S)$, then σ_S is set equal to p_X, τ is reset to zero and the entire process is repeated. If on the other hand, $\Pi(X, \sigma_S)$ is less than or equal to $\Pi(X, p_X)$, τ is reset to $\tau_D - \tau_C$, W is again sampled, X takes on this new sampled value, minimization is again carried out over the next τ_C time units..... and so on.

REMARK 4.1. Note that Σ_S is *scale independent* in that its output

σ_S remains unchanged if its performance function-weighting matrix pair (Π, W) is replaced by another performance function-weighting matrix pair $(\bar{\Pi}, \bar{W})$ satisfying $\bar{\Pi}(\bar{W}, p) = \theta\Pi(W, p)$, $p \in \mathcal{P}$, where $\theta : [0, \infty) \to \mathbb{R}$ is a positive time function. This is because for any fixed t, the values of p which minimize $\Pi(W(t), p)$ are exactly the same as the values of p which minimize $\theta(t)\Pi(W(t), p)$. ♠.

5. Exact matching. Our aim here is to analyze the closed-loop behavior of a supervisory control system for the case when $\mathcal{P}$ is finite and there is a value $p^* \in \mathcal{P}$ for which nominal transfer function $\frac{\alpha_{p^*}}{1 - \beta_{p^*}}$ matches or equals that of process model Σ_P. For the present, we assume that $\lambda = 0$ and that Σ_W and Π are defined by (4.2) and (4.3) respectively. Therefore in this case, π_p is the $\mathcal{L}^2$ performance signal

$$(5.1) \qquad \dot{\pi}_p = e_p^2, \quad p \in \mathcal{P}$$

In the sequel we call a piecewise-constant function $\sigma : [0, \infty) \to \mathcal{P}$ *admissible* if it either switches values at most once, or if it switches more than once and the set of time differences between each two successive switching times is bounded below by a positive number μ. The supremum of such values of μ is σ's *dwell time*. Because of the definition of Σ_S in §4, it is clear its output σ_S will be admissible with dwell time no smaller than that of Σ_S. This means that switching cannot occur infinitely fast and thus that existence and uniqueness of solutions to the differential equations involved, is not an issue.

It is convenient at this point to rewrite the equations describing integrating subsystem (2.2), combined controller (3.3), and feedback law (3.6), in alternative form which will prove to be especially useful in the sequel. For this define $b = [0 \quad 1]'$ and for each $q \in \mathcal{P}$, set $d_q = -[d_C' + h_q b_C' \quad 1]'$ and $\bar{c}_q = [c_q \quad 0]$. In addition, for each pair of points p and q in $\mathcal{P}$, define matrices

$$(5.2) \qquad f_{pq} = [\qquad f_p + h_p c_q \qquad\qquad g_p \qquad]$$

$$(5.3) \qquad c_{pq} = [\qquad c_p - c_q \qquad\qquad 0 \qquad]$$

$$(5.4) \qquad A_{pq} = \begin{bmatrix} A_C + d_C c_q + b_C(f_p + h_p c_q) & h_C + b_C g_p \\ -c_q & 0 \end{bmatrix}$$

Using (2.3), (3.3) and (3.6), it is straightforward to verify that

$$(5.5) \qquad \dot{x} = A_{\sigma p^*} x + d_{p^*} e_{p^*} + br$$
$$(5.6) \qquad e_p = c_{pp^*} x + e_{p^*}, \quad p \in \mathcal{P}$$
$$(5.7) \qquad u = f_{\sigma p^*} x - h_\sigma e_{p^*}$$
$$(5.8) \qquad y = \bar{c}_{p^*} x - e_{p^*}$$

where x is the composite controller state $x = [x'_C \quad z]'$. It is important to note that these equations hold for any value of $p^* \in \mathcal{P}$ whether or not p^* is actually a point at which exact matching takes place. In other words, equations (5.5) to (5.8) have nothing to do with the matching assumption and thus can be used even when the assumption is not made.

REMARK 5.1. Assumptions 5 and 6 imply that for any fixed $p, q \in \mathcal{P}$, the matrix pair (c_{pq}, A_{pq}) is detectable with stability margin[2] λ_S. To understand why this is so, first note from (5.3) and (5.4) that for any p and q in $\mathcal{P}$, there is a vector k_{pq} such that

$$(5.9) \qquad A_{pq} + k_{pq}c_{pq} = A_{pp}$$

Using (2.3), (3.2), and (5.4) it is a simple matter to verify that the characteristic polynomial of A_{pq} equals $\omega(s)$ times the numerator of the characteristic transfer function ϕ_{pq} defined in (2.4). In view of Assumptions 5 and 6, $\lambda_S I + A_{pp}$ is a stability matrix. Thus $\lambda_S I + A_{pq}$ is an output injection away from a stability matrix which proves that $(c_{pq}, \lambda_S I + A_{pq})$ is detectable as claimed. ♠

Fix set-point value r and the initial values of the process model state x_P, integrator state z, controller state x_C and supervisor's state $\{W, X, \tau, \beta, \sigma_S\}$. The exact matching assumption made above means that no matter what u is, e_{p^*} must go to zero exponentially fast. This implies that e_{p^*} is bounded on $[0, \infty)$ and has a finite $\mathcal{L}^2[0, \infty)$ norm. Thus because of (5.1),

$$\lim_{t \to \infty} \pi_{p^*}(t) \stackrel{\Delta}{=} C^* < \infty$$

Let $\mathcal{P}^*$ denote the set of $p \in \mathcal{P}$ for which $\pi_p(t) \leq C^*$ for all $t \geq 0$. Clearly $p_* \in \mathcal{P}^*$ and

$$\int_0^\infty ||e_p||^2 dt < \infty, \quad p \in \mathcal{P}^*$$

Moreover, since $\mathcal{P}$ is finite and each π_p, $p \in \mathcal{P}$ is monotone nondecreasing, there must exist a time $\bar{t}$ beyond which for each $p \notin \mathcal{P}^*$, $\pi_p(t) > C^*$. Because of the way in which Σ_S is defined, this means that $\sigma_S \in \mathcal{P}^*$, $t \geq \bar{t} + \tau_D$. Let t_0 denote the least non-negative value of t for which

$$(5.10) \qquad \sigma_S \in \mathcal{P}^*, \quad t \geq t_0$$

With c_{pq} as in (5.3), let $\{c_{p_1 p^*}, c_{p_2 p^*}, \ldots, c_{p_m p^*}\}$ be a basis for the span of $\{c_{pp^*} : p \in \mathcal{P}^*\}$. Define $C = [c'_{p_1 p^*} \quad c'_{p_2 p^*} \quad \cdots \quad c'_{p_m p^*}]'$ and

$$(5.11) \qquad \bar{e} = Cx$$

---

[2] A matrix pair (C, A) is *detectable with stability margin* $\lambda \geq 0$, if $(C, \lambda I + A)$ is a detectable pair.

These definitions together with (5.6) imply that $e_{p_i} - e_{p^*}$ is the ith entry of $\bar{e}$. Since each such entry has a finite $\mathcal{L}^2[0, \infty)$ norm, $\bar{e}$ must have a finite $\mathcal{L}^2[0, \infty)$ norm as well. Note also that the definition of C implies that there must be a bounded function $s : \mathcal{P}^* \to \mathbb{R}^{m \times 1}$ for which

$$(5.12) \qquad s(p)C = c_{pp^*}, \ p \in \mathcal{P}^*$$

In view of this and Remark 5.1, it must be that the matrix pair $(C, \lambda_S I + A_{pp^*})$ is detectable for each $p \in \mathcal{P}^*$.

To proceed, let us note that for any appropriately sized, bounded, matrix-valued function $p \longmapsto K_p$, (5.5) can be rewritten as

$$(5.13) \qquad \dot{x} = (A_{\sigma p^*} + K_\sigma C)x - K_\sigma \bar{e} + d_{p^*} e_{p^*} + br$$

Suppose that a function K_p can be shown to exist for which $A_{\sigma p^*} + K_\sigma C$ is exponentially stable. Then because br is a constant and $\bar{e}$ and e_{p^*} have finite $\mathcal{L}^2[0, \infty)$ norms, x will be bounded on $[0, \infty)$. Since e_{p^*} is bounded on $[0, \infty)$, (5.7) and (5.8) would imply boundedness of u and y as well. As a consequence x_P would have to be bounded too since Σ_P is observable. In other words, to show that x_P, z, and x_C are bounded on $[0, \infty)$, its enough to show that $A_{\sigma p^*} + K_\sigma C$ is exponentially stable for some suitably defined function K_p.

Let σ be an admissible switching signal with dwell time no smaller than τ_D. In view of (5.10), suppose that for $t \geq t_0$, $\sigma(t) \in \mathcal{P}^*$. We claim that a stabilizing function K_p exists provided τ_D is sufficiently large. To understand why this is so, suppose that k_{pp^*} is defined as in (5.9) so that $A_{pp^*} + k_{pp^*} c_{pp^*} = A_{pp}$, $p \in \mathcal{P}^*$. With $s(p)$ as in (5.12), define $K_p = k_p s(p)$ for $p \in \mathcal{P}^*$ and $K_p = 0$ for $p \notin \mathcal{P}^*$. Thus $A_{\sigma p^*} + K_\sigma C = A_{\sigma\sigma}$ for $t \geq t_0$. Therefore if ϕ and $\tilde{\phi}$ are the state transition matrices of $A_{\sigma p^*} + K_\sigma C$ and $A_{\sigma\sigma}$ respectively, then clearly $\phi(t, \mu) = \tilde{\phi}(t, \mu)$, $t \geq \mu \geq t_0$. Hence by the composition rule for state transition matrices,

$$(5.14) \qquad \phi(t, \mu) = \tilde{\phi}(t, t_0)\phi(t_0, \mu), \ t \geq \mu \geq 0$$

Now it is well known that

$$|\phi(t, \mu)| \leq e^{b(t-\mu)}, \ t \geq \mu \geq 0$$

where $b = \sup_{p \in \mathcal{P}} |A_{pp^*} + K_p C|$. Suppose that $A_{\sigma\sigma}$ is exponentially stable. Then there would be constants $\tilde{a} \geq 0$ and $\lambda > 0$ such that

$$|\tilde{\phi}(t, \mu)| \leq e^{(\tilde{a} - \lambda(t-\mu))}, \ t \geq \mu \geq 0$$

It would then follow from (5.14) that

$$|\phi(t, \mu)| \leq e^{(a - \lambda(t-\mu))}, \ t \geq \mu \geq 0$$

where $a = \tilde{a} + (b + \lambda)t_0$. In other words, $A_{\sigma p^*} + K_\sigma C$ will be exponentially stable if $A_{\sigma\sigma}$ is. We now explain how to compute a lower bound for τ_D for which the latter is true.

Since $\lambda_S I + A_{pp}$ is a stability matrix for each fixed $p \in \mathcal{P}$ {cf., Remark 5.1}, it is possible to find numbers $a_p \geq 0$ and $\lambda_p \geq \lambda_S > 0$ for which

$$(5.15) \qquad |e^{A_{pp}t}| \leq e^{(a_p - \lambda_p t)} \quad t \geq 0, \; p \in \mathcal{P}$$

Since $\frac{a_p}{\lambda_p}$ is an upper bound on the time it takes for $|e^{A_{pp}t}|$ to drop below one in value, it is perhaps not surprising that $A_{\sigma\sigma}$ will be exponentially stable provided

$$(5.16) \qquad \tau_D > \sup_{p \in \mathcal{P}} \left\{ \frac{a_p}{\lambda_p} \right\}$$

This in fact is an immediate consequence of the following lemma.

LEMMA 5.1. *Let* $\{A_p : p \in \mathcal{P}\}$ *be a closed, bounded set of real,* $n \times n$ *matrices. Suppose that for each* $p \in \mathcal{P}$, A_p *is stable and let* a_p *and* λ_p *be any finite, nonnegative and positive numbers respectively for which*

$$(5.17) \qquad |e^{A_p t}| \leq e^{(a_p - \lambda_p t)}, \quad t \geq 0$$

Suppose that τ_0 *is a number satisfying*

$$(5.18) \qquad \tau_0 > \sup_{p \in \mathcal{P}} \left\{ \frac{a_p}{\lambda_p} \right\}$$

For any admissible switching function $\sigma : [0, \infty) \to \mathcal{P}$ *with dwell time no smaller than* τ_0, *the state transition matrix of* A_σ *satisfies*

$$(5.19) \qquad |\Phi(t, \mu)| \leq e^{(a - \lambda(t - \mu))}, \quad \forall \, t \geq \mu \geq 0$$

where

$$(5.20) \qquad a \; = \; \sup_{p \in \mathcal{P}} \{a_p\}$$

$$(5.21) \qquad \lambda \; = \; \inf_{p \in \mathcal{P}} \left\{ \lambda_p - \frac{a_p}{\tau_0} \right\}$$

$$(5.22)$$

Moreover,

$$(5.23) \qquad \lambda \in (0, \lambda_p], \quad p \in \mathcal{P}$$

The lemma implies that if σ "dwells" at each of its values in $\mathcal{P}$ long enough for the norm of the state transition matrix of A_p to drop to 1 in value {i.e., at least τ_0 time units}, then A_σ will be exponentially stable with a decay rate λ no larger than the smallest of the decay rates of the A_p, $p \in \mathcal{S}$. A proof of this lemma is given in [8].

Let Σ denote the closed-loop supervisory control system consisting of process model Σ_P described by (2.1), integrating subsystem (2.2), shared controller Σ_C defined by (3.3), feedback law (3.6), switching law (3.7), performance weight generator Σ_W defined by (4.2) with $\lambda = 0$, performance function Π given by (4.3), and the dwell-time switching logic Σ_D described in §4,

THEOREM 5.1. *Let τ_D be any number satisfying (5.16). Suppose that Σ_P's transfer function equals nominal process model transfer function ν_p for some $p = p^* \in \mathcal{P}$. Then for each constant set-point value r and each initial state $\{x_P(0), z(0), x_C(0), W(0), X(0), \tau(0), \beta(0), \sigma_S(0)\}$ the sub-state response $\{x_P, z, x_C, \tau, \beta, \sigma_S\}$ of the supervisory control system Σ is bounded on $[0, \infty)$.*

We now focus on the problem of showing that $y \to r$. To do this, we first note that if x were to have a limit, then z would have a limit too since z is a component of x. y would also have a limit because of (5.8). In view of Theorem 1, these limits would be finite. As a consequence of (2.2), y would therefore have to tend to r as desired. In other words, to prove that $y \to r$, it is enough to show that x has a limit.

First suppose that $r = 0$ and assume that τ_D satisfies (5.16). Under these conditions x must have 0 as a limit. To understand why, note first that because τ_D satisfies (5.16), $A_{\sigma p^*} + K_\sigma C$ is exponentially stable. Next recall that $\bar{e}$ and e_{p^*} both have finite $\mathcal{L}^2[0, ,\infty)$ norms. From these facts and (5.13) it follows that x goes to zero as claimed.

Now suppose that $r \neq 0$. Note that if σ were to eventually stop switching and remain fixed at some value $\tilde{\sigma}$, then because of the definition of K_p, (5.13) could be written as

$$(5.24) \qquad \dot{x} = A_{\tilde{\sigma}\tilde{\sigma}} x - K_{\tilde{\sigma}} \bar{e} + d_{p^*} e_{p^*} + br$$

Since $A_{\tilde{\sigma}\tilde{\sigma}}$ is a stability matrix and e_{p^*} and $\bar{e}$ have finite $\mathcal{L}^2[0, \infty)$ norms, x would go to a limit $\tilde{x}$ for which $A_{\tilde{\sigma}\tilde{\sigma}}\tilde{x} + br = 0$. Although there is no apparent reason to assume that σ will in fact stop switching, it is nevertheless worthwhile to express in more detail just what the form of such a limit would be.

LEMMA 5.2. *Let $p, q \in \mathcal{P}$ be fixed. The value of x_{pq} for which*

$$A_{pq} x_{pq} + b = 0$$

is given by

$$x_{pq} = - \begin{bmatrix} A_C^{-1} d_C \\ 0 \end{bmatrix} - \begin{bmatrix} A_C^{-1} b_C & A_C^{-1} h_C \\ 0 & -1 \end{bmatrix} \begin{bmatrix} \Gamma_q \\ \Delta_{pq} \end{bmatrix}$$

where

$$\Delta_{pq} = \bar{\Gamma}_p \begin{bmatrix} \Gamma_q \\ -1 \end{bmatrix}$$

and Γ_q and $\bar{\Gamma}_p$ are the dc gains of nominal process model transfer function q and controller transfer matrix p respectively. Moreover,

$$(5.25) \qquad c_{pq}x_{pq} = (\Gamma_p - \Gamma_q)(c_p A_C^{-1} b_C) = c_{pq}x_{qq}$$

This lemma's proof utilizes the state space formulas for Γ_p and $\bar{\Gamma}_q$ noted previously in Remark 3.2. The proof is by direct verification and is therefore omitted.

Suppose that $r \neq 0$ and that switching stops with σ subsequently constant at some value $\tilde{\sigma} \in \mathcal{P}^*$. Then because of Lemma 5.2, $x \to rx_{\tilde{\sigma}\tilde{\sigma}}$ In addition, since $\tilde{\sigma} \in \mathcal{P}^*$, $e_{\tilde{\sigma}}$ must have a finite $\mathcal{L}^2[0, \infty)$ norm. In view of (5.6), $c_{\tilde{\sigma}p^*}x$ must have a finite $\mathcal{L}^2[0, \infty)$ norm too. This can occur only if $c_{\tilde{\sigma}p^*}x_{\tilde{\sigma}\tilde{\sigma}} = 0$, since $x \to rx_{\tilde{\sigma}\tilde{\sigma}}$. It follows from (5.25) and (3.8) that $\Gamma_{\tilde{\sigma}} = \Gamma_{p^*}$.

COROLLARY 5.1. *Let the hypotheses of Theorem 1 hold.*
- *If $r = 0$, then $x \to 0$ and $y \to 0$.*
- *If $r \neq 0$ and switching stops, then $y \to r$ and $\Gamma_\sigma \to \Gamma_{p^*}$.*

We now turn to the considerably more difficult case when switching is not presumed to stop. We will prove that even in this case $\Gamma_\sigma \to \Gamma_{p^*}$ if $r \neq 0$. We will also prove that $y \to r$.

With t_0 as in (5.10), let $t_1, t_2 \ldots$ denote the times greater than t_0 at which σ switches. Define $\sigma' : [t_0, \infty) \to \mathcal{P}$ to be the piecewise-constant switching signal whose value on $[t_{i-1}, t_i)$ is the same as the value of σ on $[t_i, t_{i+1})$. For each $p \in \mathcal{P}^*$ write $\{p\}^\Gamma$ for the equivalence class

$$\{p\}^\Gamma = \{q : \Gamma_q = \Gamma_p, \ q \in \mathcal{P}\}$$

induced by the relation $p \equiv^\Gamma q \iff \Gamma_q = \Gamma_p, \ p, q \in \mathcal{P}^*$.

THEOREM 5.2. *Let the hypotheses of Theorem 1 hold. If $r \neq 0$ then*

$$\Gamma_\sigma \quad \to \quad \Gamma_{p^*}$$
$$x_{\sigma p^*} \quad \to \quad x_{\sigma' p^*}$$

In addition there exists a time $t^* < \infty$ and a value $q^* \in \{p^*\}^\Gamma$ such that for $t \geq t^*$

$$\Gamma_\sigma \quad = \quad \Gamma_{p^*}$$
$$x_{\sigma p^*} \quad = \quad x_{q^* p^*}$$

Moreover

$$y \quad \to \quad r$$
$$x \quad \to \quad x_{q^* p^*}$$

The theorem asserts that even if switching does not stop, so long as $r \neq 0$ σ must tend to a subset of $\mathcal{P}$ on which all dc process model gains are the same as that of Σ_P. The theorem also says that the difference between any two consecutive of values of candidate limit points for x, namely $x_{\sigma p^*}$ and $x_{\sigma' p^*}$, must tend to 0 as $t \to \infty$. The theorem's second set of claims state that set-point tracking is achieved asymptotically and that the supervisor finds the correct dc process model gain equivalence class in finite time. The remainder of this section is devoted to the proof of these results.

To proceed it is helpful to use instead of (5.5) and (5.11), the "extended state space" model

$$(5.26) \qquad \dot{\bar{x}} = \bar{A}_{\sigma p^*} \bar{x} + \bar{d} e_{p^*}$$

$$(5.27) \qquad \bar{e} = \bar{C} \bar{x}$$

where $\bar{x} = [\, x' \quad r \,]'$, $\bar{d} = [\, d'_{p^*} \quad 0 \,]'$, and

$$(5.28) \qquad \bar{C} \triangleq [\, C \quad 0 \,]$$

$$(5.29) \qquad \bar{A}_{pq} \triangleq \begin{bmatrix} A_{pq} & b \\ 0 & 0 \end{bmatrix}, \quad p, q \in \mathcal{P}$$

We need to establish certain properties of such matrix pairs. For this let $\bar{\mathcal{P}}$ denote the set of $p \in \mathcal{P}$ such that c_{pp^*} is in the row span of C. Clearly $\mathcal{P}^* \subset \bar{\mathcal{P}}$. In addition, since $c_{pp^*} = [\, c_p \quad -c_{p^*} \,]$ and $p \longmapsto c_p$ is continuous on $\mathcal{P}$ {cf. Remark 3.1}, $\bar{\mathcal{P}}$ is a closed set.

For $p, k \in \bar{\mathcal{P}}$, let $\Phi_{pk}(t, -\tau_D)$ denote the state transition matrix of

$$\tilde{A}_{pk}(t) \triangleq \begin{cases} \bar{A}_{pp^*} & \text{if } t \in [-\tau_D, 0) \\ \bar{A}_{kp^*} & \text{if } t \in [0, \tau_D) \end{cases}$$

and write G_{pk} for the observability Gramian

$$G_{pk} \triangleq \int_{-\tau_D}^{\tau_D} \Phi'_{pk}(t, -\tau_D) \bar{C}' \bar{C} \Phi_{pk}(t, -\tau_D) dt$$

Observe that

$$(5.30) \qquad |\bar{C} \Phi(t, -\tau_D) \bar{d}| \le \eta, \quad t \in [-\tau_D, \tau_D]$$

where $\eta = |\bar{C}||\bar{d}| \sup_{p \in \bar{\mathcal{P}}} e^{2\tau_D |\bar{A}_{pp^*}|}$.

LEMMA 5.3. *Fix* $p, k \in \bar{\mathcal{P}}$ *and let* $\bar{n}$ *denote the size of* G_{pk}. *If* $[\, \Gamma_p \quad \Delta_{pp^*} \,] \ne [\, \Gamma_{p^*} \quad \Delta_{kp^*} \,]$ *then* $G_{pk} \ne 0$ *and*

$$(5.31) \qquad \begin{bmatrix} w \\ r \end{bmatrix}' G_{pk} \begin{bmatrix} w \\ r \end{bmatrix} \ge \mu_{pk} r^2, \quad \forall w \in \mathbb{R}^{\bar{n}-1}, \; r \in \mathbb{R}$$

where μ_{pk} *is the smallest nonzero eigenvalue of* G_{pk}.

A proof of this lemma is given in [8].

Proof of Theorem 2: For $i \geq 1$, let p_i be the value of σ on $[t_{i-1}, \ t_i)$ and let x_i denote the value of x at time $t_i - \tau_D$. It will first be shown that $\Gamma_\sigma \rightarrow \Gamma_{p^*}$ and that $\Delta_{\sigma p^*} - \Delta_{\sigma' p^*} \rightarrow 0$. To prove this is so, suppose that the contrary is true. Then there must be a positive number ϵ and an infinite subsequence of switching times $\{t_{i_1}, t_{i_2}, \ldots\}$ such that $|\Gamma_{p_{i_j}} - \Gamma_{p^*}| + |\Delta_{p_{i_j} p^*} - \Delta_{p_{i_j} + 1 p^*}| \geq \epsilon, \ \forall j \geq 1$. Let Ω_ϵ denote all pairs $(p, k) \in \bar{P} \times \bar{P}$ for which $|\Gamma_p - \Gamma_{p^*}| + |\Delta_{pp^*} - \Delta_{kp^*}| \geq \epsilon$. Since $\bar{P} \times \bar{P}$ is closed, and Γ_p and Δ_{pp^*} are continuous functions on P { cf. Remark 2.1, Lemma 5.2} Ω_ϵ must be closed as well. Moreover $(p_{i_j}, p_{i_j+1}) \in \Omega_\epsilon, \ j \geq 1$. By Lemma 5.3, $G_{pk} \neq 0, \ (p, k) \in \Omega_\epsilon$. For each such pair, let μ_{pk} denote the smallest nonzero eigenvalue of G_{pk}. Since Ω_ϵ is closed, it follows that

$$\mu^* \overset{\Delta}{=} \inf_{(p,k) \in \Omega_\epsilon} \mu_{pk}$$

is positive. In view of Lemma 5.3,

$$(5.32) \qquad \begin{bmatrix} x \\ r \end{bmatrix}' G_{p_{i_j} p_{i_j+1}} \begin{bmatrix} x \\ r \end{bmatrix} \geq \mu^*, \ \forall j \geq 1, \ x \in \mathbb{R}^{\bar{n}-1}$$

Since $\bar{e}$ and e_{p^*} have finite $\mathcal{L}^2[0, \infty)$ norms, there must be an integer m so large that

$$\mu^* m > 4 \int_{t_0}^\infty (\|\bar{e}\|^2 + (2\eta\tau_D)^2 e_{p^*}^2) dt$$

Then using (5.32) there follows

$$(5.33) \qquad \sum_{i=1}^{i_m} \begin{bmatrix} x_i \\ r \end{bmatrix}' G_{p_i p_{i+1}} \begin{bmatrix} x_i \\ r \end{bmatrix} \geq \sum_{j=1}^m \begin{bmatrix} x_{i_j} \\ r \end{bmatrix}' G_{p_{i_j} p_{i_j+1}} \begin{bmatrix} x_{i_j} \\ r \end{bmatrix} \geq m\mu^* >$$
$$4 \int_{t_0}^\infty (\|\bar{e}\|^2 + (2\eta\tau_D)^2 e_{p^*}^2) dt$$

On the other hand, from (5.26) and (5.27)

$$\bar{e}(t) = \bar{C}\Phi_{p_i p_{i+1}}(t - t_i, -\tau_D) \begin{bmatrix} x_i \\ r \end{bmatrix} + \int_{-\tau_D}^{t-t_i} \bar{C}\Phi_{p_i p_{i+1}}(t - t_i, s) \bar{d} e_{p^*}(s + t_i) ds$$

(5.34)

for $t - t_i \in [-\tau_D, \tau_D], \ i \geq 1$. With reference to (5.30), note that

$$\left\| \int_{-\tau_D}^{t-t_i} \bar{C}\Phi_{p_i p_{i+1}}(t - t_i, s) \bar{d} e_{p^*}(s + t_i) ds \right\|^2 \leq \left\| \eta \int_{-\tau_D}^{t-t_i} |e_{p^*}|(s + t_i) ds \right\|^2$$

$$\leq 2\eta^2 \tau_D \int_{-\tau_D}^{t-t_i} e_{p^*}^2(s + t_i) ds$$

$$= 2\eta^2 \tau_D \int_{t_i - \tau_D}^{t} e_{p^*}^2(s) ds$$

From this, (5.34) and the definition of $G_{p_ip_{i+1}}$ it follows that

$$\begin{bmatrix} x_i \\ r \end{bmatrix}' G_{p_ip_{i+1}} \begin{bmatrix} x_i \\ r \end{bmatrix} \leq 2\int_{t_i-\tau_D}^{t_i+\tau_D}(\|\bar{e}(t)\|^2 + 2\eta^2\tau_D\int_{t_i-\tau_D}^{t}e_{p^*}^2(s)ds)dt$$

$$\leq 2\int_{t_i-\tau_D}^{t_i+\tau_D}(\|\bar{e}(t)\|^2 + (2\eta\tau_D)^2e_{p^*}^2(t))dt$$

Therefore

$$\sum_{i=1}^{i_m}\begin{bmatrix} x_i \\ r \end{bmatrix}' G_{p_ip_{i+1}} \begin{bmatrix} x_i \\ r \end{bmatrix} = 2\sum_{i=1}^{i_m}\int_{t_i-\tau_D}^{t_i+\tau_D}(\|\bar{e}\|^2 + (2\eta\tau_D)^2e_{p^*}^2)dt$$

$$\leq 2\sum_{i=1}^{i_m}\int_{t_i-1}^{t_i+1}(\|\bar{e}\|^2 + (2\eta\tau_D)^2e_{p^*}^2)dt$$

$$\leq 4\int_{t_0}^{\infty}(\|\bar{e}\|^2 + (2\eta\tau_D)^2e_{p^*}^2)dt$$

which contradicts (5.33). Therefore $\Gamma_\sigma \to \Gamma_{p^*}$ and $\Delta_{\sigma p^*} - \Delta_{\sigma' p^*} \to 0$ as claimed. The latter together with the definition of x_{pq} in Lemma 2, imply that $x_{\sigma p^*} - x_{\sigma' p^*} \to 0$.

Now suppose that $\mathcal{P}$ is a finite set. Because $\mathcal{P}$ is finite and $\sigma \to \{p^*\}^\Gamma$, after some finite time t_1, σ must enter $\{p^*\}^\Gamma$ and remain there indefinitely.

For each $p \in \{p^*\}^\Gamma$, let $\langle p \rangle$ denote the equivalence class $\langle p \rangle \triangleq \{q : \Delta_{qp^*} = \Delta_{pp^*}, q \in \{p^*\}^\Gamma\}$. Since $\mathcal{P}$ is a finite set, there must be a positive number d such that $|\Delta_{pp^*} - \Delta_{qp^*}| > d$ if p and q are in distinct equivalence classes in $\{p^*\}^\Gamma$; i.e., if $\langle p \rangle \neq \langle q \rangle$. From this and the fact that and $\Delta_{\sigma p^*} - \Delta_{\bar{\sigma} p^*} \to 0$ it follows that beyond some finite time $t^* \geq t_1$, σ must be remain in one such class. Let $\langle q^* \rangle$ denotes this class. Therefore for $t \geq t^*$, $\Gamma_\sigma = \Gamma_{p^*}$ and $\Delta_{\sigma p^*} = \Delta_{q^* p^*}$, $t \geq t^*$. It follows from Lemma 2 that for such values of t, $x_{\sigma p^*} = x_{q^* p^*}$.

To complete the proof, it is enough to show that $x \to x_{q^* p^*}$. To establish that this is so, redefine C, much like just below (5.11), so that its rows are now a basis for the span of the vectors $\{c_{pp^*} : p \in \langle q^* \rangle\}$. Just like before and for the came reasons, (C, A_{pp^*}) is detectable for each $p \in \langle q^* \rangle$ and $\bar{e} \triangleq Cx$ has a finite $\mathcal{L}^2[0, \infty)$ norm.

In view of Lemma 2, $x_{pp^*} = x_{q^* p^*}$, $p \in \langle q^* \rangle$ since $\Delta_{pp^*} = \Delta_{q^* p^*}$ for such values of p. ¿From this and Lemma 2 it follows that $c_{pp^*}x_{q^* p^*} = c_{pp^*}x_{pp^*} = (\Gamma_p - \Gamma_{p^*})(c_p A^{-1}b_C) = 0, p \in \langle q^* \rangle$. It follows from the definition of C that $Cx_{q^* p^*} = 0$.

Let K_p be any bounded matrix on $\mathcal{P}$ which exponentially stabilizes $A_{\sigma p^*} + K_\sigma C$. Such a matrix exists because K_p can be defined (for example) so that for $t \geq t^*$, $A_{\sigma p^*} + K_\sigma C = A_{\sigma\sigma}$, just as before.

Now use the preceding and (5.5) to write

$$\frac{d\{x - x_{q^* p^*}\}}{dt} = (A_{\sigma p^*} + K_\sigma C)\{x - x_{q^* p^*}\} - K_\sigma\bar{e} + d_{p^*}e_{p^*}$$

Since $\bar{e}$ and e_{p^*} both have finite $\mathcal{L}^2[0, \infty)$ norms and $A_{\sigma p^*} + K_\sigma C$ is exponentially stable it must be that $\{x - x_{q^* p^*}\} \to 0$. This completes the proof.
□

REMARK 5.2. Note that the only requirements of the matrix function K_p used in the preceding proof are that it be bounded on $\mathcal{P}$, and that it exponentially stabilize $A_{\sigma p^*} + K_\sigma C$.

6. Performance signals. One of the problems with the preceding is that for $r \neq 0$, the weighting matrix generated by (4.2) with $\lambda = 0$ will typically not remain bounded. There are several ways to remedy this problem.

Under the exact matching hypothesis, (3.4) holds. Since ω^3 is the characteristic polynomial of A_C, by Assumption 6 all of A_C's eigenvalues have real parts no greater than $-\lambda_S$. Thus there is a non-negative constant C_0 such that $e_{p^*}^2(t) \leq C_0 e^{-\lambda_S t}$. Pick $\lambda \in (0, \lambda_S)$. Let Π and π_p be defined as in (4.1) and (4.3) respectively, but rather than using (4.2) to generate W, use the equation

$$
(6.1) \qquad \dot{W} = e^{\lambda t} \begin{bmatrix} X_C \\ -y \end{bmatrix} \begin{bmatrix} X_C \\ -y \end{bmatrix}'
$$

instead. Clearly

$$
\dot{\pi}_p = e^{\lambda t} e_p^2
$$

As defined, π_p has three crucial properties:
1. For each $p \in \mathcal{P}$, π_p is monotone nondecreasing.
2. $\lim_{t \to \infty} \pi_{p^*} \triangleq C^* \leq \pi_{p^*}(0) + \int_0^\infty C_0 e^{-(\lambda_S - \lambda)t} dt < \infty$
3. If $\mathcal{P}^*$ is again defined to be all $p \in \mathcal{P}$ for which $\pi_p(t) \leq C^*$, $\forall t > 0$, then $p^* \in \mathcal{P}^*$ and for each $p \in \mathcal{P}^*$, e_p has a finite $\mathcal{L}^2[0, \infty)$ norm.

These are precisely the properties needed to define C and $\bar{e}$ as in (5.12) so that $\bar{e}$ has a finite $\mathcal{L}^2[0, \infty)$ norm and that (C, A_{pp^*}) is detectable for each $p \in \mathcal{P}^*$. In other words, if one were to use (6.1) to generate W, then Theorems 1 and 2 and Corollary 1 would still hold.

Now consider replacing W with the "scaled" weighting matrix

$$
(6.2) \qquad \bar{W} \triangleq e^{-\lambda t} W
$$

Note that $\Pi(\bar{W}, p) = e^{-\lambda t} \Pi(W, p)$, $p \in \mathcal{P}$. In the light of the scale independence property of Σ_S noted previously in Remark 4.1, it must be that replacing W with $\bar{W}$ has no effect on σ_S so Theorem 1 still holds. The key point here is that the weighting matrix $\bar{W}$ defined by (6.2) can also be generated directly by the differential equation

$$
(6.3) \qquad \dot{\bar{W}} = -\lambda \bar{W} + \begin{bmatrix} X_C \\ -y \end{bmatrix} \begin{bmatrix} X_C \\ -y \end{bmatrix}'
$$

Moreover, since Theorem 1 asserts boundedness of y and x_C, it must be that $\bar{W}$ { and therefore X} are bounded as well.

Another possible weighting matrix, with the same essential properties as the preceding, is generated by the nonlinear equation

$$\dot{\tilde{W}} = (1 - \mathrm{tr}\{\tilde{W}\})\left(\begin{bmatrix} X_C \\ -y \end{bmatrix}\begin{bmatrix} X_C \\ -y \end{bmatrix}' - \left(\lambda + \mathrm{tr}\left\{\begin{bmatrix} X_C \\ -y \end{bmatrix}\begin{bmatrix} X_C \\ -y \end{bmatrix}'\right\}\right)\tilde{W}\right)$$

(6.4)

with initial state $\tilde{W}_0$ being any positive semidefinite matrix with trace less than 1. { Here $\mathrm{tr}\{\cdot\}$ denotes trace.} To understand why this is so we need

LEMMA 6.1. *Let $E : [0, \infty) \to \mathbb{R}^{n \times n}$ be a piecewise-continuous, positive-semidefinite, matrix valued function. For each positive semidefinite matrix X_0 with trace less than 1, the matrix*

(6.5)
$$X = \frac{\frac{e^{-\lambda t}}{1 - \mathrm{tr}\{X_0\}}X_0 + \int_0^t e^{-\lambda(t-s)}E(s)ds}{1 + \mathrm{tr}\{\frac{e^{-\lambda t}}{1 - \mathrm{tr}\{X_0\}}X_0 + \int_0^t e^{-\lambda(t-s)}E(s)ds\}}$$

uniquely solves the differential equation

(6.6)
$$\dot{X} = (1 - \mathrm{tr}\{X\})(E - (\lambda + \mathrm{tr}\{E\})X)$$

with the initial condition $X(0) = X_0$ and is bounded on $[0, \infty)$.
The proof of this lemma is by direct verification and will not be given.

Suppose that $\bar{W}$ is a weighting matrix generated by (6.3) with initial state $\bar{W}_0 = \frac{\tilde{W}_0}{1 - \mathrm{tr}\{\tilde{W}_0\}}$ Then because of (6.5), $\Pi(\tilde{W}, p) = \theta\Pi(\bar{W}, p)$, $p \in \mathcal{P}$ where θ is the positive function

$$\theta = \frac{1}{1 + \mathrm{tr}\{\frac{e^{-\lambda t}}{1 - \mathrm{tr}\{\tilde{W}_0\}}W_0 + \int_0^t e^{-\lambda(t-s)}E(s)ds\}}$$

and $E = \begin{bmatrix} X_C \\ -y \end{bmatrix}\begin{bmatrix} X_C \\ -y \end{bmatrix}'$. Again by the scale independence property of Σ_S, σ_S will remain unchanged if $\bar{W}$ is replaced by $\tilde{W}$. Moreover, because of Lemma 6.1, $\tilde{W}$ is bounded. In other words, if $\tilde{W}$ replaces $\bar{W}$, Theorems 1 and 2 and Corollary 1 will still hold and $\tilde{W}$ will be bounded as well. It is easy to see that these conclusions will be true, even if $\lambda = 0$.

While the preceding solve the problem of boundedness of weighting matrices, the real significance of these modifications is well camouflaged under the exact matching hypothesis. In fact, if the exact matching hypothesis is not made, these modifications prove to be crucial not just for the boundedness of weighting matrices, but for the boundedness of x_C and y as well. In other words, without unmodelled dynamics $\mathcal{L}^2$ performance signals such as (5.1) are in principle okay, but in the more realistic situation when unmodelled dynamics are present they are a very poor choice.

7. Fast switching. In a recent paper [13], a switching logic similar to that under consideration here has been independently proposed for switching between finite families of identically configured, but differently initialized model reference parameter adaptive controllers with the purpose of improving overall system performance. In [13] the interesting observation is made that the time between switches {i.e., the dwell time} can be arbitrarily small, without sacrificing stability, at least in the absence of unmodelled dynamics and measurement errors. What follows in the remainder of this section is prompted by that observation.

A key step in the analysis just given in §5, was to show that for the family of detectable pairs $\{(C, A_{pp^*}) : p \in \mathcal{P}^*\}$, there exists a a bounded, output injection function K_p and a dwell time τ_D for which $A_{\sigma p^*} + K_\sigma C$ is exponentially stable for any admissible switching function σ with dwell time no smaller than τ_D. In the sequel it will be shown than for *any* given positive dwell time τ_D, it is possible to find a function K_p which exponentially stabilizes $A_{\sigma p^*} + K_\sigma C$ for any admissible switching function σ with dwell time no smaller than τ_D.

It turns out that detectability of such matrix pairs is by itself *not* sufficient for the existence of a function K_p with the aforementioned property. To understand why, just consider the situation in which a family of detectable pairs of the form $\{(C, A_p) : p \in \mathcal{P}\}$ has a zero readout matrix C; in this case each A_p must be a stability matrix and $A_p + K_p C = A_p$ for all K_p. It is well known that if the A_p do not commute with each other, exponential stability of A_σ cannot in general be assured unless τ_D is large enough. In other words, there are are families of detectable pairs of the form $\{(C, A_p) : p \in \mathcal{P}\}$ for which no stabilizing function K_p exists if τ_D is too small. What's especially interesting is that if $\{(C, A_p) : p \in \mathcal{P}\}$ is a family of *observable* matrix pairs, then no matter how small τ_D is, there does in fact exist a matrix function K_p with the required stabilizing property. This is an immediate consequence of Lemma 5.1 and the following result proved previously in [14].

Squashing Lemma: *Let* (C, A) *be a fixed, constant, observable matrix pair, and let* τ_0 *be a positive number. For each positive number* δ *there exists a positive number* λ *and a constant output-injection matrix* K *for which*

(7.1) $$\left| e^{(A+KC)t} \right| \leq \delta e^{-\lambda(t-\tau_0)}, \quad t \geq 0$$

The way to construct K_p for a family of observable pairs such as $\{(C, A_p) : p \in \mathcal{P}\}$, is as follows. Pick $\delta \in (0, 1)$, set $\tau_0 = \tau_D$ and for each $p \in \mathcal{P}$ use the Squashing Lemma to find a value of K_p for which

$$\left| e^{(A_p + K_p C)t} \right| \leq \delta e^{-\lambda(t-\tau)}, \quad t \geq 0$$

This construction ensures that the hypotheses of Lemma 1 are now satisfied for the family $\{A_p + K_p C : p \in \mathcal{P}\}$ and dwell time $\tau_0 = \tau_D$. Thus K_p is as required.

Unfortunately, for the problem of interest in this paper, the matrix pairs in $\{(C, A_{pp^*}) : p \in \mathcal{P}^*\}$ cannot be assumed to be observable without a definite loss of generality. On the other hand, observability is in general sufficient for stabilizability whereas detectability is not. It is therefore clear that to make any further progress it will be necessary to look more carefully the detailed algebraic structure of the matrix pairs in $\{(C, A_{pp^*}) : p \in \mathcal{P}^*\}$.

From (5.4) and (5.2) it can be seen that

$$(7.2) \qquad\qquad A_{pp^*} = A + BF_p$$

where $B = [b'_C \quad 0]'$, $F_p = f_{pp^*}$, and

$$(7.3) \qquad\qquad A = \begin{bmatrix} A_C & h_C \\ -c_{p^*} & 0 \end{bmatrix}$$

Moreover, examination of (5.12) and (5.3) reveals that the transfer matrix of (C, A, B) will be zero just in case all of the nominal process model transfer functions in the family $\{\nu_p : p \in \mathcal{P}^*\}$ are the same. This suggests that we should consider separately, two distinct cases - one in which all of the transfer functions in $\{\nu_p : p \in \mathcal{P}^*\}$ are the same and the other when they are not. If the former is true, then $F_p = F_{p^*}$, $p \in \mathcal{P}^*$ because of Assumption 3. Therefore in this case $A_{\sigma p^*} = A_{p^* p^*}$ for $t \geq t^*$ so $A_{\sigma p^*}$ is exponentially stable {without output injection} no matter what the value of τ_D.

Now consider the case in which $\{\nu_p : p \in \mathcal{P}^*\}$ contains at least two distinct transfer functions. This means that the transfer matrix of (C, A, B) is nonzero. It will now be shown that no matter what the value of τ_D, so long as it is positive there is an output injection matrix K_p, depending on τ_D, which exponentially stabilizes $A_{\sigma p^*} + K_\sigma C$. This is a consequence of the following theorem.

Switching Theorem: *Let $\lambda_0 > 0$ and $\tau_0 > 0$ be fixed. Let $(C_{q_0 \times n}, A_{n \times n}, B_{n \times m})$ be a left invertible system. Suppose that $\{(C_p, F_p) : p \in \mathcal{P}\}$ is a closed, bounded subset of matrix pairs in $\mathbb{R}^{q \times n} \oplus \mathbb{R}^{m \times n}$ with the property that for each $p \in \mathcal{P}$, $(C_p, A + BF_p)$ is detectable with stability margin no smaller than λ_0. There exist a constant $a \geq 0$ and bounded, matrix-valued output injection functions $p \longmapsto H_p$ and $p \longmapsto K_p$ on $\mathcal{P}$ which, for any admissible switching function $\sigma : [0, \infty) \rightarrow \mathcal{P}$ with dwell time no smaller than τ_0, causes the state transition matrix of*

$$A + K_\sigma C_\sigma + H_\sigma C + BF_\sigma$$

to satisfies

$$|\Phi(t, \mu)| \leq e^{(a - \lambda_0(t - \mu))}, \quad t \geq \mu \geq 0$$

Let us note that the switching theorem[3] can be applied to the matrices $A, B, C, F_p, p \in \mathcal{P}^*$ defined just prior to the theorem's statement, by identifying $\mathcal{P}$ with $\mathcal{P}^*$, each C_p, $p \in \mathcal{P}$ with C and λ_0 with λ_S. Therefore there exists a bounded output injection matrix, namely $H_p + K_p$, which exponentially stabilizes $A_{\sigma p^*} + (H_\sigma + K_\sigma)C$.

Let Σ denote the closed-loop supervisory control system consisting of process model Σ_P described by (2.1), integrating subsystem (2.2), shared controller Σ_C defined by (3.3), feedback law (3.6), switching law (3.7), performance weight generator Σ_W defined by (4.2) with $\lambda \in (0, \lambda_S)$, performance function Π given by (4.3), and the dwell- time switching logic Σ_D described in §4. The following theorem summarizes the established properties of the closed-loop supervisory control system considered in this paper.

THEOREM 7.1. *Let τ_D be any positive dwell time. Suppose that Σ_P 's transfer function equals nominal process model transfer function ν_p for some $p = p^* \in \mathcal{P}$. Then for each constant set-point value r, and each initial state $\{x_P(0), z(0), x_C(0), W(0), X(0), \tau(0), \beta(0), \sigma_S(0)\}$, the state response $\{x_P, z, x_C, W, X, \tau, \beta, \sigma_S\}$ of the supervisory control system Σ is bounded on $[0, \infty)$. If $r \neq 0$, then Γ_σ tends to the dc gain of ν_{p^*}. Moreover if switching stops or if $\mathcal{P}$ is a finite set, then $y \to r$.*

The proof of the Switching Theorem makes use of several new ideas and results. In the sequel we discuss these and then conclude the section with a proof of the theorem.

Recall that a linear system (C, A, B) is *minimum phase* if it is left invertible and if all of its transmission zeros[4] are stable. As defined, minimum phase systems are precisely those left invertible systems which are detectable and remain detectable under transformations of both the state feedback and output injection types. *All pole* systems are minimum phase systems with no transmission zeros; they are accordingly, those left invertible systems which are observable and remain observable under both state feedback and output injection transformations. It is natural to call a left invertible linear system *completely nonminimum phase* if it is not all-pole and if all of its transmission zeros are unstable. Like minimum phase and all-pole systems, completely nonminimum phase systems remain completely nonminimum phase under both state feedback and output-injection transformations {cf. [9]}.

Note that any unobservable modes of a completely nonminimum phase system must be unstable. This implies that a detectable, completely nonminimum phase system must be observable.

---

[3] Our reason for stating the theorem in more general terms than are needed here is because this is what's required to analyze supervisory control systems in which disturbances and unmodelled dynamics are taken into consideration. This more realistic set of assumptions will be treated in the sequel to this paper.

[4] Here we use the term *transmission zero* as it was originally defined in [9].

The following lemma shows that by means of output-injection and state-feedback, it is possible to "extract" from any left-invertible linear system (C, A, B), a subsystem $(\bar{C}, \bar{A}, \bar{B})$ which is either all-pole or completely nonminimum phase, in such a way that what's left is a "completely uncontrollable," "completely unobservable" subsystem whose spectrum is the set of stable transmission zeros of (C, A, B).

LEMMA 7.1. *Let (C, A, B) be a left invertible system. There exist matrices $\bar{K}$, $\bar{F}$, and T, with T nonsingular such that*

$$(7.4) \qquad\qquad CT^{-1} = [\; \bar{C} \quad 0 \;]$$

$$(7.5) \qquad\qquad TAT^{-1} = \begin{bmatrix} \bar{A} & \bar{B}\bar{F} \\ \bar{K}\bar{C} & A_- \end{bmatrix}$$

$$(7.6) \qquad\qquad TB = \begin{bmatrix} \bar{B} \\ 0 \end{bmatrix}$$

where A_-'s spectrum is the set of stable transmission zeros of (C, A, B) and $(\bar{C}, \bar{A}, \bar{B})$ is either an all-pole or a completely nonminimum phase system.

This lemma, which generalizes earlier results [16,1,11] {see also the discussion of output feedback linearization in Chapter 5 of [5]} is a simple consequence of Theorem 4.1 of [9]. A direct proof is given in [8].

Proof of Switching Theorem: Since (C, A, B) is left invertible, so is $(C, \lambda_0 I + A, B)$. Therefore by Lemma 7.1 there exist matrices $\bar{K}$, $\bar{F}$, and T, with T nonsingular such that

$$(7.7) \qquad\qquad CT^{-1} = [\; \bar{C} \quad 0 \;]$$

$$(7.8) \qquad\qquad T(\lambda_0 I + A)T^{-1} = \begin{bmatrix} \bar{A} & \bar{B}\bar{F} \\ \bar{K}\bar{C} & A_- \end{bmatrix}$$

$$(7.9) \qquad\qquad TB = \begin{bmatrix} \bar{B} \\ 0 \end{bmatrix}$$

where A_- is a stability matrix and $(\bar{C}, \bar{A}, \bar{B})$ is either an all-pole or a completely nonminimum phase system.

Let $\bar{n}$ be the size of $\bar{A}$. Our aim is to construct bounded functions $p \longmapsto H_p$, $p \longmapsto K_p$, $p \longmapsto \bar{A}_p$, $p \longmapsto \bar{B}_p$, $p \longmapsto \bar{a}_p$, and $p \longmapsto \bar{\lambda}_p$ with codomains $\mathbb{R}^{n \times q_0}$, $\mathbb{R}^{n \times q}$, $\mathbb{R}^{\bar{n} \times \bar{n}}$, $\mathbb{R}^{\bar{n} \times (n-\bar{n})}$, $\mathbb{R}^+$, and $\mathbb{R}^+$ respectively, such that

$$(7.10) \quad \lambda_0 I + A + K_p C_p + H_p C + B F_p = T^{-1} \begin{bmatrix} \bar{A}_p & \bar{B}_p \\ 0 & A_- \end{bmatrix} T$$

$$(7.11) \qquad\qquad |e^{\bar{A}_p t}| \leq e^{(\bar{a}_p - \bar{\lambda}_p t)}, \quad t \geq 0$$

(7.12)
$$\tau_0 > \frac{\bar{a}_p}{\lambda_p}$$

for all $p \in \mathcal{P}$.

For this fix $p \in \mathcal{P}$ and partition C_pT^{-1} and F_pT^{-1} as $C_pT^{-1} = [\bar{C}_p \quad \tilde{C}_p]$ and $F_pT^{-1} = [\bar{F}_p \quad \tilde{F}_p]$ respectively. Note that

$$\left(\begin{bmatrix} \bar{C} \\ \tilde{C}_p \end{bmatrix}, \quad \bar{A}, \quad \bar{B} \right)$$

is either all-pole or completely nonminimum phase since $(\bar{C}, \bar{A}, \bar{B})$ is and since the set of transmission zeros of the former is a subset of the set of transmission zeros of the latter.

Note next that $\lambda_0 I + A + BF_p$ can be written as

$$\lambda_0 I + A + BF_p = T^{-1} \begin{bmatrix} \bar{A} + \bar{B}\bar{F}_p & B(\bar{F} + \tilde{F}_p) \\ \bar{K}\bar{C} & A_- \end{bmatrix} T$$

By hypothesis, $(C_p, A+BF_p)$ is detectable with stability margin no smaller than λ_0. Therefore $(C_p, \lambda_0 I + A + BF_p)$ is detectable. Clearly

$$\left(\begin{bmatrix} C \\ C_p \end{bmatrix}, \lambda_0 I + A + BF_p \right)$$

is detectable as well. Since detectability is invariant under both state-coordinate and output-injection transformations, the matrix pair

$$\left(\begin{bmatrix} \bar{C} & 0 \\ \bar{C}_p & \tilde{C}_p \end{bmatrix}, \begin{bmatrix} \bar{A} + \bar{B}\bar{F}_p & B(\bar{F} + \tilde{F}_p) \\ 0 & A_- \end{bmatrix} \right)$$

is detectable. This in turn implies detectability of $\left(\begin{bmatrix} \bar{C} \\ \bar{C}_p \end{bmatrix}, \bar{A} + \bar{B}\bar{F}_p \right)$.

But as noted previously, $\left(\begin{bmatrix} \bar{C} \\ \tilde{C}_p \end{bmatrix}, \quad \bar{A}, \quad \bar{B} \right)$ is either all-pole or completely nonminimum phase, so $\left(\begin{bmatrix} \bar{C} \\ \tilde{C}_p \end{bmatrix}, \bar{A} + \bar{B}\bar{F}_p \right)$ is actually observable.

Pick $\delta \in (0,1)$ and use the Squashing Lemma to find an output-injection matrices $\bar{H}_p$ and $\bar{K}_p$ and a positive number $\bar{\lambda}_p$ for which

(7.13)
$$|e^{(\bar{A}+\bar{B}\bar{F}_p+\bar{K}_p\bar{C}_p+\bar{H}_p\bar{C})t}| \leq e^{(\bar{a}_p - \bar{\lambda}_p t)}, \quad t \geq 0$$

where $\bar{a}_p = \log(\delta) + \bar{\lambda}\tau_0$. Clearly $\log(\delta) < 0$ so

$$\tau_0 > \frac{\bar{a}_p}{\bar{\lambda}_p}$$

It follows from this, that (7.10) to (7.12) will hold if functions $H_p, K_p, \bar{A}_p$, and $\bar{B}_p$ are defined at by

$$H_p = T^{-1}\begin{bmatrix} \bar{H}_p \\ -\bar{K} \end{bmatrix}$$

$$K_p = T^{-1}\begin{bmatrix} \bar{K}_p \\ 0 \end{bmatrix}$$

$$\bar{A}_p = \bar{A} + \bar{B}\bar{F}_p + \bar{H}_p\bar{C} + \bar{K}_p\bar{C}_p$$

$$\bar{B}_p = \bar{B}(\bar{F} + \tilde{F}_p) + \bar{K}_p\bar{C}_p$$

Now suppose that σ is an admissible function with dwell time no smaller than τ_0. In view of (7.10), the state transition matrix of $\lambda_0 I + A + K_\sigma C_\sigma + H_\sigma C + BF_\sigma$ is of the form $T^{-1}\tilde{\Phi}(t,\mu)T$ where $\tilde{\Phi}(t,\mu)$ is the state transition matrix of

$$\tilde{A}(t) = \begin{bmatrix} \bar{A}_\sigma & \bar{B}_\sigma \\ 0 & A_- \end{bmatrix}$$

From the structure of $\tilde{A}(t)$, it is clear that

$$\tilde{\Phi}(t,\mu) = \begin{bmatrix} \bar{\Phi}(t,\mu) & \int_\mu^t \bar{\Phi}(t,s)\bar{B}_{\sigma(s)}e^{A_-(s-\mu)}ds \\ 0 & e^{A_-(t-\mu)} \end{bmatrix}$$

where $\bar{\Phi}(t,\mu)$ is the state transition matrix of $\bar{A}_\sigma$. Pick $\lambda \in (0,\tilde{\lambda})$ and define

$$a = \log\left\{ |T^{-1}||T|\left(e^{\bar{a}} + e^{a^*} + \frac{\bar{b}}{(\tilde{\lambda}-\lambda)}e^{(\bar{a}+a^*)} \right) \right\}$$

where $\bar{b} = \sup_{p\in\mathcal{P}}|\bar{B}_p|$ It follows that

$$|T^{-1}\tilde{\Phi}(t,\mu)T|$$

$$\leq |T^{-1}||T|(|\bar{\Phi}(t,\mu)| + |e^{A_-(t-\mu)}| + |\int_\mu^t \bar{\Phi}(t,s)\bar{B}_{\sigma(s)}e^{A_-(s-\mu)}ds|)$$

$$\leq |T^{-1}||T|(e^{(\bar{a}-\tilde{\lambda}(t-\mu))} + e^{(a^*-\tilde{\lambda}(t-\mu))} + \int_\mu^t |\bar{\Phi}(t,s)||\bar{B}_{\sigma(s)}||e^{A_-(s-\mu)}||ds)$$

$$\leq |T^{-1}||T|(e^{(\bar{a}-\lambda(t-\mu))} + e^{(a^*-\lambda(t-\mu))} + \bar{b}\int_\mu^t e^{(\bar{a}-\tilde{\lambda}(t-s))}e^{(a^*-\tilde{\lambda}(s-\mu))}ds)$$

$$= |T^{-1}||T|(e^{(\bar{a}-\lambda(t-\mu))} + e^{(a^*-\lambda(t-\mu))} + \bar{b}(t-\mu)e^{(\bar{a}+a^*-\tilde{\lambda}(t-\mu))})$$

$$\leq |T^{-1}||T|(e^{(\bar{a}-\lambda(t-\mu))} + e^{(a^*-\lambda(t-\mu))} + \frac{\bar{b}}{(\tilde{\lambda}-\lambda)}e^{(\bar{a}+a^*-\lambda(t-\mu))})$$

$$= e^{(a-\lambda(t-\mu))}$$

Since the state transition matrix of $A + K_\sigma C_\sigma + H_\sigma C + B F_\sigma$ equals $e^{-\lambda_0(t-\mu)}$ times $T^{-1}\tilde{\Phi}(t,\mu)T$, which in turn is the state transition matrix of $\lambda_0 I + A + K_\sigma C_\sigma + H_\sigma C + B F_\sigma$, it follows that

$$|\Phi(t,\mu)| \leq e^{(a-(\lambda_0+\lambda)(t-\mu))} \leq e^{(a-\lambda_0(t-\mu))}$$

and thus that the theorem is true.   □

8. Concluding remarks. This paper has demonstrated the feasibility of using a supervisor to orchestrate the switching of a sequence of linear position or set-point controllers into feedback with a siso process so as to cause the output of the process to approach and track a constant reference input. It has been shown that in the absence of unmodelled process dynamics and disturbances, the proposed supervisor can successfully perform its function provided $\mathcal{P}$ is a finite set. In the sequel to this paper we will analyze the performance of the same supervisor in the presence of unmodelled process dynamics and disturbances.

A proof that $y \to r$ if $\mathcal{P}$ is infinitely large has so far eluded us. While this issue may be of theoretical interest, as a practical matter its probably not very important because in most applications only approximate process models are needed for control purposes. Moreover, even if $\mathcal{P}$ is infinitely large, the transfer functions in $\mathcal{C}_P$ can almost certainly be approximated to any desired degree of accuracy by transfer functions from a sub-class $\mathcal{C}'_P$ whose underlying parameter space $\mathcal{P}'$ is finite {cf. [17], [15]}. What's really important, is that the 'complexity' of a supervisory control system as defined here, depends primarily on the dimension of the linear space which contains its underlying parameter space and *not* on the number of points in that parameter space. Therefore by defining $\mathcal{P}'$ with enough points, it should be possible to approximate $\mathcal{C}_P$ with $\mathcal{C}'_P$ as closely as desired without ending up with a control system of Gargantuan size.

Acknowledgments. The author thanks David Mayne, George Zames, Felipe Pait, and Brian Anderson for useful discussions which have contributed to this work.

REFERENCES

[1] C. I. BYRNES AND J. C. WILLEMS, *Adaptive stabilization of multivariable linear systems*, in Proceedings of the 23rd Conference on Decision and Control, Control Systems Society, IEEE, 1984, pp. 1574–1577.

[2] S. J. CUSUMANO AND POOLA, *Adaptive control of uncertain systems: A new approach*, in Proceedings of the American Automatic Control Conference, 1988, pp. 355–359.

[3] J. C. DOYLE AND G. STEIN, *Multivariable feedback design - concepts for a classical modern synthesis*, IEEE Transactions on Automatic Control, (1981), pp. 4–11.

[4] M. FU AND B. R. BARMISH, *Adaptive stabilization of linear systems via switching controls*, IEEE Transactions on Automatic Control, (1986), pp. 1079–1103.

[5] A. ISIDORI, *Nonlinear Control Systems*, Springer-Verlag, 1989.

[6] B. MARTENSSON, *The order of any stabilizing regulator is sufficient a priori information for adaptive stabilization*, Systems and Control Letters, 6 (1985), pp. 87–91.

[7] D. E. MILLER, *Adaptive Control of Uncertain Systems*, PhD thesis, University of Toronto, 1989.

[8] A. S. MORSE, *Supervisory control of families of linear set-point controllers - part 1: Exact matching.* submitted for publication.

[9] ———, *Structural invariants of linear multivariabe systems*, SIAM J. Control, 11 (1973), pp. 446–465.

[10] ———, *System invariants under feedback and cascade control*, in Mathematical System Theory, Springer-Verlag, 1975, pp. 61–74.

[11] ———, *A three-dimensional universal controller for the adaptive stabilization of any strictly proper minimum-phase system with relative degree not exceeding two*, IEEE Transactions on Automatic Control, 30 (1985), pp. 1188–1191.

[12] ———, *Towards a unified theory of parameter adaptive control - part 2: Certainty equivalence and implicit tuning*, IEEE Transactions on Automatic Control, 37 (1992), pp. 15–29.

[13] K. S. NARENDRA AND J. BALAKRISHNAN, *Improving transient response of adaptive control systems using multiple models and switching*, tech. rep., Yale University, 1992.

[14] F. M. PAIT AND A. S. MORSE, *A cyclic switching strategy for parameter-adaptive control*, IEEE Transactions on Automatic Control, (1994). to appear.

[15] L. Y. WANG AND L. LIN, *Complexity and entropy of identification and adaptation*, in Proceedings of the IEEE Conference on Decision and Control, 1992, pp. 38–43.

[16] J. C. WILLEMS AND C. I. BYRNES, *Global adaptive stabilization in the absence of information on the sign of the high frequency gain*, in Proceedings of the Sixth International Conference on Analysis and Optimization of Systems, INRIA, Springer-Verlag, 1984, pp. 58–68.

[17] G. ZAMES, *On the metric complexity of causal linear systems: ϵ-entropy and ϵ-dimension for continuous time*, IEEE Transactions on Automatic Control, (1979), pp. 222–229.

POTENTIAL SELF-TUNING ANALYSIS OF STOCHASTIC ADAPTIVE CONTROL*

KARIM NASSIRI-TOUSSI[†] AND WEI REN[†]

Abstract. In this paper we present the first stage of a unified two-stage approach to analyzing stochastic adaptive control. In the first stage, we study the issue of *potential self-tuning* where we ask the question whether a certainty-equivalence adaptive control scheme achieves the same control objective as the ideal control design at the potential convergence points of the estimation algorithm. We exploit the fact that this important property can be analyzed independent of the estimation method that is used, without restoring to complicated convergence analysis. For linear time-invariant systems, this reduces to simply studying two identifiability equations; the Identifiability Equation for Internal Excitation (IEIE) and the Identifiability Equation for External Excitation (IE^3) whose solutions determine the potential convergence points of the parameter estimates. Sufficient conditions and necessary conditions are then derived for potential self-tuning and identifiability of general control schemes. Applications of these general results to specific adaptive control policies then show that regardless of the external excitation, the certainty-equivalence adaptive control based on generalized Minimum-Variance, generalized predictive, and pole-placement control are potentially self-tuning. On the other hand, the LQG feedforward and feedback control designs are shown to require sufficient external excitation. In the next stage, we will show how to proceed from potential self-tuning to asymptotic self-tuning.

Key words. Certainty-equivalence adaptive control, self-tuning, identifiability, persistency of excitation.

1. Introduction. Since the pioneering work of Astrom and Wittenmark (1973), the important issue of the asymptotic behaviour of stochastic certainty-equivalence adaptive control systems has been the subject of extensive study and research, the results of which have been published in numerous papers, see for example Goodwin and Sin (1984), Astrom and Wittenmark (1989), Chen and Guo (1991), Ren and Kumar (1991) and Lai and Ying (1991) and references therein. However, much of the analysis regarding the main issue, that is self-tuning or convergence of the controller to its ideal design, has been specific to particular estimation algorithms and control schemes, while a unified analysis of the subject from a general perspective has been rare.

The certainty-equivalence-based algorithms are based on the idea that the plant parameter estimates, viewed as the correct parameter values, be used at each time instant to design the controller. The motivation is of course the simplicity and ingenuousness of the resulting adaptation schemes which can be simply decomposed into an identifier algorithm and a controller design mapping. The main question however remains whether the adaptive controller is asymptotically self-tuning, *i.e.* is capable of asymp-

---

* This research has been supported by the NSF grant FD92-11025-REN.

† EECS Dept., University of California at Berkeley, Berkeley, CA, 94720, U.S.A. email addresses: *karim@eclair.eecs.berkeley.edu, ren@control.eecs.berkeley.edu*

totically achieving the same control objective (such as cost minimization) as the ideal control design. We remind the reader that in general, self-tuning might hold without the parameter estimates being consistent or even convergent. We will answer this question for stochastic systems by using a two-stage approach which is applicable to a broad range of plants and control designs.

In the first stage, which is the aim of this paper, we investigate the issue of *potential self-tuning, i.e.* whether at the *potential* convergence points of the identification algorithm, the adaptive control scheme achieves the ideal control objective. This study therefore separates the characteristics of the controller design and plant structure from the properties of the adaptation algorithm that are carried over to the asymptotic behaviour. In the case of linear systems in particular, potential self-tuning analysis can be accomplished through simple algebraic operations over polynomials of the plant and controller transfer functions. The potential self-tuning analysis differs from the traditional analysis of closed-loop identifiability in two ways. First is the obvious distinction that in the potential self-tuning analysis, the controller is not fixed a priori but depends on the estimates. Second, potential self-tuning is weaker than identifiability since there may exist a set of models, different from the true model, which also yield the desired controller.

Clearly, it is highly desirable to have a potentially self-tuning adaptive control scheme. Because in that case, establishing stability and convergence of the parameter estimates is sufficient to prove asymptotic self-tuning, one of the main objectives of adaptive control. Conversely, simulations show that when the potential self-tuning property does not hold, even with almost surely convergent estimates, the control algorithm may not be asymptotically self-tuning.

An early attempt to analyze the potential self-tuning property was made by Lin, Kumar and Seidman (1985), where it was shown that for a first order system, the LQG adaptive control design is not potentially self-tuning in absence of external excitation. Along the same line of work, but for deterministic adaptive control systems, are the works of Polderman (1986, 1987, 1989) where a more detailed analysis of this issue was carried out. But in the absence of internal and external excitations, his results consequently could not provide a basic identifiability relation which may determine the self-tuning property of general adaptive control systems. Other recent works related to the subject of potential self-tuning are those of Van Schuppen (1991), Vorchik (1988, 1990) and Kogan an Neimark (1989, 1992). Although these papers give formulations of this problem somewhat similar to ours, they do not offer general results or a comprehensive treatment of the problem.

In this paper, we study the potential self-tuning issue by considering the identifiability equations which are derived simply from equating the estimated closed-loop transfer function and the actual closed-loop transfer function with certainty-equivalence-based control parameters. This is equivalent to equating the certainty-equivalence prediction of the output based on the parameter estimates with its optimal prediction. For linear systems these result in separate Identifiability Equations for the Internal and the External Excitations, referred to hereafter as IEIE and IE[3].

It is well-known that for some estimation methods based on prediction error, such as the Extended Least-Squares (ELS) and Stochastic Gradient (SG) algorithms, the solutions of these identifiability equations include the convergence points of the parameter estimates. This for example can be shown by using averaging techniques to approximate the recursive estimator of the stochastic adaptive algorithm with an associated ordinary differential equation, see Ljung and Soderstrom (1983).

The question of potential self-tuning thus will be equivalent to whether each solution of the identifiability equations yields the desired controller transfer function corresponding to the true parameters, which can be answered by using simple algebraic operations.

Now, it is well-known that self-tuning can not generally be achieved in absence of external excitation in the deterministic case where the system noise has been ignored. However, as will be shown in this paper, there are many situations where the stochastic disturbance inherent to the stochastic system provides adequate excitation for the adaptive control scheme to be self-tuning. This inspires us to delineate the effects of the internal and external excitations. Thus, we would like to find out whether an adaptive control algorithms is potentially self-tuning in absence of external excitation, and if not, what levels of persistency of the external excitation ensure this property.

In §4, we derive general results concerning potential self-tuning and identifiability in absence of external excitation. We then use these results to prove that the generalized minimum-variance control and pole-placement, as well as generalized predictive control subject to certain assumptions, are potentially self-tuning, while the LQG control designs in general require sufficiently persistent external excitation.

In the second stage, which due to lack of space will be the subject of future papers, we shall deal with the more demanding issue of asymptotic self-tuning. We shall discuss the connection between potential and asymptotic self-tuning, and assuming certain sets of properties for the estimation algorithm, will derive sufficient conditions on the plant and controller structure, for asymptotic self-tuning. In particular, we shall establish that with possibly stronger assumptions, the same self-tuning results hold asymptotically for the specific adaptive controllers that were mentioned in the last paragraph.

The organization of the paper is as follows: In the next section, we explain the problem setup, the plant, the controller design, the estimation algorithm and the required assumptions. In §3 we present formal definitions of potential self-tuning and identifiability and derive the identifiability equations; IEEE and IE³. The main results are presented in §4. §5 is dedicated to the application of these results to some widely-used certainty-equivalence adaptive control schemes.

2. Problem formulation

A. Plant:

For the sake of simplicity, we concentrate on the important class of SISO linear time-invariant systems in this paper.

Hence, the plant subject to our discussion will have the following AR-MAX model realization:

$$(2.1) \qquad A(q^{-1})y_k = B(q^{-1})u_{k-d} + C(q^{-1})\epsilon_k \qquad \text{for } k \geq 0,$$

where q^{-1} is the backward shift operator, and y, u, r and ϵ are respectively, the one-dimensional output, control input, reference signal and unobserved disturbance. All the processes are assumed to be adapted to the increasing family of σ-fields $\{\mathcal{F}_k\}$ with respect to the underlying probability space $(\Omega, \mathcal{F}, P)$, and

$$
\begin{aligned}
A &= 1 + a^1 q^{-1} + \ldots + a^\nu q^{-\nu}, \\
B &= b^0 + b^1 q^{-1} 1 + \ldots + b^{\nu-d} q^{-\nu+d}, \qquad d \geq 1 \\
C &= 1 + c^1 q^{-1} + \ldots + c^\nu q^{-\nu}.
\end{aligned}
$$

Henceforth, when there is no risk of confusion, we shall use A in place of $A(q^{-1})$ etc, and let ∂A stand for the degree of A. $\nu = \max\{\partial A, \partial B + d, \partial C\}$, which is given, is greater than or equal to the minimal order of the plant.

H1) *We assume that in the minimal realization of (2.1), C is a stable polynomial, i.e. $C(z) \neq 0$ for all $|z| \leq 1$, and that the transfer function from u to y is stabilizable and detectable.*

Let the system parameter vector consist of all the variable coefficients of A, B and C. Then, (2.1) can be represented in a regression form linear in term of the system parameter vector θ^*:

$$(2.2) \qquad \mathcal{L}(\theta^*): \qquad\qquad y_k = \theta^{*T} \phi_k^\epsilon + \epsilon_k,$$

where

$$\theta^* := [a^1, \ldots, a^\nu, b^0, \ldots, b^{\nu-d}, c^1, \ldots, c^\nu]^T \in \Theta \qquad \text{and}$$

$$\phi_k^\epsilon := [-y_{k-1}, \ldots, -y_{k-\nu}, u_{k-d}, \ldots, u_{k-\nu}, \epsilon_{k-1}, \ldots, \epsilon_{k-\nu}]^T \in \mathcal{F}_{k-1}.$$

We shall assume θ^* to be unknown with independent elements in general, and denote the parameter set in which θ^* is known to be located by Θ. Also, using loose notation, we shall henceforth identify θ^* with the polynomials (A, B, C) and each $\overline{\theta} \in \Theta$ with $(\overline{A}, \overline{B}, \overline{C})$.

We should point out that for potential self-tuning, the principal notations and definitions which will be introduced hereafter are applicable to a much larger class of systems.

B. Certainty-Equivalence Adaptive Controller:

We consider the class of linear certainty-equivalence adaptive controllers $\widehat{C}$ given by the following equation:

$$\widehat{C}: \quad R(q^{-1}, \widehat{\theta}_{k-d'})u_k = -S(q^{-1}, \widehat{\theta}_{k-d'})y_k + T(q^{-1}, \widehat{\theta}_{k-d'})r_k, \quad \forall k \geq 0$$
(2.3)

where $\widehat{\theta}_k$ is the estimate of the unknown parameter vector θ^* at time k using an appropriate estimation method, and r is the reference or probing signal which constitutes the external excitation. The time-varying controller polynomials are computed from the estimate of θ^* with a delay of $d' \geq 0$ steps, by using a well-defined time-independent controller mapping $\mathcal{K}$ from the parameter set Θ to the space of polynomials:

$$\mathcal{K}: \quad (\overline{\theta} \equiv (\overline{A}, \overline{B}, \overline{C})) \mapsto (R(q^{-1}, \overline{\theta}), S(q^{-1}, \overline{\theta}), T(q^{-1}, \overline{\theta})).$$

We assume that the polynomial $R(\cdot)$ has always nonzero leading coefficient so that u is well-defined and use the simplified notations $(R^*, S^*, T^*) = \mathcal{K}(\theta^*)$ and $(\overline{R}, \overline{S}, \overline{T}) = \mathcal{K}(\overline{\theta})$, for every $\overline{\theta} \in \Theta$.

Definition and Notations:

We use the following notations associated with the *frozen* closed-loop systems consisting of a plant $\mathcal{L}(\theta)$ of the form (2.2), and a *frozen* linear time-invariant controller corresponding to a *frozen* parameter-estimate vector $\overline{\theta}$ given by

(2.4) $\quad C[\mathcal{K}(\overline{\theta})]: \qquad \overline{R}\, u_k = -\overline{S}\, y_k + \overline{T}\, r_k, \qquad (\overline{R}, \overline{S}, \overline{T}) = \mathcal{K}(\overline{\theta}).$

— $y(\overline{\theta})$, $u(\overline{\theta})$: "Frozen" output and control input in the "frozen" closed-loop system consisting of the plant $\mathcal{L}(\theta^*)$ and the controller $C[\mathcal{K}(\overline{\theta})]$. $\phi(\overline{\theta})$ is similarly defined by replacing y and u with $y(\overline{\theta})$ and $u(\overline{\theta})$. In particular, $y(\theta^*)$ and $u(\theta^*)$ are the desired output and control input.

By the above definition, $u(\overline{\theta})$ and $y(\overline{\theta})$ satisfy

(2.5) $\qquad\qquad A\, y_k(\overline{\theta}) \;=\; q^{-d} B\, u_k(\overline{\theta}) + C\, \epsilon_k$

and

(2.6) $\qquad\qquad \overline{R}\, u_k(\overline{\theta}) \;=\; -\overline{S}\, y_k(\overline{\theta}) + \overline{T}\, r_k.$

— $\mathbf{M}_c(\theta, \mathcal{K}(\overline{\theta}))$: the transfer function from (r, ϵ) to (y, u), for the "frozen" closed-loop system consisting of plant $\mathcal{L}(\theta)$ and "frozen" controller $C[\mathcal{K}(\overline{\theta})]$ as in Figure 1, *i.e.* $[y \;\; u]^T = \mathbf{M}_c(\theta, \mathcal{K}(\overline{\theta})) \left([r \;\; \epsilon]^T\right).$

— We say the above closed-loop system is *stable* if $\mathbf{M}_c$ is BIBO stable.

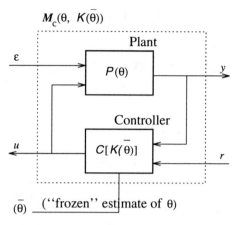

(2.1) and (2.4) imply the following description for the frozen closed-loop system corresponding to $\mathbf{M}_c(\theta^o, \mathcal{K}(\theta))$, where $\theta^o \equiv (A^o, B^o, C^o)$. Clearly, $A^o \overline{R} + q^{-d} B^o \overline{S}$ is the characteristic polynomial of $\mathbf{M}_c(\theta^o, \mathcal{K}(\theta))$:

$\overline{(\theta)}$ ("frozen" estimate of θ)

FIGURE 1. The "frozen" adaptive control system

$$(2.7) \quad \mathbf{M}_c(\theta^o, \mathcal{K}(\theta)) = \left(A^o \overline{R} + q^{-d} B^o \overline{S}\right)^{-1} \begin{bmatrix} q^{-d} B^o \overline{T} & C^o \overline{R} \\ A^o \overline{T} & -C^o \overline{S} \end{bmatrix}.$$

C. Identification Algorithm:

Consider (2.2), the linear regression representation of the plant. Let the estimation algorithm for identifying the system parameter vector θ^* be of the following general recursive form:

$$(2.8) \qquad \widehat{\theta}_k = \widehat{\theta}_{k-1} + \gamma_k\, f_k(y^k, u^k, \widehat{\theta}_{k-1}), \qquad \text{where} \qquad \gamma_k > 0$$

with the constraint that $\widehat{\theta}_k \in \Theta$ the parameter set. We use $\widehat{\theta}_k$ to compute the instantaneous controller transfer function in the certainty-equivalence adaptive control algorithm (2.3). As mentioned in introduction, our goal at this stage is to analyze the potential of the adaptive control algorithm for self-tuning independent of the specific estimation method.

Nonetheless, our analysis is based on the implicit assumption that the estimation algorithm attempts to globally minimize the second moment of the prediction error, which is the error between the actual output and the minimum-variance one-step-ahead output prediction based on the parameter estimates.

Definition:

The *frozen* prediction error $\hat{\epsilon}(\overline{\theta})$ is the estimate of the unobserved noise ϵ from input-output observations based on the parameter estimate $\overline{\theta}$. Thus considering (2.2) and (2.1), $\hat{\epsilon}(\overline{\theta})$ is given by

$$\hat{\epsilon}_k(\overline{\theta}) := y_k(\overline{\theta}) - \overline{\theta}^T \phi_k^\epsilon\left(y^{k-1}(\overline{\theta}), u^{k-1}(\overline{\theta}), \hat{\epsilon}^{k-1}(\overline{\theta})\right) := y_k(\overline{\theta}) - \overline{\theta}^T \widehat{\phi}_k(\overline{\theta}),$$

and satisfies

$$(2.9) \qquad\qquad \overline{C}\, \hat{\epsilon}_k(\overline{\theta}) = \overline{A}\, y_k(\overline{\theta}) - q^{-d}\overline{B}\, u_k(\overline{\theta}).$$

3. Potential self-tuning: preliminaries and definitions. First, in order to simplify the discussion of potential self-tuning, let us make the following non-restrictive assumptions about the input processes:

I1) ϵ *is a wide sense stationary, second-order ergodic stochastic process with zero mean and strictly positive power spectrum $S_\epsilon(\omega) \geq \delta > 0$ for all $\omega \in \mathcal{R}$.*

The above assumption implies that ϵ is infinitely persistently exciting. Note that we do not assume knowledge of the spectrum of ϵ.

I2) *r the reference signal is bounded, independent of ϵ, and can be written as a linear combination of deterministic <u>periodic</u> signals and wide sense stationary, second-order ergodic stochastic processes.*

I2 means that r is path-wise second-order ergodic.

Hence, if $\mathbf{H}(q)$ and $\mathbf{G}(q)$ are two stable transfer functions, then we know from stationarity or periodicity of ϵ and r that

$$(3.1) \quad \lim_{N \to \infty} \frac{1}{N} \sum_{k=0}^{N} E\left[(\mathbf{H}(q)\,\epsilon_k + \mathbf{G}(q)\,r_k)^2\right] = 0 \quad \Longleftrightarrow \quad \mathbf{H}(q) = 0$$
$$\text{and} \quad \mathbf{G}(q)\,r = 0$$

Now, it is well-known that for the class of estimation algorithms (2.8), if in addition to **I1–I2**, the input processes satisfy certain *mixing* condition, the adaptive system is stable, the plant and controller model satisfy certain regularity assumptions, and $\sum_{k=0}^{\infty} \gamma_k = \infty$, $\sum_{k=0}^{\infty} \gamma_k^\alpha < \infty$ for some $\alpha > 1$, then every convergence point of $\hat{\theta}_k$, $\bar{\theta}$, satisfies

$$(3.2) \quad \lim_{N \to \infty} \frac{1}{N} \sum_{k=0}^{N} E\left[f_k(y^k(\bar{\theta}), u^k(\bar{\theta}), \bar{\theta})\right] = 0.$$

See Ljung and Soderstrom (1983) for more details.

But as mentioned in the last section, we use identification algorithms whose goal is to globally minimize $E\left[\hat{\epsilon}(\bar{\theta})^2\right]$. Indeed, for the plant model (2.1) and identification algorithms such as ELS and SG, it is easy to show that if the polynomial C satisfies certain SPR (Strictly Positive Real) conditions, then (3.2) is equivalent to[1]

$$(3.3) \quad \lim_{N \to \infty} \frac{1}{N} \sum_{k=0}^{N} E\left[(\hat{\epsilon}(\bar{\theta}) - \epsilon)^2\right] = 0.$$

Therefore, based on the above discussion it is natural to assume that the following statement is true. Note that we require the stability of $\mathbf{M}_c(\theta, \mathcal{K}(\bar{\theta}))$ to ensure that the limits in (3.2–3.3) are well-defined:

----

[1] One obtains the same equivalence for the Prediction Error methods if one can remove the possibility of existence of local minima. No SPR condition is however required in this case.

H2) *The set of convergence points of the parameter estimate vector $\hat{\theta}_k$ is a subset of*

$$\mathcal{D}_\infty := \left\{ \bar{\theta} \in \Theta \mid \mathbf{M}_c(\theta, \mathcal{K}(\bar{\theta})) \quad \text{is stable} \quad \text{and} \quad \hat{\epsilon}(\bar{\theta}) \overset{m.s.}{=} \epsilon \right\}.$$

Notation: $x \overset{m.s.}{=} y$ stands for $\lim_{N \to \infty} \sum_{k=0}^{N} E\left[(x_k - y_k)^2 \right] = 0.$[2]

Note that by **H1** and (9), $\theta^* \in \mathcal{D}_\infty$ if $\mathbf{M}_c(\theta^*, \mathcal{K}(\theta^*))$ is stable. The above assumption means that we need only to examine $\mathcal{D}_\infty$ in order to determine whether a certainty-equivalent control design has potential for self-tuning, or if the plant subject to this control design has potential for identifiability. We point out that $\mathcal{D}_\infty$ as defined here is the set of equilibria of the ODE associated with the adaptive control system under the ELS or SG estimation algorithms if certain SPR conditions are satisfied.

Now, one can easily find from (2.9) that (3.3) implies

$$(\theta^* - \bar{\theta})^T \phi^\epsilon(\bar{\theta}) = \overline{A}\, y(\bar{\theta}) - q^{-d}\overline{B}\, u(\bar{\theta}) - \overline{C}\epsilon \overset{m.s.}{=} 0.$$

Since $\mathbf{M}_c(\theta^*, \mathcal{K}(\bar{\theta}))$ is stable for every $\bar{\theta} \in \mathcal{D}_\infty$, one can ignore the exponentially decaying effects of the initial conditions on $y(\bar{\theta})$ and $u(\bar{\theta})$. Hence by using (3.1), one can write the above equations as exact equalities with probability 1. In other words, we have

$$(3.4) \qquad \overline{A}\, y_k(\bar{\theta}) - \overline{B}\, u_{k-d}(\bar{\theta}) - \overline{C}\epsilon_k = 0 \qquad a.s. \qquad \forall \bar{\theta} \in \mathcal{D}_\infty.$$

But $y(\bar{\theta})$ and $u(\bar{\theta})$ also satisfy the true plant equation in (2.5), as well as the frozen controller equation (2.6), for all $\bar{\theta} \in \Theta$. These 3 equations can equivalently be expressed as

$$[y(\bar{\theta}) \quad u(\bar{\theta})] = \mathbf{M}_c(\theta^*, \mathcal{K}(\bar{\theta}))\Big([r \quad \epsilon]\Big) = \mathbf{M}_c(\bar{\theta}, \mathcal{K}(\bar{\theta}))\Big([r \quad \epsilon]\Big),$$
$$\forall \bar{\theta} \in \mathcal{D}_\infty \qquad a.s.$$

But for a linear system, $\mathbf{M}_c(\cdot, \mathcal{K}(\cdot))\,([r \quad \epsilon]) = \mathbf{M}_r(\cdot, \mathcal{K}(\cdot))\,(r) + \mathbf{M}_\epsilon(\cdot, \mathcal{K}(\cdot))$ (ϵ). Thus, considering the result (3.1), we find that the above equation is equivalent to the following fundamental identifiability equations:

$$\begin{cases} \left(\mathbf{M}_r(\theta^*, \mathcal{K}(\bar{\theta})) - \mathbf{M}_r(\bar{\theta}, \mathcal{K}(\bar{\theta}))\right) r = 0 \quad a.s. & (\mathrm{IE}^3), \\[2mm] \mathbf{M}_\epsilon(\theta^*, \mathcal{K}(\bar{\theta})) = \mathbf{M}_\epsilon(\bar{\theta}, \mathcal{K}(\bar{\theta})) & (\mathrm{IEEE}), \end{cases} \qquad \forall \bar{\theta} \in \mathcal{D}_\infty$$

Clearly, one would further get $\mathbf{M}_c(\theta^*, \mathcal{K}(\bar{\theta})) = \mathbf{M}_c(\bar{\theta}, \mathcal{K}(\bar{\theta}))$ if r were sufficiently persistent. The first equation (IE^3) characterizes the effect of

---

[2] Using **I1–I2**, that is ergodicity of ϵ and r, one can show that for stable $\mathbf{M}_c(\theta^*, \mathcal{K}(\bar{\theta}))$, $\epsilon(\bar{\theta}) \overset{m.s.}{=} \epsilon$ is equivalent to $\dfrac{1}{N}\sum_{k=0}^{N}(\hat{\epsilon}_k(\bar{\theta}) - \epsilon_k)^2 = 0$ almost surely.

the external excitation and hence is called the Identifiability Equation for External Excitation. The second equation (IEIE), on the other hand, characterizes the effect of the internal excitation and is called the Identifiability Equation for Internal Excitation. IEIE indicates that the closed-loop transfer function from ϵ to (y, u) is always identifiable.

Using the description of $\mathbf{M}_c$ given in (2.7), one can show that these equations can be written in terms of the system polynomials as

$$(3.5) \quad (\overline{A}\,B - \overline{B}\,A)\,\overline{T}\,r \; = 0 \qquad a.s., \qquad (\text{IE}^3)$$

$$(3.6) \quad (\overline{A}\,\overline{R} + q^{-d}\overline{B}\,\overline{S})\,C \; = (A\,\overline{R} + q^{-d}B\,\overline{S})\,\overline{C}, \qquad (\text{IEIE})$$

for all $\overline{\theta} \in \mathcal{D}_\infty$. Note that $\overline{R}$ is nonzero by assumption.

Definition: *We say a controller mapping $\mathcal{K}$ is stabilizing over a set if the estimated closed-loop transfer function $\mathbf{M}_c(\theta, \mathcal{K}(\theta))$ is stable for every θ in that set.*

LEMMA 3.1. *Suppose that the controller mapping $\mathcal{K}$ is stabilizing over the set $\mathcal{S} = \{\, \overline{\theta} \in \Theta \mid \overline{A} \text{ and } \overline{B} \text{ have no unstable common factor} \,\}$. Then, $\mathcal{K}$ can be defined over $\Theta \setminus \mathcal{S}$ such that when $\nu = \max\{\partial A,\, d + \partial B,\, \partial C\}$, $\mathcal{D}_\infty \subset \mathcal{S}$ and*

$$(3.7) \quad \begin{aligned} \mathcal{D}_\infty : \; &= \{\overline{\theta} \in \Theta \mid \overline{A}\,y_k(\overline{\theta}) - \overline{B}\,u_{k-d}(\overline{\theta}) - \overline{C}\epsilon_k = 0 \quad a.s. \} \\ &= \{\overline{\theta} \in \Theta \mid \text{IE}^3 \text{ and } \text{IEIE are satisfied.} \}. \end{aligned}$$

Proof.   Clearly, $\mathcal{D}_\infty \subset \{\overline{\theta} \in \Theta \mid \overline{A}\,y_k(\overline{\theta}) - \overline{B}\,u_{k-d}(\overline{\theta}) - \overline{C}\epsilon_k = 0$ $a.s.\,\}$. For the other direction, let $\overline{\theta} \in \Theta$ be such that $\overline{A}\,y_k(\overline{\theta}) - \overline{B}\,u_{k-d}(\overline{\theta}) - \overline{C}\epsilon_k = 0$ almost surely.

First assume that $\overline{A}$ and $\overline{B}$ have no unstable common factor. Hence, $\mathbf{M}_c(\overline{\theta}, \mathcal{K}(\overline{\theta}))$ is stable by assumption which is equivalent to saying that its characteristic polynomial $\overline{A}\,\overline{R} + q^{-d}\overline{B}\,\overline{S}$ is a stable polynomial. As explained, $\overline{\theta}$ also satisfies the identifiability equations (3.5). Using IEIE (3.6) and **H1** (C being stable), one can then conclude that $A\,\overline{R} + q^{-d}B\,\overline{S}$ and $\overline{C}$ are stable. Therefore, $\mathbf{M}_c(\theta^*, \mathcal{K}(\overline{\theta}))$ is BIBO stable. Furthermore, one infers from (2.9) and stability of $\overline{C}$, after ignoring the effects of initial conditions, that $\widehat{\epsilon}(\overline{\theta}) = \epsilon$.

Next, consider the points $\overline{\theta} \in \Theta$ which correspond to $\overline{A}$ and $\overline{B}$ polynomials with unstable common factor and modify $\mathcal{K}$ at that points such that they do not satisfy the identifiability equations. For example let $\mathcal{K}(\overline{\theta})$ for any such point be equal to a fixed coprime pair (R_0, S_0) such that $\partial R_0 > \partial B + \partial C$ or $d + \partial S_0 > \partial A + \partial C$. Then if such $\overline{\theta}$ were to satisfy IEIE (3.6), one would get $A^{-1}[B\ C] = \overline{A}^{-1}[\overline{B}\ \overline{C}]$, which contradicts **H1** and the hypothesis. This proves the first equality of (3.7). We showed the second identity when deriving the identifiability equations. $\qquad\square$

Therefore, based on lemma 3.1 and since most practical certainty-equivalence control designs are stabilizing, we are justified in letting $\mathcal{D}_\infty$ be defined from now on by (3.7).[3]

Notation: Let $\mathcal{D}_{opt} := \{ \bar{\theta} \in \Theta \mid u(\bar{\theta}) \stackrel{m.s.}{=} u(\theta) \}$.

$\mathcal{D}_{opt}$ is the set of parameter vectors which yield the desired or correct controller transfer function. It is easy to verify from (2.5–2.6) that for every $\bar{\theta} \in \mathcal{D}_{opt}$, $y(\bar{\theta}) \stackrel{m.s.}{=} y(\theta^*)$ as well.

Considering (2.6), (2.7) and (3.1), we can show that for every point $\bar{\theta} \in \mathcal{D}_{opt}$,

$$\overline{R}^{-1}\overline{S} = R^{*-1}S^* \quad \text{and} \quad (R^*\overline{T} - \overline{R}T^*)r \stackrel{m.s.}{=} 0,$$

where R^*, S^* and T^* are the <u>desired</u> controller polynomials corresponding to θ^*. Since one can always simplify the controller design for a specific reference input r such that the controller transfer function from r to u is identifiable, we can assume without any loss of generality that

$$\begin{aligned} \mathcal{D}_{opt} &:= \{ \bar{\theta} \in \Theta \mid \mathcal{K}(\bar{\theta}) \equiv \mathcal{K}(\theta^*) \} \\ &= \{ \bar{\theta} \in \Theta \mid \overline{R}^{-1}(\overline{S}\ \overline{T}) = R^{*-1}(S^*\ T^*) \}. \end{aligned}$$

$\mathcal{D}_{opt}$, defined as such, is then the set of parameter estimates which yield the ideal controller design under the certainty-equivalence approach.

Now using the above notations, we may formally define potential self-tuning:

Definition:

> We call a certainty-equivalence adaptive control scheme *potentially self-tuning* if $\mathcal{D}_\infty \subseteq \mathcal{D}_{opt}$.

For stochastic systems, we are particularly interested in self-tuning regardless of persistency of external excitation, hence:

> We say an adaptive control algorithm is potentially self-tuning in absence of external excitation if $\mathcal{D}_\infty \big|_{\{r\equiv 0\}} \subseteq \mathcal{D}_{opt}$.

As explained in the introduction, it is desirable to have a potentially self-tuning adaptive controller. In that case, with appropriate estimation method, stability and convergence of the parameter estimates guarantees convergence of the control input to its desired or correct value, regardless of the inconsistency of the parameter estimates.

---

[3] As we shall see in §5, control designs such as generalized minimum-variance and generalized predictive control, which make use of certain generalized minimum-phase assumptions, are not stabilizing for all possible parameter vectors. However, one can still show in these cases that $\mathbf{M}_c(\theta^*, \mathcal{K}(\bar{\theta}))$ is stable for every $\bar{\theta} \in \Theta$ such that $\overline{A}\, y_k(\bar{\theta}) - \overline{B}\, u_{k-d}(\bar{\theta}) - \overline{C}\epsilon_k = 0$. Thus, (3.7) is valid in these cases if we assume that $\overline{C}$ is stable for every $\bar{\theta} \in \Theta$.

A stronger property than potential self-tuning is potential identifiability as defined in the following. It is clear that potential identifiability implies potential self-tuning:

Definition:

> We call a plant *potentially identifiable*, under some *certainty-equivalence adaptive control scheme*, if $\mathcal{D}_\infty = \{\theta\}$.

Strictly speaking, from the point of view of the control objective, potential identifiability has no advantage over potential self-tuning. However, if one considers the issue of robustness of the controller design, then it is well-known that consistent adaptive systems possess certain robustness qualities. See Sastry and Bodson (1987), §5, for more details.

Based on the above arguments, we are therefore interested in classifying the controller mappings, or the plant structures, for which potential self-tuning in absence of external excitation, or potential identifiability, are guaranteed. We would also like to examine popular certainty-equivalence adaptive control algorithms for the mentioned properties. In the next section, we present our main results in this regard.

To further illustrate the subject, we end this section with an example. In the following example, we consider an stable **ARMAX** model where in addition to the unobserved noise ϵ, an observed disturbance w also enters the plant. Hence, the plant equation in this case is given by $Ay_k = q^{-d} Bu_k + q^{-\gamma} Dw_k + C\epsilon_k$ with the unknown parameter $\theta^* \equiv (A, B, D, C)$. The control objective is to use a *feedforward* strategy to compensate the effect of the disturbance on output. All previous definitions will be modified accordingly:

Example 3.2 As an example, let us consider the feedforward minimum variance self-tuning regulation of the simple SISO finite impulse response plant model

$$y_k = b^0 u_{k-1} + b^1 u_{k-2} + d^0 w_{k-1} + \epsilon_k,$$

where y and u are the output and control input, and w and ϵ, the observed and unobservable disturbances, are independent stationary stochastic processes satisfying **I1**. If the parameters b^0, b^1 and d^0 were known, then the "stabilized" feedforward MV control policy u^*, which minimizes the cost function

$$J = \lim_{\lambda \to 0} \limsup_{N \to \infty} \frac{1}{N} E\left[y_k^2 + \lambda u_k^2\right],$$

would be equal to $\quad u^* := \frac{s_0}{t_0 + t_1 q^{-1}} w, \quad$ where the controller parameters (s_0, t_0, t_1) are computed from the system parameters $\theta^* = (b^0, b^1, d^0)$ by using the controller mapping $\mathcal{K} : \bar{\theta} = (\bar{b}_o, \bar{b}^1, \bar{d}^0) \mapsto (\bar{s}_0, \bar{t}_0, \bar{t}_1)$ such that:

$$\bar{s}_0 = \bar{d}^0 \frac{\bar{b}^0}{\bar{t}_0}, \qquad \begin{cases} (\bar{t}_0, \bar{t}_1) = (\bar{b}^0, \bar{b}^1) & \text{if } |\bar{b}^0| \geq |\bar{b}^1| \\ (\bar{t}_0, \bar{t}_1) = (\bar{b}^1, \bar{b}^0) & \text{if } |\bar{b}^1| > |\bar{b}^0|. \end{cases}$$

See for example Nassiri-Toussi and Ren (1994a). Notice that other than the disturbance w, no external excitation is present. The certainty-equivalence adaptive design then consists of estimating b^0, b^1 and d^0 by an appropriate estimation method and using the estimates to design the controller. The frozen controller equation based on the frozen parameter estimate $\bar{\theta}$ thus becomes $\bar{t}_0 u_k + \bar{t}_1 u_{k-1} = \bar{s}_0 w_k$.

Since w is of infinite degree of persistency of excitation, the corresponding $\mathcal{D}_\infty$ set in this case will be the set of points $\bar{\theta}$ for which $\mathbf{M}_w(\theta^*, \mathcal{K}(\bar{\theta})) = \mathbf{M}_w(\bar{\theta}, \mathcal{K}(\bar{\theta}))$ and $\mathbf{M}_\epsilon(\theta^*, \mathcal{K}(\bar{\theta})) = \mathbf{M}_\epsilon(\bar{\theta}, \mathcal{K}(\bar{\theta}))$. These form the identifiability equations IEIE. After routine algebraic manipulations, one can then show that $\mathcal{D}_\infty = \Gamma_1 \cup \Gamma_2 \cup \Gamma_3$, where

$$\Gamma_1 = \left\{ \bar{b}^0, \quad \bar{d}^0 = d^0 \right\}, \qquad \text{and}$$

$$\Gamma_2 = \left\{ \frac{\bar{b}^1}{\bar{b}^0} = \frac{\bar{b}^0 - b^0}{\bar{b}^1 - b^1}, \quad \bar{d}^0 = d^0 \frac{\bar{b}^1}{b^1}, \quad |\bar{b}^1| > |\bar{b}^0| \right\} \qquad \text{and}$$

$$\Gamma_3 = \begin{cases} \{\kappa(b^0, b^1, d^0), \quad \forall \kappa \in \mathcal{R}\} & \text{if } |b^0| > |b^1|, \\ \emptyset & \text{if } |b^0| < |b^1|. \end{cases}$$

One can also show that $\mathcal{D}_{opt}$ is the union of two lines:

$$\mathcal{D}_{opt} = \left\{ \kappa\left(b^0, b^1, d^0\right), \quad \forall \kappa \in \mathcal{R} \right\} \cup \left\{ \kappa\left(b^{1^2}/b^0, b^1, d^0\right), \quad \forall \kappa \in \mathcal{R} \right\}.$$

Figure 2 shows the graph of these sets in $\mathcal{R}^3$ for $(|b^0| > |b^1|)$ and $(|b^1| > |b^0|)$ cases. It is evident that not every point in these sets yields the optimal controller design. In particular, we observe that for every $\bar{\theta} \in \Gamma_1$, the corresponding control input is equal to zero. Indeed, simulations show that for certain initial conditions and with positive probability the parameter estimates converge to $\mathcal{D}_\infty \setminus \mathcal{D}_{opt}$, including the set Γ_1. See Nassiri-Toussi and Ren (1994a) for more details.

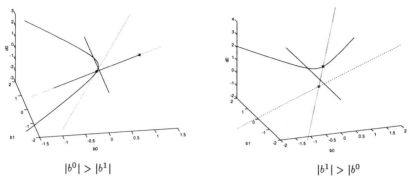

$$|b^0| > |b^1| \qquad\qquad\qquad\qquad |b^1| > |b^0|$$

FIGURE 2.   The equilibrium set $\mathcal{D}_\infty$ and the set of points yielding the optimal controller design $\mathcal{D}_{opt}$

$\mathcal{D}_\infty$:   Solid line, $\mathcal{D}_{opt}$:   Dotted line.

4. Potential self-tuning: the main results.

In this section, we attempt to answer the question of when a certainty-equivalence adaptive control system is potentially self-tuning or identifiable. Theorem 1 which is deduced from IE[3] and IEIE, provides two sufficient conditions and a necessary condition for potential self-tuning in absence of external excitation. The first sufficient condition assumes that under the desired control design, the l-step-ahead predictor (in closed-loop) of $P(q^{-1})y$, some combination of the output and its past samples, can be realized independent of the system parameters. As we shall see in §5, this condition is in particular applicable to minimum-variance and generalized predictive control.

The second sufficient condition requires uniqueness of solutions for a certain Diophantine equation associated with IEIE (3.6) and in particular applies to pole-placement control designs.

The necessary condition states that for an adaptive control system to be potentially self-tuning, certain characteristics of the frozen closed-loop system must be invariant in $\mathcal{D}_\infty$.

Notation: Let $x_{k|k-l} := E\left[x_j \mid y^{k-l}, r^k\right]$ denote the l-step-ahead prediction of x_k.

Theorem 4.1. Conditions for potential self-tuning:
Consider the plant (2.1) subject to the general certainty-equivalence adaptive control algorithm (2.3–2.4), based on the identification algorithm (2.8). Assume that **H1–H2** *and* **I1–I2** *are satisfied.   Then, the following statements hold:*

i) *Consider the closed-loop system* $[y \quad u] = \mathbf{M}_c(\theta, \mathcal{K}(\theta)) \left(\begin{bmatrix} r & \epsilon \end{bmatrix}\right)$ *and suppose that the l-step-ahead predictor of $P(q^{-1})y$ for some $l \geq d$ has a realization, given by the following equation, which is independent of θ:*

$$(4.1) \qquad \left(P(q^{-1})y\right)_{k|k-l} = Q(q^{-1})u_{k-d} + G(q^{-1})r_k + W(q^{-1})\epsilon_{k-l}.$$

Here P, Q, G and W are polynomials independent of θ such that P is monic and $PB - QA$ is stable, while $u_{k-d} \in \mathcal{F}_{k-l}$.
Then, the adaptive control algorithm is potentially self-tuning regardless of the degree of persistency of r.

ii) Suppose that IEIE, the identifiability equation for internal excitation, results in the following equation:

$$(4.2) \qquad X(\bar{\theta}, \theta^*)(R^* - \bar{R}) + Z(\bar{\theta}, \theta^*)(S^* - \bar{S}) = 0,$$

where $\Pi_{X,Z}(\bar{\theta}, \theta^*)$, the Sylvester matrix of X and Z,[4] satisfies

$$\inf_{\bar{\theta} \in \Theta} \det\left(\Pi_{X,Z}^T \Pi_{X,Z}\right) > 0.$$

Then, provided that $T(\theta) = T^*$ for all $\theta \in \Theta$, the adaptive control algorithm is potentially self-tuning regardless of the degree of persistency of r.

iii) Suppose that the adaptive control algorithm is potentially self-tuning and assume that $\bar{A}\bar{R} + q^{-d}\bar{B}\bar{S}$, the characteristic polynomial of $\mathbf{M}_c(\bar{\theta}, \mathcal{K}(\bar{\theta}))$, is divisible by $\bar{C}$ for every $\bar{\theta} \in \Theta$. Then, we will have

$$(4.3) \qquad \bar{A}\bar{R} + q^{-d}\bar{B}\bar{S} = D^*\bar{C} \qquad\qquad \forall\, \bar{\theta} \in \mathcal{D}_\infty,$$

where D^* is a function of θ^* and the controller mapping $\mathcal{K}$.
Moreover, ignoring the effects of initial conditions and assuming that $r \equiv 0$, there exists a polynomial $P^*(q^{-1})$ and a realization of the d-step-ahead predictor of $P^* y(\bar{\theta})$ that is given by the following equation, for every $\bar{\theta} \in \mathcal{D}_\infty$:

$$(4.4) \qquad \left(P^*(q^{-1})\, y(\bar{\theta})\right)_{k|k-d} = Q^*(q^{-1})\, u(\bar{\theta})_{k-d} + W^*(q^{-1})\, \epsilon_{k-d}.$$

Here, P^*, W^* and Q^* depend only on θ^* and the controller mapping $\mathcal{K}$,
$BP^* - AQ^* = D^*$ and $P^*C = F^*A + q^{-d}(W^* + S^*)$ such that $\partial F^* < d$.

See the appendix for the proof.   □

Remarks: Statements i) and ii) give the sufficient conditions in absence of r, while iii) gives the necessary condition. As to the assumption in statement iii), the characteristic polynomial of the closed-loop system is dividable by the C polynomial if and only if the control design can be

---

[4] Every polynomial Diophantine equation, $XA + YB = 0$, with X and Y unknown polynomials of specific degrees, can be written in the form of. $\Pi_{X,Y}\,\Psi = 0$, where $\Pi_{X,Y}$ is the Sylvester matrix of X and Y and Ψ is the vector of coefficients of X and Y.

realized by a Kalman filter and a dynamic state feedback. A similar, though more complex, necessary condition is found for the general case, as well. In practice, the controller mapping is C^∞ almost everywhere. Hence, we conjecture from iii) that in such cases, one can achieve potential self-tuning in absence of external excitation, only if the plant is potentially identifiable in a neighborhood of θ^*, that is if θ^* is an isolated point of $\mathcal{D}_\infty$, or if the control design is of the pole-placement or model-matching type in a neighborhood of θ^*. Note that (4.4) (or (4.1)) can be regarded as a model matching objective, equivalent to minimizing $E\left[(P^*y_{k+d} - Q^*u_k - W^*\epsilon_k)^2 \mid \mathcal{F}_k\right]$. We shall discuss the model-matching and pole-placement adaptive control designs in §5.

In the next theorem, we collect a set of results concerning some sufficient conditions and a necessary condition for potential identifiability:

Theorem 4.2. Conditions for potential identifiability:
Consider the plant (2.1) subject to the general certainty-equivalence adaptive control algorithm (2.3–2.4), based on the identification algorithm (2.8). Assume that **H1-H2** *and* **I1-I2** *are true and the plant is identifiable, that is to say* $\nu = \max\{\partial A, \partial B + d, \partial C\}$ *is the minimal order of the plant. Then, the following statements hold:*

i) Let (A, B) be a coprime pair, r be persistently exciting of at least degree $\partial A + \partial B + 1$, and $T(\theta) = T^*$ for all $\theta \in \Theta$ such that T^*r has the same degree of persistency as r. Then, the plant is potentially identifiable.

ii) The plant is potentially identifiable regardless of the degree of persistency of r if $d > \partial A + \partial C$ and $\inf_{\theta \in \Theta} \|\frac{1}{r_0(\theta)}S(\theta)\|_2 > 0$[5], where $r_0(\theta)$ is the first coefficient of $R(\theta)$.

iii) Suppose that the controller mapping $\mathcal{K}$ is continuously differentiable in a neighborhood of θ^*. Then, in order to have a potentially identifiable plant regardless of the degree of persistency of r, it is necessary that for some θ in each neighborhood of θ^*,

(4.5) $$\max\{\partial R(\theta) - \partial B, \, d + \partial S(\theta) - \partial A\} > 0.$$

On the other hand, a sufficient condition for potential identifiability, regardless of the degree of persistency of r, is that for all $\theta \in \Theta$,

(4.6) $$\max\{\partial R(\theta) - \partial B - \partial C, \, d + \partial S(\theta) - \partial A - \partial C\} > 0,$$

where we have assumed without loss of generality that $R(\theta)$ and $S(\theta)$ are coprime for each $\theta \in \Theta$. In particular, if $\partial C = 0$, then (4.5) corresponds to both a sufficient and a necessary condition for potential identifiability.

---

[5] For a polynomial A, $\|A\|_2^2 := (a^0)^2 + \ldots + (a^{\partial A})^2$.

Moreover, if the adaptive controller is potentially self-tuning, then a sufficient condition for potential identifiability is that

$$(4.7) \qquad \max\{\partial R^* - \partial B - \partial C, \, d + \partial S^* - \partial A - \partial C\} > 0,$$

See the appendix for the proof.   □

Remarks: Statement $i)$ presents only a sufficient degree of persistency of excitation for r. In most practical cases however, the plant is potentially identifiable with external excitations that are persistently exciting of a degree less than $\partial A + \partial B + 1$. Statement $ii)$ means that sufficiently large delay in control input-to-output transfer function results in identifiability. This is of course a trade-off between controllability and identifiability since larger delay from control input to output means also less controllability. Statement $iii)$ suggests that larger complexity of the controller design is also beneficial towards identifiability. The special case of $\partial C = 0$ case is particularly noteworthy. Note that we have assumed that all common factors between $R(\theta)$ and $S(\theta)$ have been cancelled. Another point worth pointing out is that from statement $iii)$, we can see that potential identifiability conditions are particularly easy to satisfy for the special cases of ARX and FIR models.

5. Applications to some control designs. In this section, we apply the general results of §4 to some widely-used adaptive control schemes. The plant model is assumed to be described by the ARMAX equation (2.1).

Remark: The results established in this paper for the ARMAX model (1) trivially extend to the more general case of ARIMAX model

$$(5.1) \qquad A(q^{-1})y_k = B(q^{-1})u_{k-d} + C(q^{-1})\frac{\epsilon_k}{\Delta(q^{-1})},$$

where $\Delta(q^{-1})$ is a known polynomial (usually $1 - q^{-1}$), with possible roots outside the unit disk or on the unit circle in the complex plane. Indeed, we obtain similar results if only we replace the A polynomial with $A\Delta$ and u with $\tilde{u}$ in the definitions of $\mathbf{M}_c$, $\mathcal{D}_\infty$ and $\mathcal{D}_{opt}$, and in the identifiability equations IEIE and IE^3.

A. Self-Tuning Model Matching or Generalized Minimum-Variance (GMV) Control: The GMV control policy u^* for the plant (2.1) is defined as

$$u^* := \arg\min_{u|u_k \in \mathcal{F}_k^y} \limsup_{N\to\infty} \frac{1}{N} E\left[(P\,y_k - G\,r_k - Q\,u_{k-d})^2 \mid \mathcal{F}_{k-d}\right],$$

where r is the deterministic reference signal. See for example Wellstead and Zarrop (1991).

$P(q^{-1})$, $Q(q^{-1})$ and $G(q^{-1})$ here are given polynomials such that P is monic. $\mathcal{F}_k^y$ is the σ-field generated by observations of the output y up to time k. One obtains the standard MV control if $P = G = 1$ and $Q = 0$.

Under the optimal GMV control policy, y and u satisfy the following equation:

$$(5.2) \qquad P(q^{-1})y_k = Q\,u_{k-d} + G\,r_k + F^*\epsilon_k,$$

where F^* with $\partial F^* < d$ is the $(d-1)$-degree truncation of $A^{-1}PC$.[6] The closed-loop system characteristic polynomial under the GMV control law is $PB - QA$. Therefore, in order to have a stabilizing control policy u^*, the following assumption must be satisfied:

GMV1) $\quad PB + QA$, i.e. the characteristic polynomial of $\mathbf{M}_c(\theta^*, \mathcal{K}_{GMV}(\theta^*))$ the closed-loop system subject to the GMV law, is stable.

Comparing (5.2) with (4.1) and noticing that P, Q and G are independent of θ^*, one observes that statement i) of Theorem 4.1 is applicable, provided that **GMV1** is true and $u_k \in \mathcal{F}_k^y$ is well-defined.
Let the corresponding controller mapping be denoted by $\mathcal{K}_{GMV}: \bar{\theta} \to (\overline{R}, \overline{S}, \overline{T})$ where $\overline{R}$, $\overline{S}$ and $\overline{T}$ are the controller polynomials of the certainty-equivalent frozen controller equation $\overline{R}u_k(\bar{\theta}) = -\overline{S}y_k(\bar{\theta}) + \overline{T}r_k$. Then, one can show that $\mathcal{K}_{GMV}$, and $u_k(\bar{\theta})$, are well-defined for every parameter vector $\bar{\theta}$ such that the first coefficient of $\overline{B} - Q$ is nonzero. But by Lemma 3.1, one can always define the mapping at other points such that they are excluded from $\mathcal{D}_\infty$. Thus, we have the following assertion:

- **Assertion 5.1.** Let **GMV1** in addition to the main assumptions; **H1-H2** and **I1-I2**, be satisfied. Then, the certainty-equivalence adaptive GMV control scheme based on the controller mapping $\mathcal{K}_{GMV}$ is potentially self-tuning regardless of the degree of persistency of the external excitation.

Moreover, by using statement iii) of Theorem 4.2 and the equations defining $\mathcal{K}_{GMV}$, we can show that:

If R^* and S^* are coprime, the following inequalities present a sufficient condition and a necessary condition for potential identifiability of the plant under the GMV law:

$$\begin{cases} \max\{d - 1 - \partial C, \quad \partial P - \partial A\} > 0 & \text{is a sufficient condition,} \\ & \text{while} \\ \max\{d - 1, \quad \partial P + \partial C - \partial A\} > 0 & \text{is a necessary condition.} \end{cases}$$

B. Self-Tuning Pole-Placement Control: Suppose ν is equal to the minimal order of the plant and that (A, B) is a coprime pair. Then, u^*, the pole-placement control policy for the plant model (2.1) satisfies

$$(5.3) \qquad R^*\,u_k^* := -S^*\,y_k + r_k,$$

---

[6] The $(d-1)$-degree truncation of $A^{-1}(q^{-1})B(q^{-1})$ is a polynomial $F(q^{-1})$ of degree $d-1$ such that $B - FA$ is divisible by q^{-d}.

where the controller polynomials are the unique solutions of either of the following Diophantine equations:

$$(5.4) \quad \begin{cases} PPL1: & A R^* + q^{-d} B S^* = D^* C, \qquad \text{or} \\ PPL2: & A R^* + q^{-d} B S^* = D^*, \end{cases}$$

such that $\partial S^* < \partial A$. $D^*(q^{-1})$ here is a given stable monic polynomial.

Diophantine equations (5.4) induce controller mappings that are used by the certainty-equivalence pole-placement control algorithm to calculate the controller polynomials from the system parameters. These equations do not have, however, unique solutions on the set $\mathcal{S}' = \{\bar{\theta} \mid \bar{A} \text{ and } \bar{B} \text{ are not coprime.}\}$. However, by the assumptions and the same argument used in the proof of Lemma 3.1, we can define the controller polynomials over $\mathcal{S}'$ such that $\mathcal{D}_\infty \cap \mathcal{S}' = \emptyset$.

Thus, the controller mapping for the *PPL1* pole-placement design is given over $\Theta \setminus \mathcal{S}'$ by

$$\mathcal{K}_{PPL1}: \quad (\bar{A}, \bar{B}, \bar{C}) \mapsto (\bar{R}, \bar{S}) \qquad \text{such that} \qquad \bar{A}\,\bar{R} + q^{-d}\bar{B}\,\bar{S} = D^*\bar{C}.$$

Clearly, we have to assume that $\bar{C}$ is stable for all $\bar{\theta}$ in order to guarantee stability of $\mathbf{M}_c(\bar{\theta}, \mathcal{K}_{PPL1}(\bar{\theta}))$ over $\mathcal{D}_\infty$. From IEIE (14b) then, we can get

$$A(\bar{R} - R^*) + q^{-d} B(\bar{S} - S^*) = 0 \qquad \forall \bar{\theta} \in \mathcal{D}_\infty.$$

Potential self-tuning regardless of the degree of persistency of r then follows from statement *ii)* of Theorem 4.1 upon noting that since A and B are coprime and $\partial S(\theta) < \partial A$, the Sylvester matrix corresponding to A and B is invertible.

We can not make the same conclusion about the *PPL2* pole-placement design, unless we assume that $\bar{C}$ and D^* are coprime for all $\bar{\theta}$. Alternatively, we may modify the *PPL2* mapping over $\Theta \setminus \mathcal{S}'$ as in the following:

$$\mathcal{K}_{PPL2}: \quad (\bar{A}, \bar{B}, \bar{C}) \mapsto (\bar{R}, \bar{S}) \qquad \text{such that}$$

$$\begin{cases} \bar{A}\,\bar{R} + q^{-d}\bar{B}\,\bar{S} = D^* & \text{if } D^* \text{and } \bar{C} \text{are coprime,} \\ \bar{A}\,\bar{R} + q^{-d}\bar{B}\,\bar{S} = \frac{D^* + \delta}{1 + \delta} & \text{otherwise.} \end{cases}$$

Here, $\delta > 0$ is small enough such that $D^* + \delta$ is stable. Note that if C and D^* are coprime, then $\mathcal{K}_{PPL2}$ remains unchanged at θ^*. With the above modification, IEIE implies

$$\bar{C} = C \qquad \text{and} \qquad A\,\bar{R} + q^{-d} B\,\bar{S} = A R^* + q^{-d} B S^*.$$

Potential self-tuning regardless of r then follows.

- **Assertion 5.2.** *Let the main assumptions be satisfied. Further assume that $\nu = \max\{\partial A, d + \partial B, \partial C\}$ is the minimal order of the plant and that A and B are coprime. Then, the certainty-equivalence adaptive pole-placement control schemes, based on the controller equation (25) and the controller mappings K_{PPL1}, or K_{PPL2}, are potentially self-tuning regardless of the degree of persistency of the external excitation, provided that for the PPL1 design, $\overline{C}$ is stable for every $\overline{\theta} \in \Theta$.*

Moreover, by using statement iii) of theorem 4.2 and (24), we can show that:

If R^* and S^* are coprime, a sufficient condition for potential identifiability of the plant under the PPL1, or PPL2, pole-placement law is given by:

$$
\begin{cases}
\max\{d - 1 - \partial C, \quad \partial D^* - \partial A - \partial B\} > 0 \\
\qquad\qquad\qquad\qquad \text{for the } PPL1 \text{ design,} \quad \text{or} \\
\max\{d - 1 - \partial C, \quad \partial D^* - \partial A - \partial B - \partial C\} > 0 \\
\qquad\qquad\qquad\qquad \text{for the } PPL2 \text{ design.}
\end{cases}
$$

C. Self-Tuning Generalized Predictive Control (GPC):

See Clarke, Mohtadi and Tuffs (1987) or Bitmead, Gevers and Wertz (1990) for more details and alternative formulations. Let

$$
\overline{Y}_k = [(y - r)_k \quad \cdots \quad (y - r)_{k+N}]^T \quad \text{and} \quad U_k = [u_k \quad \cdots \quad u_{k+M}]^T,
$$

where $M \leq N$ and $\Sigma > 0$, $\Lambda > 0$ are given. r here refers to the deterministic reference signal. Then, the generalized predictive control policy u^* is

$$
u_k^* \ := \ [1\ 0\ \ldots\ 0]_{1 \times (M+1)} \times U_k^* \qquad\qquad \text{where}
$$

$$
U_k^* \ := \ \underset{U_k \in \mathcal{F}_k^y}{\arg\min} E\left[\overline{Y}_{k+d}^T \Sigma \overline{Y}_{k+d} + U_k^T \Lambda U_k \mid \mathcal{F}_k\right]
$$

such that $\qquad u_{k+M+1} = \ldots = u_{k+N} = 0.$

The above criterion is of course a generalization of the standard GPC criterion

$$
u_k^* := [1\ 0\ \ldots\ 0] \underset{u_k^{k+M} \in \mathcal{F}_k^y}{\arg\min} E\left[\sum_{i=0}^{N}(y_{k+d+i} - r_{k+d+i})^2 + \lambda \sum_{i=0}^{N} u_{k+i}^2 \mid \mathcal{F}_k\right]
$$

based on the same constraint. Traditionally, the Generalized Predictive Control is applied to the plants described by the **ARIMAX** model (5.1), with u being replaced with $\tilde{u} = \Delta(q^{-1}) u$ in (5). See Bitmead *et al.* (1990) for example. The self-tuning properties of the resulting certainty-equivalence adaptive system are however similar to those discussed here.

Under the optimal GPC policy, y and u satisfy the following equation:

(5.5)
$$\sum_{i=0}^{N} \alpha_i^* q^{i-N} y_k + \left(q^{-N} - \sum_{i=0}^{N} \alpha_i^* q^{i-N} G_i^*\right) u_{k-d}$$
$$= \sum_{i=0}^{N} \alpha_i^* q^{i-N} r_k + \sum_{i=0}^{N} \alpha_i^* q^{i-N} F_i^* \epsilon_k,$$

where G_i^* is the i-degree truncation of $A^{-1}B$, F_i^* is the $(i+d-1)$-degree truncation of $A^{-1}C$ and α_i's are given by the following:

$$\begin{bmatrix} \alpha_0^* \\ \vdots \\ \alpha_N^* \end{bmatrix} = \Sigma \mathbf{G}^* (\mathbf{G}^{*T} \Sigma \mathbf{G}^* + \Lambda)^{-1} \begin{bmatrix} 1 \\ 0 \\ \vdots \\ 0 \end{bmatrix} \quad \text{with} \quad \mathbf{G}^* = \begin{bmatrix} g_0 & & 0 \\ \vdots & \ddots & 0 \\ g_M & \cdots & g_0 \\ \vdots & \vdots & \vdots \\ g_N & \cdots & g_{N-M} \end{bmatrix}$$

and g_i the $(i+1)$-st Markov parameter of the transfer function $A^{-1}B$ (last coefficient of G_i^*). One also finds the closed-loop characteristic polynomial in this case to be
$$\left[A + \sum_{i=0}^{N} \alpha_i^* q^i (B - G_i^* A)\right].$$
Thus, in order to have closed-loop stability, we must have the following assumption:

GPC1) $A_c = A + \sum_{i=0}^{N} \alpha_i^* q^i (B - G_i^* A)$, *i.e. the characteristic polynomial of* $\mathbf{M}_c(\theta^*, \mathcal{K}_{GPC}(\theta^*))$ *the closed-loop system subject to the GPC law, is stable.*

The controller polynomials corresponding to the certainty-equivalent frozen controller equation are given by $\mathcal{K}_{GPC}(\hat{\theta})$, where the Generalized Predictive controller mapping $\mathcal{K}_{GPC}$ is proved to be well-defined for all possible parameter vectors. Unfortunately, the certainty-equivalence adaptive GPC is not potentially self-tuning in the general case, one can easily show this for example for the $N = M = 0$ case. However, the algorithm becomes potentially self-tuning if we assume some prior information about the system parameters. Specifically, we need to assume that:

GPC2) *The first $N+1$ Markov parameters of the transfer function $A^{-1}B$ and the first $N+d$ Markov parameters of $A^{-1}C$ are known a priori. This is equivalent to prior knowledge of G_i^* and F_i^* polynomials for $i = 0, \ldots, N$.*

A similar requirement, but only for deterministic systems, has already been used in order to prove stability and asymptotic optimality of adaptive GPC algorithms, see Wang and Henriksen (1993) or Ortega and Sanchez-Galindo (1989).

To incorporate this information into the estimation algorithm, one has to transform the ARMAX model into a regression model of the form $y_k = G_N{}^* + F_N{}^* \epsilon_k + \theta^* \phi_k^\epsilon$, where θ^* is now the reduced parameter vector. This assumption implies that all the parameters and polynomials in (5.5) are computable prior to estimation. Thus, in the new reduced parameterization, (5.5) is independent of the unknown system parameters. Comparing with (4.1), then one may invoke statement $i)$ of Theorem 4.1 to conclude the potential self-tuning of the adaptive GPC scheme regardless of r. It is worth pointing out that over this parameter space, the generalized predictive control policy is equivalent to matching the plant model to the fixed model (5.5).

- **Assertion 5.3.** *The certainty-equivalence adaptive GPC algorithm, based on the controller mapping $\mathcal{K}_{GPC}$, is not in general potentially self-tuning in absence of external excitation. However if GPC1 and GPC2 in addition to the main assumptions are true, then the control scheme is potentially self-tuning regardless of the degree of persistency of the external excitation.*

It should also be pointed out however that **GPC2** is a strong assumption. For example if a sufficient number of Markov parameters are known, then the exact plant model is in fact known. Therefore, **GPC2** will make sense only if N is small compared to ν.

D. Self-Tuning LQG (Feedback) Control: The LQG control policy for the plant model (2.1) is given by

$$u^* := \arg\min_{u \,|\, u_k \in \mathcal{F}_k^y} \limsup_{N \to \infty} \frac{1}{N} E\left[y_k^2 + \lambda u_k^2 \mid \mathcal{F}_k \right] \qquad \text{for some } \lambda > 0.$$

u^* satisfies the controller equation $R^* u_k = -S^* y_k$, where $(R^*, S^*) = \mathcal{K}_{LQG}(\theta^*)$.   One can show that $\mathcal{K}_{LQG}$ the LQG controller mapping is well-defined (and continuous) over the set of points $\bar{\theta}$ for which $\overline{A}$ and $\overline{B}$ have no common unstable factor. Nevertheless, Lin, Kumar and Seidman (1985) showed by a simple first-order example ($\partial A = 1$, $\partial B = 0$, $C = 1$) that the certainty-equivalence adaptive control scheme based on the above controller mapping is not potentially self-tuning.

E. Self-Tuning Feedforward LQG Regulator: Let the plant in this case be described by

$$(5.6) \qquad A(q^{-1}) y_k = B(q^{-1}) u_{k-d} + D(q^{-1}) w_k + C(q^{-1}) \epsilon_k,$$

where w is a stationary stochastic process satisfying **I1** and independent of ϵ and r. Assume that A and C are stable and monic. The parameter vector θ^* in this case consists of the variable coefficients of A, B, C and D and the linear regression formulation is obtained accordingly.

Then, the feedforward LQG control u^* is the control policy satisfying

$$u^* := \underset{u|(u_k - r_k) \in \mathcal{F}_k^w}{\arg\min} \ \limsup_{N \to \infty} \frac{1}{N} E\left[y_k^2 + \lambda u_k^2 \mid \mathcal{F}_k \right] \qquad \text{for some } \lambda > 0.$$

The controller equation for u^* is given by

$$(5.7) \qquad\qquad R^*(q^{-1})\,(u_k^* - r_k) = -V^*(q^{-1})\,w_k,$$

where $(R^*, V^*) = \mathcal{K}_{FFLQG}(\theta^*)$. $\mathcal{K}_{FFLQG}$ is well-defined if Θ is limited to points $\bar{\theta}$ such that $\overline{A}$ is stable, but it can be extended to the set of all possible parameter vectors. See Nassiri-Toussi and Ren (1994a) for details.

Because of mutual independence of r, w and ϵ, there are 3 identifiability equations describing the set $\mathcal{D}_\infty$ in this case. It is easy to check that the identifiability equations corresponding to ϵ and w (IEIE) are

$$\overline{A}\,C - \overline{C}\,A = 0 \qquad \text{and} \qquad \overline{A}\,(q^{-d}B\,\overline{V} - D\overline{R}) = A\,(q^{-d}\overline{B}\,V - \overline{D}\,R).$$

This means that the transfer function $A^{-1}C$ is potentially identifiable for every external excitation r. However, as shown in Example 3.2, this adaptive control scheme is not in general potentially self-tuning in absence of r. The sufficient degree of persistency of r which guarantees potential self-tuning in these situations, can however be as low as 1, depending on the structure of the plant. In Nassiri-Toussi and Ren (1994a), certain sufficient conditions and a necessary condition on the structure of the plant, for its potential identifiability in absence of external excitation, were established.

- **Assertion 5.4.** *In general, the certainty-equivalence LQG adaptive control schemes, based on the controller mappings $\mathcal{K}_{LQG}$ or $\mathcal{K}_{FFLQG}$ are not potentially self-tuning in absence of external excitation. The degree of persistency of the external excitation which is required for potential self-tuning depends on the structure of the plant.*

Remark.   Note that by (5.7) and assumption, the frozen control input $u(\bar{\theta})$ is independent of ϵ. Hence, ignoring the ϵ-to-y dynamics, one may instead transform (5.6) into the output-error form $\quad y_k = \theta^{*T} \phi_k^x + \nu_k$, where

$$\theta^* = \left[a^1, \ldots, a^\nu, b^0, \ldots, b^{\nu-d}, d^0, \ldots, d^\nu \right]^T,$$

$$\phi_k^x = \left[-x_{k-1}, \ldots, -x_{k-\nu}, u_{k-d}, \ldots, u_{k-\nu}, w_k, \ldots, w_{k-\nu} \right]^T$$

$$\text{and} \quad \begin{cases} A(q^{-1})\,x_k = B(q^{-1})\,u_{k-d} + D(q^{-1})\,w_k \\ \quad \text{with same initial conditions as in (5.6),} \\ A(q^{-1})\,\nu_k = C(q^{-1})\,\epsilon_k \\ \quad \text{with zero initial conditions.} \end{cases}$$

The parameter space in this case corresponds to the variable coefficients of A, B and D. Let $\hat{x}(\bar{\theta})$ satisfy $\overline{A}\,\hat{x}_k(\bar{\theta}) = q^{-d}\overline{B}\,u_k(\bar{\theta}) + \overline{D}\,w_k$, and $\hat{\phi}_k(\bar{\theta}) = \phi_k^x(\bar{\theta})\Big|_{x(\bar{\theta})\to\hat{x}(\bar{\theta})}$. It is then easy to realize that the prediction error in this case is $\hat{\nu}_k(\bar{\theta}) = y_k - \bar{\theta}^T\hat{\phi}_k(\bar{\theta})$. Also, with the same certainty-equivalence control policy and in absence of external excitation, $\mathcal{D}_\infty$ will be described by the sole identifiability equation:

$$\mathbf{M}_w(\theta^*,\mathcal{K}(\bar{\theta})) = \mathbf{M}_w(\bar{\theta},\mathcal{K}(\bar{\theta})) \iff \overline{A}\,(q^{-d}B\,\overline{V}-D\overline{R}) = A\,(q^{-d}\overline{B}\,\overline{V}-\overline{D}\,R),$$

where $\mathbf{M}_w$ is the w-to-x transfer function. Obviously, the adaptive feedforward LQG regulator based on the new scheme is not in general potentially self-tuning, either. In fact, the new parameterization renders the plant "less" identifiable.

Let us now summarize the above results in the following table:

TABLE 1.   External excitation requirements for some certainty-equivalence adaptive control algorithms.

Control Policy for ARMAX Model (2.1)	Does potential self-tuning require external excitation? (Min. required degree of p.e.?)
Generalized Minimum Variance	No
Pole-Placement	No
Generalized Predictive Control	No[†]
LQG (Feedback or Feedforward)	Yes (depends on the plant structure)

[†] No provided that **GPC2** holds, otherwise depends on the plant structure

6. Conclusions.
We presented the first stage in a two-stage unified approach to analyzing the self-tuning of certainty-equivalence adaptive control systems. The concepts of potential self-tuning and identifiability were introduced in order to distinguish the self-tuning and identifiability properties that are inherent to the control design and plant structure, from the properties of the specific estimation algorithm. In addition, the roles of internal and external excitations in potential self-tuning and identifiability were delineated. The analysis of PST was conducted by studying two identifiability equations, IE[3] and IEIE. General conditions for potential self-tuning and identifiability were then derived and applied to some specific control designs. In particular, we conjecture that among practical control designs, only those that locally, may be formulated as model-matching or pole-placement control policies, can yield potentially self-tuning adaptive controllers for every plant structure and every external excitation.

While our results were limited to the SISO linear time-invariant discrete-time plants and an ideal set-up, the approach is equally applicable to a much larger class of systems, including continuous-time, MIMO (see Nassiri-Toussi and Ren (1994b) and nonlinear sytems.

In the next stage, we shall discuss the connection between potential and asymptotic self-tuning, and by assuming certain sets of properties for the estimation algorithm and the control design, we will derive general conditions for asymptotic self-tuning.

7. Appendix. **A. Proof of Theorem 4.1:**

i)   For every $\bar{\theta} \in \mathcal{D}_\infty$, we have by definition

$$\left[y(\bar{\theta}) \quad u(\bar{\theta})\right] = \mathbf{M}_c(\theta, \mathcal{K}(\bar{\theta}))\left(\begin{bmatrix} r & \epsilon \end{bmatrix}\right) = \mathbf{M}_c(\bar{\theta}, \mathcal{K}(\bar{\theta}))\left(\begin{bmatrix} r & \epsilon \end{bmatrix}\right). \qquad a.s.$$

Therefore for every $\bar{\theta} \in \mathcal{D}_\infty$ including θ, $y(\bar{\theta})$ and $u(\bar{\theta})$ satisfy (16), while they satisfy (2.5) for every $\bar{\theta} \in \Theta$.

Now, consider the following Diophantine equations:

$$(7.1) \qquad \begin{cases} CP = F^*A + q^{-l}\Gamma^* & \text{such that } \partial F^* < l, \text{and} \\[2mm] 1 = E^*C + q^{-l}H^* & \text{such that } \partial E^* < l. \end{cases}$$

The degree conditions guarantee that (F^*, Γ^*) and (E^*, H^*) are unique solutions to the above Diophantine equations for all (A, C). Multiplying (2.5) by $E^* F^*$, we thus find

$$P\,y_k(\bar{\theta}) = \left[E^*F^*B\,u_{k-d}(\bar{\theta}) + (E^*\Gamma^* + H^*P)\,y_{k-l}(\bar{\theta}) - H^*F^*\epsilon_{k-l}\right] + F^*\epsilon_k,$$

Note that by assumption, the first term of RHS is $\mathcal{F}_{k-l}$-measurable, while $F^*\epsilon_k$ is a linear combination of $\epsilon_k, \dots, \epsilon_{k-l+1}$. Thus from the above equation and the hypothesis, we find the l-step-ahead-prediction of $P\,y_k(\bar{\theta})$ to be

$$\left(P\,y(\bar{\theta})\right)_{k|k-l} = \\ \left[E^*F^*B\,u_{k-d}(\bar{\theta}) + (E^*\Gamma^* + H^*P)y_{k-l}(\bar{\theta}) - H\,F^*\epsilon_{k-l} - G\,r_k\right].$$

Comparing with (4.1), it is clear that the following equation must hold:

$$(7.2) \qquad P\,y_k(\bar{\theta}) = Q\,u_{k-d}(\bar{\theta}) + G\,r_k + W\,\epsilon_{k-l} + F^*\epsilon_k.$$

Note that $F^*(q^{-1})$ is the $(d-1)$-degree truncation of $A^{-1}PC$ and independent of Q, G or $\bar{\theta}$. In particular, this shows that regardless of equation (4.1),

$$(7.3) \qquad P\,y_k(\bar{\theta}) - \left(P\,y(\bar{\theta})\right)_{k|k-l} = F^*\epsilon_k \qquad \forall \bar{\theta} \in \mathcal{D}_\infty,$$

as long as P is monic and independent of $\bar{\theta}$, and $u_{k-d} \in \mathcal{F}_{k-l}$.

From (2.5) and (7.2), we get the following closed-loop system equations for every $\bar{\theta} \in \mathcal{D}_\infty$, including θ:

$$\begin{cases} (BP - AQ)y_k(\bar{\theta}) &= BGr_k + ((F^* + q^{-l}W)B - QC)\epsilon_k, \\ (BP - AQ)u_k(\bar{\theta}) &= AGr_{k+d} + (W - \Gamma^*)\epsilon_{k-l+d}, \end{cases}$$

This means that on $\mathcal{D}_\infty$, $u(\bar{\theta}) = u(\theta^*)$ and $y(\bar{\theta}) = y(\theta^*)$. As explained in §3, we can assume this to be equivalent to $\mathcal{K}(\bar{\theta}) = \mathcal{K}(\theta^*)$ for all $\bar{\theta} \in \mathcal{D}_\infty$. In other words, $\mathcal{D}_\infty \subset \mathcal{D}_{opt}$ for every r. Note that $BP - AQ$, the characteristic polynomial of $\mathbf{M}_c(\theta^*, \mathcal{K}(\bar{\theta}))$ for every $\bar{\theta} \in \mathcal{D}_\infty$, is stable by assumption.

   ii)   (4.2) follows from IEIE (3.6) by assumption. But the Diophantine equation (4.2) can be written as

$$\Pi_{X,Z}(\bar{\theta}, \theta^*)\Psi = 0$$

where $\Pi_{X,Z}$ is the corresponding Sylvester matrix and Ψ is the vector of coefficients of $R^* - \bar{R}$ and $S^* - \bar{S}$. Since it is assumed that $\Pi_{X,Z}(\bar{\theta}, \theta^*)$ has full column rank uniformly in Θ, the only solution of the above equation is $\Psi = 0$, i.e. $\bar{R} = R^*$ and $\bar{S} = S^*$. Along with $T(\bar{\theta}) = T^*$ for all $\bar{\theta} \in \Theta$, this implies potential self-tuning. Since we did not make use of $IE^3(3.5)$, the identifiability equation for external excitation, potential self-tuning is achieved regardless of the degree of persistency of r.

   iii)   Let $\mathcal{D}_\infty \subseteq \mathcal{D}_{opt}$. By the definitions of $\mathcal{D}_\infty$ and $\mathcal{D}_{opt}$, this means that

$$\mathbf{M}_c(\theta^*, \mathcal{K}(\theta^*))\begin{pmatrix} [r & \epsilon] \end{pmatrix} = \mathbf{M}_c(\theta^*, \mathcal{K}(\bar{\theta}))\begin{pmatrix} [r & \epsilon] \end{pmatrix} = \mathbf{M}_c(\bar{\theta}, \mathcal{K}(\bar{\theta}))\begin{pmatrix} [r & \epsilon] \end{pmatrix} \quad a.s.$$

for all $\bar{\theta} \in \mathcal{D}_\infty$. Let without loss of generality $R(\bar{\theta})$ and $S(\bar{\theta})$ be coprime for all $\bar{\theta}$. Then, $\bar{R} = R^*$ and $\bar{S} = S^*$ for all $\bar{\theta} \in \mathcal{D}_\infty$, and using **I1**, **I2** and the description of $\mathbf{M}_c$ given by (2.7), we get

$$(\bar{A}\bar{R} + q^{-d}\bar{B}\bar{S})C = (A\bar{R} + q^{-d}B\bar{S})\bar{C} = (AR^* + q^{-d}BS^*)\bar{C} \qquad \forall\bar{\theta} \in \mathcal{D}_\infty.$$

Letting $(AR^* + q^{-d}BS^*)/C = D^*$, we arrive at (4.3). Now, let $r \equiv 0$. Then, ignoring the effects of the initial conditions, we find that $\mathbf{M}_c(\theta^*, \mathcal{K}(\bar{\theta}))$ can be described by

$$D^*y_k(\bar{\theta}) = R^*\epsilon_k \qquad \text{and} \qquad D^*u_k(\bar{\theta}) = -S^*\epsilon_k$$

for each $\bar{\theta} \in \mathcal{D}_\infty \subseteq \mathcal{D}_{opt}$. The rest of the statement is proved by using these equalities and the following fact:

Lemma A.1. Let $AR^* + q^{-d}BS^* = D^*C$.   Then, there exist polynomials P^*, Q^*, W^* and F^* such that $\partial F^* < d$,

$$P^*C = F^*A + q^{-d}(W^* + S^*), \qquad BP^* - AQ^* = D^*,$$

$$\text{and} \qquad P^*R^* + q^{-d}Q^*S^* = D^*(F^* + q^{-d}W^*).$$

Using Lemma A.1, we can thus write $R^* P^* y_k(\overline{\theta}) + Q^* S^* y_{k-d}(\overline{\theta}) = (F^* + q^{-d} W^*) D^* y_k(\overline{\theta})$. But for every $\overline{\theta} \in \mathcal{D}_\infty \subseteq \mathcal{D}_{opt}$, $y(\overline{\theta})$ and $u(\overline{\theta})$ satisfy the controller equation $R^* u_k(\overline{\theta}) = -S^* y_k(\overline{\theta})$. Therefore,

$$R^* \left(P^* y_k(\overline{\theta}) - Q^* u_{k-d}(\overline{\theta}) \right) = (F^* + q^{-d} W^*) D^* y_k(\overline{\theta}) = R^* (F^* + q^{-d} W^*) \epsilon_k.$$

Ignoring the effects of the initial conditions, this implies (4.4). □

Proof of Lemma A.1: With no loss of generality, we assume that A, B and C have no common factor. Let M be the greatest common factor of A, B and hence, D^*. Take (P^o, Q^o) to be any solution of the Diophantine equation $P^o B - Q^o A = D^*$. Then, by assumption we have

$$B(P^o C - q^{-d} S^*) = D^* C - q^{-d} B S^* + Q^o A C = A(R^* + Q^o C).$$

Since A/M and B/M are coprime, A/M must divide $P^o C - q^{-d} S^*$ or $P^o C - q^{-d} S^* = \frac{A}{M} Y$ for some polynomial Y. Let V and X solve $Y = M V + C X$ and set

$$P^* = P^o - \frac{A}{M} X, \qquad Q^* = Q^o - \frac{B}{M} X, \qquad \text{and} \qquad F^* + q^{-d} W^* = V.$$

where $\partial F^* < d$. Then, it is easy to verify that P^*, Q^*, W^* and F^* satisfy the statement of the lemma. □

B. Proof of Theorem 4.2:

i) Consider IE^3(3.5). We have

$$(\overline{A} B - A \overline{B}) (T(\overline{\theta}) r) = 0 \qquad a.s. \qquad \Longrightarrow \qquad \overline{A} B - A \overline{B} = 0$$

because $\overline{A} B - A \overline{B}$ is only of degree $\partial A + \partial B$ while $T(\overline{\theta}) r = T^* r$ is persistently exciting of degree $\partial A + \partial B + 1$. But by coprimeness of A and B, $A^{-1} B$ is globally identifiable and $\overline{A} = A$ and $\overline{B} = B$. Applying these equalities to IEIE, we get

$$(A \overline{R} + q^{-d} B \overline{S})(\overline{C} - C) = 0 \qquad \Longrightarrow \qquad \overline{C} = C.$$

ii) IEIE can be rewritten as

(7.4) $$(\overline{A} C - A \overline{C}) \overline{R} + q^{-d} (\overline{B} C - B \overline{C}) \overline{S} = 0.$$

If $d > \partial A + \partial C$, then one can easily verify that $r_0(\overline{\theta})(\overline{A} C - A \overline{C}) = 0$. Dividing by $r_0(\overline{\theta})$, we get

$$\overline{A} C - A \overline{C} = 0 \qquad \text{and} \qquad \frac{1}{r_0(\overline{\theta})} \overline{S}(\overline{B} C - B \overline{C}) = 0.$$

By the identifiability assumption, these imply that $\overline{A} = A$, $\overline{B} = B$ and $\overline{C} = C$. Note that we did not make use of IE^3.

iii)   If $\mathcal{K}$ is a C^1 function in a neighborhood of θ^*, then one can regard IEIE (7.4) as the nonlinear equation

$$\Phi_{\theta^*}(\overline{\theta}) = 0 \qquad \text{where} \qquad \Phi_{\theta^*} : \mathcal{R}^{\partial A + \partial B + \partial C + 1} \to \mathcal{R}^{\overline{n}},$$

with $\overline{n} = \max_{\theta \in \Theta} \max\{\partial R(\theta) + \partial A, \ d + \partial S(\theta) + \partial B\} + \partial C$, is a C^1 function in the neighborhood of θ^*. Note that $\partial A + \partial B + \partial C + 1$ is the dimension of Θ and that the first coefficient of the RHS of (7.4) is equal to 0.

Now suppose there exists a neighborhood of θ^* such that for every $\overline{\theta}$ in that neighborhood, the following inequality is satisfied:

$$\max\{\partial R(\overline{\theta}) + \partial A, \ d + \partial S(\overline{\theta}) + \partial B\} + \partial C < \partial A + \partial B + \partial C + 1,$$

or equivalently the necessary condition (4.5) is not satisfied. This implies that in a neighborhood of θ^*,

$$\text{rank}\left(\frac{\partial}{\partial \theta}\Phi_{\theta^*}(\theta^*)\right) < \partial A + \partial B + \partial C + 1, \qquad \text{while} \qquad \Phi_{\theta^*}(\theta^*) = 0.$$

Let us assume without loss of generality that $\frac{\partial}{\partial \theta}\Phi_{\theta^*}$ has constant rank in a neighborhood of θ^*. Then, we can use the implicit function theorem to conclude that $\Phi_{\theta^*}(\overline{\theta}) = 0$ on a manifold passing through θ^*. This is equivalent to saying that in this case, the plant is not potentially identifiable regardless of r. (On the other hand, if $\text{rank}\left(\frac{\partial}{\partial \theta}\Phi_{\theta^*}(\theta^*)\right) = \partial A + \partial B + \partial C + 1$ in a neighborhood of θ^*, then the inverse function theorem will imply that θ^* is an isolated point of $\mathcal{D}_\infty$.)

Now let $R(\overline{\theta})$ and $S(\overline{\theta})$ be coprime and (4.6) be true for all $\overline{\theta} \in \Theta$. Then, regarding (7.4) as a Diophantine equation of the form $X\,R(\overline{\theta}) + Y\,q^{-d}S(\overline{\theta}) = 0$, one may conclude that

$$X = \overline{A}\,C - A\,\overline{C} = 0 \qquad \text{and} \qquad Y = \overline{B}\,C - B\,\overline{C} = 0.$$

Potential identifiability then follows from the identifiability assumption. The last statement, sufficiency of (4.7), follows trivially from the definition of potential self-tuning and identifiability. $\square$

REFERENCES

[1] K.J. ASTROM AND B. WITTENMARK, *On self-tuning regulators*, Automatica 9 (1973), pp. 185–199.

[2] K.J. ASTROM, AND B. WITTENMARK, *Adaptive Control*, Addison-Wesley, 1989.

[3] R.R. BITMEAD, M. GEVERS AND V. WERTZ, *Adaptive Control, the Thinking Man's GPC*, Prentice-Hall, 1990.

[4] H.F. CHEN AND L. GUO, *Identification and Stochastic Adaptive Control*, Birkhäuser, 1991.

[5] D.W. CLARKE, C. MOHTADI, AND P.S. TUFFS, *Generalized predictive control— Parts I & II: the basic algorithm & extensions and interpretations*, Automatica, **23**(2) (1987), pp. 137–160.

[6] N.M. KOGAN AND Y.I. NEIMARK, *Study of identifiability in adaptive control systems by averaging method*, Automation and Remote Control, **50**(3) (1989), pp. 374–380.

[7] N.M. KOGAN AND Y.I. NEIMARK, *Adaptive control of a stochastic system with unobservable state under conditions of unidentifiability*, Automation and Remote Control, **53**(6) (1992), pp. 884–891.

[8] T.L. LAI AND Z.L. YING, *Parallel recursive algorithms in asymptotically efficient adaptive control of linear stochastic systems*, SIAM J. Control & Optimization, **29**(5) (1991), pp. 1091–1127.

[9] W. LIN, P.R. KUMAR AND T.I. SEIDMAN, *Will the self-tuning approach work for general criteria?*, Systems & Control Letters, **6** (1985), pp. 77–85.

[10] L. LJUNG AND T. SODERSTROM, *Theory and Practice of Recursive Identification*, MIT Press, 1983.

[11] A.S. MORSE, *Towards a unified theory of parameter adaptive control: tunability*, IEEE Trans. Aut. Control, **AC-35**(9) (1990), pp. 1002–1012.

[12] A.S. MORSE, *Towards a unified theory of parameter adaptive control—part II: certainty equivalence and implicit tuning*, IEEE Trans. Aut. Control, **AC-37**(1) (1992), pp. 15–29.

[13] K. NASSIRI-TOUSSI AND W. REN, *On asymptotic properties of the LQG feedforward self-tuner*, Int. J. Control (to appear). Also in *Proceedings of the 1993 American Control Conference*, (1993), pp. 1354–1358.

[14] K. NASSIRI-TOUSSI AND W. REN, *Indirect adaptive pole-placement control of MIMO stochastic systems: self-tuning results* (To be presented at the 33rd IEEE Conference on Decision and Control, (1994)).

[15] R. ORTEGA AND G. SANCHEZ-GALINDO, *Globally convergent multistep receding horizon adaptive controller*, Int. J. Control, **49**(5) (1989), pp. 1655–1664.

[16] J.W. POLDERMAN, *A note on the structure of two subsets of the parameter space in adaptive control problems*, Tech. Report OS-R8509, Center for Mathematics and Computer Science, The Netherlands (1986).

[17] J.W. POLDERMAN, *Adaptive Control and Identification: Conflict or Conflux?*, Centrum voor Wiskunde en Informatica, CWI Tract, **67**, 1987.

[18] J.W. POLDERMAN AND C. PRAAGMAN, *The closed-loop identification problem in indirect adaptive control*, Proc. of the 1989 IEEE Conference on Decision and Control, Tampa, FL (1989), pp. 2120–2124.

[19] W. REN AND P.R. KUMAR, *Stochastic adaptive system theory: recent advances and a reappraisal*, Foundations of Adaptive Control (P.V. KOKOTOVIC, ED.,) Springer-Verlag, 1991, pp. 269–307.

[20] S. SASTRY AND M. BODSON, *Adaptive Control: Stability, Convergence, and Robustness*, Prentice-Hall, 1989.

[21] J.H. VAN SCHUPPEN, *Tuning of gaussian stochastic control systems*, Technical Report BS-R9223, Centrum voor Wiskunde en Informatica (1992).

[22] B.G. VORCHIK, *Limit properties of adaptive control systems with identification (Using the identifiability equations), I. One-input one-output plants*, Automation and Remote Control, **49**(6) (1988), pp. 765–777.

[23] B.G. VORCHIK AND O.A. GAISIN, *Limit properties of adaptive control systems with identification (using the identifiability equations) II: Multivariable plants*, Automation and Remote Control, **51**(4) (1990), pp. 495–506.

[24] W. WANG AND R. HENRIKSEN, *Direct adaptive generalized predictive control*, Modeling Identification and Control, **14**(4) (1993), pp. 181–191.

[25] P.E. WELLSTEAD AND M.B. ZARROP, *Self-Tuning Systems: Control and Signal Processing*, John Wiley, 1991.

STOCHASTIC ADAPTIVE CONTROL*

B. PASIK-DUNCAN[†]

Abstract. The objective of this paper is to present some identification problems and adaptive control problems for continuous time linear and nonlinear stochastic systems that are completely or partially observed. For continuous time linear of stochastic systems the consistency of a family of least squares estimates of some unknown parameters is verified. The unknown parameters appear in the linear transformations of the state and the control. An approach to the verification of the consistency associates a family of control problems to the identification problem and the asymptotic behavior of the solutions of a family of algebraic Riccati equations from the control problems implies a persistent excitation property for the identification problem. The theorem of locally asymptotically normal experiment is used to test hypotheses about the parameters of a controlled linear stochastic system. The tests are formulated for both continuous and sampled observations of the input and the output.

An adaptive control problem will be described and solved for continuous time linear stochastic systems using a diminishing excitation control to show a strong consistency of a family of least squares estimates and using switchings to show self-optimizing property.

We shall investigate the ergodic control of a multidimensional diffusion process described by a nonlinear stochastic differential equation that has unknown parameters appearing in the drift. For $\epsilon > 0$ it is required to find an adaptive control such that the ergodic cost for this control is within ϵ of the optimal ergodic likelihood estimation procedure that was used by Kumar and Becker. An adaptive control is constructed from a discretization of the range of this family of estimates using the certainty equivalence principle and this control is verified to be almost self-optimizing.

We shall also consider adaptive control problem of a discrete time Markov process that is completely observed in a fixed recurrent domain and partially observed elsewhere. An almost self-optimal strategy is constructed for this problem. Finally, some numerical examples and simulation results will be presented.

Part I
Adaptive Control of Continuous Time Linear Stochastic Systems

1. Introduction. In this paper an adaptive control problem is formulated and solved for a completely observed, continuous time, linear stochastic system with an ergodic (or long run average) quadratic cost criterion. By the solution of an adaptive control problem we mean exhibiting a strongly consistent family of estimators of the unknown parameters and constructing a self-optimizing adaptive control.

For the solution of the identification problem a family of least squares estimators are given that are strongly consistent using a diminishing excitation control (or dither) that is asymptotically negligible for an ergodic quadratic cost criterion. This method has been previously used successfully in identification problems for ARMAX systems [2,3]. If the linear stochastic differential equation for the system is described by the triple

---

* Research partially supported by NSF Grants ECS-9102714 and ECS-9113029.

† Department of Mathematics, University of Kansas, Lawrence, KS 66045.

(A, B, C) where A is the linear transformation of the state, B is the linear transformation of the control and C is the linear transformation of the (white, Gaussian) noise, then for strong consistency it is only assumed that A is stable and (A, C) is controllable. It is required to estimate the linear transformations A and B with only the aforementioned requirements on A and no requirements on B. In particular there are no boundedness assumptions on the family of unknown linear transformations (A, B). The control at time t is required to be measurable with respect to the past of the state process up to time $t - \Delta$ where $\Delta > 0$ is arbitrary but fixed. This assumption accounts for some natural delay in processing the information for the construction of the control.

For the adaptive control, the certainty equivalence control for the ergodic cost functional is used with a switching to the zero control. These switchings are determined by a family of stopping times. It is shown that for almost all sample paths there is a finite number of switchings with the certainty equivalence control used lastly for an infinite time. This adaptive control is shown to be self-optimizing. For the self-optimizing property of this adaptive control it is assumed that A is stable, (A, B) and (A, C) are controllable and (A, D) is observable where D is a square root of the nonnegative, symmetric transformation for the state in the quadratic cost functional.

For discrete time linear systems, specifically ARMAX models, results related to ours are given in [3] where a diminishing excitation control and switchings are used to establish strong consistency for a family of least squares estimates and the self-optimizing property for a certainty equivalence adaptive control. For continuous time linear stochastic systems, some results for strong consistency and self-tuning without dither and switchings are given in [6]. Another approach to strong consistency without dither for continuous time linear systems is given in [7,10]. For discrete time linear stochastic systems, especially ARMAX models, there is a significant amount of work on consistency and self-tuning (e.g., [8,9]). The recent monograph [4] describes these results with extensive references.

The complete proofs of the main results of strong consistency and self-optimality are given in [1].

2. Preliminaries. The model for the adaptive control problem is given by the controlled diffusion $(X(t), t \geq 0)$ that is the solution of the stochastic differential equation

$$dX(t) = AX(t)dt + BU(t)dt + CdW(t)$$
(2.1)
$$X(0) = X_0$$

where $X(t) \in \mathbb{R}^n$, $U(t) \in \mathbb{R}^m$ and $(W(t), t \geq 0)$ is a standard p-dimensional Wiener process. The probability space is $(\Omega, \mathcal{F}, P)$ and $(\mathcal{F}_t, t \geq 0)$ is an increasing family of sub-σ-algebras of $\mathcal{F}$ such that $\mathcal{F}_0$ contains all P-null sets, $(W(t), \mathcal{F}_t, t \geq 0)$ is a continuous martingale and $X(t) \in \mathcal{F}_t$ for all

$t \geq 0$. A, B and C are suitable linear transformations. It is assumed that the triple (A, B, C) is unknown. Since the adaptive control does not depend on C it is only necessary to estimate the pair (A, B). For notational simplicity we let $\theta^T = [A, B]$ and we suppress the dependence of $X(t)$ on U and (A, B).

An adaptive control problem is formulated and solved for (2.1) where it is desired to minimize the ergodic cost functional

$$(2.2) \qquad \limsup_{t \to \infty} J(t, u) = \limsup_{t \to \infty} \frac{1}{t} \int_0^t X^T(s)Q_1 X(s) + U^T(s)Q_2 U(s) ds$$

where $Q_1 \geq 0$ and $Q_2 > 0$ and U is an admissible control. The family of admissible controls is subsequently specified.

Initially some results are described that are used in the verification of the strong consistency and self-optimality.

LEMMA 2.1. *Let $\Delta > 0$ be fixed and let $(M(t), t \geq 0)$ be an $\mathcal{L}(\mathbb{R}^k, \mathbb{R}^\ell)$-valued process such that $M(t) \in \mathcal{F}_{(t-\Delta)\vee 0}$ for all $t \geq 0$ and let $(f(t), t \geq 0)$ be an $\mathbb{R}^k$-valued process such that $f(t) \in \mathcal{F}_t$, $E[f(t)|\mathcal{F}_{(t-\Delta)\vee 0}] = 0$, $E[\|f(t)\|^{2+\delta}] \leq c$ for some $\delta > 0$ and all $t \geq 0$ and*

$$(2.3) \qquad \|E[f(t)f^T(s)|\mathcal{F}_{((t\vee s)-\Delta)\vee 0}]\| \leq c \qquad a.s.$$

for $|t - s| \leq \Delta$ where c is a constant. Let $E[f(t)|\mathcal{F}_s]$ be independent of $\mathcal{F}_{(t-\Delta)\vee 0}$ for all positive numbers s and t. Then

$$(2.4) \qquad \left\| \int_0^t M(s)f(s)ds \right\| = 0\left(\left(1 + \int_0^t \|M(s)\|^2 ds \right)^{1/2+\eta} \right) \qquad a.s.$$

for all $\eta > 0$ as $t \to \infty$.

It is easy to verify that the estimate (2.4) can be strengthened to

$$(2.5) \qquad \begin{aligned} \left\| \int_0^t M(s)f(s)ds \right\| &= 0\left(\left(1 + \int_0^t \|M(s)\|^2 ds \right)^{1/2} \right. \\ &\left. \log^{1/2+\eta}\left(1 + \int_0^t \|M(s)\|^2 ds \right) \right). \end{aligned}$$

Define the $\mathbb{R}^n$-valued processes $(\xi(t), t \geq 0)$ and $(\eta(t), t \geq 0)$ by the equations

$$(2.6) \qquad \xi(t) = \int_0^t e^{(t-s)A} C dW(s)$$

and

$$(2.7) \qquad \eta(t) = \int_0^t e^{(t-s)\phi} C dW(s)$$

where A, C and $(W(t), t \geq 0)$ are given in (2.1) and $\phi \in \mathcal{L}(\mathbb{R}^n, \mathbb{R}^n)$.

Let $\Gamma \in \mathcal{L}(\mathbb{R}^n, \mathbb{R}^n)$ be stable and let $(g(t), t \geq 0)$ be the $\mathbb{R}^n$-valued process that satisfies

$$(2.8) \qquad g(t) = \int_{(t-\Delta)\vee 0}^{t} e^{\Gamma(t-s)} C dW(s).$$

Now it is shown that $(g(t), t \geq 0)$ satisfies the assumptions on $(f(t), t \geq 0)$ in Lemma 1. Let $|t - s| \leq \Delta$ and consider only the case where $t \geq s \geq \Delta$ because the other case follows by similar arguments.

$$\begin{aligned}
&\|E[g(t)g^T(s)|\mathcal{F}_{(t-\Delta)}]\| \\
&= \left\| E\left[g(t) \left(\int_{s-\Delta}^{t-\Delta} e^{\Gamma(s-r)} C dW(r) + \int_{t-\Delta}^{s} e^{\Gamma(s-r)} C dW(r) \right)^T |\mathcal{F}_{t-\Delta}\right] \right\| \\
&= \left\| E\left[E\left[\left(\int_{t-\Delta}^{s} e^{\Gamma(t-r)} C dW(r) + \right.\right.\right.\right. \\
&\qquad\qquad \left.\left.\left.\left. + \int_{s}^{t} e^{\Gamma(t-r)} C dW(r) \right) \left(\int_{t-\Delta}^{s} e^{\Gamma(s-r)} C dW(r) \right)^T |\mathcal{F}_s \right] |\mathcal{F}_{t-\Delta} \right] \right\| \\
&= \left\| E\left[\int_{t-\Delta}^{s} e^{\Gamma(t-r)} C dW(r) \left(\int_{t-\Delta}^{s} e^{\Gamma(s-r)} C dW(r) \right)^T |\mathcal{F}_{t-\Delta} \right] \right\| \\
&\leq (trC^T C) e^{-\rho(t+s)} \int_{t-\Delta}^{s} e^{2\rho r} dr
\end{aligned}$$
$$(2.9)$$

where $\rho > 0$ is determined from Γ.

If A and ϕ are stable then the processes defined by (2.6) and (2.7) each have a unique invariant measure and the ergodic theorem is satisfied. Thus we have the following lemma.

LEMMA 2.2. *If A and ϕ are stable, $(\xi(t), t \geq 0)$ and $(\eta(t), t \geq 0)$ satisfy (2.6) and (2.7) respectively and $\Delta > 0$ is fixed then*

$$\lim_{t\to\infty} \frac{1}{t} \int_0^t (\eta(s) - e^{\Delta\phi}\eta((s-\Delta)\vee 0))(\xi(s) - e^{\Delta A}\xi((s-\Delta)\vee 0))^T ds$$
$$(2.10)$$

$$= \int_0^\Delta e^{s\phi} C C^T e^{sA^T} ds \quad a.s.$$

$$(2.11)\lim_{t\to\infty} \frac{1}{t}(\xi(s) - e^{\Delta A}\xi((s-\Delta)\vee 0))\eta^T((s-\Delta)\vee 0)ds = 0 \quad a.s.$$

$$\lim_{t\to\infty} \frac{1}{t} \int_0^t \int_{(s-\Delta)\vee 0}^s e^{(s-r)\phi} F \int_{(r-\Delta)\vee 0}^r e^{(r-q)\phi} C dW(q) dr$$
$$(2.12)$$

$$\left(\int_{(s-\Delta)\vee 0}^s e^{(s-r)A} C dW(r) \right)^T ds$$

$$(2.13) \qquad = \int_0^{\Delta} e^{s\phi} F \int_0^{\Delta - s} e^{r\phi} CC^T e^{rA^T} dr \, e^{sA^T} ds := L \qquad a.s.$$

where $F \in \mathcal{L}(\mathbb{R}^n, \mathbb{R}^n)$.

Now a stability property is described for the solution of a stochastic differential equation of a special form that is important for the adaptive control of (2.1).

LEMMA 2.3. *Let $(Y(t), t \geq 0)$ satisfy the stochastic differential equation*

$$(2.14) \qquad dY(t) = \phi(t)Y(t)dt + f(t)dt + CdW(t) + F(\xi(t) -$$
$$-e^{\Delta A} \xi((t - \Delta) \vee 0))dt$$
$$Y(0) = Y_0$$

where $(\xi(t), t \geq 0)$ is given in (2.6), $Y(t) \in \mathbb{R}^n$, A is stable, $F \in \mathcal{L}(\mathbb{R}^n, \mathbb{R}^n)$, $f(t) \in \mathcal{F}_t$ for all $t \geq 0$, $\int_0^t \|f(s)\|^2 ds = 0(t)$ a.s. as $t \to \infty$, $\phi(t) \in \mathcal{F}_t$ for all $t \geq 0$ and

$$(2.15) \qquad \lim_{t \to \infty} \phi(t) = \phi \qquad a.s.$$

where ϕ is a stable matrix. Then

$$(2.16) \qquad \lim_{t \to \infty} \frac{Y(t)Y^T(t)}{t} = 0 \qquad a.s.$$

$$\lim_{t \to \infty} \frac{1}{t} \int_0^t Y(s)Y^T(s)ds = \int_0^{\infty} s^{s\phi}(NF^T + FN^T + CC^T)e^{s\phi^T} ds = Q \qquad a.s.$$
$$(2.17)$$
where

$$(2.18) \qquad N = \int_0^{\Delta} e^{s\phi} CC^T e^{sA^t} ds + L$$

and L is given by (2.13).

3. Parameter Estimation. A family of least squares estimates $(\theta(t), t \geq 0)$ is used to estimate the unknown $\theta = [AB]^T$. The estimate $\theta(t)$ is given by

$$(3.1) \qquad \theta(t) = \Gamma(t) \int_0^t \phi(s)dX^T(s) + \Gamma(t)\Gamma^{-1}(0)\theta(0)$$

$$(3.2) \qquad \Gamma(t) = \left(\int_0^t \phi(s)\phi^T(s)ds + aI \right)^{-1}$$

where $a > 0$ is fixed and $\phi(s) = [X^T(s)U^T(s)]^T$. It is known [5] that

$$(3.3) \qquad \|\theta - \theta(t)\|^2 = 0\left(\frac{\log r(t)}{\lambda_{min}(t)} \right)$$

where $r(t) = e + \int_0^t \|\phi(s)\|^2 ds$, $e = 2.718...$ and $\lambda_{min}(t)$ is the minimum eigenvalue of $\Gamma^{-1}(t)$.

Let $(U^d(t), t \geq 0)$ be the desired control. It is assumed that $U^d(t) \in \mathcal{F}_{(t-\Delta)\vee 0}$ for all $t \geq 0$ where $\Delta > 0$ is fixed.

It is quite natural to assume that the information processing for the desired control introduces a delay Δ.

In addition to the desired control a diminishing excitation control is used. Let $(\epsilon_n, n \in \mathbb{N})$ be a sequence of $\mathbb{R}^m$-valued independent, identically distributed random variables that is independent of the Wiener process $(W(t), t \geq 0)$. It is assumed that $E(\epsilon_n) = 0$, $E[\epsilon_n \epsilon_n^T] = I$ for all $n \in \mathbb{N}$ and there is a $\sigma > 0$ such that $\|\epsilon_n\|^2 \leq \sigma$ a.s. for all $n \in \mathbb{N}$. Let $\epsilon \in (0, \frac{1}{2})$ and fix it. Define the $\mathbb{R}^m$-valued process $(V(t), t \geq 0)$ as

$$(3.4) \qquad V(t) = \sum_{n=0}^{[t/\Delta]} \frac{\epsilon_n}{n^{\epsilon/2}} 1_{[n\Delta,(n+1)\Delta)}(t)$$

Clearly we have

$$(3.5) \qquad \lim_{t\to\infty} V(t) = 0 \qquad a.s.$$

and

$$(3.6) \qquad \begin{aligned} &\lim_{t\to\infty} \frac{1}{t^{1-\epsilon}} \int_0^t V(s)V^T(s)ds \\ &= \lim_{t\to\infty} \frac{1}{t^{1-\epsilon}} \sum_{i=1}^{[t/\Delta]} \frac{\epsilon_i \epsilon_i^T}{i^\epsilon} \Delta + o(1) \\ &= \Delta^\epsilon(1-\epsilon)I \qquad a.s. \end{aligned}$$

It is assumed that $\epsilon_n \in \mathcal{F}_{n\Delta}$ and ϵ_n is independent of $\mathcal{F}_s$ for $s < n\Delta$ for $n \in \mathbb{N}$.

Let $(f(t), t \geq 0)$ be defined by $f(t) = \epsilon_i$, $i\Delta \leq t < (i+1)\Delta$. Then this process satisfies the assumptions of Lemma 2.1. Clearly we have $E[f(t)|\mathcal{F}_{t-\Delta}] = 0$ and

$$E[f(t)|\mathcal{F}_s] = \begin{cases} f(t) & s \geq i\Delta \\ 0 & s < i\Delta \end{cases}$$

Thus $E[f(t)|\mathcal{F}_s]$ is independent of $\mathcal{F}_{t-\Delta}$ for all $s, t \geq 0$. To verify (2.3) let $t \geq s$. For $t - \Delta \leq s \leq t$, $f(s)$ either is $f(t)$ or is $\mathcal{F}_{t-\Delta}$ measurable. Thus $\|E[f(t)f^T(s)|\mathcal{F}_{t-\Delta}]\|$ is bounded by c.

The diminishingly excited control is

$$(3.7) \qquad U(t) = U^d(t) + V(t)$$

for $t \geq 0$.

The following lemma is important in the verification of the strong consistency of (3.1).

LEMMA 3.1. *Let $(U(t), t \geq 0)$ satisfy the conditions in Theorem 3.2 and let $(V(t), t \geq 0)$ be given by (3.4). The following equations are satisfied:*

$$(3.8) \qquad \lim_{t \to \infty} \frac{1}{t^\alpha} \int_0^t U^d(s) V^T(s) ds = 0 \text{ a.s.}$$

$$(3.9) \qquad \lim_{t \to \infty} \frac{1}{t^\alpha} \int_0^t e^{sA} \int_0^s e^{-rA} B U^d(r) dr V^T(s) ds = 0 \qquad a.s.$$

$$(3.10) \qquad \lim_{t \to \infty} \frac{1}{t^\alpha} \int_0^t \xi(s) V^T(s) ds = 0 \quad a.s.$$

$$(3.11) \qquad \lim_{t \to \infty} \frac{1}{t^\alpha} \int_0^t V(s) \int_0^s V^T(r) B^T e^{(s-r)A^T} dr \, ds$$
$$- \int_0^t V(s) V^T(s) ds C(\Delta) = 0 \qquad a.s.$$

where $(\xi(t), t \geq 0)$ is given by (2.6),

$$(3.12) \qquad C(\Delta) = B^T (A^T)^{-2} (e^{\Delta A^T} - I - \Delta A^T)$$

and $\alpha \in (\frac{1+\delta}{2}, 1 - \epsilon)$ is arbitrary where δ and ϵ are given in Theorem 3.1.

The following result verifies the strong consistency of $(\theta(t), t \geq 0)$ defined by (3.1).

THEOREM 3.2. *Let $\epsilon \in (0, \frac{1}{2})$ be given from the definition of $(V(t), t \geq 0)$ in (3.4). Consider the stochastic system given by (2.1). If A is stable, (A, C) is controllable and the control $(U(t), t \geq 0)$ is given by (3.7) where $U^d(t) \in \mathcal{F}_{(t-\Delta) \vee 0}$ for $t \geq 0$ and*

$$(3.13) \qquad \int_0^t \|U^d(s)\|^2 ds = 0(t^{1+\delta}) \qquad a.s.$$

for some $\delta \in [0, 1 - 2\epsilon)$, then

$$(3.14) \qquad \|\theta - \theta(t)\|^2 = 0 \left(\frac{\log t}{t^\alpha} \right) \qquad a.s.$$

as $t \to \infty$ for each $\alpha \in (\frac{1+\delta}{2}, 1 - \epsilon)$ where $\theta = [AB]^T$ and $\theta(t)$ satisfies (3.1).

To verify (3.14) using (3.3) it suffices to show that

$$(3.15) \qquad \liminf_{t \to \infty} \frac{1}{t^\alpha} \lambda_{\min} \left(\int_0^t \varphi(s) \varphi^T(s) ds \right) > 0 \qquad a.s.$$

where $\alpha \in (\frac{1+\delta}{2}, 1 - \epsilon)$ is fixed $\lambda_{\min}(X)$ denotes the minimum eigenvalue of the symmetric matrix X. The inequality (3.15) is verified by contradiction using a sequence of random times that converge (almost surely) to infinity and Lemmas 2.1,2.2, 2.3 and 3.1.

Remark. If $U^d(t) \equiv 0$ then we have that

$$\lim_{t \to \infty} \frac{1}{t} \int_0^t X(s)X^T(s)ds = \int_0^\infty e^{sA}CC^T e^{sA^T} ds > 0$$

and $\frac{\lambda_{\max}(t)}{\lambda_{\min}(t)} = ct^\epsilon + 0(1)$ a.s. where $c > 0$ and $\lambda_{\max}(t)$ is the maximum eigenvalue of $\Gamma^{-1}(t)$. Thus the system does not satisfy a persistent excitation condition but there is strong consistency of the family of least squares estimates.

4. Adaptive Control. In this section, a self-optimizing adaptive control is constructed for the linear stochastic system (2.1) with the quadratic ergodic cost functional (2.2) Throughout this section it is assumed that $(\mathcal{F}_t, t \geq 0)$ is right continuous.

Initially the family of admissible controls $\mathcal{U}(\Delta)$ is defined:

$$\mathcal{U}(\Delta) = \{U : U(t) = U^d(t) + U^1(t), U^d(t) \in \mathcal{F}_{(t-\Delta)\vee 0}$$
$$\text{and } U^1(t) \in \sigma(V(s), (t-\Delta) \vee 0 \leq s \leq t) \text{ for all } t \geq 0,$$
$$\|X(t)\|^2 = o(t)a.s. \text{ and } \int_0^t (\|U(s)\|^2 +$$
$$(4.1) \qquad\qquad +\|X(s)\|^2)ds = 0(t) \quad a.s. \text{ as } t \to \infty\}.$$

Let $Q_1 = D^T D$ where Q_1 is given in (2.2) and D is a square root of Q_1. If the triple (A, B, D) is controllable and observable (i.e., minimal) then it is well known that there is a unique solution P of the algebraic Riccati equation

$$A^T P + PA - PBQ_2^{-1}B^T P + Q_1 = 0$$

in the family of positive definite linear transformations. Using this equation and applying Itô's formula to $(X^T(t)PX(t), t \geq 0)$ we have that

$$X^T(t)PX(t) + \int_0^t X^T(s)Q_1 X(s) + U^T(s)Q_2 U(s)ds$$
$$(4.2) \quad = t \, tr(C^T PC) + 2 \int_0^t X^T(s)PCdW(s)$$
$$+ \int_0^t (U(s) + Q_2^{-1}B^T PX(s))^T Q_2(U(s) + Q_2^{-1}B^T X(s))ds.$$

Define the process $(\hat{X}(t), t \geq \Delta)$ by the equation

$$(4.3) \qquad \hat{X}(t) = e^{\Delta A}X(t-\Delta) + \int_{t-\Delta}^t e^{(t-s)A}V(s)ds.$$

Clearly for $t \geq \Delta$

$$X(t) = \hat{X}(t) + \int_{t-\Delta}^{t} e^{(t-s)A} C dW(s).$$

By Lemmas 2.2 and 2.3 and (4.2) we have that for any $U \in \mathcal{U}(\Delta)$

$$
\limsup_{t \to \infty} J(t, U) = tr(C^T P C)
$$

$$
+ \limsup_{t \to \infty} \frac{1}{t} \int_0^t (U(s) + Q_2^{-1} B^T P \hat{X}(s)
$$

(4.4)
$$
+ Q_2^{-1} B^T P \int_{(s-\Delta)\vee 0}^s e^{(s-r)A} C dW(r)^T Q_2
$$

$$
(U(s) + Q_2^{-1} B^T P \hat{X}(s) + Q_2^{-1} B^T P \int_{(s-\Delta)\vee 0}^s e^{(s-r)A} C dW(r)) ds
$$

$$
\geq tr(C^T P C) + tr(B^T P R(\Delta) P B Q_2^{-1}) \qquad a.s.
$$

where $J(t, U)$ is given by (2.2) and $R(\Delta)$ is given by $R(\Delta) = \int_0^\Delta e^{sA} C C^T e^{sA^T}\, ds$. By Lemma 2.3 and (4.4) it is clear that $-Q_2^{-1} B^T P \hat{X}(\cdot) \in \mathcal{U}(\Delta)$ and it minimizes the ergodic cost functional (2.2) for $U \in \mathcal{U}(\Delta)$.

Define $\phi \in \mathcal{L}(\mathbb{R}^n, \mathbb{R}^n)$ by the equation

(4.5)
$$\phi = A - B Q_2^{-1} B^T P.$$

It is well known that ϕ is stable.

Define the $\mathbb{R}^m$-valued process $(U^0(t), t \geq 0)$ by the equation

(4.6)
$$
U^0(t) = -Q_2^{-1} B^T((t-\Delta)\vee 0) P((t-\Delta)\vee 0)(e^{\Delta A((t-\Delta)\vee 0)}
$$
$$
X((t-\Delta)\vee 0) + \int_{(t-\Delta)\vee 0}^t e^{(t-s)A((t-\Delta)\vee 0)} U^d(s) ds)
$$

where $A(t)$ and $B(t)$ are the least squares estimates for A and B given by (3.1) and $P(t)$ is the solution of the algebraic Riccati equation

(4.7)
$$A^T(t) P(t) + P(t) A(t) - P(t) B(t) Q_2^{-1} B^T(t) P(t) + Q_1 = 0$$

if $(A(t), B(t), D)$ is controllable and observable and otherwise $P(t) = 0$. The process $(U^d(t), t \geq 0)$ is defined subsequently. It will be clear that $U^0 \in \mathcal{U}(\Delta)$.

Define two sequences of stopping times $(\sigma_n, n = 0, 1, ..)$ and $(\tau_n, n = 1, 2, ...)$ as follows:

$$\sigma_0 \equiv 0$$

$$\sigma_n = \sup\{t \geq \tau_n : \int_0^s \|U^0(r)\|^2 dr \leq s\tau_n^\delta, (A(s), B(s), D)$$

(4.8)
$$\text{is controllable and observable for all } s \in [\tau_n, t)\}$$

$$\tau_n = \inf\{t > \sigma_{n-1} + 1 : \int_0^t \|U^0(r)\|^2 dr \le \frac{1}{2} t^{1+\delta},$$

$(A(t), B(t), D)$ is controllable and observable

(4.9)   and $\|X(t)\|^2 \le t^{1+\delta/2}\}.$

It is clear that $(\tau_n - \sigma_{n-1}) \ge 1$ on $\{\sigma_{n-1} < \infty\}$ for all $n \ge 1$.

Define the adaptive control $(U^*(t), t \ge 0)$ by the equation

(4.10) $U^*(t) = U^d(t) + V(t)$

where

(4.11) $U^d(t) = \begin{cases} 0 & \text{if } t \in [\sigma_n, \tau_{n+1}) \quad \text{for some } n \ge 0 \\ U^0(t) & \text{if } t \in [\tau_n, \sigma_n) \quad \text{for some } n \ge 1 \end{cases}$

$U^0(t)$ satisfies (4.6) and $V(t)$ satisfies (3.4).

The adaptive control $(U^*(t), t \ge 0)$ is self-optimizing as the following result describes.

THEOREM 4.1. *If A is stable, (A, C) is controllable, (A, B, D) is controllable and observable where $Q_1 = D^T D$, then the adaptive control $(U^*(t), t \ge 0)$ given by (4.10) is an element of $\mathcal{U}(\Delta)$ and is self-optimizing, that is,*

(4.12)
$$\inf_{U \in \mathcal{U}(\Delta)} \limsup_{t \to \infty} J(t, U) = \lim_{t \to \infty} J(t, U^*)$$
$$= tr(C^T PC) + tr(B^T PR(\Delta)PBQ_2^{-1}) \qquad a.s.$$

where $J(t, U)$ satisfies (2.2).

The proof of this theorem uses the switchings from the two sequences of stopping times $(\sigma_n, n \in \mathbb{N})$ and $(\tau_n, n \in \mathbb{N})$ to verify stability and self-optimality for the adaptive control.

REFERENCES

[1] H.F. CHEN, T.E. DUNCAN AND B. PASIK-DUNCAN, *Stochastic adaptive control for continuous time linear systems with quadratic cost*, to appear in Journal of Applied Mathematics and Optimization.

[2] H.F. CHEN AND L. GUO, *Optimal adaptive control with quadratic index*, Int J. Control 43 (1986), pp. 869–881.

[3] H.F. CHEN AND L. GUO, *Optimal adaptive control and parameter estimates for ARMAX model with quadratic cost*, SIAM J. Control Optim. 25 (1987), pp. 845–867.

[4] H.F. CHEN AND L. GUO, *Identification and Stochastic Adaptive Control*, Birkhauser, Boston, 1991.

[5] H.F. CHEN AND J.B. MOORE, *Convergence rate of continuous time ELS parameter estimation*, IEEE Trans. Autom. control. AC-32 (1987), pp. 267–269.

[6] T.E. DUNCAN AND B. PASIK-DUNCAN, *Adaptive control of continuous-time linear stochastic systems*, Math. Control Signals Systems 3 (1990), pp. 45–60.

[7] T.E. DUNCAN, P. MANDL AND B. PASIK-DUNCAN, *On least squares estimation in continuous time linear stochastic systems*, Kybernetika 28 (1992), pp. 169–180.

[8] O.B. HIJAB, *The adaptive LQG problem, Part 1*, IEEE Trans. Autom. Control AC-28 (1983), pp. 171–178.

[9] P.R. KUMAR, *Optimal adaptive control of linear-quadratic-Gaussian systems*, SIAM J. Control Optim. 21 (1983), pp. 163–178.

[10] P. MANDL, T.E. DUNCAN AND B. PASIK-DUNCAN, *On the consistency of a least squares identification procedure*, Kybernetika 24 (1988), pp. 340–346.

[11] B. PASIK-DUNCAN, T.E. DUNCAN, AND H.F. CHEN, *Continuous Time Adaptive LQG Control*, Proceedings of the 31st IEEE Control and Decision Conference, (1992), pp. 3227–3232.

Part II
Parameter Estimation in Higher Order Stochastic Systems with Discrete Observations

1. Introduction. In the theory of parameter estimation for continuous time linear systems it is often assumed that the observation of the state trajectory is continuous as well. It is natural to expect that the consistency of the estimates or their asymptotic normality are approximately valid if the observations are discrete and the sampling interval is sufficiently small, that is, there is a continuity property for these asymptotic properties as the sampling interval tends to zero. However it is shown here that this is not true in higher order systems where for discrete time observations the derivatives that are required to determine the estimates for the continuous time observations are replaced by finite differences. For the discretization of the equations that define a family of least squares estimates based on the continuous time observations, it is shown that for these discretized equations an additional term is required to ensure strong consistency of the family of estimates based on sampled observations as the sampling interval approaches zero. This correction term is related to the error in estimating the local variance matrix of a Wiener process by the quadratic variation using only a family of discretizations of the output of a dth order, linear, stochastic differential equation.

Let $(X(t), t \geq 0)$ be an $\mathbb{R}^n$-valued process that is the solution of the following stochastic differential equation of order d

$$(1.1) \quad dX^{(d-1)}(t) = \left(\sum_{i=1}^{d} f_i X^{(i-1)}(t) + gU(t) \right) dt + dW(t) \quad X^{(i)}(0) = X_0^{(i)}$$

for $i = 0, ..., d-1$ where $t \geq 0$, $X^{(i)}(t) = \frac{dX^{(i-1)}}{dt}$ for $i = 1, 2, ..., d-1$, $X^{(0)}(t) = X(t)$, $(f_1, f_2, ... f_d, g)$ are constant matrices, $U(t) \in \mathbb{R}^q$, $(W(t), t \geq 0)$ is an $\mathbb{R}^n$-valued Wiener process with local variance matrix h, that is, $dW(t) \, dW'(t) = h \, dt$ and prime denotes matrix transpose. The $\mathbb{R}^q$-valued process $(U(t), t \geq 0)$ is the solution of the linear stochastic differential equation

$$(1.2) \qquad dU(t) = cU(t)dt + dW_0(t) \qquad U(0) = U_0$$

where $t \geq 0$, c is a constant matrix and $(W_0(t), t \geq 0)$ is an $\mathbb{R}^q$-valued
Wiener process with local variance matrix h_0 that is independent of $(W(t)$,
$t \geq 0)$.

A first order system of linear stochastic differential equations is obtained from (1.1-1.2) by defining the following vector and matrices in block form

$$(1.3) \qquad \mathbb{X}(t) = \begin{pmatrix} X^0(t) \\ X^1(t) \\ \vdots \\ X^{(d-1)}(t) \\ U(t) \end{pmatrix}$$

$$(1.4) \qquad F = \begin{pmatrix} 0 & I & & & & \\ & 0 & 1 & & & \text{\Large O} \\ & \text{\Large O} & & \ddots & I & 0 \\ & & & & 0 & \\ f_1 & f_2 & \cdots & f_d & g \\ 0 & 0 & \cdots & 0 & c \end{pmatrix}$$

$$(1.5) \qquad H = \begin{pmatrix} \text{\Large O} & \text{\Large O} \\ \text{\Large O} & \begin{matrix} h & 0 \\ 0 & h_0 \end{matrix} \end{pmatrix}$$

where I is the identity in $\mathbb{R}^n$ and the blocks in F and H correspond to the blocks in $\mathbb{X}$. Thus (1.1-1.2) can be expressed as a system of first order equations as

$$(1.6) \qquad d\mathbb{X}(t) = F\mathbb{X}(t)dt + d\mathbb{W}(t) \qquad \mathbb{X}(0) = \mathbb{X}_0$$

where $t \geq 0$, $\mathbb{X}(t)$ is given by (1.3), F is the constant matrix (1.4), $(\mathbb{W}(t), t \geq 0)$ is an $\mathbb{R}^{dn+q-}$ valued Wiener process with local variance matrix H given by (1.5).

The following assumption is made on F.

(A1) F is a stable linear transformation, that is, the spectrum of F is contained in the open left half plane.

If (A1) is satisfied then $(X(t), t \geq 0)$ has a limiting Gaussian distribution with zero mean and variance matrix R that

$$(1.7) \qquad R = E[\mathbb{X}\mathbb{X}^1]$$

where $\mathbb{X}$ is a random variable with the limiting distribution and R satisfies the Lyapunov equation

$$(1.8) \qquad FR + RF^1 + H = 0$$

The variance matrix R is partitioned into blocks that correspond to the block components $X^0(t), ..., X^{(d-1)}(t), U(t)$ of $\mathbb{X}(t)$ as follows

$$(1.9) \qquad\qquad R = (r_{ij})$$

for $i, j \in \{1, ..., d+1\}$ where $r_{ij} = E[X^{(i)}X^{(j)1}]$ for $i, j \in \{1, ..., d\}$, $r_{i,d+1} = E[X^{(i)}U^1] = r^1_{d+1,i}$ for $i \in \{1, ..., d\}$, $r_{d+1,d+1} = E[UU^1]$ and E is expectation with respect to the limiting distribution. The assumption (A1) of the stability of F ensures the validity of the subsequent applications of the Law of Large Numbers.

It is assumed that there are discrete observations of $(X(t), t \geq 0)$ and $(U(t), t \geq 0)$ with the uniform sampling interval $\delta > 0$. This sampling yields the following random variables

$$(1.10) \qquad (X(m\delta), U(m\delta), m = 0, 1, ..., N + d - 1)$$

The derivatives $(X^{(j)}(m\delta), m = 0, ..., N + d - 1 - j)$ are approximated by the forward differences

$$(1.11) \qquad\qquad X^{(i)}_{m,\delta} = (X^{(i-1)}_{m+1,\delta} - X^{(i-1)}_{m,\delta})/\delta$$

for $i = 1, 2, ..., d - 1$. For subsequent notational convenience let $X^{(0)}_{m,\delta} = X(m\delta)$ for $m = 0, 1, ..., n + d - 1$. Since $X^{(i)}_{m,\delta}$ is not $X^{(i)}(m\delta)$, the ith derivative of $(X(t), t \geq 0)$, a bias for some asymptotic computation is introduced that does not converge to zero as δ tends to 0. For a scalar second order equation the bias that is caused by the sampled observations is computed in [1].

2. A Quadratic Variation Estimate. The well known quadratic variation formula for (1.1) is

$$(2.1) \qquad \lim_{\delta \to 0} \frac{1}{T} \sum_{m=0}^{[T/\delta]} (X^{(d-1)}((m+1)\delta) - X^{(d-1)}(m\delta))$$

$$(X^{(d-1)}((m+1)\delta) - X^{(d-1)}(m\delta))^1 = h$$

where $T > 0$ is fixed and the limit can be taken in $L^2(P)$. The family of random variables for the limit on the left hand side of (2.1) suggest the following family of estimates for h

$$(2.2) \quad h^*(N, \delta) = \frac{1}{N\delta} \sum_{m=0}^{N-1} (X^{(d-1)}_{m+1,\delta} - X^{(d-1)}_{m,\delta})(X^{(d-1)}_{m+1,\delta} - X^{(d-1)}_m)^1$$

that are based on the observations (1.10) where $N \in IN$ and $\delta > 0$.

The following proposition shows that the family of estimates $(h^*(N, \delta)$, $N \in IN$, $\delta > 0)$ does not converge to h as $N \to \infty$ and $\delta \to 0$ but it converges to $C(d)$ h where $C(d)$ is a nontrivial, explicit constant that depends on the order d of the system.

Proposition 1. Assume that (A1) is satisfied. Let $(X^{(d-1)}(t), t \geq 0)$ satisfy (1.1) and let $h^*(N, \delta)$ for $N \in IN$ and $\delta > 0$ be given by (2.2). The following equality is satisfied

$$(2.3) \qquad \lim_{\delta \to 0} \lim_{N \to \infty} h^*(N, \delta) = C(d)h \qquad a.s.$$

where

$$(2.4) \qquad C(d) = \frac{(-1)^d}{(2d-1)!} \sum_{j=1}^{d} (-1)^j j^{2d-1} \begin{pmatrix} 2d \\ d-j \end{pmatrix}$$

for $d = 2, 3, \ldots$

COROLLARY 2.1. *For $i < 2d$, $(F^i R)_{11}$ is symmetric for i even and skew symmetric for i odd.*

The verification of this corollary follows from the arguments at the end of the proof of the Proposition 1.

3. Parameter Estimation. Now it is assumed that (1.1) contains a p-dimensional unknown parameter $\alpha = (\alpha^1, \ldots, \alpha^p)$ so that (1.1) is expressed as

$$(3.1) \quad dX^{(d-1)}(t) = \left(\sum_{i=1}^{d} f_i(\alpha) X^{(i)} d(t) + g(\alpha) U(t) \right) dt + dW(t)$$

where

$$(3.2) \qquad f_i(\alpha) = f_{i0} + \sum_{j=1}^{p} \alpha^j f_{ij}$$

for $i = 1, \ldots, d$,

$$(3.3) \qquad g(\alpha) = g_0 + \sum_{j=1}^{p} \alpha^j g_j$$

$(f_{ij}, i \in \{1, \ldots, d\}, j \in \{1, \ldots, p\})$ and $(g_j, j \in \{1, \ldots, p\})$ are known fixed matrices. The true parameter value is denoted α_0. It is assumed that (A1) is satisfied with

$$(3.4) \qquad f_i = f_i(\alpha_0)$$

for $i = 1, 2, \ldots, d$ and

$$(3.5) \qquad g = g(\alpha_0)$$

Using the identifications that are made in (3.4, 3.5) other equations from Section 1 are used in this section.

A least square estimate of α_0 is obtained from the observations $(X(t), t \in [0, T])$ by minimizing the quadratic functional

$$(3.6) \quad \int_0^T \left[\left(X^{(d)} - \sum_{i=1}^d f_i(\alpha) X^{(i-1)} - g(\alpha) U \right)^1 \ell \left(X^{(d)} - \sum_{i=1}^d f_i(\alpha) X^{(i-1)} \right. \right.$$
$$\left. \left. - g(\alpha) U \right) - X^{(d)1} \ell X^{(d)} \right] dt$$

where ℓ is a positive, semidefinite matrix. In (3.6) the undefined term $X^{(d)1} \ell X^{(d)}$ is cancelled and $X^{(d)} dt = dX^{(d-1)}$. The minimization of (3.6) yields the following family of equations for the least squares estimate, $\alpha^*(T) = (\alpha^{*1}(T), ..., \alpha^{*p}(T))$, of α_0

$$\sum_{k=1}^p \frac{1}{T} \int_0^T \left(\sum_{i=1}^d f_{ij} X^{(i-1)} + g_j U \right)^1 \ell \left(\sum_{i=1}^d f_{ik} X^{(i-1)} + g_k U \right) dt \alpha^{*k}(T)$$
$$= \frac{1}{T} \int_0^T \left(\sum_{i=1}^d f_{ij} X^{(i-1)} + g_j U \right)^1 \ell \left(dX^{(d-1)} - \sum_{i=1}^d f_{i0} X^{(i-1)} - g_0 U \right) dt$$
(3.7)

for $j = 1, 2, ..., p$. Using (3.1) with $\alpha = \alpha_0$, (3.7) can be rewritten as

$$(3.8) \quad \sum_{k=1}^p \frac{1}{T} \int_0^T \left(\sum_{i=1}^d f_{ij} X^{(i-1)} + g_j U \right)^1 \ell \left(\sum_{i=1}^d f_{ik} X^{(i-1)} + \right.$$
$$\left. + g_k U \right) dt (a^{*k}(T) - \alpha_0^k)$$
$$= \frac{1}{T} \int_0^T \left(\sum_{i=1}^d f_{ij} X^{(i-1)} + g_j U \right) \ell dW$$

for $j = 1, 2, ..., p$ of (A1) is satisfied then

$$(3.9) \quad \lim_{T \to \infty} \frac{1}{T} \int_0^T \left(\sum_{i=1}^d f_{ij} X^{(i-1)} + g_j U \right)^1$$
$$\ell \left(\sum_{i=1}^d f_{ij} X^{(i-1)} + g_j U \right) dt = tr(F_j^1 \ell F_k R)$$

where $tr(\cdot)$ is the trace, $F_j = (f_{ij}, ..., f_{dj}, g_j)$ for $j = 1, ..., d$ and R satisfies the Lyapunov equation (1.8).

The following assumption is used subsequently

(A2) The matrix $Q = (tr(F_j^1 \ell F_k R))$ for $j, k \in \{1, ..., p\}$ is nonsingular.

Since the right hand side of (39) converges to zero a.s. as $T \to \infty$, it follows from (A2) that

$$(3.10) \qquad \lim_{T \to \infty} \alpha^*(T) = \alpha_0 \qquad a.s.$$

Thus the family of least squares estimates $(\alpha^*(T), T > 0)$ is strongly consistent.

Let $G_j(T)$ for $j = 1, 2, ..., p$ be the random variable on the right hand side of (3.7). The following equality is satisfied

$$(3.11) \qquad \lim_{T \to \infty} G_j(T) = \sum Q_{jk} \alpha_0^k$$

for $j = 1, ..., p$ where Q is given in (A2).

For the proof of the next proposition it is useful to decompose the right hand side of (3.7) and compute the limits separately. By an integration by parts it follows that

$$\lim_{T \to \infty} \frac{1}{T} \int_0^T \left(\sum_{i=1}^{d-1} f_{ij} X^{(i-1)} \right)^1 \ell dX^{(d-1)} = -E \left[\left(\sum_{i=1}^{d-1} f_{ij} X^{(i)} \right)^1 \ell dX^{(d-1)} \right]$$

(3.12)

Furthermore

$$(3.13) \quad \lim_{T \to \infty} \frac{1}{T} \int_0^T \left(\sum_{i=1}^{d} f_{ij} X^{(i-1)} + g_j U \right)^1 \ell \left(\sum_{i=1}^{d} f_{i0} X^{(i-1)} + g_0 U \right) dt$$

$$= E \left[\left(\sum_{i=1}^{d} f_{ij} X^{(i-1)} + g_j U \right)^1 \ell \left(\sum_{i=1}^{d} f_{i0} X^{(i-1)} + g_0 U \right) \right]$$

Using (1.1, 1.7) it follows that

$$(3.14) \qquad \lim_{T \to \infty} \frac{1}{T} \int_0^T X^{(d-1)} dX^{(d-1)1} = \lim_{T \to \infty} \frac{1}{T} \int_0^T X^{(d-1)} \left(dW + \right.$$

$$\left. + \left(\sum_{i=1}^{d} f_i X^{(i-1)} + g U \right)^1 dt \right)$$

$$= \sum_{i=1}^{d} r_{di} f_i^1 + r_{d,d+1} g^1$$

and

$$(3.15) \qquad \lim_{T \to \infty} \frac{1}{T} \int_0^T U dX^{(d-1)1} = \sum_{i=1}^{d} r_{d+1,i} f_i^1 + r_{d+1,d+1} g^1$$

so that

$$(3.16) \quad \lim_{T \to \infty} \frac{1}{T} \int_0^T (f_{dj} X^{(d-1)} + g_j U)^1 \ell dX^{(d-1)}$$

$$= f_{dj} \left(\sum r_{di} f_i^1 + r_{d,d+1} g^1 \right) + g_j \left(\sum r_{d+1,i} f_i^1 + r_{d+1,d+1} g^1 \right)$$

Now it is assumed that instead of the continuous observation of the state $X(t), t \geq 0$ and the input $(U(t), t \geq 0)$ there is only the sampled observations (10) from which the approximate derivatives (1.11) are computed. The equation (3.7) is replaced by a modified discrete analogue as follows

$$(3.17) \quad \frac{1}{N\delta} \sum_{k=1}^p \delta \sum_{m=0}^{n-1} \left(\sum_{i=1}^d f_{ij} X_{m,\delta}^{(i-1)} + g_j U(m\delta) \right)^1 \ell \left(\sum_{i=1}^d f_{ik} X_{m,\delta}^{(i-1)} + \right.$$

$$\left. + g_k U(m\delta) \right) \hat{\alpha}_{N\delta}^k$$

$$\frac{1}{N\delta} \sum_{m=0}^{n-1} \left(\sum_{i=1}^d f_{ij} X_{m,\delta}^{(i-1)} + g_j U(m\delta) \right)^1 \ell (X_{m+1,\delta}^{(d-1)} - X_{m,\delta}^{(d-1)})$$

$$-\delta \left(\sum_{i=1}^d f_{i0} X_{m,\delta}^{(i-1)} + g_0 U(m\delta) \right)$$

$$-\frac{1}{N\delta} \frac{D_d}{C_d} \sum_{m=0}^{N-1} (X_{m+1,\delta}^{(d-1)} - X_{m,\delta}^{(d-1)}) f_{dj}^1 \ell (X_{m+1,\delta}^{(d-1)} - X_{m,\delta}^{(d-1)})$$

for $j = 1, 2, ..., p$ where C_d is given by (15), $\hat{\alpha}_{N\delta} = (\hat{\alpha}_{N\delta}^1, ..., \hat{\alpha}_{N\delta}^p)$ is an estimate of α_0 and D_d is given by

$$(3.18) \quad D_d = \frac{(-1)^{d-1}}{(2d-1)!} \sum_{j=1}^{d-1} (-1)^j j^{2d-1} \binom{2d-1}{d-j-1}$$

The following result verifies that the family of estimates $(\hat{\alpha}_{N\delta}, N \in \mathbb{N}, \delta > 0)$ is consistent.

Proposition 2. Assume that (A1) is satisfied for $\alpha = \alpha_0$ and (A2) is satisfied. For $N \in \mathbb{N}$ and $\delta > 0$ let $\hat{\alpha}_{N\delta} = (\hat{\alpha}_{N\delta}^1, ..., \hat{\alpha}_{N\delta}^p)$ be the solution of (48). Then

$$(3.19) \quad \lim_{\delta \to 0} p \lim_{N \to \infty} \hat{\alpha}_{N\delta} = \alpha_0$$

where $\lim_{N \to \infty} \hat{\alpha}_{N\delta}$ is the nonrandom limit in probability.

The numerical values of C_d and D_d for small values of d are given in the following table.

TABLE 3.1

d	C_d	D_d
2	0.66667	0.25000
3	0.55000	0.40909
4	0.47937	0.54305
5	0.43042	0.66166

REFERENCES

[1] T.E. DUNCAN, P. MANDL AND B. PASIK-DUNCAN, *On statistical sampling for system testing*, IEEE Trans. Auto. Control. 36, No. 1, (1994), pp. 118–122.
[2] T.E. DUNCAN, P. MANDL AND B. PASIK-DUNCAN, *Parameter estimation in higher order stochastic systems with discrete observations*, preprint.

Part III
Adaptive Control of Continuous Time Nonlinear Stochastic Systems

1. Introduction. In this part the ergodic control of a multidimensional diffusion process described by a stochastic differential equation that has unknown parameters appearing in the drift is investigated. For $\epsilon > 0$ it is required to find an adaptive control such that the ergodic cost for this control is within ϵ of the optimal ergodic cost that is obtained if the system is known. An estimation scheme is obtained from a biased maximum likelihood estimation procedure that was used by Kumar and Becker [Ref. 1]. An adaptive control is constructed from a discretization of the range of this family of estimates using the certainty equivalence principle and this control is verified to be almost self-optimizing. To verify that an adaptive control is almost self-optimizing it is necessary to establish some properties of the invariant measure for the diffusion process as a function of the control and the unknown parameters. The invariant measure is exhibited by the well known method of establishing ergodic properties of an embedded Markov chain.

Borkar [Ref. 2] has considered the adaptive control of diffusion processes. We provide a brief comparison between our results and the results in [Ref. 2]. We consider a more general class of controlled diffusions that are solutions of stochastic differential equations. Specifically the drift term in our model is Lipschitz only in the unknown parameter while in [Ref. 2] the drift term is Lipschitz in both the unknown parameter and the state. Borkar [Ref. 2] uses a Hamilton-Jacobi-Bellman equation that has not been studied for our problem. Instead we use probabilistic methods. We show the continuity of the invariant measures in variation norm that should be of independent interest. Borkar [Ref. 2] uses a particular version of the optimal controls that is measurable with respect to both variables. We use a family of almost optimal controls which is weaker from the viewpoint of optimality but provides a more feasible procedure. The methods of proof that are used here are significantly different from the methods in [Ref. 2].

Previous work on the approximate self-optimizing adaptive control for a discrete time stochastic system is given in [Ref. 3].

2. Preliminaries. Let $(X(t; \alpha, u), t \geq 0)$ be a controlled diffusion process that satisfies the following stochastic differential equation

(2.1a)
$$dX(t; \alpha, u) = f(X(t; \alpha, u))dt + h(X(t; \alpha, u), \alpha, u)dt$$
$$+ \sigma(X(t; \alpha, u))dW(t)$$

(2.1b) $$X(0; \alpha, u) = x$$

where $X(t; \alpha, u) \in \mathbb{R}^n$, $(W(t), t \geq 0)$ is a standard $\mathbb{R}^n$-valued Wiener process, $u(t) \in U \subset \mathbb{R}^m$ and U is a compact set and $\alpha \in \mathcal{A} \subset \mathbb{R}^q$ and $\mathcal{A}$ is a compact set. The functions f and σ satisfy a global Lipschitz condition, $\sigma(x)\sigma^*(x) \geq c > 0$ for all $x \in \mathbb{R}^n$ and h is a bounded, Borel function on $\mathbb{R}^n \times \mathcal{A} \times U$. The family $\mathcal{U}$ of admissible controls is

(2.2) $$\mathcal{U} = \{u : u : \mathbb{R}^n \to U \text{ is Borel measurable}\}.$$

The probability space for the controlled diffusion is denoted $(\Omega, \mathcal{F}, P)$. The solution of the stochastic differential equation is a weak solution that can be obtained by absolutely continuous transformation of the measure of the solution of

(2.3a) $$dY(t) = f(Y(t))dt + \sigma(Y(t))dW(t)$$

(2.3b) $$Y(0) = x.$$

which has one and only one strong solution by the Lipschitz continuity of f and σ. Often it is notationally convenient to express an expectation of a function of $X(t; \alpha, u)$ as $E_x^{\alpha, u}(g(X(t)))$ instead of as $E_x(g(X(t; \alpha, u)))$.

For a Borel set A, let T_A be the first hitting time of A, that is

(2.4) $$T_A = \begin{cases} \inf\{s > 0 : X(s) \in A\} \\ +\infty \quad \text{if the above set if empty} \end{cases}$$

For notational convenience the dependence on (α, u) has been suppressed. Let Γ_1 and Γ_2 be two spheres in $\mathbb{R}^n$ with centers at 0 and radii $0 < r_1 < r_2$ respectively. Let τ be defined by the equation

(2.5) $$\tau = T_{\Gamma_2} + T_{\Gamma_1} \circ \theta_{T_{\Gamma_2}}$$

where $\theta_{T_{\Gamma_2}}$ is the positive time shift by T_{Γ_2} that acts on $C(\mathbb{R}_+, \mathbb{R}^n)$. The random variable τ is first time that the process $(X(t), t \geq 0)$ hits Γ_1 after hitting Γ_2. This definition of τ is used throughout this paper unless specified otherwise.

The following assumptions are selectively used in this paper:

(A1)
$$\sup_{\alpha \in \mathcal{A}} \sup_{u \in \mathcal{U}} \sup_{x \in \Gamma_1} E_x^{\alpha, u}[\tau^2] < \infty$$

where τ is given by (2.5).

(A2)   There is an $L_h > 0$ such that for all $\alpha, \beta \in \mathcal{A}$

$$\sup_{x \in \mathbb{R}^n} \sup_{v \in U} |h(x, \alpha, v) - h(x, \beta, v)| \leq L_h |\alpha - \beta|.$$

(A3)   For each $(x, \alpha, u) \in \mathbb{R}^n \times \mathcal{A} \times \mathcal{U}, E_x^{\alpha, u}[T_{\Gamma_1}] < \infty$.

Some sufficient conditions for (A1) are given in Refs. 2 and 4.

A family of measures $(m_x(\cdot; \alpha, u); x \in \mathbb{R}^n, \alpha \in \mathcal{A}, u \in \mathcal{U})$ on the Borel σ-algebra of $\mathbb{R}^n, \mathcal{B}(\mathbb{R}^n)$ is defined by the equation

$$(2.6) \qquad m_x(D; \alpha, u) = E_x^{\alpha, u}\left[\int_0^\tau 1_D(X(s))ds\right]$$

where 1_D is the indicator function of D and τ is given by (5) assuming (A1). The measure $m_x(\cdot; \alpha, u)$ is well defined for each $(\alpha, u) \in A \times \mathcal{U}$.

If (A1) is satisfied then it is well known [Ref. 5] that there is an invariant measure $\mu(\cdot; \alpha, u)$ on $\mathcal{B}(\mathbb{R}^n)$ for the process $(X(t; \alpha, u), t \geq 0)$ that is given by the equation

$$(2.7) \quad \mu(D; \alpha, u) = \int_{\Gamma_1} m_x(D; \alpha, u)\eta(dx; \alpha, u) \left(\int_{\Gamma_1} E_x^{\alpha, u}(\tau)\eta(dx; \alpha, u)\right)^{-1}$$

where $D \in \mathcal{B}(\mathbb{R}^n)$ and $\eta(\cdot; \alpha, u)$ is an invariant measure for the embedded Markov chain $(x_0 \in \Gamma_1, X(\tau_n; \alpha, u), n \in \mathbb{N})$ where

$$(2.8) \qquad \tau_{n+1} = \tau_n + \tau \circ \theta_{\tau_n}$$

where $n \geq 1$ and $\tau_1 = \tau$.

Initially it is verified that the family of invariant measures $(\mu(\cdot; \cdot, u), u \in \mathcal{U})$ is uniformly equicontinuous on $\mathcal{A}$.

THEOREM 2.1. *If (A1) and (A2) are satisfied then for each $\epsilon > 0$ there is a $\delta > 0$ such that if $\alpha, \beta \in \mathcal{A}$ and $|\alpha - \beta| < \delta$ then*

$$(2.9) \qquad \sup_{u \in \mathcal{U}} \|\mu(\cdot; \alpha, u) - \mu(\cdot; \beta, u)\| < \epsilon$$

where $\| \cdot \|$ is the variation norm and $\mu(\cdot; \cdot, \cdot)$ is given by (2.7).

The adaptive control problem includes the control of the diffusion process $(X(t; \alpha, u), t \geq 0)$ where the control is $u \in \mathcal{U}$ and the unknown parameter is $\alpha \in \mathcal{A}$. Since $\alpha \in \mathcal{A}$ is unknown it is necessary to estimate it. An optimal control is a control from $\mathcal{U}$ that minimizes the ergodic cost functional

$$(2.10) \qquad J(u; x, \alpha) = \limsup_{t \to \infty} t^{-1} E_x^{\alpha, u} \left[\int_0^\tau k(X(s), u(X(s))) ds \right]$$

where $k : \mathbb{R}^n \times U \to \mathbb{R}$ is a fixed, bounded Borel function. An adaptive control is constructed that is almost optimal with respect to this cost functional (2.10).

LEMMA 2.2. *If (A1), (A2) and (A3) are satisfied and $k : \mathbb{R}^n \times U \to \mathbb{R}$ is a bounded, Borel function then for each $\alpha \in \mathcal{A}$, $x \in \mathbb{R}^n$ and $u \in \mathcal{U}$*

$$\limsup_{t \to \infty} t^{-1} E_x^{\alpha, u} \left[\int_0^t k(X(s), u(X(s))) ds \right]$$
$$= \int_{\mathbb{R}^n} k(z, u(z)) \mu(dz; \alpha, u)$$

where μ is defined by (2.7).

LEMMA 2.3. *If g is a bounded, Borel function then the map from Γ_1 to $\mathbb{R}$ given by*

$$(2.11) \qquad x \mapsto E_x \left[\int_0^\tau g(X(s)) ds \right]$$

is continuous.

Let $J^* : \mathcal{A} \to \mathbb{R}$ be defined as

$$(2.12) \qquad J^*(\alpha) = \inf_{u \in \mathcal{U}} J(u; x, \alpha)$$

where $J(u; x, \alpha)$ is given by (24). J^* does not depend on x by (2.11).

The following continuity property of the optimal cost with respect to the parameter α follows from the continuity property that is verified in Theorem 2.1 (cf. Corollary 3, [Ref. 3]).

Proposition 2.4. Assume that (A1), (A2) and (A3) are satisfied. For $\alpha \in \mathcal{A}$ let $J^*(\alpha)$ be given by (2.13). Then

$$(2.13) \qquad \lim_{\alpha_n \to \alpha} J^*(\alpha_n) = J^*(\alpha).$$

By Theorem 2.1 it follows that for each $\epsilon > 0$ there is a $\delta(\epsilon) > 0$ such that if $|\alpha - \beta| < \delta(\epsilon)$ and $\alpha, \beta \in \mathcal{A}$ then

$$(2.14) \qquad \sup_{u \in \mathcal{U}} \|\mu(\cdot; \alpha, u) - \mu(\cdot; \beta, u)\| \leq \epsilon/4\|k\|$$

where $\|k\| = \sup_{x \in \mathbb{R}^n} \sup_{u \in \mathcal{U}} |k(x,u)|$ and k is given in (2.10). In the subsequent discussion when $\epsilon > 0$ is given then $\delta(\epsilon)$ always denotes a $\delta(\epsilon) > 0$ that is chosen to satisfy (2.15). Sometimes for notational convenience the dependence of δ on ϵ is suppressed.

To establish almost optimality for an adaptive control it is necessary to relate almost optimal controls for different values of the parameter. This relation is formalized in the following proposition.

Proposition 2.5. If (A1), (A2) and (A3) are satisfied and given $\epsilon > 0$ and $a\bar{u} \in \mathcal{U}$ that is $\epsilon/2$-optimal for (2.1, 2.10) with $\alpha \in \mathcal{A}$ then $\bar{u}$ is ϵ-optimal for (2.1, 2.10) with $\beta \in \mathcal{A}$ where $|\alpha - \beta| < \delta(\epsilon)$.

A family of functions, $w(\cdot;\cdot,\cdot)$ is introduced in the following lemma that is used in the subsequent analysis.

LEMMA 2.6. *Assume that (A1) is satisfied. For $\alpha \in \mathcal{A}$ and $u \in \mathcal{U}$ let $w(\cdot;\cdot,\cdot) : \Gamma_1 \times \mathcal{A} \times U \to \mathbb{R}$ be defined as*

$$w(x;\alpha,u) = \lim_{n \to \infty} E_x^{\alpha,u} \left[\int_0^{\tau_n} k(X(s)), U(x(s)) - \int_{\mathbb{R}^n} k(z, u(z)) \mu(dz; \alpha, u) ds \right].$$
(2.15)

Then the family of functions $(w(x;\alpha,u), x \in \Gamma_1, \alpha \in \mathcal{A}, u \in \mathcal{U})$ is well defined and

$$\sup_{\alpha \in \mathcal{A}} \sup_{u \in \mathcal{U}} \sup_{x \in \Gamma_1} |w(x;\alpha,u)| < \infty \tag{2.16}$$

3. Identification and Adaptive Control. An identification procedure is defined by a biased maximum likelihood method [Ref. 1] where the estimates are changed at random times. The parameter set $\mathcal{A}$ is covered by a finite, disjoint family of sets. The adaptive control is constructed by choosing from a finite family of controls each of which corresponds to an almost optimal control for a distinguished point from one of the sets of the cover of $\mathcal{A}$. It is assumed that (A1), (A2) and (A3) are satisfied throughout this section.

Fix $\epsilon > 0$ and choose $\delta(\epsilon) > 0$ to satisfy (2.15). By the compactness of $\mathcal{A}$ there is a finite cover of $\mathcal{A}$, $(B(\alpha_i, \delta), i = 1, 2, ..., r)$, where $\alpha_i \notin B(\alpha_j, \delta)$ for $i \neq j$ and $B(\alpha, \delta)$ is the open ball with center α and radius $\delta > 0$. Define $(A_i(\epsilon), i = 1, ..., r)$ by the equations

$$A_i(\epsilon) = (B(\alpha_i; \delta) \setminus \bigcup_{j=1}^{i-1} A_j(\epsilon)) \cap \mathcal{A}. \tag{3.1}$$

where $i = 2, ..., r$ and $A_1(\epsilon) = B(\alpha_1, \delta) \setminus \mathcal{A}$. Clearly

$$\mathcal{A} = \bigcup_{j=1}^{r} A_j(\epsilon)$$

Let $e : \mathcal{A} \to \{\alpha_1, ..., \alpha_r\}$ be defined by

$$e(\alpha) = \alpha_j \quad \text{if } \alpha \in A_j(\epsilon) \tag{3.2}$$

and let $\lambda : \mathcal{A} \to \mathbb{R}$ be defined by

$$(3.3) \qquad\qquad \lambda(\alpha) = J^*(e(\alpha))$$

where J^* is defined by (2.13).

By modifying the definition of $(A_1(\epsilon), ..., A_r(\epsilon))$ on the boundaries of these sets it can be assumed that λ is lower semicontinuous. It is assumed that this has been done but the same notation $(A_1(\epsilon), ..., A_r(\epsilon))$ is used for these modified sets.

Given $\epsilon > 0$, choose $N \in \mathbb{N}$ such that

$$(3.4) \qquad\qquad 2 \sup_{\alpha \in \mathcal{A}} \sup_{u \in \mathcal{U}} \|w(\cdot; \alpha, u)\| \leq \frac{\epsilon}{4} m \cdot N$$

where w is defined by (2.16) and $m > 0$ satisfies

$$(3.5) \qquad\qquad m \leq \inf_{x \in \Gamma_1} \inf_{u \in \mathcal{U}} \inf_{\alpha \in \mathcal{A}} E_x^{\alpha u}(\tau)$$

Define a sequence of stopping times $(\sigma_n, n \in \mathbb{N})$ as

$$(3.6) \qquad\qquad \sigma_{n+1} = \sigma_n + \tau_N \circ \theta_{\sigma_n}$$

where $n \geq 1$, $\sigma_1 = \tau_N$, τ_N is given by (2.8) and N is given in (3.4).

The unknown parameter α^0 is estimated at the random times $(\sigma_n, n \in \mathbb{N})$ by a biased maximum likelihood method, that is, $\hat{\alpha}(\sigma_n)$ is a maximizer of

$$(3.7) \qquad L_n(\alpha) = ln\overline{M}(\sigma_n; \alpha, \alpha^0, \eta) + z(\sigma_n)ln(\lambda(\alpha^0)/\lambda(\alpha))$$

where $\overline{M}(\sigma_n; \alpha, \alpha^0, \eta) = dP^\alpha/dP^{\alpha^0}$ is the likelihood function evaluated at time σ_n with the control η, $z : \mathbb{R} \to \mathbb{R}_+$ and $z(t)/t \to 0$ and $z(t)/t^\beta \to \infty$ for some $\beta \in (1/2, 1)$ as $t \to \infty$. By lower semicontinuity of λ, continuity of h with respect to $\alpha \in \mathcal{A}$ and compactness of $\mathcal{A}$ there is an element of $\mathcal{A}$ that achieves the supremum of $L_n(\cdot)$. If this maximizer is not unique then we assume that there is some rule for selecting one from this set of maximizers. The use of terms involving α^0 in (3.7) is valid because these do not affect the maximization.

The family of estimates $(\hat{\alpha}(t), t \geq 0)$ is defined as follows: Choose $\overline{\alpha} \in \mathcal{A}$ and let

$$(3.8) \qquad\qquad \hat{\alpha}(t) = \hat{\alpha}(0) = \overline{\alpha} \quad \text{for } 0 \leq t < \sigma_1.$$

and for $n \geq 1$ let

$$(3.9) \qquad\qquad \hat{\alpha}(\sigma_n) = \arg\max L_n(\alpha)$$
$$(3.10) \qquad\qquad \hat{\alpha}(t) = \hat{\alpha}(\sigma_n) \quad \text{for } \sigma_n \leq t < \sigma_{n+1}$$

Clearly we have

$$(3.11) \qquad\qquad L_n(\hat{\alpha}(\sigma_n)) \geq 0 \qquad a.s.$$

Using this family of estimates, $(\hat{\alpha}(t), t \geq 0)$, and an approximate certainty equivalence principle we define an adaptive control as

$$(3.12) \qquad\qquad \eta(s; \epsilon) = u_{e(\hat{\alpha}(s))}(X(s))$$

where u_{α_i} for $i = 1, 2, ..., r$ is a fixed $\epsilon/2$-optimal control corresponding to the value α_i and $e(\cdot)$ is given by (3.2).

The main result for the ϵ-optimality of the adaptive control (3.12) is given now.

THEOREM 3.1. *If (A1), (A2) and (A3) are satisfied then for each* $\epsilon > 0$

$$(3.13) \qquad \limsup_{t \to \infty} \frac{1}{t} \int_0^t k(X(s), \eta(s; \epsilon)) ds \leq J^*(\alpha^0) + 2\epsilon \qquad a.s.$$

where η is the approximate certainty equivalence control (3.12) and $J^*(\alpha^0)$ satisfies (2.13).

REFERENCES

[1] P.R. KUMAR AND A. BECKER, *A new family of optimal adaptive controllers for Markov chains*, IEEE Transactions on Automatic Control, 27 (1982), 137–146.

[2] V.S. BORKAR, *Self-tuning control of diffusions without the identifiability condition*, Journal of Optimization Theory and Applications, 68 (1991), 117–138.

[3] L. STETTNER, *On nearly selfoptimizing strategies for a discrete time uniformly ergodic adaptive model*, to appear in Journal of Applied Mathematics and Optimization.

[4] V.S. BORKAR AND M.K. GHOSH, *Ergodic control of multidimensional diffusions I: the existence results*, SIAM Journal on Control and Optimization, 26 (1988), 112–126.

[5] R.Z. KHASMINSKII, *Stochastic Stability of Differential Equations*, (translation from Russian), Sigthoff and Noordhoff, Alphen aan den Rijn, Holland, 1980.

[6] D.W. STROCK AND S.R.S. VARADHAN, *Multidimensional Diffusion Processes*, Springer-Verlag, New York, 1979.

[7] A. BENSOUSSSAN, *Perturbation Methods in Optimal Control*, J. Wiley, New York, 1988.

[8] V. BORKAR AND A. BAGCHI, *Parameter estimation in continuous-time stochastic processes*, Stochastics, 8 (1982), 193–212.

[9] N.V. KRYLOV, *On an approach to controlled diffusion processes*, Probability Theory and Applications, 31 (1986), 685–709.

[10] T.E. DUNCAN, B. PASIK-DUNCAN, L. STETTNER, *Almost self-optimizing strategies for the adaptive control of diffusion processes*, Journal of Optimization Theory and Applications, 81(3)(1994), 479–507.

Part IV
Adaptive Control of a Partially Observed Discrete Time Markov Process

1. Introduction and Announcement of the Result. An adaptive control problem for a discrete time Markov process is formulated and its solution is described. The Markov process is completely observed in a fixed recurrent domain and partially observed in the complement of this domain. The study of the adaptive control of partially observed Markov processes seems to be quite limited. In [2] a special maintenance mode is investigated and in [1] there is an approach to self optimality for the adaptive control of partially observed Markov chains.

Consider a controlled Markov process $(X_n, n \in \mathbb{N})$ on a (measurable) state space $(E, \mathcal{E})$ with transition operator $P^{a_i \alpha^0}(x_i, dy)$ evaluated at time i where a_i is the control at time i that is a $\mathcal{Y}_i$-measurable U-valued random variable. The process $(X_n, n \in \mathbb{N})$ is completely observed in a fixed recurrent domain Γ and is partially observed in Γ^c. The observations are denoted $(Y_n, n \in \mathbb{N})$ and $\mathcal{Y}_n = \sigma(Y_i, i \leq n)$. The control $u = (a_n, n \in \mathbb{N})$ is adapted to $(\mathcal{Y}_n, n \in \mathbb{N})$ and takes values in the compact set U. The transition operator $P^{a\alpha^0}(x, dy)$ depends on an unknown parameter α^0 that is an element of a compact set $A \subset \mathbb{R}^k$. The cost $c : E \times U \to \mathbb{R}$ at each time is a bounded, positive, continuous function. It is desired to minimize the ergodic cost functional

$$(1.1) \qquad J((a_i)_{i \in \mathbb{N}}) \limsup_{n \to \infty} \frac{1}{n} \sum_{i=0}^{n-1} c(X_i, a_i).$$

In this paper the state space E is a closed subset of $\mathbb{R}^d$, though all the results can be easily modified if E is countable.

Let

$$P^{a\alpha^0}(x, B) = \int_B p(x, y, a, \alpha^0) dy$$

for $B \in \mathcal{E}$ where $a \in U$, $a^0 \in A$ and the (assumed) density p is a measurable function of all of its arguments.

Assume that there is a compact set Γ such that the observation Y_i of X_i satisfies the following relation

$$P(Y_i \in B | X_i, \mathcal{Y}_{i-1}) =$$
$$= 1_{B \cap \Gamma}(X_i) + 1_{\Gamma^c}(X_i) \int_{B \cap \Gamma^c} r(X_i, y) dy$$

for $B \in \mathcal{E}$, $i \in \mathbb{N}$ where $\mathcal{Y}_0 = \{\emptyset, \Omega\}$, $\Gamma^C = E \backslash \Gamma$ and 1_B is the indicator function of the set B. The equality (3) shows that in Γ there is the complete observation of X_i while in Γ^c there is a noisy observation of X_i with the conditional density $r(X_i, \cdot)$.

Given the initial law μ of X_0, the filter process $(\pi_i^{\alpha^0}, i \in \mathbb{N})$ corresponding to the observations $(\mathcal{Y}_i, i \in \mathbb{N})$ is defined as follows

$$\pi_0^{\alpha^0}(B) = \mu(B)$$
$$\pi_i^{\alpha^0}(B) = P_\mu(X_i \in B | \mathcal{Y}_i)$$

for $i \geq 1$ and $B \in \mathcal{E}$. By Lemma 1 of [6] the filter process $(\pi_i^{\alpha^0}, i \in \mathbb{N})$ has the following representation

$$\pi_{i+1}^{\alpha^0}(B) = 1_{B \cap \Gamma}(Y_{i+1}) + 1_{\Gamma^c} M^{\alpha^0}(Y_{i+1}, \pi_i^{\alpha^0}, a_i)(B)$$

for $B \in \mathcal{E}$ where M^{α^0} is given by

$$M^{\alpha^0}(y, v, a)(B) = \frac{\int_{B \cap \Gamma} r(z, y) p(v, z, a, \alpha^0) dz}{\int_{\Gamma} r(z, y) p(v, z, a, \alpha^0) dz}$$

for $y \in E$, $v \in \mathcal{P}(E)$, $a \in U$ and

$$p(v, z, a, \alpha^0) = \int_E p(x, z, a, \alpha^0) v(dx).$$

Since the parameter α^0 is unknown, to minimize the ergodic cost functional (1) it is necessary to estimate α^0 using the control at time i based on the estimate of α^0 at time i. Under suitable assumptions an adaptive strategy is constructed that is almost self-optimal. This construction is based on the particular observation structure that is used in this paper and an adaptive control method that is used in [3,4,6].

The main result on almost self-optimality can be stated as follows:

THEOREM 1.1. *Under suitable assumptions there is a $K \in \mathbb{R}$ that does not depend on ϵ such that*

$$J((\hat{a}_i)) \leq \lambda^{\alpha^0} + K\epsilon \qquad a.e.P$$

so that in particular

$$J_\mu^{\alpha^0}((\hat{a}_i)) \leq \lambda^{\alpha_0} + K\epsilon.$$

The assumptions and the proof of the theorem are given in [5].

REFERENCES

[1] E. FERNANDEZ-GAUCHERAND, A. ARAPOSTATHIS AND S. MARCUS, *Analysis of an adaptive control scheme for a partially observed controlled Markov chain,* IEEE Trans. Auto. Control, to appear.

[2] E. FERNANDEZ-GAUCHERAND, A. ARAPOSTATHIS AND S. MARCUS, *A methodology for the adaptive control of Markov chains under partial state information,* Proc. 31st Conf. on Decision and control, Tucson, 1992, 2750–2751.

[3] T.E. DUNCAN, B. PASIK-DUNCAN AND L. STETTNER, *Almost self-optimizing strategies for the adaptive control of diffusion processes*, J. Optim. Th. Appl., 81(3)(1994), 407–507.

[4] T.E. DUNCAN, B. PASIK-DUNCAN AND L. STETTNER, *On the ergodic and the adaptive control of stochastic differential delay systems*, J. Optim. Th. Appl., 81(3)(1994), 509–531.

[5] T.E. DUNCAN, B. PASIK-DUNCAN AND L. STETTNER, *Adaptive control of a partially observed discrete time Markov process*, in preparation.

[6] L. STETTNER, *On nearly self-optimizing strategies for a discrete time uniformly ergodic adaptive model*, Appl. Math. Optim., to appear.

[7] L. STETTNER, *Ergodic control of Markov processes with mixed observation structure*, preprint.

OPTIMALITY OF THE ADAPTIVE CONTROLLERS

MILOJE S. RADENKOVIC* AND B. ERIK YDSTIE[†]

Abstract. Results presented in [1] can be considered as the fundamental one in the adaptive control theory. Essentially, it is shown that if the system is exactly modelled, a self-tuning controller provides the same performance as the minimum-variance controller. It turns out that similar results are valid even in the presence of a modelling error. In this paper it is proved that in the presence of unmodelled dynamics, adaptive controller guarantees in a certain sense the same performance as the best non-adaptive controller. Despite the fact that the estimation algorithm has a vanishing gain sequence, uniform boundedness of all signals is established.

1. Introduction. One of the original motivations behind development of the adaptive control theory is to provide a way of handling system uncertainty by adjusting the controller parameters on-line to optimize system performance. In the absence of modeling imperfections it is shown that the asymptotical performance of the adaptive systems are the same as the performance of the corresponding non-adaptive optimal controller [1]. Since adaptive controllers are tuned to the uncertain system, it is expected that their performance will be better compared to a fixed robust controller, even in the presence of modeling errors. In this paper we consider direct adaptive control with the objective to minimize the output tracking error. For a given controller parametric structure, there exists such controller transfer function so that the tracking error is minimal. In general, this transfer function depends on the external system signals. Therefore, among all possible controllers with the same prespecified order, one is optimal in the sense that it generates the smallest tracking error. In [2] this controller is referred to as a tuned controller, while in [3] it is named centered controller. Actually, centered controller is the best reduced order controller corresponding to a given performance index and it is related to the tuned (centered) nominal system model [2,3]. Later one is defined as the best reduced order model so that discrepancies between the model and the physical system are minimal. The centered controller is, unfortunately, unrealizable since it requires prior knowledge of the actual system. The fact that the adaptive controller can guarantee the same or similar performance as the unknown centered controller, so far has not been verified in the adaptive control theory. The reason for this is that in the past 12 years attention of the adaptive control community mainly has been focused on a robustness problem. This became a focal research point and ever since has been demonstrated that small unmodelled dynamics can cause instability in the adaptive system [4,5]. In order to provide global

* Department of Electrical Engineering, University of Colorado at Denver, Denver, Colorado 80217-3364.

† Department of Chemical Engineering, Carnegie Mellon University, Doherty Hall, Pittsburgh, PA 15213.

stability in the face of modeling errors, a variety of modifications of the algorithms originally designed for the ideal system model were proposed. These include σ-modification, parameter projection, ϵ_1-modification, dead-zone techniques, etc. (see for example [6]).

Existing robust adaptive control results proved boundedness of the signals in the adaptive loop. The upper bound on the tracking error is established and it depends on the algorithm gain and some generic constants with unclear size and nature. There is no precise indication of how well a resulting performance is close to the case of exactly known nominal system model.

In this paper we consider adaptive algorithm with the vanishing gain sequence, without using σ-modification or parameter projection. Reference signal is assumed to be persistently exciting. It is proven that the mean-square tracking error is the same (in a certain sense defined in section 2) as in the case of known centered parameters, and it does not depend on the algorithm gain or any design parameter. The basis for this conclusion is the result presented in section 3 and it states that $\tilde{\theta}(t)^T \phi(t)$ and $e(t+1)$ are "uncorrelated," i.e.

$$(1.1) \qquad \lim_{N \to \infty} \frac{1}{N} \sum_{t=1}^{N} \tilde{\theta}(t)^T \phi(t) e(t+1) = 0$$

where $e(t)$ is the tracking error, $\phi(t)$ is the signal vector and $\tilde{\theta}(t)$ is the parameter estimation error. Regardless of the fact that the proposed estimation algorithm has a vanishing gain sequence, uniform boundedness of all signals in the adaptive loop is established.

1.1. Notation and Terminology. For a discrete-time function x : $\mathbf{T} \to \mathbf{R}^+$, we define the following seminorm:

$$(1.2) \qquad n_x(t) = \left\{ \sum_{j=1}^{t} \lambda^{t-j} x(j)^2 \right\}^{1/2} , \quad 0 < \lambda < 1$$

where $\mathbf{T}$ is the set of positive integers, while $\mathbf{R}^+$ is the set of non-negative real numbers. When $n_x(t)$ is uniformly bounded over all $t \geq 0$, x is said to be in l_2^λ. H^∞ will denote the space of transfer functions $T(z)$ which are analytic and bounded outside and on the unit circle in the z plane. S^λ is the operator defined by

$$(1.3) \qquad S^\lambda T(z) = T(\lambda^{1/2} z)$$

for a fixed parameter $\lambda, 0 < \lambda \leq 1$. $S^\lambda H^\infty$ is the space of transfer functions $T(z)$ such that $S^\lambda T(z) \in H^\infty$. In other words, $T(z) \in S^\lambda H^\infty$ if $T(z)$ is analytic and bounded outside and on the circle $|z| = \lambda^{1/2}$ in the z plane. For $T(z) \in H^\infty$, the H^∞ norm is defined by

$$(1.4) \qquad \|T(z)\|_{H^\infty} := \max_{|z|=1} |T(z)|.$$

Likewise, the norm of the $S^\lambda H^\infty$ space is defined by

$$(1.5) \qquad \|T(z)\|_{H\infty}^\lambda := \|S^\lambda T(z)\|_{H\infty} = \max_{|z|=1} |T(\lambda^{1/2}z)|.$$

This norm is induced by the l_2^λ norm of the input and output signals of $T(z)$.

When performing majorizations, in order to account for initial conditions we will use non-negative generic functions in the present form

$$(1.6) \qquad \xi(t) = c\lambda_\xi^t, 0 \le c < \infty, 0 < \lambda_\xi < 1$$

2. Problem Formulation and Major Assumptions. We consider the following discrete time SISO system with unmodelled dynamics

$$(2.1) \quad A(q^{-1})y(t+1) = B(q^{-1})u(t) + A(q^{-1})\Delta(q^{-1})u(t) + w(t+1)$$

where $\{y(t)\}, \{u(t)\}$ and $\{w(t)\}$ are output, input and disturbance sequences, respectively, while q^{-1} represents the unit delay operator. The polynomials $A(q^{-1})$ and $B(q^{-1})$ describe the nominal system model and are given by

$$A(q^{-1}) = 1 + a_1 q^{-1} + ... + a_{n_A} q^{-n_A}, B(q^{-1}) = b_0 + b_1 q^{-1} + ... + b_{n_B} q^{-n_B}, b_0 \ne 0$$
$$(2.2)$$

In eq. (2.1) $\Delta(q^{-1})$ denotes additive system perturbation and it is given by

$$(2.3) \qquad \Delta(q^{-1}) = P(q^{-1}) - P_n(q^{-1}), P_n(q^{-1}) = B(q^{-1})/A(q^{-1})$$

where $P(q^{-1})$ is the transfer operator describing dynamics of a real physical system and it is of higher order than $P_n(q^{-1})$.

The objective is to simultaneously stabilize the input-output behavior of the system (2.1) and minimize the criterion

$$(2.4) \qquad J = \lim_{N \to \infty} \frac{1}{N} \sum_{t=1}^{N} (y(t) - y^*(t))^2$$

where $y^*(t)$ is the given reference signal. Concerning disturbance $w(t)$ and the reference signal $y^*(t)$ we assume:

$$(2.5) \qquad |w(t)| \le k_w < \infty, |y^*(t)| \le k_{y^*} < \infty, \forall t \ge 0$$

To define adaptive controller it is convenient to write system (2.1) in the form

$$(2.6) \quad \frac{1}{|b_0|}e(t+1) = (sgnb_0)u(t) + \theta^T\phi(t) + \frac{\gamma(t)}{|b_0|}; e(t+1) = y(t+1) - y^*(t+1)$$

where

(2.7)
$$\theta^T = \left[-\frac{1}{|b_0|}; -\frac{a_1}{|b_0|}, ..., -\frac{a_{n_A}}{|b_0|}, ..., \frac{b_{n_B}}{|b_0|}\right]$$

(2.8) $\phi(t)^T = [y^*(t+1); y(t), ..., y(t-n_A+1); u(t-1), ..., u(t-n_B)]$

and

(2.9)
$$\gamma(t) = A(q^{-1})\Delta(q^{-1})u(t) + \omega(t+1)$$

If the nominal system model is minimum phase and $\gamma(t) \equiv 0$, from (2.6) it is obvious that the control law optimal in the sense of (2.4) is given by $(\text{sgn } b_0)u(t) = -\theta^T\phi(t)$. In the adaptive case we use the certainty equivalence type controller

(2.10)
$$(\text{sgn } b_0)u(t) = -\theta(t)^T\phi(t)$$

where $\theta(t)$ is an estimate of θ. From (2.6) and (2.10) it follows that the closed-loop adaptive system is given by

(2.11) $e(t+1) = -|b_0|z(t) + \gamma(t); z(t) = \tilde{\theta}(t)^T\phi(t); \tilde{\theta}(t) = \theta(t) - \theta.$

From (2.10) and (2.11) it is not difficult to obtain

(2.12) $B(q^{-1})u(t) - q[A(q^{-1}) - 1]y(t) = y*(t+1) - |b_0|z(t)$

wherefrom by (2.11) we can derive

(2.13) $B(q^{-1})u(t) = A(q^{-1})[-|b_0|z(t) + y^*(t+1)] + [A(q^{-1}) - 1]\gamma(t)$

Substituting $u(t)$ from (2.13) into (2.9) we get [7,8]

(2.14) $\gamma(t) = -H_0(\theta, q^{-1})[|b_0|z(t) - y^*(t+1)] + H_1(\theta, q^{-1})\omega(t+1)$

where

$$H_0(\theta, q^{-1}) = \frac{A(q^{-1})\Delta(q^{-1})A(q^{-1})}{B(q^{-1}) - A(q^{-1})\Delta(q^{-1})[A(q^{-1}) - 1]}$$

and

(2.15) $H_1(\theta, q^{-1}) = \dfrac{B(q^{-1})}{B(q^{-1}) - A(q^{-1})\Delta(q^{-1})[A(q^{-1}) - 1]}$

When denoting transfer operators $H_0(.,.)$ and $H_1(.,.)$ we use argument θ in order to emphasize that $H_0(.,.)$ and $H_1(.,.)$ depend on the nominal model $B(q^{-1})/A(q^{-1})$, characterized by the parameter vector θ. Let us assume for a moment that the nominal parameters θ are known. Then in (2.10) we can set $\theta(t) = \theta$ and obtain the following closed-loop system

$$
\begin{aligned}
e(t+1) &= \gamma(t), z(t) = 0, \gamma(t) = \\
&H_0(\theta, q^{-1})y^*(t+1) + H_1(\theta, q^{-1})w(t+1)
\end{aligned}
$$
(2.16)

which follows from (2.11) and (2.14). For given orders n_A and n_B of the polynomials $A(q^{-1})$ and $B(q^{-1})$, there exists the nominal system model $B^*(q^{-1})/A^*(q^{-1})$, so that the H^∞-norm of $H_0(\theta, q^{-1})$ is minimal. Following ideas presented in [3], let us define the "centered" parameters as follows

$$
\theta^* = arg \min_{\theta \in \Theta} ||H_0(\theta, z)||_{H^\infty}
$$
(2.17)

The parameters θ^* are called "centered" because they have been tuned to give optimal performance and minimize sensitivity with respect to the truncation (modelling) errors. In this paper we will refer to θ^* as a vector of optimal control parameters, and it is given by

$$
\theta^* = \left[-\frac{1}{|b_0^*|}; -\frac{a_1^*}{|b_0^*|}, \dots, -\frac{a_{n_A}^*}{|b_0^*|}; \frac{b_1^*}{|b_0^*|}, \dots, \frac{b_{n_B}^*}{|b_0^*|}\right]
$$
(2.18)

Obviously, the centering problem is closely related to the problem of finding the best reduced order (nominal system) model [9], and with the "tuned parameters" concept introduced in [2]. In practical situations it is difficult to know centered parameters θ^* and therefore optimal performance cannot be obtained. In this paper we show that it is possible to design an adaptive controller which will provide in a certain sense the same performance as the centered non-adaptive controller. For future reference it is convenient to rewrite (2.10), (2.11) and (2.14) in terms of centered parameters, i.e.,

(2.19) $(\text{sgn} b_0^*)u(t) = -\theta(t)^T \phi(t); e(t+1) = -|b_0^*|z(t) + \gamma(t)$

(2.20) $z(t) = \tilde{\theta}^*(t)^T \phi(t); \tilde{\theta}^*(t) = \theta(t) - \theta^*$

and

(2.21) $\gamma(t) = -H_0^*(q^{-1})[|b_0^*|z(t) - y^*(t+1)] + H_1^*(q^{-1})w(t+1)$

where $H_0^*(q^{-1}) = H_0(\theta^*, q^{-1})$ and $H_1^*(q^{-1}) = H_1(\theta^*, q^{-1})$. In (2.19) $\theta(t)$ is an estimate of θ^*. Before we propose an algorithm which will estimate θ^*, let us introduce the following assumption:

Assumption A_1:
(i) polynomial $B^*(z^{-1})$ and the transfer functions $H_0^*(z^{-1})$ and $H_1^*(z^{-1})$ are stable,
(ii) the sign of the high-frequency gain b_0^* and an upper bound $b_{0,max}$ of $|b_0^*|$ are known. Without loss of generality we assume that $b_0^* > 0$.

Estimates of θ^* will be generated by the following algorithm

$$(2.22) \quad \theta(t+1) = \theta(t) + \frac{\bar{a}}{r(t)}\phi(t)[y(t+1) - y^*(t+1)], 0 < \bar{a} \leq \frac{1}{b_{0,max}}$$

where $\phi(t)$ is given by (2.8), while

$$(2.23) \qquad\qquad r(t) = \max\left\{\max_{1\leq\tau\leq t} \|\phi(\tau)\|^2; f(t)\right\}$$

and $f(t)$ is any function satisfying

$$(2.24) f(t) > 0, f(t) \geq f(t-1), \forall t \geq 0; \lim_{t\to\infty} f(t) = +\infty; \lim_{t\to\infty} \frac{f(t)}{r_0(t) + t} = 0$$

where $r_0(t) = \sum_{k=1}^t \|\phi(k)\|^2$. For example, $f(t)$ can be given by $f(t) = t^{1-\epsilon}$ or $f(t) = r_0(t)^{1-\epsilon}, 0 < \epsilon < 1$.

It is well known that the above adaptive algorithm is not stable. The problem is due to the presence of an unstable manifold along which the parameter estimates may diverge. Since we are not using parameter projection or leakage techniques, the stability problem will be solved by using external excitation. We take the same approach as in [7] where it is assumed that the reference signal $y^*(t)$ is persistently exciting. Note that by using (2.19)-(2.21), the measurement vector can be written in the form

$$(2.25) \qquad\qquad \phi(t) = \phi^*(t) + \phi_\omega(t) + \phi_z(t) + \phi_\gamma(t)$$

where

$$\phi^*(t)^T =$$
$$\left[y^*(t+1); y^*(t), ..., y^*(t-n_A+1); \frac{A^*}{B^*}y^*(t), ..., \frac{A^*}{B^*}y^*(t-n_B+1)\right]$$
(2.26)
$$\phi_\omega(t)^T =$$
$$\left[0, \omega(t), ..., \omega(t-n_A+1); \frac{A^*-1}{B^*}\omega(t), ..., \frac{A^*-1}{B^*}\omega(t-n_B+1)\right]$$
(2.27)
$$\phi_z(t)^T =$$
$$\left[0, -b_0^*z(t-1), ..., -b_0^*z(t-n_A); \frac{A^*}{B^*}b_0^*z(t-1), ..., -\frac{A^*}{B^*}b_0^*z(t-n_B)\right]$$
(2.28)

$$\phi_\gamma(t)^T = \left[0, \gamma(t-1), ..., \gamma(t-n_A); \frac{A^*-1}{B^*}\gamma(t-1), ..., \frac{A^*-1}{B^*}\gamma(t-n_B)\right]$$
(2.29)

Now we can introduce the assumption related to the frequency content of the reference signal $y^*(t)$ and the structure of the centered nominal model $B^*(q^{-1})/A^*(q^{-1})$.

Assumption A_2: For all sufficiently large N,

$$\sum_{t=1}^{N} \lambda^{N-t} \phi^*(t)\phi^*(t)^T \geq \sigma_1^* I, \sigma_1^* > 0, 0 < \lambda < 1$$

with $\phi^*(t)$ given by (2.26).

In order to quantify the size of the admissible modeling errors we define the following H^∞ norms:

(2.30)
$$C_{AB} = \left\| \frac{A^*(z)}{B^*(z)} \right\|_{H\infty}^{\lambda}, C_A = \left\| \frac{A^*(z)-1}{B^*(z)} \right\|_{H\infty}^{\lambda},$$
$$C_\gamma = \| H_0^*(z) \|_{H\infty}^{\lambda}, C_\omega = \| H_1^*(z) \|_{H\infty}^{\lambda}$$

where $H_0^*(q^{-1})$ and $H_1^*(q^{-1})$ are given by (2.15), when $\theta = \theta^*$, with θ^* defined by (2.18).

Assumption A_3: *(concerning the size of modeling errors C_γ and external disturbances k_ω)*

(i) $\rho_1 = 1 - \frac{\mu}{2} - (1-\mu)C_\gamma - \frac{\mu}{2}C_\gamma^2 > 0, \mu = \overline{a}b_0^*$

(ii) $\sigma_2^* = \frac{\sigma_1^*}{2} - \left\{ n_1 b_0^* \Sigma_\gamma + n_2 \left[b_0^* C_\gamma \Sigma_\gamma + \frac{C_\gamma k_{y^*} + (1+C_\omega)k_\omega}{(1-\lambda)^{1/2}} \right] \right\}^2 > 0$

where σ_1^* is given by assumption (A_2), while

$$\Sigma_\gamma^2 = \max \left[16(1 - \mu + \mu C_\gamma)/\rho_1^2; 2\mu/\rho_1 \right] \frac{(C_\gamma k_{y^*} + C_\omega k_\omega)^2}{1 - \lambda}$$

and

$$n_1 = \left[\sum_{i=1}^{n_A} \lambda^{-i} + C_{AB}^2 \sum_{i=1}^{n_B} \lambda^{-i} \right]^{1/2}, n_2 = \left[\sum_{i=1}^{n_A} \lambda^{-i} + C_A^2 \sum_{i=1}^{n_B} \lambda^{-i} \right]^{1/2}$$

As it is stated in [7], the first part of the above assumption implies that the intensity of the unmodelled dynamics C_γ satisfies $C_\gamma < 1$. The second part of this assumption is more complicated and essentially means the following: C_γ and the disturbance upper bound k_ω should be small compared with the level of the external excitation.

Let us define the criterion according to which performance of the adaptive and centered non-adaptive controller will be compared. From (2.19) and (2.21) it follows that

(2.31)
$$e(t+1) = -b_0^* \left[1 - H_0^*(q^{-1}) \right] z(t) +$$
$$H_0^*(q^{-1})y^*(t+1) + H_1^*(q^{-1})\omega(t+1)$$

or

$$(2.32) \qquad \frac{1}{1 - H_0^*(q^{-1})} e(t+1) = -b_0^* z(t) + d(t)$$

where

$$(2.33) \qquad d(t) = \frac{H_0^*(q^{-1})}{1 - H_0^*(q^{-1})} y^*(t+1) + \frac{H_1^*(q^{-1})}{1 - H_0^*(q^{-1})} w(t+1)$$

Relation (2.32) implies

$$(2.34) \; e(t+1) \frac{1}{1 - H_0^*(q^{-1})} e(t+1) = -b_0^* z(t) e(t+1) + e(t+1) d(t)$$

In the case when the centered parameters θ^* are known, from (2.20) it follows that the optimal non-adaptive controller will generate $z(t) = 0$, which together with (2.34) gives

$$(2.35) \; \lim_{N \to \infty} \frac{1}{N} \sum_{t=1}^{N} e(t+1) \frac{1}{1 - H_0^*(q^{-1})} e(t+1) = \lim_{N \to \infty} \frac{1}{N} \sum_{t=1}^{N} e(t+1) d(t)$$

where $d(t)$ is defined by (2.33). We will show that the proposed adaptive controller provides the same performance as the centered non-adaptive controller in the sense that relation (2.35) is valid. Actually, we will prove that in the adaptive case

$$(2.36) \qquad \lim_{N \to \infty} \frac{1}{N} \sum_{t=1}^{N} z(t) e(t+1) = 0$$

i.e., $z(t)$ and the tracking error $e(t+1)$ are "uncorrelated." It is not difficult to recognize that in the "ideal" adaptive control theory [1], relation (2.36) implies global stability and optimality of the adaptive system.

3. Technical Results. In this section we present the technical result which will be used to establish performance of the considered adaptive system. The following lemma states that all signals are bounded by the l_2^λ norm of the error signal $z(t)$ given by (2.20).

LEMMA 3.1. *Let the assumption* (A_1) *hold. Then*
1)

$$(3.1) \qquad n_\gamma(t) \le C_\gamma b_0^* n_z(t) + n_v(t)$$

with

$$(3.2) \qquad n_v(t) = (C_\gamma k_{y^*} + C_w k_w)/(1 - \lambda)^{1/2} + \xi(t)$$

where $n_\gamma(t)$ *and* $n_z(t)$ *are a given by (1.2) when* $x(t) = \gamma(t)$ *and* $x(t) = z(t)$, *respectively, with* $\gamma(t)$ *and* $z(t)$ *defined as in (2.20) and (2.21).*

2)

$$(3.3) \quad n_\phi(t) \leq C_{\phi1}b_0^* n_z(t-1) + (C_{\phi1}^* k_{y^*} + C_{\phi2}k_\omega)/(1-\lambda)^{1/2} + \xi(t)$$

where $n_\phi(t)$ is defined by (1.2), with $x(t) = \|\phi(t)\|^2$, while

$$(3.4) \qquad \begin{aligned} C_{\phi1} &= C_1[1 + (1+C_A)C_\gamma + C_{AB}], \\ C_{\phi2} &= C_1 C_\omega(1+C_A), C_{\phi1}^* = \lambda^{-1}C_{\phi1} + 1 \end{aligned}$$

where $C_1 = \left(\sum_{i=0}^{\overline{n}} \lambda^{-i}\right)^{1/2}$ and $\overline{n} = \max(n_A; n_B)$, while C_A, C_{AB}, C_γ and C_ω are given by (2.30).

Proof. Proof of the lemma is given in the Appendix □

From (2.25) it is obvious that whenever the l_2^λ norm of $z(t)$ is small relative to the level of the external excitation σ_1^* defined by the assumption (A_3), frequency content of the regressor $\phi(t)$ is dominated by $\phi^*(t)$. In other words, signal vector $\phi(t)$ will be persistently exciting, which is stated by the next lemma.

LEMMA 3.2. Let the assumptions $(A_1) \div (A_3)$ hold. Then on the subsequence $\{N_k\}$ where $n_z(N_k) \leq \sum_\gamma^2 + \xi(N_k)$, the following holds

$$(3.5) \qquad \lambda_{\min}\left\{\sum_{t=1}^{N_k} \lambda^{N_k - t}\phi(t)\phi(t)^T\right\} \geq \sigma_2^* - \rho_0, 0 < \rho_0 \ll \sigma_2^*,$$

for sufficiently large k. Constants $\sum_\gamma$ and σ_2^* are defined in the assumption (A_3).

Proof. The proof is given in the Appendix. □

Intuitively, it is clear that on the subsequence where the PE conditions are satisfied in the adaptive loop, estimator (2.22) guarantees small parameter estimation error. This fact is formulated by the next lemma.

LEMMA 3.3. Let the assumptions $(A_1) - (A_3)$ hold. Then on the subsequence $\{N_k\}$ where $n_z(N_k) \leq \sum_\gamma^2 + \xi(N_k)$, parameter estimation error is given by

$$(3.6) \qquad \|\theta(N_k+1) - \theta^*\| \leq \sum_\theta + \xi(t)$$

where

$$(3.7) \quad \sum_\theta = C_\phi\left\{\sum_\gamma + \overline{a}[b_0^*(1+C_\gamma)\sum_\gamma + \frac{C_\gamma k_{y^*} + C_\omega k_\omega}{(1-\lambda)^{1/2}}\right\}/(\sigma_2^* - \rho_0)$$

with $0 < \rho \ll \sigma_2^*$ and

$$(3.8) \qquad C_\phi = C_{\phi1}b_0^*\lambda^{-1}\sum_\gamma + (C_{\phi1}^* k_{y^*} + C_{\phi2}k_\omega)/(1-\lambda)^{1/2}$$

where the constants $C_{\phi1}^*$, $C_{\phi i}$, $i = 1, 2$ are given by (3.4). $\sum_\gamma$ and σ_2^* are defined in the assumption (A_3).

Proof. Proof of the lemma is given in the Appendix. □

Note that Lemmas 3.2 and 3.3 state that on the subsequence where $n_z(t)$ is small, PE conditions are satisfied in the adaptive loop and parameter estimation error is bounded by the $\sum_\theta$. Careful examination of (3.7) shows that small modeling errors C_γ and small k_w (disturbance upper bound) imply small $\sum_\theta$. For the purpose of future analysis, by the next assumption we further restrict admissible modeling errors C_γ and disturbance bound k_w.

Assumption A_4: $\lambda^{-1} \sum_\theta C_{\phi 1} b_0^* < 1$, where $\sum_\theta$ and $C_{\phi 1}$ are given by (3.7) and (3.4) respectively.

4. Performance of the Adaptive Controller. In this section we will prove that (2.36) and consequently (2.35) holds, and all signals in the adaptive loop are uniformly bounded despite the fact that the estimator (2.22) has vanishing gain sequence. Let us determine the Liapunov type difference inequality which describes the behavior of the parameter estimator. From the estimation algorithm (2.22) we have

$$V(t+1) \leq V(t) + \frac{2\overline{a}\tilde{\theta}^*(t)^T \phi(t)e(t+1)}{r(t)} + \frac{\overline{a}^2 \|\phi(t)\|^2 e(t+1)^2}{r(t)^2}; V(t) = \|\tilde{\theta}^*(t)\|^2$$
(4.1)

where $\tilde{\theta}^*(t)$ is given by (2.20). Substituting $e(t+1)$ from (2.19) into (4.1), and after simple majorization, we obtain

$$(4.2) \quad V(t+1) \leq V(t) - 2\mu\left(1 - \frac{\mu}{2}\right)\frac{z(t)^2}{r(t)} + \frac{2\mu(1-\mu)|z(t)\gamma(t)| + \mu^2\gamma(t)^2}{r(t)}$$

where μ is defined in the assumption $(A_3 - i)$. Global stability of the above difference inequality can be proved by using ideas presented in [7,8] where the behavior of the following function is examined

$$(4.3) \qquad\qquad S(t+1) = V(t+1) + \frac{W(t+1)}{r(t)}$$

with

$$(4.4) \quad \begin{aligned} W(t+1) = \mu\sum_{j=1}^{t} \lambda^{t-j} \Big\{ &\big[1 - \tfrac{\mu}{2} + (1-\mu)C_\gamma + \tfrac{\mu}{2}C_\gamma^2\big] z(j)^2 \\ &- 2(1-\mu)|z(j)\gamma(j)| - \mu\gamma(j)^2 \Big\} \end{aligned}$$

THEOREM 4.1. *Let the assumptions $(A_1) - (A_4)$ hold. Then the adaptive algorithm (2.19), (2.22)-(2.24), provide*
1)

$$(4.5) \qquad\qquad \limsup_{t \to \infty} \|\theta(t) - \theta^*\| \leq \sum_\theta$$

where $\sum_\theta$ is given by (3.7).

2)

$$(4.6) \limsup_{t \to \infty} n_\phi(t) \leq (C_{\phi 1} k_{y*} + C_{\phi 2} k_w)/(1 - \lambda)^{1/2}(1 - \sum_\theta C_{\phi 1} b_0^* \lambda^{-1})$$

where $n_\phi(t)$ is given by (1.2), when $x(t) = \|\phi(t)\|^2$.
3) relations (2.35) and (2.36) hold.

Proof. First statement of the theorem will be derived by examining "burst recovery" effect and self-stabilization in the adaptive loop [7,8]. Following the concept presented in [8], let us define the sequences τ_k and $\sigma_k, k \geq 1$ as follows

$$(4.7) \qquad 1 \stackrel{\Delta}{=} \tau_1 < \sigma_1 < \tau_2 < ... < \tau_k < \sigma_k < \tau_{k+1} < ...$$

so that for $W(t+1)$ given by (4.4), the following holds

$$(4.8) \qquad W(t+1) \leq 0, \forall t \in Q_k \text{ and } W(t+1) > 0, \forall t \in T_k$$

where the time intervals T_k and Q_k are defined by

$$(4.9) \qquad Q_k = [\tau_k, \sigma_k) \text{ and } T_k = [\sigma_k, \tau_{k+1}), k \geq 1$$

If $W(2) > 0$, we set $\tau_1 = 0$, $\sigma_1 = 1$ and Q_k is defined for $k \geq 2$. Regarding the sequences τ_k and σ_k, we consider three possible cases:
(i) for all finite k, we have $\tau_k < \infty$ and $\sigma_k < \infty$,
(ii) there exists a finite k_0 such that $\tau_{k_0} < \infty$ and $\sigma_{k_0} = +\infty$,
(iii) there exists a finite k_1 such that $\sigma_{k_1} < \infty$ and $\tau_{k_1+1} = +\infty$.
Let us first consider the case when $\tau_k < \infty$ and $\sigma_k < \infty$ for all finite k. Since in the time intervals $Q_k, W(t+1) \leq 0$, from (3.1) and (4.4), after simple majorizations we obtain $\forall t \in Q_k$

$$(4.10) \qquad \rho_1 n_z(t)^2 \leq 2[1 - \mu + \mu C_\gamma] n_z(t) n_v(t) + \mu n_v(t)^2$$

where ρ_1 is defined by assumption $(A_3 - i)$. Previous relation implies

$$(4.11) \; \rho_1 n_z(t)^2 \leq 2 \max \left[2(1 - \mu + \mu C_\gamma) n_z(t) n_v(l); n_v(t)^2 \right], \forall t \in Q_k$$

from where by (3.2) we can derive

$$(4.12) \qquad n_z(t)^2 \leq \sum_\gamma^2 + \xi(t), \forall t \in Q_k$$

with $\sum_\gamma$ given in assumption $(A_3 - ii)$. By using Lemma 3.3 and (4.12), we conclude that for sufficiently large k,

$$(4.13) \qquad V(t) \leq \sum_\theta^2 + \xi(t), \forall t \in [\tau_k + 1, \sigma_k]$$

where $V(t)$ is defined by (4.1), while $\sum_\theta$ is given by (3.7). Next we analyze the time intervals T_k, where $W(t+1) > 0$. From (4.2) and (4.4) it follows that

$$(4.14)\; V(t+1) + \frac{W(t+1)}{r(t)} \le V(t) + \frac{W(t)}{r(t-1)} - \mu\rho_1 \frac{z(t)^2}{r(t)}, \quad \forall t \in T_k$$

where ρ_1 is defined by the assumption $(A_3 - i)$. Obviously $\forall t \in T_k$ function $S(t)$ given by (4.3) is nonincreasing. After summation from $t = \sigma_k + 1$ to $N < \tau_{k+1}$, we obtain from (4.14)

$$(4.15)\quad V(N+1) + \frac{W(N+1)}{r(N)} \le V(\sigma_k) - \mu\rho_1 \sum_{t=\sigma_k}^{N} \frac{z(t)^2}{r(t)}, \forall N \in T_k$$

where we used the fact that from (4.8), $W(\sigma_k) \le 0$. Previous relation implies that $V(t) \le V(\sigma_k), \forall t \in [\sigma_k + 1, \tau_{k+1}]$ which together with (4.13) gives the first statement of the theorem. Statement (4.6) can be obtained from (2.20) and (4.5). Since from (2.20), $|z(t)| \le \|\tilde{\theta}^*(t)\| \cdot \|\psi(t)\|$, by (4.5) we have

$$(4.16)\qquad \limsup_{t\to\infty} n_z(t) \le \Sigma_\theta \limsup_{t\to\infty} n_\phi(t)$$

Substituting (4.16) into (3.3) we get

$$\limsup_{t\to\infty} n_\phi(t) \le \Sigma_\theta C_{\phi 1} b_0^* \lambda^{-1} \limsup_{t\to\infty} n_\phi(t) + (C_{\phi 1}^* k_{y*} + C_{\phi 2} k_w)/(1-\lambda)^{1/2}$$
$$(4.17)$$
from where statement (4.6) directly follows. In the case when there exists a finite k_0 so that $\tau_{k_0} < \infty$ and $\sigma_{k_0} = +\infty$, relation (4.13) is valid $\forall t \ge \tau_{k_0}+1$ and consequently (4.6) holds for all $t \ge \tau_{k_0} + 1$.

Next we analyze the more complicated case when in (4.7) there exists a finite k_1 so that $\sigma_{k_1} < \infty$ and $\tau_{k_1+1} = +\infty$. Since now $W(t+1) > 0, \forall t \ge \sigma_{k_1}$, from (4.15) we derive

$$(4.18)\qquad \mu\rho_1 \sum_{t=\sigma_{k_1}}^{N} \frac{z(t)^2}{r(t)} \le V(\sigma_{k_1}) < \infty, \forall N \ge \sigma_{k_1}$$

and by the Kronecker's Lemma $\lim_{N\to\infty}(\sum_{t=1}^{N} z(t)^2)/r(N) = 0$, from where by (3.3) it follows that

$$(4.19)\qquad \lim_{N\to\infty} \frac{1}{N} \sum_{t=1}^{N} z(t)^2 = 0 \text{ or } \lim_{N\to\infty} \frac{1}{N} \sum_{t=1}^{N} n_z(t)^2 = 0$$

Let us show that the above relation implies statements (4.5) and (4.6). We define the following function

$$(4.20)\qquad W_1(t+1) = \mu\rho_1 \left[n_z(t)^2 - (1-\lambda)\Sigma_\gamma^2 \sum_{j=1}^{t} \lambda^{t-j} \right]$$

It is not difficult to see that there exists subsequence $\{t_p\}$, so that $W_1(t_p + 1) \leq 0$. If there is no such subsequence, then $W_1(t + 1) > 0, \forall t \geq \sigma_{k_1}$, and (4.20) gives

$$(4.21) \qquad \liminf_{N \to \infty} \frac{1}{N} \sum_{t=\sigma_{k_1}}^{N} n_z(t)^2 \geq \Sigma_\gamma^2$$

which contradicts (4.19). Therefore, a subsequence $\{t_p\}$ exists, i.e., $W_1(t)$ changes its sign and we can define the time intervals D_i and L_i so that

$$(4.22) \qquad W_1(t + 1) \leq 0, \forall t \in D_i, \quad W_1(t + 1) > 0, \forall t \in L_i$$

where

$$(4.23) \qquad D_i = [\alpha_i, \beta_i) \text{ and } L_i = [\beta_i, \alpha_{i+1}), i \geq 1$$

with $\alpha_1 < \beta_1 < \alpha_2 < ... < \alpha_i < \beta_i < \alpha_{i+1} < ...;$ and $\alpha_1 \geq \sigma_{k_1}$. If $W_1(\sigma_{k_1} + 1) > 0$, we set $\beta_1 = \sigma_{k_1}$ and D_i is defined for $i \geq 2$. From (4.20) and (4.23) we conclude that $n_z(t)^2 \leq \Sigma_\gamma^2, \forall t \in D_i$ and from Lemma 3.3 it follows that

$$(4.24) \qquad \limsup_{i \to \infty} \sup_{t \in [\alpha_i + 1, \beta_i]} V(t) \leq \Sigma_\theta, \alpha_i \geq \sigma_{k_1}$$

Since $W_1(t + 1) > 0, \forall t \in L_i$ and $W(t + 1) > 0, \forall t \geq \sigma_{k_1}$, from (4.14) we derive $\forall t \in L_i$

$$(4.25) \qquad \begin{aligned} V(t + 1) + \tfrac{W(t+1)+W_1(t+1)}{r(t)} &\leq \\ V(t) + \tfrac{W(t)+W_1(t)}{r(t-1)} &\leq ... \leq V(\beta_i) + \tfrac{W(\beta_i)}{r(\beta_i - 1)} \end{aligned}$$

where we used the fact that by (4.22), $W_1(\beta_i) \leq 0$. On the other hand, $W_1(\beta_i) \leq 0$ gives $n_z(\beta_i) \leq \Sigma_\gamma^2$, which together with (2.23) and (4.4) implies $\lim_{i \to \infty} W(\beta_i)/r(\beta_i - 1) = 0$. Then from (4.24) and (4.25) we obtain

$$(4.26) \qquad \limsup_{i \to \infty} \sup_{t \in [\beta_i + 1; \alpha_{i+1}]} V(t) \leq \Sigma_\theta$$

Relations (4.24) and (4.26) constitute the proof of the statement (4.5). Second statement of the theorem follows from (4.16) and (4.17). Thus we show that for all possible scenarios regarding sequences τ_k and σ_k, relations (4.5) and (4.6) hold. Only the third statement of the theorem is left to be proved. From (4.1) we have

$$\begin{aligned} V(t + 1)r(t) &= V(t)r(t - 1) + V(t)(r(t) - r(t - 1)) + 2\bar{a}\tilde{\theta}^*(t)^T \phi(t)e(t + 1) \\ &\quad + \bar{a}^2 \|\phi(t)\|^2 e(t + 1)^2/r(t) \end{aligned}$$

$$(4.27)$$

which gives

$$(4.28) \quad \begin{aligned} V(N+1)r(N) &= V(1)r(0) + \sum_{t=1}^{N} V(t)(r(t) - r(t-1)) + \\ &+ 2\bar{a}\sum_{t=1}^{N} \tilde{\theta}^*(t)^T \phi(t)e(t+1) + \bar{a}^2 \sum_{t=1}^{N} \|\phi(t)\|^2 e(t+1)^2/r(t) \end{aligned}$$

It is obvious (2.23), (2.24), (4.5) and (4.6) imply

$$(4.29) \quad \lim_{N\to\infty} \frac{1}{N}\sum_{t=1}^{N} V(t)(r(t) - r(t-1)) \leq \lim_{N\to\infty} \frac{1}{N}\Sigma_\theta(r(N) - r(0)) = 0$$

Similarly, by the Stoltz's theorem, from (4.24) and (4.6) we get

$$\lim_{N\to\infty} \frac{1}{N}\sum_{t=1}^{N} \|\phi(t)\|^2 e(t+1)^2/r(t) = \lim_{N\to\infty} \|\phi(N)\|^2 e(N+1)^2/r(N) = 0$$
(4.30)

Finally, (4.28) - (4.30) give

$$(4.31) \quad \lim_{N\to\infty} \frac{1}{N}\sum_{t=1}^{N} \tilde{\theta}^*(t)^T \phi(t)e(t+1) = 0$$

where we used that by (4.6), $\lim_{N\to\infty} V(N+1)r(N)/N = 0$. The third statement of the theorem follows from (2.20), (2.34) and (4.31). Thus the theorem is proved. □

Remark 1: The upper bound of the tracking error $e(t)$ can be established by using (2.34), from where we have

$$e(t+1)^2 = e(t+1)\frac{H_0^*(q^{-1})}{1 - H_0^*(q^{-1})}e(t+1) - b_0^* z(t)e(t+1) + e(t+1)d(t)$$
(4.32)

with $d(t)$ given by (2.33). After simple majorizations, from (4.31) and (4.32) we derive

$$\limsup_{N\to\infty} \frac{1}{N}\sum_{t=1}^{N} e(t+1)^2 \leq$$

$$(4.33) \quad \left\{ \frac{K_\gamma}{1 - K_\gamma}\left[\limsup_{N\to\infty} \frac{1}{N}\sum_{t=1}^{N} y^*(t+1)^2\right]^{\frac{1}{2}} + \right.$$

$$\left. + \frac{K_\omega}{1 - K_\gamma}\left[\limsup_{N\to\infty} \frac{1}{N}\sum_{t=1}^{N} w(t+1)^2\right]^{\frac{1}{2}} \right\}^2$$

where $K_\gamma = \left\|\frac{H_0^*(z)}{1-H_0^*(z)}\right\|_{H^\infty}$ and $K_\omega = \left\|\frac{H_1^*(z)}{1-H_0^*(z)}\right\|_{H^\infty}$, and it is assumed that $K_\gamma < 1$.

In the case when $\omega(t) = 0$, $\forall t \geq 0$, from (2.33) and (4.33) it is obvious that the adaptive controller minimizes the upper bound of the tracking error. When there is no unmodelled dynamics, i.e., when $\Delta(q^{-1}) = 0$, from (2.15) we have $H_0^*(q^{-1}) = 0$ and $H_1^*(q^{-1}) = 1$. Then $K_\gamma = 0$ and $K_\omega = 1$, and from (4.33) it is clear that the upper bound of the mean-square tracking error is minimal.

Remark 2: Let us mention that by using external excitation (assumption A_2, global stabilization is possible without using projection or leakage techniques. In addition, during the adaptation, parameter estimation error can be kept small, which is stated in (4.5). This implies that only small bursts are possible in the adaptive loop. At the same time, external excitation provides uniform boundedness of all signals, regardless of the fact that the algorithm has vanishing gain sequence. All these properties together with the optimality result and satisfactory disturbance rejection, suggest that the estimator with the vanishing gain, can have practically acceptable performance. From (4.28) and (4.30) it is obvious that the non-vanishing gain estimators ($r(t) < \infty$) do not guarantee relation (4.31). With such algorithms, meansquare tracking error unavoidably will depend on the algorithm gain.

On the other hand, in the case of time-varying parameters, a non-vanishing gain estimator has to be used. This means that in the presence of the parameter's time variations, relation (4.31) cannot be obtained.

5. Conclusion. Based on a'priori system information, in certain practical situations non-adaptive robust controllers can be designed. Existing results in robust adaptive control theory do not provide clear indication of whether or not an adaptive system has better performance than a fixed robust controller. The reason is that all results in robust adaptive control theory give performance upper bound in terms of certain generic constants with undefined size and nature. In this paper it is shown that in the presence of modeling errors, an adaptive system guarantees similar performance as the best non-adaptive controller. Since the best fixed controller is unrealizable, results presented in this paper can be considered as a strong motivation for using an adaptive system concept in solving practical control problems.

Appendix

Proofs of the Lemmas 3.1, 3.2 and 3.3 follow the same steps as the proofs of Lemmas 4.1, 4.2 and 4.3 in [7]. There is a slight technical difference due to the fact that in [7] sign of the high frequence gain, b_0^* is assumed to be unknown. For the sake of completeness and clarity, we present these proofs in the following text.

Proof of Lemma 3.1: Statement (1) of the lemma follows from (2.21).

Let us prove the second statement of the lemma. From (2.8) it follows that

(A.1) $$n_\phi(t) \leq C_1[n_y(t) + n_u(t)] + n_{y^*}(t+1) + \xi(t)$$

where C_1 is given with (3.4). Note that from (2.11) and (3.1), we can obtain

(A.2) $$n_y(t) \leq (1 + C_\gamma)[b_0^* n_z(t-1) + n_{y^*}(t)] + C_\omega n_\omega(t) + \xi(t)$$

Similarly, from Eq. (2.13) and (3.1) we derive

$$n_u(t) \leq (C_{AB} + C_A C_\gamma)[b_0^* n_z(t) + n_{y^*}(t+1)] + C_A C_\omega n_\omega(t+1) + \xi(t)$$
(A.3)

Substituting (A.2) and (A.3) into (A.1), the second statement of the lemma follows directly.

Proof of Lemma 3.2: From (2.28) and by the condition of the lemma, it is not difficult to obtain

(A.4) $$\sum_{t=1}^{N_k} \lambda^{N_k - t} [\eta^T \phi_z(t)]^2 \leq n_1^2 (b_0^*)^2 \Sigma_\gamma^2 + \xi(t)$$

where n_1 is given by the assumption (A_3), and η is any vector satisfying $\|\eta\| = 1$. Similarly, from (2.29) and statement (1) of the Lemma 3.1, we can obtain

$$\sum_{t=1}^{N_k} \lambda^{N_k - t} [\eta^T \phi_\gamma(t)]^2 \leq n_2^2 \left\{ C_\gamma b_0^* \Sigma_\gamma + [C_\gamma m_1 + C_\omega k_\omega]/(1 - \lambda)^{1/2} \right\}^2 + \xi(t)$$
(A.5)

where n_2 and Σ_γ are defined by the assumption (A_3). From (2.27) it follows that

(A.6) $$\sum_{t=1}^{N_k} \lambda^{N_k - t} [\eta^T \phi_\omega(t)]^2 \leq n_2^2 k_\omega^2/(1 - \lambda) + \xi(t)$$

Relations (A.4) - (A.6) yield

$$\sum_{t=1}^{N_k} \lambda^{N_k - t} [\eta^T \phi_1(t)]^2 \leq$$

$$\left\{ n_1 b_0^* \sum_\gamma + n_2 [b_0^* C_\gamma \sum_\gamma + (C_\gamma m_1 + (1 + C_\omega) k_\omega)/(1 - \lambda)^{1/2}] \right\}^2 + \xi(t)$$
(A.7)

where $\phi_1(t)$ is the component of the regressor which is not directly affected by the reference signal, i.e., $\phi_1(t) = \phi_z(t) + \phi_\gamma(t) + \phi_\omega(t)$. From (2.25) it follows that

(A.8) $$[\eta^T \phi(t)]^2 \geq \frac{1}{2}[\eta^T \phi^*(t)]^2 - [\eta^T \phi_1(t)]^2$$

which together with (A.7) and assumptions (A_2) and (A_3) give

$$(A.9) \qquad \lambda_{\min}\left\{\sum_{t=1}^{N_k}\lambda^{N_k-t}\phi(t)\phi(t)^T\right\} \geq \sigma_2^* - \xi(N_k)$$

Using the fact that for sufficiently large N_k, $\xi(N_k) \leq \rho_0$, $0 < \rho_0 \ll \sigma_2^*$, the statement of the lemma directly follows. Thus the lemma is proved. □

Proof of Lemma 3.3: From (2.19) and statement (1) of Lemma 3.1, we derive

$$(A.10)\, n_e(t) \leq b_0^*(1+C_\gamma)n_z(t) + C_\gamma n_{y*}(t+1) + C_w n_w(t+1) + \xi(t)$$

where $n_e(t)$ is given by Eq. (1.2), when $x(t) = e(t+1)$. By the condition of the lemma and (A.10) we can obtain

$$(A.11)\quad n_e(N_k)^2 \leq [b_0^*(1+C_\gamma)\Sigma_\gamma + (C_\gamma m_1 + C_w k_w)/(1-\lambda)^{1/2}]^2 + \xi(t)$$

Similarly, from statement (2) of Lemma 3.1, we conclude that

$$(A.12) \qquad n_\phi(N_k) \leq C_\phi + \xi(t)$$

where $C\phi$ is given by (3.8). From (2.22) it follows that

$$\tilde{\theta}^*(t+1)^T p(t)^{-1} = \lambda\tilde{\theta}^*(t)^T p(t-1)^{-1} + z(t)\phi(t)^T + \frac{\bar{a}\phi(t)^T p(t)^{-1}}{r(t)}e(t+1)$$
(A.13)

where $p(t)^{-1}$ is given by $p(t)^{-1} = \lambda p(t-1)^{-1} + \phi(t)\phi(t)^T$, $p(0)^{-1} = p_0 I$, $p_0 > 0$ and $\tilde{\theta}^*(t)$ is defined by (2.20). It is not difficult to see that relation (A.13) implies

$$\tilde{\theta}^*(N+1)^T = \lambda^N\tilde{\theta}^*(1)p_0 p(N) + \left(\sum_{t=1}^{N}\lambda^{N-t}z(t)\phi(t)^T\right)p(N)+$$
(A.14)
$$+\bar{a}\left(\sum_{t=1}^{N}\lambda^{N-t}\frac{p(t)^{-1}}{r(t)}e(t+1)\right)p(N)$$

Note that

$$(A.15)\quad \left\|\sum_{t=1}^{N}\lambda^{N\pm t}\frac{\phi(t)^T p(t)^{-1}}{r(t)}e(t+1)\right\| \leq n_e(N)n_\phi(N)/(1-\lambda)$$

and

$$(A.16) \qquad \left\|\sum_{t=1}^{N}\lambda^{N-t}z(t)\phi(t)^T\right\| \leq n_z(N)n_\phi(N)$$

From (A.14), by using relations (A.15) and (A.16) we obtain

$$\|\tilde{\theta}^*(N+1)\| \leq \|p(N)\|\left\{p_0\lambda^N\|\tilde{\theta}^*(1)\| + [n_z(N) + \frac{\bar{a}}{1-\lambda}n_e(N)]n_\phi(N)\right\}.$$
(A.17)

Note that by Lemma 3.2, for sufficiently large k, on the subsequence $\{N_k\}$ we have

(A.18)
$$\|p(N_k)\| \leq \frac{1}{\sigma_2^* - \rho_0}$$

The statement of the lemma directly follows from (A.17) by using relations (A.11), (A.12), and (A.18). Thus the lemma is proved.   □

REFERENCES

[1] G.C. Goodwin, P.J. Ramadge and P.E. Caines, *Discrete time stochastic adaptive control*, SIAM J. Contr. and Optimization, 19 (1981) pp. 829–853.

[2] R.L. Kosut and B. Friedlander, *Robust adaptive control: conditions for global stability*, IEEE Trans. Aut. Contr., AC-30 (1985) pp. 610–624.

[3] B.E. Ydstie, *Transient performance and robustness of direct adaptive control*, IEEE Trans. Aut. Contr., AC-37 (1992) pp. 1092–1105.

[4] E. Egardt, *Stability of adaptive controllers* (Springer-Verlag, New York, 1979).

[5] C.E. Rohrs., L. Valavani, M. Athans and G. Stein, *Analytical verification of undesirable properties of direct model reference adaptive control algorithm*, Proc. 20th IEEE Conf. Decision and Contr., 2 (1981) pp. 1272–1284.

[6] P.A. Ioannou and J. Sun, *Theory and design of robust direct and indirect adaptive-control schemes*, Int. J. Contr., 47 (1988) 775–813.

[7] M.S. Radenkovic and B. Ydstie, *Using persistent excitation with fixed energy to stabilize adaptive controllers and obtain hard bounds for the parameter estimation error*, approved for publication in SIAM J. Contr. and Optimization, 1994.

[8] M.S. Radenkovic and A.N. Michel, *Robust adaptive systems and self stabilization*, IEEE Trans. Aut. Contr., AC-37 (1992) pp. 1355–1369.

[9] K. Glover, *All optimal Hankel norm approximations of linear multivariable systems and their L^∞-error bounds*, Int. J. Contr., 39 (1984) pp. 1145–1193.

UNCERTAIN REAL PARAMETERS WITH BOUNDED RATE OF VARIATION*

ANDERS RANTZER[†]

Abstract. This paper treats stability robustness for systems with uncertain time-varying parameters, exploiting "bandwidth constraints" on the parameters. This closes the gap between μ-analysis for time-invariant uncertainty and circle criteria for time-varying uncertainty.

1. Notation. Let $\mathbf{RH}_\infty$ be the set of proper (bounded at infinity) rational functions with real coefficients and without poles in the closed right half plane. The set of $m \times n$ matrices with elements in $\mathbf{RH}_\infty$ will be denoted $\mathbf{RH}_\infty^{m \times n}$. The Fourier transform of u is denoted by $\hat{u}$.

2. Introduction and main result. Recent progress in computation of structured singular values has significantly improved our ability to do stability analysis for linear systems containing uncertain constant parameters [FTD91,You93]. However, the assumption that the uncertain parameters are constant is crucial and there is a large gap between these results and those that can be obtained for arbitrarily time-varying parameters using the circle criterion.

In the late sixties there were several papers devoted to linear time-invariant convolution operators cascaded in a feedback loop with a positive time-varying gain $\delta(t)$ having bounded rate of variation. For example Brockett and Forys [BF64], Gruber and Willems [GW66] and Sundareshan and Thathachar [ST72] used bounds on $d\delta/dt$ to relax the circle criterion. Recently these results were generalized by Megretski [Meg93a].

In this paper, we shall introduce a new type of bound on the variations in δ, putting constraints on the support of the Fourier transform $\hat{\delta}$, (which is defined in the sense of a temperate distribution [Hör85, Definition 7.1.7] when $\delta \in L^\infty$). Slow time-variations in δ are then modelled as a small bandwidth, i.e. $\hat{\delta}$ having support in a small interval.

Standard upper bounds for structured singular values [FTD91] are obtained in the limit as the support of $\hat{\delta}$ approaches zero. Conversely, unlimited bandwidth of δ gives the small gain theorem with constant scalings.

It is worth noting that there is a corrspondence between bandwidth constraints and derivative constraints. In fact, if $\delta \in \mathbf{L}_\infty(\mathbf{R})$ and $\operatorname{supp} \hat{\delta} \in [-a, a]$, then there is an entire function on $\mathbf{C}$ that equals δ a.e. on the real axis, see e.g. [Hör85, Theorem 7.1.14]. Furthermore, we have

$$\sup_t |\delta'(t)| < a \sup_t |\delta(t)| \quad \text{(Bernstein's inequality)}.$$

* This work was supported by the Swedish Natural Science Research Council.

† Department of Automatic Control, University of Lund, Box 118, S-221 00 Lund, Sweden.

For proof of this and other properties of entire functions, see [Boa54].

Assuming that $\delta(t)$ is a scalar function with bandwidth constraints, our main result gives a sufficient condition for $\mathbf{L_2(R)}$-stability of the differential equation

$$(2.1) \qquad \dot{x}(t) = [A + B\delta(t)C]x(t),$$

i.e. that $\int_0^\infty |x(t)|^2 dt < \infty$ for any solution of the equation. The exact statement is as follows.

THEOREM 1. *Suppose that* $\delta \in \mathbf{L_\infty(R)}$, $\|\delta_i\|_\infty \leq 1$ *and* $\operatorname{supp} \widehat{\delta}_i \subset [-a, a]$. *Let* $G(s) = C(sI - A)^{-1}B + D \in \mathbf{RH_\infty^{n \times n}}$. *If there exists a* $\Psi \in \mathbf{L_\infty^{2n \times 2n}}(j\mathbf{R})$ *such that*

$$\begin{bmatrix} G(j\omega) \\ G(j\omega) \end{bmatrix}^* \Psi(j\omega) \begin{bmatrix} G(j\omega) \\ G(j\omega) \end{bmatrix} < \begin{bmatrix} G(j\omega) \\ I \end{bmatrix}^* \Psi(j\omega + ja) \begin{bmatrix} G(j\omega) \\ I \end{bmatrix}$$

$$\Psi(j\omega) = \Psi(j\omega)^* = \begin{bmatrix} \Psi_{11} & \Psi_{12} \\ -\Psi_{12} & \Psi_{22} \end{bmatrix}(j\omega) > 0, \quad \omega \in [0, \infty]$$

$$\Psi(j\omega_1) \geq \Psi(j\omega_2), \quad 0 \leq \omega_1 \leq \omega_2 \leq \infty$$

then $\int_0^\infty |x(t)|^2 dt < \infty$ *for any solution* x *of (2.1).*

The proof is given in section 3.

Application of the theorem requires a procedure to solve the inequalities for Ψ. This is convex but infinite-dimensional feasability problem. Finite-dimensional restrictions of this problem can be efficiently treated by algorithms based on linear matrix inequalities.

It is straightforward, but notationally less convenient to formulate a corresponding theorem for several uncertain parameters $\delta_1, \ldots, \delta_m$ with different bandlimits $a_1, \ldots, a_m$. The proof remains essentially unchanged.

3. Proofs. The proof will be based on the following two lemmata. The first lemma is a standard argument for invertibility of a linear operator. An stronger formulation used in a similar context can be found in [Meg93b].

LEMMA 2. *Let* $\Pi \in \mathbf{L_\infty^{2n \times 2n}}(j\mathbf{R})$ *and suppose* $\delta \in \mathbf{L_\infty(R)}]$ *with*

$$\int_{-\infty}^{\infty} \begin{bmatrix} \widehat{u}(j\omega) \\ \widehat{v}(j\omega) \end{bmatrix}^* \Pi(j\omega) \begin{bmatrix} \widehat{u}(j\omega) \\ \widehat{v}(j\omega) \end{bmatrix} d\omega \geq 0$$

for all $u \in \mathbf{L_2^n(R)}$ *and* $v(t) = \tau\delta(t)u(t)$, $\tau \in [0, 1]$. *If*

$$(3.1) \qquad \begin{bmatrix} G(j\omega) \\ I \end{bmatrix}^* \Pi(j\omega) \begin{bmatrix} G(j\omega) \\ I \end{bmatrix} < -\epsilon I$$

for $\omega \in \mathbf{R}$ *and some* $\epsilon > 0$, *then* $\int_0^\infty |x(t)|^2 dt < \infty$ *for any solution* x *of (2.1).*

Proof. Introduce the linear operators F_G and T_δ on $\mathbf{L_2(R)}$ defined by

$$F_G u(t) = \int_{-\infty}^t C e^{A(t-\tau)} B u(\tau) d\tau$$
$$T_\delta v(t) = \delta(t) v(t).$$

The main step of the proof will be to show that the operator $I - T_\delta F_G$ has a bounded inverse. Then, given $x(0)$, the unique solution of (2.1) can be constructed for $t \geq 0$ as

$$x(t) = e^{At} x(0) + \int_0^t e^{A(t-\tau)} B u(\tau) d\tau,$$

where

$$u = (I - T_\delta F_G)^{-1} f$$
$$f(t) = \begin{cases} 0 & \text{for } t < 0 \\ \delta(t) C e^{At} x(0) & \text{for } t \geq 0. \end{cases}$$

To prove the invertibility of $I - T_\delta F_G$, let τ_0 be the smallest number $\tau \geq 0$ such that no bounded inverse $I - \tau T_\delta F_G$ exists. Then $\tau_0 > 0$ and

$$\|(I - \tau T_\delta F_G)^{-1}\| \to \infty \text{ as } \tau \to \tau_0 - 0.$$

We will prove that this is impossible for $\tau_0 \leq 1$. Introduce the notation

$$\sigma(u, v) = \int_{-\infty}^\infty \begin{bmatrix} \widehat{u}(j\omega) \\ \widehat{v}(j\omega) \end{bmatrix}^* \Pi(j\omega) \begin{bmatrix} \widehat{u}(j\omega) \\ \widehat{v}(j\omega) \end{bmatrix} d\omega$$

for $u, v \in \mathbf{L_2(R)}$. Then, by the boundedness of Π and F_G, there is a number $C > 0$ such that

$$|\sigma(F_G v, v - w) - \sigma(F_G v, v)| \leq C(\|v\| \cdot \|w\| + \|w\|^2)$$

for all $v, w \in \mathbf{L_2(R)}$. Furthermore (3.1) implies the existence of $\epsilon > 0$ such that $\sigma(F_G v, v) \leq -\epsilon \|v\|^2$ for $v \in \mathbf{L_2(R)}$. With $w = (I - \tau T_\delta F_G)v$, $\tau \in [0, 1]$, this gives

$$\begin{aligned}
0 &\leq \sigma(F_G v, \tau T_\delta F_G v) \\
&= \sigma(F_G v, v - w) \\
&\leq \sigma(F_G v, v) + C(\|v\| \cdot \|w\| + \|w\|^2) \\
&\leq -\epsilon \|v\|^2 + C(\|v\| \cdot \|w\| + \|w\|^2) \\
&= -0.5\epsilon \|v\|^2 + (0.5C^2/\epsilon + C)\|w\|^2 - 0.5\epsilon(\|v\| - C\|w\|/\epsilon)^2
\end{aligned}$$

and

$$\|(I - \tau T_\delta F_G)v\|^2 \geq \frac{\epsilon^2}{C^2 + 2C\epsilon} \|v\|^2.$$

This completes the proof. $\qquad\square$

LEMMA 3. *Suppose that* $\Psi \in \mathbf{L}_\infty^{2n \times 2n}(j\mathbf{R})$ *satisfies* (2.2) *and* (2.2). *Then*

$$\int_0^\infty \left\{ \left[\begin{array}{c} \widehat{u}(j\omega) \\ \widehat{u}(j\omega) \end{array} \right]^* \Psi(j\omega) \left[\begin{array}{c} \widehat{u}(j\omega) \\ \widehat{u}(j\omega) \end{array} \right] - \left[\begin{array}{c} \widehat{u}(j\omega) \\ \widehat{v}(j\omega) \end{array} \right]^* \Psi(j\omega + ja) \left[\begin{array}{c} \widehat{u}(j\omega) \\ \widehat{v}(j\omega) \end{array} \right] \right\} d\omega$$

is non-negative for $u \in \mathbf{L}_2^n(\mathbf{R})$ *and* $v(t) = \delta(t)u(t)$, *with* $\delta \in \mathbf{L}_\infty(\mathbf{R})$, $\|\delta\|_\infty \le 1$ *and* supp $\widehat{\delta}_i \subset [-a, a]$.

Proof. Consider first the special case that Ψ has the form of a constant matrix multiplied by the characteristic function of the interval $[0, \nu]$. Let $u_0 \in \mathbf{L}_2^n(\mathbf{R})$ have the Fourier transform $\widehat{u}_0(j\omega) = \widehat{u}(j\omega)$ for $|\omega| \le \nu + a$ and $\widehat{u}_0(j\omega) = 0$ elsewhere. Then

$$\int_0^\nu \left[\begin{array}{c} \widehat{u} \\ \widehat{v} \end{array} \right]^* \Psi \left[\begin{array}{c} \widehat{u} \\ \widehat{v} \end{array} \right] d\omega = \int_0^\nu \left[\begin{array}{c} \widehat{u}_0 \\ (\widehat{\delta} * \widehat{u}_0) \end{array} \right]^* \Psi \left[\begin{array}{c} \widehat{u}_0 \\ (\widehat{\delta} * \widehat{u}_0) \end{array} \right] d\omega$$

$$\le \frac{1}{2} \int_{-\infty}^\infty \left[\begin{array}{c} u_0(t) \\ \delta(t)u_0(t) \end{array} \right]^* \Psi \left[\begin{array}{c} u_0(t) \\ \delta(t)u_0(t) \end{array} \right] dt$$

$$= \frac{1}{2} \int_{-\infty}^\infty \left(u_0^* \Psi_{11} u_0 + \delta^2 u_0^* \Psi_{22} u_0 \right) dt$$

$$\le \frac{1}{2} \int_{-\infty}^\infty \left[\begin{array}{c} u_0 \\ u_0 \end{array} \right]^* \Psi \left[\begin{array}{c} u_0 \\ u_0 \end{array} \right] dt$$

$$= \int_0^{\nu+a} \left[\begin{array}{c} \widehat{u} \\ \widehat{u} \end{array} \right]^* \Psi \left[\begin{array}{c} \widehat{u} \\ \widehat{u} \end{array} \right] d\omega.$$

This proves the lemma in the special case. In the general case, let $\Psi_{k,\varepsilon} = \Psi(jk\varepsilon + ja) - \Psi(j(k+1)\varepsilon + ja)$ for $\epsilon > 0$ and $k = 1, 2, \dots$ Introduce also $\Psi_\infty = \lim_{\omega \to \infty} \Psi(j\omega)$. Then

$$\int_0^\infty \left[\begin{array}{c} \widehat{u}(j\omega) \\ \widehat{v}(j\omega) \end{array} \right]^* \Psi(j\omega + ja + j\varepsilon) \left[\begin{array}{c} \widehat{u}(j\omega) \\ \widehat{v}(j\omega) \end{array} \right] d\omega$$

$$\le \int_0^\infty \left[\begin{array}{c} \widehat{u} \\ \widehat{v} \end{array} \right]^* \Psi_\infty \left[\begin{array}{c} \widehat{u} \\ \widehat{v} \end{array} \right] d\omega + \sum_{k=1}^\infty \int_0^{k\varepsilon} \left[\begin{array}{c} \widehat{u} \\ \widehat{v} \end{array} \right]^* \Psi_{k,\varepsilon} \left[\begin{array}{c} \widehat{u} \\ \widehat{v} \end{array} \right] d\omega$$

$$\le \int_0^\infty \left[\begin{array}{c} \widehat{u} \\ \widehat{u} \end{array} \right]^* \Psi_\infty \left[\begin{array}{c} \widehat{u} \\ \widehat{u} \end{array} \right] d\omega + \sum_{k=1}^\infty \int_0^{k\varepsilon+a} \left[\begin{array}{c} \widehat{u} \\ \widehat{u} \end{array} \right]^* \Psi_{k,\varepsilon} \left[\begin{array}{c} \widehat{u} \\ \widehat{u} \end{array} \right] d\omega$$

$$\le \int_0^\infty \left[\begin{array}{c} \widehat{u}(j\omega) \\ \widehat{u}(j\omega) \end{array} \right]^* \Psi(j\omega) \left[\begin{array}{c} \widehat{u}(j\omega) \\ \widehat{u}(j\omega) \end{array} \right] d\omega$$

for arbitrarily small ε. This proves the lemma in the general case. □

Proof of Theorem 1 The theorem follows by combination of Lemma 2 and Lemma 3 using

$$\Pi(j\omega) = \left[\begin{array}{cc} (\Psi_{11} + \Psi_{22})(j\omega) & 0 \\ 0 & 0 \end{array} \right] - \Psi(j\omega + ja \operatorname{sign} \omega).$$

□

4. Acknowledgements. The author is grateful to the Institute for Mathematics and its Applications (IMA) for providing an extremely stimulating environment during the preparation of this manuscript. The author is particularly greatful to A. Megretski for comments on the proof of Lemma 2.

REFERENCES

[BF64] R.W. Brockett and L.J. Forys. On the stability of systems containing a time-varying gain. In *Proc. 2nd Allerton Conf. Circuit and System Theory*, pages 413–430, 1964.

[Boa54] R.P. Boas. *Entire functions*. Academic Press, 1954.

[FTD91] M.K.H. Fan, A.L. Tits, and J.C. Doyle. Robustness in presence of mixed parametric uncertainty and unmodeled dynamics. *IEEE Transactions on Automatic Control*, 36(1):25–38, January 1991.

[GW66] M. Gruber and J.L. Willems. On a generalization of the circle criterion. In *Proc. 4th Allerton Conf. Circuit and System Theory*, pages 827–848, 1966.

[Hör85] L. Hörmander. *The analysis of linear partial differential operators I*. Springer-Verlag, 1985.

[Meg93a] A. Megretski. Frequency domain criteria of robust stability for slowly time-varying systems. Submitted for publication in IEEE Transactions on Automatic Control, 1993.

[Meg93b] A. Megretski. Power distribution approach in robust control. In *Proceedings of IFAC Congress*, 1993.

[ST72] M.K. Sundareshan and M.A.L. Thathachar. L_2-stability of linear time-varying systems—conditions involving noncasual multipliers. *IEEE Transaction on Automatic Control*, 17(4):504–510, 1972.

[You93] Peter M. Young. Robustness with parametric and dynamic uncertainty. Technical report, Caltech, 1993. PhD thesis.

AVERAGING METHODS FOR THE ANALYSIS OF ADAPTIVE ALGORITHMS

VICTOR SOLO*

Abstract. We give a brief review of some ideas in averaging methods for analysing adaptive algorithms. The 3 types of first order averaging result are described and a mention is given as to how averaging can be used to get second order results in particular perturbation expansions for invariant measures.

1. Introduction. Averaging methods were introduced into Adaptive Control and Signal Processing in the 1970's initially for long memory algorithms in a stochastic framework and then deterministic averaging methods were introduced to analyse short memory algorithms in the early 1980's [5], [10], [11].

A long memory adaptive algorithm has its gain sequence $\rightarrow 0$ as time $\rightarrow \infty$ whereas a short memory algorithm has gain $>$ constant > 0 for all time. Since only short memory algorithms can track time varying parameters (which is the justification for adaptive algorithms) we only consider those.

Our aim here is to give a brief overview of the method of averaging and its use in analysing adaptive algorithm. There is some utility in this because there is a lot of confusion (even amongst those supposedly knowledgeable) about what averaging methods can do and how they work. We give a brief illustration of averaging applied to a simple adaptive control problem and refer to recent further work of the author.

2. An example. Consider the following simple (certainly equivalent) adaptive control problem (in which (y, u) is an output, input pair; y^* a desired Signal; a a parameter; w_k a white noise)

Plant: $y_k = a_k y_{k-1} + u_{k-1} + \omega_k$

Control: $u_{k-1} = y_k^* - \hat{a}_{k-1} y_{k-1}$

Tracking error: $e_k = y_k - [\hat{a}_{k-1} y_{k-1} + u_{k-1}] = y_k - y_k^*$

Parameter estimator: $\delta \hat{a}_k = \hat{a}_k - \hat{a}_{k-1} = \frac{\mu y_{k-1}}{1 + \mu y_{k-1}^2} e_k$

Combining these equations leads to the following error system.

$$(2.1) \qquad \delta \tilde{a}_k = \frac{\mu(e_{k-1} - \omega_{k-1} + y_{k-1}^* + \omega_{k-1})}{1 + \mu(e_{k-1} + y_{k-1}^*)^2}(e_k - \omega_k + \omega_k) - \delta a_k$$

$$(2.2) \qquad e_k - \omega_k = -\tilde{a}_{k-1}(e_{k-1} - \omega_{k-1}) - \tilde{a}_{k-1}(y_{k-1}^* + \omega_{k-1})$$

where, $\tilde{a}_k = \tilde{a}_k - a_k$. Now we are interested to analyse the behaviour of this simple adaptive control system. That is, we ask whether $\tilde{a}_k$ remains

---

* Department of Statistics, Macquarie University, Sydney NSW 2109, Australia.

bounded and if so, what is the performance of the system. We might expect
for example that

$$(2.3) \qquad \lim_{N \to \infty} N^{-1} \sum_{1}^{N} e_k^2 = \sigma_w^2 + \text{ learning cost}$$

To do the analysis it seems necessary to say something about how a_k varies
with time, so we consider the following large amplitude slowly time varying
model.

$$(2.4) \qquad \text{parameter model:} \qquad a_k = a_0(\epsilon k), \quad \varepsilon << 1$$

Note that $\delta a_k \cong \epsilon \dot{a}_0(\epsilon k)$ hence the speed of change is slow.

3. Averaging setup. To understand how averaging analysis can be
used to analyse (2.1) and (2.2), it is convenient to get some perspective and
observe that (2.1) and (2.2) have the general form of a mixed time scale
system with

Slow state $\delta z_k = \mu f(k, z_{k-1}, y_{k-1}, \epsilon k, \mu) - \delta z_0(\epsilon k)$

Fast state $y_k = A(\epsilon k, z_{k-1})y_{k-1} + h(k, z_{k-1}, \epsilon k) + \mu g(k, z_{k-1}, y_{k-1}, \epsilon k, \mu)$

If we could show (z_k, y_k) is uniformly bounded in k with fixed μ, ϵ we
could hope this markov system has a steady state measure π so that

$$N^{-1} \sum_{1}^{N} e_k^2 \to E_\pi(e_0^2), \text{ as } N \to \infty$$

and for any continuous function $f(\cdot)$

$$N^{-1} \sum_{1}^{N} f(y_k) \to E_\pi(y_0), \text{ as } N \to \infty$$

So if we can calculate π then we can answer at least a performance question
such as (2.3). Unfortunately even in the simplest cases calculating π seems
hopeless. The best we can do is to develop a perturbation expansion. This
nontrivial calculation might work as follows. Setting $\mu = \epsilon$ (so that we
assume we know the speed of change in (2.4)) let us denote z_k by z_k^μ to
remind ourselves of the μ dependence. Suppose we can show:

(i) z_k^μ has steady state z_∞^μ (i.e. z_∞^μ has distribution π) uniformly in μ.
(ii) z_k^μ converges weakly as $\mu \to \infty$ to an averaged process z_∞^0
(iii) z_k^0 has a steady state z_∞^0.
(iv) z_∞^μ converges weakly as $\mu \to 0$ to the same z_∞^0 as in (ii).

Then z_∞^0 provides a first order approximation to the steady state random variable z_∞^μ but importantly we need only find the averaged process
z_k^0 in order to get z_∞^0.

A simple (non adaptive) example may clarify these ideas. Consider the
slow autoregression

$$\xi_k^\mu = (1 - \mu)\xi_{k-1}^\mu + [W(\mu k) - W(\mu(k-1))]$$

where $W(t)$ is a Brownian notion of variance $\sigma_\nu^2 t$. Then ξ_k^μ has a steady state variance

$$\text{var}\,(\xi_\infty^\mu) = \frac{\sigma_\nu^2}{2-\mu} \to \frac{\sigma_\nu^2}{2} = \text{var}\,(\xi_\infty^0), \text{ as } \mu \to 0$$

Also let $\xi(t)$ be the continuous autoregression

$$d\xi = -\xi dt + dW$$

then $\xi(t)$ has steady state variance

$$\text{var}\,(\xi_\infty) = \sigma_\nu^2/2$$

One can show that $\xi_{\mu k}^\mu - \xi(\mu k) \Rightarrow 0$ (where $\Rightarrow$ means converges weakly).

Referring ahead to section 4 we may say that (i), (iii) require a stability analysis while (ii), (iv) require a hovering theorem. For more details see [17].

4. Averaging review. Here we give a quick review of salient features of averaging. To keep the discussion simple we work with a single time scale system

$$(4.1) \qquad \delta z_k = \mu f(k, z_{k-1})$$

We start with some simple heuristics which show how to find the averaged system. Sum (5.1) to find

$$z_{N+m} = z_m + \mu \sum_{m+1}^{m+N} f(s, z_{s-1})$$

If μ is small, z_s changes little over a short time window so

$$z_{N+m} \simeq z_m + \mu \sum_{m+1}^{m+N} f(s, z_m).$$

If we now introduce the averaged function

$$(4.2) \qquad f_{av}(z) = \lim_{N\to\infty} \sum_{m+1}^{m+N} f(s, z)N^{-1}$$

where the limit exists independently of m then if N is not too large we have

$$z_{N+m} \simeq z_m + \mu N f_{av}(z_m)$$

Now reversing our argument we have

$$z_{N+m} \simeq z_m + \mu \sum_{m+1}^{m+N} f_{av}(z_{s-1})$$

and this leads us to introduce the averaged system

$$(4.3) \qquad\qquad \delta \bar{z}_k = \mu f_{av}(\bar{z}_{k-1})$$

The question now is how to link (4.3) rigorously to (4.1). In doing this it does not seem to be properly understood that there are 3 types of averaging result. We discuss them in deterministic form but stochastic versions are available in [17].

(i) Finite Time Averaging = 'Trajectory Locking'
    This type of result says, under certain technical conditions

$$(4.4) \qquad\qquad \sup_{1 \le k \le T\mu^{-1}} \|z_k - \bar{z}_k\| \le C_T(\mu), \quad 0 \le \mu \le \mu_T$$

$$(4.5) \qquad\qquad C_T(\mu) \to 0, \quad \text{as } \mu \to 0$$

That is, given $T > 0$, there is a function $C_T(\mu)$ which for fixed μ in $[0, \mu_T]$ obeys (4.4) and has the property (4.5). A simple proof of such a result is given below. So as μ gets smaller the trajectories of $\bar{z}_k$ lock on to those of z_k. Of course (4.4) gives no information about stability.
    To be very clear here note finally that (4.4) is valid for μ fixed.

(ii) Infinite Time Averaging - Hovering Theorem
    This result extends (4.4) to an infinite time interval. Under certain regularity conditions the chief of which is
        A_∞: The averaged system (4.3) is exponentially stable
        (ES) to the equilibrium point (EP) z_e of (4.3). But z_e
        need not be an EP of (4.1).
Then there exists $c(\mu)$ such that

$$\sup_{k \ge 1} \|z_k - \bar{z}_k\| \le c(\mu), 0 \le \mu \le \mu_0, c(\mu) \to 0, \text{ as } \mu \to 0$$

Since (4.1) need not have any EP's this result says that the primary system hovers forever near an EP of (4.3) if μ is small enough. This is not a stability result (it does not for instance imply Lyapunov stability) but is nevertheless very useful.

(iii) Infinite Time Averaging - Stability
    This result delivers stability. Under certain regularity conditions; for fixed μ, if the averaged system (4.3) is ES to z_e and z_e is an EP of (4.1) and (4.1) starts near z_c (the nearness does not depend on μ) then z_k is ES to z_e.
    Proofs of results (ii), (iii) can be found in [17] in various settings. The so-called ode method [9], [10] can be deduced from (ii). With (4.3) we associate an ode

$$d\bar{z}(\tau)/d\tau = f_{av}(\bar{z}(\tau))$$

Then under certain regularity conditions

$$\sup_{0 \leq \tau \leq T} \|\bar{z}_{[\tau\mu^{-1}]} - \bar{z}(\tau)\| \leq c_T(\mu), 0 \leq \mu \leq \mu'_T, c_T(\mu) \rightarrow 0 \text{ as } \mu \rightarrow 0$$

and together with (4.5) this yields

$$\sup_{0 \leq \tau \leq T} \|z_{[\tau\mu^{-1}]} - \bar{z}(\tau)\| \leq c_T(\mu), \quad 0 \leq \mu \leq \min(\mu_T, \mu'_T)$$

Similarly weak convergence results [9] can be linked in, although are not as satisfactory because they do not deliver results that are valid for fixed μ rather only for $\mu \rightarrow 0$.

It may be remarked that although averaging has a long history in Applied Mathematics (as a special kind of perturbation analysis) stretching back into the nineteenth century a rigorous foundation was supplied only in the 1930's chiefly by Bogoliubov [1], [4].

For a method which is so widespread and important it is surprising that there is no definitive discussion in textbook form. Instead there seems to be a plethora of methods and a lot of confusion in the various literatures.

In fact, there are only 3 proof techniques that this author is aware of. All others seem to be versions (sometimes more complicated!) of these. Bogoliubov's original method delivers the 3 types of averaging result mentioned above but is very cumbersome and does not seem to extend to stochastic problems. There is another method due to Gikhman [6], [7] which even now is not completely developed (although some use of it is made in [9]) yet it is not only much simpler than the Bogoliubov method but also yields results under weaker regularity conditions. The author has used the method recently to analyse an adaptive algorithm in the presence of time varying parameters [16]. Finally there is the method of Papanicoloau [3] which has been pursued and extended (outside the markov setting) by [9] and [17]. It seems to work nearly as well as Gikhman's method but deals more easily with stability and is much better developed for doing second order averaging. Further discussion and references are available in [17]. It may be remarked that the 'Poisson equation' method developed in [2] in a Markov setting is essentially a discrete time version of [3] although strangely [3] is not referenced in [2]: in a similar vein [9] does not reference [7]. Perhaps mention should be made of [12] which has some interesting history and a good intuitive development of some deterministic averaging ideas. Unfortunately many proofs are faulty and technical conditions missing.

To sum up, it cannot be said that there is anything like a definitive set of theorems ready for application (especially in the stochastic case).

5. A simple proof of first order averaging. To show the clever idea behind averaging we give a proof of finite time averaging in a simple case. Subtract (4.3) from (4.1) and sum to find

$$\Delta_k = z_k - \bar{z}_k$$
$$= \Delta_0 + \mu \sum_1^k [f(s, z_{s-1}) - f(s, \bar{z}_{s-1})] + \mu J_k$$
$$J_k = \sum_1^k [f(s, \bar{z}_{s-1}) - f_{av}(\bar{z}_{s-1})]$$

let us assume $f(s, z)$ obeys a Lipschitz condition with constant L so

$$\|\Delta_k\| \le \|\Delta_0\| + \mu L \sum_1^k \|\Delta_{s-1}\| + \mu \|J_k\|$$

Now apply the discrete Bellman - Gronwall lemma to get

$$\|\Delta_k\| \le (\|\Delta_0\| + \mu \max_{1 \le k \le T/\mu} \|J_k\|) e^{\mu k L}$$

where T is fixed. Thus

$$\max_{1 \le k \le T/\mu} \|\Delta_k\| \le (\|\Delta_0\| + \mu \max_{1 \le k \le T/\mu} \|J_k\|) e^{TL}$$

We need a bound on J_k to get RHS $\to 0$ as $\mu \to 0$. If we bound J_k in a crude way we cannot achieve this and it is here that the subtle part of the averaging method enters. Introduce the frozen perturbation

$$p(k, z) = \sum_1^k (f(s, z) - f_{av}(z)), \quad p(k, 0) = 0$$

And assume a Lipschitz condition

(5.1) $$\|p(k, z) - p(k, z')\| \le L\|z - z'\|$$

Below, we indicate when this might reasonably be satisfied. Then the idea is to approximate J_k by $P_k = p(k, \bar{z}_k)$. Consider that

$$P_k - P_{k-1} = p(k, \bar{z}_k) - p(k-1, \bar{z}_{k-1})$$
$$= p(k, \bar{z}_k) - p(k, \bar{z}_{k-1}) + [p(k, \bar{z}_{k-1}) - p(k-1, \bar{z}_{k-1})]$$
$$= \eta_k + [f(k, \bar{z}_{k-1}) - f_{av}(\bar{z}_{k-1})]$$
$$\Rightarrow \quad P_k = \sum_1^k \eta_s + J_k$$

but from the assumption (5.1)

$$\|\eta_s\| \le L\|\bar{z}_s - \bar{z}_{s-1}\|$$
$$= L\mu \|f_{av}(\bar{z}_{s-1})\|$$
$$\le L\mu h$$

where we suppose for $1 \leq k \leq T/\mu$

(5.2) $$\|f_{av}(\bar{z}_k)\| \leq h, \|p(k,0)\| \leq h, \|\bar{z}_k\| \leq h$$

(this can be justified). Thus

$$
\begin{aligned}
\|J_k\| &\leq \|P_k\| + Lk\mu h \\
&\leq L\|\bar{z}_k\| + h + Lhk\mu \\
&\leq (L+1)h + Lhk\mu \\
\to \mu \max_{1 \leq k \leq T/\mu} \|J_k\| &\leq \mu((L+1)h + LhT)
\end{aligned}
$$

So we get

$$\max_{1 \leq k \leq T/\mu} \|\Delta_k\| \leq (\|\Delta_0\| + \mu a)e^{TL}$$

and if we start the averaged system at the same point as the primary system (4.1) then we get the required finite time averaging bound.

Further examples and extensions of these theorems in both deterministic and stochastic settings and under weaker regularity conditions can be found in [17]. Let us briefly remark that if $f(k,z)$ is periodic then (5.1) is easy to justify, but can hold also when periodicity fails. For instance in the LMS algorithm $f(k,z) = -x_k x_k^T z$ where x_k is an exogenous signal and if x_k is periodic with $R_x = $ average of $x_k x_k^T$ over a cycle then (5.1) holds because for instance $\sum_1^k \cos(k\theta) = 0(1)$.

6. Brief example. If we carry out a first order averaging analysis of our system (2.1), (2.2) it is somewhat more complicated than the single time scale system (see [17]) but the heuristic development leads to an averaged system (when $\mu = \epsilon$)

$$\delta \bar{a}_k = -\mu R(\bar{a}_{k-1})\bar{a}_{k-1} - \delta a_k$$

where

$$R(\bar{a}) = f_{-\pi}^{\pi} \frac{F_{y^*}(\omega) + \sigma_w^2}{|1 + \bar{a}e^{-jw}|^2} \frac{d\omega}{2\pi}$$

and the frozen state is

$$e_k - w_k = \frac{-\bar{a}(y_{k-1}^* + w_{k-1})}{1 + q^{-1}\bar{a}}$$

and $F_{y^*}(\omega)$ is the spectrum of y_k^* and σ_w^2 is the variance of w_k. We see that $\bar{a}_k$ tracks a_k as a first order system. A second order averaging analysis of the system in section 2 has been sketched in [14] where an expression for the learning cost in (2.3) is obtained.

7. Summary and extensions. We have pointed out here the utility of averaging methods and mentioned 3 types of first order averaging results. Using these methods the author has recently reproduced results of [8, section 10.14] on the LMS algorithm, in a new way [15]. This is useful because the methods used in [8] seem suited only to linear problems whereas averaging is not so limited. Steady state behaviour can also be obtained by averaging methods (there seems to be no other way in general) and a simple example has been mentioned [14]. A general procedure is given in [17] and a special case was developed (using more cumbersome methods) in [13].

REFERENCES

[1] M. Balachandra, P.R. Sethna, *A Generalization of the Method of Averaging for Systems with Two Time Scales*, Archiv. Rat. Mech. Anal., Vol. 58 (1975).

[2] A. Benveniste, M. Metivier and P. Priouret, *Adaptive Algorithms and Stochastic Approximations*, Springer-Verlag (1990).

[3] G. Glankenship and G.C. Papanicoloau, *Stability and Control of Stochastic Systems with Wide-band Noise Disturbances*, SIAM J. Appl. Math., 34 (1978), pp. 437–476.

[4] N.N. Bogoliubov, Y. Mitropolsky, *Asymptotic Methods in Nonlinear Mechanics*, Gordon and Breach, New York (1961).

[5] B.P. Deveritskii and A. Fradkov, *Two models for analysing the dynamics of adaptive algorithms*, Automatikha i Telemekhanica, 1 (1974), pp. 66–75.

[6] M.I. Freidlin, A.D. Wentzell, *Random Perturbations of Dynamical Systems*, Springer-Verlag (1984).

[7] I.I. Gikhman, *Poporodu Odnoi Teoremy N.I. Bogolybova*, Ukrain. Mat. Zh., 4 (1952), pp. 215–218.

[8] L. Guo and H.F. Chen, *Identification and Stochastic Adaptive Control*, Birkhauser, Boston (1991).

[9] H. Kushner, *Approximation and Weak Convergence Methods for Random Process*, MIT Press (1984).

[10] L. Ljung and T. Soderstrom, *Theory and Practice of Recursive Identification*, MIT Press (1983).

[11] B.D. Riedle, P.V. Kokotovic, *Integral Manifolds of Slow Adaptation*, IEEE Trans. Autom. Contr., Vol. AC-31 (1986), pp. 316–324.

[12] J.A. Sanders and F. Verhulst, *Averaging Methods in nonlinear dynamical systems*, Springer-Verlag, Berlin (1985).

[13] V. Solo, *The error variance of LMS with time-varying weights*, IEEE Trans. Sig. Proc., 40 (1992), pp. 803–813.

[14] V. Solo, *Adaptive Control Performance with time varying parameters*, Proc. 12th World Congress of IFAC, 18-23 July, Sydney, Australia, Vol. 10 (1993), pp. 173–175.

[15] V. Solo, *Averaging Analysis of the LMS algorithm*, in Advances in Control and Dynamic Systems, Academic Press, Vol. 65, C.T. Leondes, editor (1994).

[16] V. Solo, *Deterministic Adaptive Control with slowly time varying parameters: an averaging analysis*, submitted to Int. J. Control (1993).

[17] V. Solo and X. Kong, *Adaptive Signal Processing: Stability and Performance*, Prentice-Hall, to appear (1994).

A MULTILINEAR PARAMETRIZATION APPROACH FOR IDENTIFICATION OF PARTIALLY KNOWN SYSTEMS*

JING SUN†

Abstract. This paper deals with an identification problem which arises in adaptive control for partially known systems. The linear systems under consideration can be represented by a *nonlinear parametric* model which is polynomial in the unknown physical parameters, as opposed to the linear parametric model used in most black-box identification problem. A multilinear parametrization approach is proposed and an identification algorithm based on the multilinear model is developed. The properties of the multilinear identification algorithm are explored and analyzed. Simulation results are also presented to demonstrate the effectiveness of the proposed algorithm.

1. Introduction. The research work described in this paper was motivated by some problems encountered in adaptive control of partially known systems. In general, the classical adaptive control theory utilizes very limited information about the physical plant, normally only the knowledge about the order of the plant and the assumption of minimum phase (if the model reference adaptive control scheme is used) are used in designing the estimation and control algorithm. However, in dealing with real world adaptive control problems, most of the time the physical systems we try to control are partially known in the sense that information other than the order of the plant are made available for the purpose of controller design. Without fully utilizing such *a priori* information, the classical adaptive control theory usually results in an overparametrized controller with slow convergence rate and extra long transient period [1,2,3]. To take full advantages of the available information and achieve faster convergence rate and better transient performance, the black-box approach adopted by the classical adaptive control theory which ignores other knowledge about the plant except its order needs to be reconsidered and revised. The major difficulty is, however, that the mathematical model with all available information embodied is usually no longer linear with respect to unknown parameters [4,5,6]. Therefore, standard algorithms for parameter identification, such as least squares and gradient algorithms, cannot be used in conjunction with the certainty equivalence principle for adaptive control design. To solve the problems in adaptive control of partially known systems, we need to develop efficient identification algorithms for nonlinear parametric models.

In this paper, we present an identification algorithm for a class of partially known systems which have a transfer function representation with coefficients as polynomial functions of a single unknown parameter θ. This class of partially known systems is representative and interesting for several

---

* This research was supported by National Science Foundation under grant ECS-9110984.

† ECE Department, Wayne State University, Detroit, MI 48202.

reasons. First, it arises in many physical problems where a single parameter affects several components. Second, since a general nonlinear function can be approximated by a polynomial function to any given accuracy, the proposed algorithm can be applied even though the original model is not polynomial in θ. Third, the proposed algorithm can be extended to cases where more than one unknown parameter need to be identified.

The paper is organized as follows: We present a parametric model for a class of partially known continuous-time systems and formulate the identification problem in section 2. In section 3, an identification algorithm for a multilinear parametric model is proposed, and the properties of the identification algorithm are also explored. In section 4, we connect the original identification problem formulated in section 2 with the multilinear identification problem, and discuss algorithms to obtain a single parameter estimate from the estimates of the multilinear model. Example and simulation results will be presented and discussed in section 5 before we reach our conclusion in section 6.

2. A polynomial parametric model for partially known systems. We consider a class of partially known continuous-time systems whose transfer function can be represented by a parametric model which is polynomial in the unknown physical parameter θ, i.e.:

$$(2.1) \quad y = G(s)u, \quad G(s) = \frac{g_0(s) + \theta g_1(s) + \theta^2 g_2(s) + \cdots + \theta^r g_r(s)}{f_0(s) + \theta f_1(s) + \theta^2 f_2(s) + \cdots + \theta^r f_r(s)}$$

where s denotes the Laplace transform operator in the transfer function formulation and the differential operator in the time domain representation. It is assumed that g_i, f_i are proper stable rational functions, and the sign of θ is known. Without loss of generality, we assume that $\theta > 0$. In most practical problems, the sign of θ can be determined from the physical content of the parameter, and therefore this assumption does not impose a restrictive condition from practical stand point of view. Our goal is to identify θ using the input and output measurements u and y.

To develop an identification algorithm, we write (2.1) in the parametric model form as:

$$(2.2) \qquad z = \theta\psi_1 + \theta^2\psi_2 + \cdots + \theta^r\psi_r$$

where

$$z \stackrel{\triangle}{=} f_0(s)y - g_0(s)u, \quad \psi_i \stackrel{\triangle}{=} g_i(s)u - f_i(s)y, \quad i = 1, 2, \cdots, r$$

(2.2) is a nonlinear-in-unknown-parameter model for which standard parameter identification techniques cannot be applied. One way to deal with the polynomial in unknown parameter model in the form of (2.2) is using linear overparametrization. That is, we assume $\theta, \theta^2, \cdots, \theta^r$ are independent parameters and write (2.2) as

$$(2.3) \qquad z = K^{\mathsf{T}} \Psi, \quad K = \begin{bmatrix} k_1 \\ k_2 \\ \vdots \\ k_r \end{bmatrix}, \quad \Psi = \begin{bmatrix} \psi_1 \\ \psi_2 \\ \vdots \\ \psi_r \end{bmatrix}$$

where $k_1 = \theta, k_2 = \theta^2, \cdots, k_r = \theta^r$ are new parameters to be identified. On one hand, this leads to a linear parametric model for which many algorithms, such as least squares and gradient, can be used for the purpose of identification. But on the other hand, the *apriori* information which relates the parameters $k_1, k_2, \cdots, k_r$ is essentially wasted.

Another approach for solving the identification problem of (2.2) is discussed in [6,7], where the identification problem is formulated as a nonlinear optimization problem. The solution of the optimization problem leads to a finite dimensional recursive algorithm which is nonlinearly interlaced, and its implementation requires a total number of $2r$ integrators.

In this paper, we present a novel approach which differs from approaches discussed in literature. We first construct a multilinear parametric model which under simple linear constraints is equivalent to the polynomial model (2.2). We show that such a multilinear parametric model is amenable to the gradient technique and thus a simple identification algorithm can be designed for the purpose of parameter estimation. Then we derive an estimate for the original identification problem by incorporating these simple constraints. The advantage of the proposed approach is that the resulting algorithm is very intuitive and simple compared to that used in [6,7].

3. Identification of a multilinear parametric model. In this section, we introduce a multilinear parametric model which is equivalent to the original polynomial parametric model (2.2) under simple linear constraints. This multilinear model allows us to use simple gradient-type algorithms for parameter estimation.

Consider the following fictitious multilinear parametric model:

$$(3.1) \qquad z = \theta_1 \psi_1 + \theta_1 \theta_2 \psi_2 + \cdots + \theta_1 \theta_2 \cdots \theta_r \psi_r = \sum_{i=1}^{r} \left(\prod_{j \leq i} \theta_j \right) \psi_i$$

where z and $\psi_i, i = 1, \cdots, r$ are defined in the same way as in (2.2) and therefore can be obtained from the measurements of u, y. We also assume that $\theta_i > 0, i = 1, 2, \cdots, r$. It is obvious that the parametric model (3.1) is equivalent to (2.2) with the constraints that $\theta_1 = \theta_2 = \cdots = \theta_r$.

Let $\hat{\theta}_1, \hat{\theta}_2, \cdots, \hat{\theta}_r$ be an estimate of $\theta_1, \theta_2, \cdots, \theta_r$ respectively. We define the corresponding estimated observation $\hat{z}$ and the output estimation error (or observation error as being referred to in some literatures) ϵ as:

$$(3.2) \quad \hat{z} \overset{\triangle}{=} \hat{\theta}_1 \psi_1 + \hat{\theta}_1 \hat{\theta}_2 \psi_2 + \hat{\theta}_1 \hat{\theta}_2 \hat{\theta}_3 \psi_3 + \cdots + \hat{\theta}_1 \hat{\theta}_2 \cdots \hat{\theta}_r \psi_r = \sum_{i=1}^{r} \left(\prod_{j \leq i} \hat{\theta}_j \right) \psi_i$$

$$(3.3) \qquad\qquad\qquad\qquad \epsilon \overset{\triangle}{=} \hat{z} - z.$$

Then we have the following lemma which relates the output estimation error with the parameter error:

LEMMA 3.1. *For the multilinear parametric model (3.1), the estimation error ϵ can be expressed in terms of the parameter error $\tilde{\theta}_i \overset{\triangle}{=} \hat{\theta}_i - \theta_i$ as*

$$(3.4) \qquad \epsilon = \tilde{\theta}_1 \phi_1 + \theta_1 \tilde{\theta}_2 \phi_2 + \theta_1 \theta_2 \tilde{\theta}_3 \phi_3 + \cdots + \theta_1 \theta_2 \cdots \theta_{r-1} \tilde{\theta}_r \phi_r$$

where

$$(3.5) \qquad \begin{aligned} \phi_r &\overset{\triangle}{=} \psi_r \\ \phi_i &\overset{\triangle}{=} \psi_i + \hat{\theta}_{i+1} \phi_{i+1}, \quad i = 1, 2, \cdots, r-1. \end{aligned}$$

Proof: Starting from the definition of ϕ_i and $\tilde{\theta}_i$, we can write:

$$(3.6) \qquad \phi_i = \psi_i + \tilde{\theta}_{i+1} \phi_{i+1} + \theta_{i+1} \phi_{i+1}, \quad i = 1, 2, \cdots, r-1.$$

Using the expression of $\hat{z}$, one can verify using simple algebra that

$$(3.7) \qquad\qquad \hat{z} = \hat{\theta}_1 \phi_1 = \tilde{\theta}_1 \phi_1 + \theta_1 \phi_1.$$

Substituting (3.6) consecutively for ϕ_i in the expression of $\hat{z}$ given by (3.7), we have

$$(3.8) \quad \hat{z} = \tilde{\theta}_1 \phi_1 + \theta_1 \tilde{\theta}_2 \phi_2 + \theta_1 \theta_2 \tilde{\theta}_3 \phi_3 + \cdots + \theta_1 \theta_2 \cdots \theta_{r-1} \tilde{\theta}_r \phi_r + \sum_{i=1}^{r} \left(\prod_{j \leq i} \theta_j \right) \psi_i.$$

Therefore, (3.4) follows immediately from the definition of ϵ. $\square$

(3.4) is the key equation that we will use to develop an identification algorithm.

Remark 1: Lemma 1 shows that even though the parametric model (3.1) appears to be nonlinear in the unknown parameters, the error equation (3.4) is virtually linear in terms of the parameter errors. This is due to the special multilinear form of (3.1). As we will show later, this error equation enables us to derive a simple identification algorithm similar to

the the gradient algorithm for the linear parametric model and allows us
to perform ananlysis using Lyapunov arguments. $\triangledown$

The following theorem gives an identification algorithm for the multilinear parametric model (3.1) and its properties:

THEOREM 3.1. *For the multilinear parametric model (3.1), if the following estimation algorithm:*

$$(3.9) \qquad \dot{\hat{\Theta}} = -\gamma\epsilon\Phi, \quad \hat{\Theta} = \begin{bmatrix} \hat{\theta}_1 \\ \hat{\theta}_2 \\ \vdots \\ \hat{\theta}_r \end{bmatrix}, \quad \Phi = \begin{bmatrix} \phi_1 \\ \phi_2 \\ \vdots \\ \phi_r \end{bmatrix}$$

is used, where $\gamma > 0$ is a constant, then:

(i) $\hat{\Theta}$ is uniformly bounded;

(ii) If Ψ is uniformly bounded, then $\dot{\hat{\Theta}}, \epsilon \in \mathcal{L}_2$;

*(iii) If Ψ is uniformly continuous and bounded, then $\epsilon(t), \dot{\hat{\Theta}}(t) \to 0$ as
$t \to \infty$.*

*(iv) Furthermore, if Ψ is persistently exciting, i.e., there exist α_1, α_2,
$T > 0$ such that the inequality:*

$$(3.10) \qquad \alpha_1 I \leq \frac{1}{T}\int_t^{t+T} \Psi(\tau)\Psi^{\mathsf{T}}(\tau)d\tau \leq \alpha_2 I$$

holds for all $t \geq t_0$ for some $t_0 \geq 0$, then Φ is PE and $\hat{\Theta}(t) \to \Theta$ as $t \to \infty$.

Proof: Consider the function:

$$V = \frac{1}{2}\tilde{\Theta}^{\mathsf{T}}\Gamma\tilde{\Theta}$$

where $\tilde{\Theta} \overset{\Delta}{=} \hat{\Theta} - \Theta$, $\Gamma \overset{\Delta}{=} diag\{1, \theta_1, \theta_1\theta_2, \cdots, \theta_1\theta_2\cdots\theta_{r-1}\}$. Since $\Gamma = \Gamma^{\mathsf{T}} > 0$ which is due to our assumption that $\theta_i > 0$, V, as a function of $\tilde{\Theta}$, is positive definite and it qualifies as a Lyapunov candidate. Taking time derivative of V along the trajectory of (3.9), we have

$$(3.11) \qquad \dot{V} = -\tilde{\Theta}^{\mathsf{T}}\Gamma\epsilon\Phi.$$

From the definition of Γ, i.e., Γ is a diagonal matrix, we have

$$\tilde{\Theta}^{\mathsf{T}}\Gamma\Phi = \tilde{\theta}_1\phi_1 + \theta_1\tilde{\theta}_2\phi_2 + \theta_1\theta_2\tilde{\theta}_3\phi_3 + \cdots + \theta_1\theta_2\cdots\theta_{r-1}\tilde{\theta}_r\phi_r = \epsilon$$

thus (3.11) implies that

$$(3.12) \qquad \dot{V} = -\epsilon^2 \leq 0.$$

Therefore, V is a monotonically decreasing function which implies that $V, \tilde{\Theta}$ and $\hat{\Theta}$ are all uniformly bounded regardless of the boundedness of other signals, thus (i) is proved. From (3.13), it follows immediately that

$$\int_0^\infty \epsilon^2(t)dt = V(0) - V(\infty) < \infty$$

which implies that $\epsilon \in \mathcal{L}_2$. If Ψ is uniformly bounded, from the boundedness of $\hat{\Theta}$ and the definition of Φ, we have that Φ is uniformly bounded. Thus, $\epsilon \in \mathcal{L}_2$ implies that $\dot{\hat{\Theta}} \in \mathcal{L}_2$.

If Ψ is uniformly continuous, we can conclude that Φ is also uniformly continuous and therefore it follows from Barbalas's lemma [8] that $\epsilon, \dot{\hat{\Theta}} \in \mathcal{L}_2$ implies $\epsilon(t), \dot{\hat{\Theta}}(t) \to 0$ as $t \to \infty$.

To prove the first part of (iv), i.e., to show that (3.10) implies the existence of α_1', α_2', T' such that the inequality

$$(3.13) \qquad \alpha_1' I \leq \frac{1}{T} \int_t^{t+T'} \Phi(\tau)\Phi^\top(\tau)d\tau \leq \alpha_2' I$$

holds for all $t \geq t_0'$ and some $t_0' > 0$, we write $\Phi(t)$ as:

$$(3.14) \qquad \Phi = F(t)\Psi$$

where

$$F(t) = \begin{bmatrix} 1 & \hat{\theta}_2 & \hat{\theta}_2\hat{\theta}_3 & \cdots & & \hat{\theta}_2\hat{\theta}_3\cdots\hat{\theta}_{r-1} & \hat{\theta}_2\hat{\theta}_3\cdots\hat{\theta}_r \\ 0 & 1 & \hat{\theta}_3 & \cdots & & \hat{\theta}_3\cdots\hat{\theta}_{r-1} & \hat{\theta}_3\cdots\hat{\theta}_r \\ \vdots & & \ddots & & \ddots & \vdots & \vdots \\ 0 & 0 & & \cdots & 1 & \hat{\theta}_{r-1} & \hat{\theta}_{r-1}\hat{\theta}_r \\ 0 & 0 & & \cdots & 0 & 1 & \hat{\theta}_r \\ 0 & 0 & & \cdots & 0 & 0 & 1 \end{bmatrix}.$$

It is clear from the properties (i)-(iii) that $F(t)$ is uniformly bounded, nonsingular and $\dot{F}(t) \to 0$ as $t \to \infty$. Therefore, the existence of an upper bound in (3.13) can be easily established. To establish a lower bound, we consider $\eta^\top \int_t^{t+T} \Phi(\tau)\Phi^\top(\tau)d\tau\eta$ where $\eta \in \mathcal{R}^n$ is an arbitrary constant vector, then:

$$\eta^\top \int_t^{t+T} \Phi(\tau)\Phi^\top(\tau)d\tau\eta = \eta^\top \int_t^{t+T} F(\tau)\Psi(\tau)\Psi^\top(\tau)F^\top(\tau)d\tau\eta$$

$$= \eta^\top F(t) \int_t^{t+T} \Psi(\tau)\Psi^\top(\tau)d\tau F^\top(t)\eta$$

$$(3.15) \qquad + \eta^\top \int_t^{t+T} (F(\tau) - F(t))\Psi(\tau)\Psi^\top(\tau)(F^\top(\tau) + F^\top(t))d\tau\eta$$

$$= \eta^\mathsf{T} F(t) \int_t^{t+T} \Psi(\tau)\Psi^\mathsf{T}(\tau)d\tau F^\mathsf{T}(t)\eta$$

$$+ \eta^\mathsf{T} \int_t^{t+T} \left(\int_t^\tau \dot{F}(\sigma)d\sigma \right) \Psi(\tau)\Psi^\mathsf{T}(\tau)(F^\mathsf{T}(\tau) + F^\mathsf{T}(t))d\tau\eta.$$

Since $\Psi(t)$, $F(t)$ are uniformly bounded, and Ψ satisfies (3.10), we have the inequality:

$$\eta^\mathsf{T} \int_t^{t+T} \Phi(\tau)\Phi^\mathsf{T}(\tau)d\tau\eta \geq T\alpha_1\eta^\mathsf{T} F(t)F^\mathsf{T}(t)\eta - f_0 \max_{\sigma\in[t,t+T]} \|\dot{F}(\sigma)\| \int_t^{t+T}(\tau-t)d\tau\eta^\mathsf{T}\eta$$

$$(3.16) \qquad = T\alpha_1\eta^\mathsf{T} F(t)F^\mathsf{T}(t)\eta - \frac{1}{2}f_0 T^2 \max_{\sigma\in[t,t+T]} \|\dot{F}(\sigma)\|\eta^\mathsf{T}\eta$$

where $f_0 = \sup_{\tau,t} \|\Psi(\tau)\Psi^\mathsf{T}(\tau)(F^\mathsf{T}(\tau)+F^\mathsf{T}(t))\|$ and $\|M\|$ for a $r \times r$ matrix M denotes the induced Euclidean norm of M in the space of $\mathcal{R}^r$. Since $F(t)$ is nonsingular, $F(t)F^\mathsf{T}(t)$ is a positive definite matrix and there exists a constant $\alpha > 0$ such that $F(t)F^\mathsf{T}(t) \geq \alpha I$. Thus we can prove that the inequality:

$$\frac{1}{T}\eta^\mathsf{T} \int_t^{t+T} \Phi(\tau)\Phi^\mathsf{T}(\tau)d\tau\eta \geq \left(\alpha_1\alpha - \frac{f_0 T \max_{\sigma\in[t,t+T]} \|\dot{F}(\sigma)\|}{2} \right) \eta^\mathsf{T}\eta$$

(3.17)
holds for any $t \geq t_0$. Note that since $\dot{F}(t) \to 0$ as $t \to \infty$, there exists a $t_1 > 0$ such that for any $t \geq t_1$, we have $\|\dot{F}(t)\| \leq \frac{f_0 T}{\alpha_1\alpha}$. Therefore, for any $t \geq t_0' \triangleq max\{t_0, t_1\}$,

$$(3.18) \qquad \frac{1}{T}\eta^\mathsf{T} \int_t^{t+T} \Phi(\tau)\Phi^\mathsf{T}(\tau)d\tau\eta \geq \frac{\alpha_1\alpha}{2} . \eta^\mathsf{T}\eta$$

Note that (3.18) holds for any $\eta \in \mathcal{R}^n$ which implies

$$\frac{1}{T} \int_t^{t+T} \Phi(\tau)\Phi^\mathsf{T}(\tau)d\tau \geq \frac{\alpha_1\alpha}{2} I$$

and the existence of a lower bound in (3.13) is proved with $\alpha_1' = \frac{\alpha_1\alpha}{2}, T' = T$, hence Φ being PE is established.

Following the standard procedure of convergence proof in adaptive control (see [9] for example), we can show that Φ being PE implies that $\hat{\Theta}(t) \to \Theta$ as $t \to \infty$. $\qquad\square$

Theorem 1 shows that the identification algorithm (3.9), which is very similar to the gradient algorithm used in standard identification and adaptive control schemes for a linear model, have the same properties for a multilinear parametric model as that of the gradient algorithm for the linear parametric model. If the signal Ψ is persistently exciting, a property which can be guaranteed by choosing u to be sufficiently rich as we have done in the standard adaptive control design [9,10], then parameter convergence can be assured and the convergence rate can be designed arbitrarily by properly selecting the adaptive gain γ.

Remark 2: Some of the conditions in Theorem 1 are sufficient rather than necessary. For example, the uniform continuity condition imposed on Ψ for establishing convergence of $\epsilon, \dot{\theta}$ is because of the Barbalas lemma that we have used in the proof. It can be relaxed to Ψ being pointwise continuous if other more sophisticated arguments are used in establishing the convergence properties from the $\mathcal{L}_2$ properties.

Remark 3: Here we assume that all θ_i's have the same positive sign. If the signs of θ_i's are known but otherwise arbitrary, the adaptive law (3.9) can be modified to become:

$$(3.19) \qquad \dot{\Theta} = -\gamma S \epsilon \Phi$$

where $S = diag\{1, sign(\theta_1), sign(\theta_1\theta_2), \cdots, sign(\theta_1\theta_2\cdots\theta_{r-1})\}$ is a diagonal $r \times r$ matrix whose diagonal elements is either 1 or -1. The same properties can be established for (3.19) by performing the similar analysis, with Γ being replaced by $\Gamma' = diag\{1, |\theta_1|, |\theta_1\theta_2|, \cdots, |\theta_1\theta_2\cdots\theta_{r-1}|\}$. $\bigtriangledown$

4. Identification of the polynomial parametric model. In this section, we make use of the identification algorithm developed in the previous section for the multilinear parametric model to solve our identification problem discussed in section 2. It is obvious that the polynomial parametric model

$$(4.1) \qquad z = \theta\psi_1 + \theta^2\psi_2 + \cdots + \theta^r\psi_r$$

is equivalent to the multilinear parametric model:

$$(4.2) \qquad z = \theta_1\psi_1 + \theta_1\theta_2\psi_2 + \cdots + \theta_1\theta_2\cdots\theta_r\psi_r$$

under the constraints $\theta_1 = \theta_2 = \cdots = \theta_r$. Therefore, (4.2) is an alternative way of overparametrization for the polynomial parametric model (4.1). Thus, an estimate $\hat{\theta}$ of θ for the model (4.1) can be derived as:

$$(4.3) \qquad \dot{\Theta} = -\gamma\epsilon\Phi, \quad \hat{\theta} = \frac{\sum_{i=1}^r \hat{\theta}_i}{r}$$

where $\hat{\theta}$ minimizes the mean square error $f(\theta_e) \triangleq \sum_{i=1}^r (\theta_e - \hat{\theta}_i)^2$.

In general, the estimates $\hat{\theta}_i$ obtained from the multilinear estimation scheme will be different for different i, and the constraints $\theta_1 = \theta_2 = \cdots = \theta_r$ are not satisfied at each time t. The estimate $\hat{\theta}$ of θ obtained in (4.3) by averaging $\hat{\theta}_i$ is optimal in the sense that $\sum_{i=1}^r (\theta_e - \hat{\theta}_i)^2$ is minimized at

$\theta_e = \hat{\theta}$. However, from the system identification point of view, ultimately we are interested in obtaining an estimate $\hat{\theta}$ which matches the polynomial parametric model (4.1). To this end, we define an alternative observation error:

$$(4.4) \qquad e_\theta \triangleq \hat{z}_\theta - z, \quad \hat{z}_\theta \triangleq \hat{\theta}\psi_1 + \hat{\theta}^2\psi_2 + \cdots + \hat{\theta}^r\psi_r$$

which provides a measurement of the closeness of $\hat{\theta}$ to the actual unknown parameter θ for any given input-output data $\{\psi_1, \cdots, \psi_r, z\}$. Using e_θ as an error signal to drive the estimation scheme, we can derive the following modified identification algorithm for the polynomial parametric model:

$$(4.5) \qquad \dot{\hat{\Theta}} = -\gamma|e_\theta|sign(\epsilon)\Phi, \quad \hat{\theta} = \frac{\sum_{i=1}^r \hat{\theta}_i}{r}$$

where Φ, ϵ are defined in the same way as in (3.2), (3.3) and (3.5), $sign(\epsilon)$ is the sign function which is $+1$ for positive ϵ and -1 for negative ϵ. Comparing (4.5) with (4.3), we note that $|e_\theta|sign(\epsilon)$ is used to replace ϵ in order to accommodate the polynomial parametric model and address the non-uniqueness problem which is resulted from over-parametrization. The algorithm (4.5) tends to minimize the estimation error e_θ. Following the similar arguments used in proving Theorem 1, we can show that the following properties hold for (4.5):

(i) $\hat{\theta}$ is uniformly bounded;

(ii) $\int_0^\infty |e_\theta \epsilon|d\tau < \infty$.

(iii) If $\dot{\hat{\Theta}}(t) \to 0$ as $t \to \infty$ and $\|\Psi(t)\| > c$ for some constant c, then $e_\theta(t)sign(\epsilon) \to 0$ as $t \to \infty$.

It is quite straightforward to verify the properties (i)-(iii). For example, one can consider the same Lyapunov-type function $V = \frac{1}{2}\tilde{\Theta}^T\Gamma\tilde{\Theta}$. Taking derivative of V along the trajectories of (4.5), we have:

$$\dot{V} = -|e_\theta|\tilde{\Theta}^T\Gamma\Phi sign(\epsilon) = -|e_\theta\epsilon|$$

which implies that V is monotonically decreasing and therefore $V, \hat{\Theta}, \hat{\theta}$ are uniformly bounded.

Remark 4: For each $\hat{\theta}_i$ obtained from the multilinear estimation scheme, one can define the observation error:

$$e_i \triangleq \hat{z}_{\theta_i} - z, \quad \hat{z}_{\theta_i} \triangleq \hat{\theta}_i\psi_i + \hat{\theta}_i^2\psi_2 + \cdots + \hat{\theta}_i^r\psi_r$$

to reflect the parameter error $\tilde{\theta}_i \triangleq \hat{\theta}_i - \theta$. There are several mechanisms with which we can make use of this additional information to improve the performance of the identification for the polynomial parametric model (4.1). One way to utilize these signals is to replace e_θ in (4.5) by $\bar{e}_\theta = \sqrt{\sum_{i=1}^r e_i^2}$. As a consequence, we will have $\int_0^\infty |e_i\epsilon|d\tau < \infty$, and the modified identification algorithm tends to drive e_i to zero for $i = 1, 2, \cdots, r$. Since smaller e_i usually implies smaller parameter error $\tilde{\theta}$ and thus better quality of estimation, another way of incorporating this information is to use e_i to put different weighting factors on $\hat{\theta}_i$ when calculating $\hat{\theta}$ from $\hat{\theta}_i$.

Remark 5: In implementing (4.5), one may face some problems caused by the discontinuity in the function $sign(\epsilon)$. An easy fix is to replace $sign(\epsilon)$ by its approximation. For example, one can use:

$$\dot{\Theta} = -\gamma|e_\theta|\frac{\epsilon}{|\epsilon| + \epsilon_0}\Phi, \quad \hat{\theta} = \frac{\sum_{i=1}^r \hat{\theta}_i}{r}$$

where $f(\epsilon, \epsilon_0) \triangleq \frac{\epsilon}{|\epsilon| + \epsilon_0}$ with $\epsilon_0 > 0$ being a small number is an approximate for the function $sign(\epsilon)$. Various other approximations for the non-smooth discontinuous function $|e_\theta|sign(\epsilon)$ are available and may have different numerical characteristics.

Remark 6: Another effective approach to improve the convergence rate of the multilinear identification scheme is to enforce the constraints $\theta_1 = \theta_2 = \cdots = \theta_r$ by resetting the parameters in the multilinear identification algorithm. That is, we use

$$(4.6) \qquad \dot{\Theta}(t) = -\gamma|e_\theta|f(\epsilon, \epsilon_0)\Phi, \quad \hat{\Theta}(t_r) = \hat{\theta}(t_r^-)E, \quad \hat{\theta} = \frac{\sum_{i=1}^r \hat{\theta}_i}{r}$$

where $E = [1, 1, \cdots 1]^T \in \mathcal{R}^r$ and t_r is the resetting time. The original estimates $\hat{\theta}_i$ are reset so that all estimated parameters in the multilinear model are equal to $\hat{\theta}$ and the constraints are satisfied at $t = t_r$. Several schemes for selecting resetting time are discussed in [11]. In all of our simulation results, even without persistent excitation, we observed parameter converging to the true value with very simple resetting scheme.

5. Example and simulations. As an example, we consider a second order system with repeated poles, i.e.,

$$G(s) = \frac{4}{(s + \theta)^2}$$

where θ is the unknown parameter to be identified. If the blackbox approach is adopted, then two parameters will be estimated and the information that "the poles are repeated" would be wasted. Using the algorithm proposed in this paper, we write

$$G(s) = \frac{\frac{4}{\Lambda(s)}}{\frac{s^2}{\Lambda(s)} + \theta\frac{2s}{\Lambda(s)} + \theta^2\frac{1}{\Lambda(s)}}$$

where $\Lambda(s) = s^2 + \lambda_1 s + \lambda_0$ can be chosen to be any Hurwitz polynomial and the signal Ψ and z can be generated from stable filters as:

$$\Psi = \begin{bmatrix} -\frac{2s}{\Lambda(s)}y \\ -\frac{1}{\Lambda(s)}y \end{bmatrix}, \quad z = \frac{s^2}{\Lambda(s)}y - \frac{4}{\Lambda(s)}u.$$

Using our multilinear identification scheme, two estimated parameter $\hat{\theta}_1, \hat{\theta}_2$ are updated directly and $\hat{\theta}$, which is an estimate for the original unknown parameter, is then computed from $\hat{\theta}_1, \hat{\theta}_2$. The following algorithms are simulated with different inputs:

- Algorithm I:

$$\begin{bmatrix} \dot{\hat{\theta}}_1 \\ \dot{\hat{\theta}}_2 \end{bmatrix} = -\gamma\epsilon \begin{bmatrix} \phi_1 \\ \phi_2 \end{bmatrix}, \quad \hat{\theta} = \frac{\hat{\theta}_1 + \hat{\theta}_2}{2}$$

- Algorithm II:

$$\begin{bmatrix} \dot{\hat{\theta}}_1 \\ \dot{\hat{\theta}}_2 \end{bmatrix} = -\gamma|e_\theta|\frac{\epsilon}{|\epsilon| + \epsilon_0} \begin{bmatrix} \phi_1 \\ \phi_2 \end{bmatrix}, \quad \hat{\theta} = \frac{\hat{\theta}_1 + \hat{\theta}_2}{2}$$

- Algorithm III

$$\begin{bmatrix} \dot{\hat{\theta}}_1 \\ \dot{\hat{\theta}}_2 \end{bmatrix} = -\gamma(|e_1| + |e_2|)\frac{\epsilon}{|\epsilon| + \epsilon_0} \begin{bmatrix} \phi_1 \\ \phi_2 \end{bmatrix}, \quad \hat{\theta} = \frac{\hat{\theta}_1 + \hat{\theta}_2}{2}$$

where $\epsilon = \hat{\theta}_1\psi_1 + \hat{\theta}_1\hat{\theta}_2\psi_2 - z$, $e_\theta = \hat{\theta}\psi_1 + \hat{\theta}^2\psi_2 - z$, $e_1 = \hat{\theta}_1\psi_1 + \hat{\theta}_1^2\psi_2 - z$, $e_2 = \hat{\theta}_2\psi_1 + \hat{\theta}_2^2\psi_2 - z$ are different observation errors. The results of the simulation are shown in Figures 5.1–5.5.

Figures 5.1 and 5.2 give the results of the algorithm I when the input is a unit step function and a sinusoidal function ($u = sin0.5t$) respectively. From Figure 5.1, one can see that parameters do not converge to their true value because of the lack of persistent excitation in the input, while Figure 5.2 shows that the multilinear estimation scheme guarantees convergence for a persistently exciting signal $sin0.5t$. For the same step input, the result in Figure 5.3 shows that algorithm II gives the estimate $\hat{\theta}$ that converges to the true value even though $\hat{\theta}_1, \hat{\theta}_2$ do not converge to their true value. When algorithm III is used, not only $\hat{\theta}$ converges to θ, $\hat{\theta}_1, \hat{\theta}_2$ also converge to θ due to the incorporation of the observation error e_1, e_2.

Figure 5.5 shows the results of algorithm I when it is combined with parameter resetting. The resetting time is selected based on the performance of signal ϵ. When ϵ is sufficiently small ($|\epsilon| \leq 0.001$), the multilinear estimation is reset such that $\hat{\theta}_1 = \hat{\theta}_2$. The effect of resetting is equivalent to restartting the identification process from a point that is closer to the true parameter than our initial condition. In most of our simulation, only a couple of resetting can result in the estimated parameter error converging to zero.

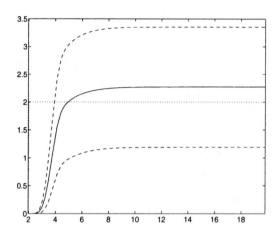

FIG. 5.1. *Results of algorithm I with unit step input:* $\hat{\theta}$ *(solid line),* $\hat{\theta}_1, \hat{\theta}_2$ *(dashed line).*

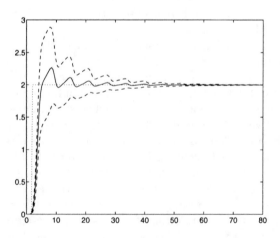

FIG. 5.2. *Results of algorithm I with input* $u = \sin 0.5t$: $\hat{\theta}$ *(solid line),* $\hat{\theta}_1, \hat{\theta}_2$ *(dashed line).*

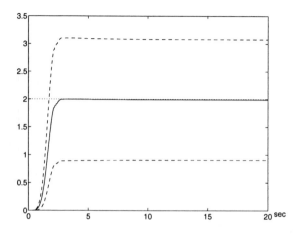

FIG. 5.3. *Results of algorithm II with unit step input:* $\hat{\theta}$ *(solid line),* $\hat{\theta}_1, \hat{\theta}_2$ *(dashed line).*

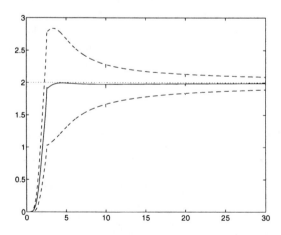

FIG. 5.4. *Results of algorithm III with unit step input:* $\hat{\theta}$ *(solid line),* $\hat{\theta}_1, \hat{\theta}_2$ *(dashed line).*

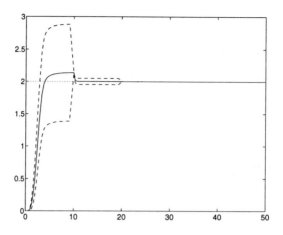

FIG. 5.5. *Results of algorithm I with unit step input and resetting:* $\hat{\theta}$ *(solid line),* $\hat{\theta}_1, \hat{\theta}_2$ *(dashed line).*

6. Conclusion. In this paper, we present a multilinear parameter estimation algorithm which can be used for identification of partially known systems. The algorithm is very simple and intuitive, good performance has been observed in all our simulation results. The results presented here are limited to identification problems. Even though we have tested this algorithm in adaptive control of partially known systems and obtained good performance, further theoretical work needs to be performed to understand its impacts on stability and performance for the closed-loop adaptive control systems. These issues related to using multilinear parametric model for adaptive control are now under investigation.

REFERENCES

[1] Carlos A Canudas de Wit, *Adaptive Control for Partially Known Systems*, Elsevier Science Publishers B. V., 1988.

[2] B. Wittenmark, "Adaptive stability augmentation", *Automatica*, Vol. 26, No. 4, 1990.

[3] E. W. Bai and S. S. Sastry, " Parameter identification using priori information", *Int. J. Control*, Vol. 44, No. 2, 1986.

[4] S. Dasgupta, "Adaptive Identification of Systems with Polynomial Parametrization", *IEEE Transaction on Circuits and Systems*, May 1988.

[5] S. Dasgupta, B. D. O. Anderson and R. J. Kaye, "Output-error identification for partially known systems", *Int. J. Control*, Vol. 43, No. 1, 1986.

[6] P. J. Gawthrop, R. W. Jones and S. A. Macjebzie, "Identification of Partially Known Systems", *Automatica*, Vol. 28, No. 4, 1992.

[7] M. T. Nihtila, "Optimal finite dimentional recursive identification in a polynomial output mapping class", *System & Control Letter*, Vol. 3, December, 1983.

[8] V. M. Popov, *Hyperstability of Control Systems*, Springer-Verlag, 1973, p. 211.

[9] K. S. Narendra and A. M. Annaswamy, *Stable Adaptive Systems*, Prentice-Hall, 1989.

[10] S. Sastry and M. Bodson, *Adaptive Control: Stability, Convergence and Performance*, Prentice-Hall, 1989.

[11] J. Sun, "A Multilinear Identification for Partially Known Systems", to appear in the proceedings of the 32nd CDC, 1993.

ADAPTIVE FILTERING WITH AVERAGING*

G. YIN[†]

Abstract. Adaptive filtering algorithms are considered in this work. The main effort is devoted to improve the performance of such algorithms. Two classes of algorithms are given. The first one uses averaging in the approximation sequence obtained via slowly varying gains, and the second one utilizes averages in both the approximation sequence and the observed signals. Asymptotic properties–convergence and rate of convergence are developed. Analysis to one of the algorithms is presented. It is shown that the averaging approach gives rise to asymptotically optimal performance and results in asymptotically efficient procedures.

Key words. adaptive filtering, averaging, asymptotic optimality.

AMS(MOS) subject classifications. 93E11, 93E25, 60G35, 60F05.

1. Introduction. The purpose of this work is to study two classes of stochastic recursive algorithms, which can be utilized in a wide range of applications in adaptive signal processing and many other related fields. The main effort is placed on improving the asymptotic performance of the algorithms.

The problem under consideration is to recursively update an approximating sequence to the vector $\theta \in \mathbb{R}^r$ that minimizes the estimation error of a random signal, $y \in \mathbb{R}$, from an observation vector $\varphi \in \mathbb{R}^r$. The calculations are done without knowing the statistics of y and φ, on the basis of a sequence of observations $\{(\varphi_n, y_n)\}$. Throughout the paper, we shall assume the sequence $\{(\varphi_n, y_n)\}$ to be stationary and

$$(1.1) \qquad E\varphi_n\varphi_n' = R > 0, \ E\varphi_n y_n = q,$$

where $R > 0$ means that the matrix R is symmetric positive definite. it is easily seen that θ is the unique solution of the Wiener-Hopf equation $R\theta = q$.

A standard algorithm for approximating θ is of the form:

$$(1.2) \qquad \theta_{n+1} = \theta_n + a_n\varphi_n(y_n - \varphi_n'\theta_n),$$

where $\{a_n\}$ is a sequence of positive scalars satisfying $\sum_n a_n = \infty$, $a_n \xrightarrow{n} 0$, and z' denotes the transpose of z.

Many algorithms for adaptive filtering, adaptive array processing, adaptive antenna systems (cf. [1] and the references therein), adaptive equalization (cf. [2]), adaptive noise cancellation (cf. [1]), pattern recognition

---

* This research was supported in part by the National Science Foundation under Grant DMS-9224372, in part by the IMA with fund provided by the National Science Foundation, and in part by Wayne State University.

† Department of Mathematics, Wayne State University, Detroit, MI 48202.

and learning (cf. [3]) etc. have been or can be recast into the same form as (1.2), with only signal, training sequence and/or reference signals varying from applications to applications. An extensive list of references on the applications mentioned above can be found for example in [1], [4] etc. For related problems in adaptive systems, consult [5], [6] among others.

Algorithm (1.2) and its variations have been studied extensively for many years, various results of convergence and rates of convergence have emerged, and numerous successful applications have been reported (cf. [7], [8], [9], [10] and the references therein).

In contrast with these developments, the efficiency issue (asymptotic optimality) is the main focus here. Our primary concern is to design asymptotically efficient and easily implementible algorithms with asymptotically optimal convergence speed so as to improve the performance of the algorithm.

The rest of the work is arranged as follows. Discussions on asymptotic optimality is given next. The precise problem formulation is presented in Section 3. Then Section 4 is devoted to the convergence and asymptotic normality of the algorithm, from which the asymptotic optimality is obtained. A number of further remarks are made in Section 5.

2. Asymptotic optimality. It was shown in the literature that under appropriate conditions, $\theta_n \xrightarrow{n} \theta$ with probability one or weakly, and $(1/\sqrt{a_n})(\theta_n - \theta)$ converges in distribution to a normal random vector with covariance Σ. The scaling factor $\sqrt{a_n}$ together with the covariance Σ is a measure of rate of convergence.

It has been a long time effort to improve the rate of convergence and reduce the variance in the adaptive estimation problems. The investigation of obtaining asymptotic optimality can be traced back to the early 50's. As was noted in [11], this is closely linked to an optimization problem.

To review the development in this direction, we digress a little, and begin with a related problem. Consider the following one dimensional, stochastic approximation algorithm

$$(2.1) \qquad x_{n+1} = x_n + \frac{\Gamma}{n}(f(x_n) + \tilde{\zeta}_n),$$

where $\{\tilde{\zeta}_n\}$ is a sequence of random disturbances, and Γ is a parameter to be specified later. Under appropriate conditions, it can be shown that $x_n \to x^0$ w.p.1 (where x^0 is such that $f(x^0) = 0$) and $\sqrt{n}(x_n - x^0) \sim N(0, \Sigma)$ with the asymptotic variance given by

$$(2.2) \qquad \Sigma = \Sigma(\Gamma) = \frac{\Gamma^2 \Sigma_0}{2\Gamma H + 1},$$

where $H = f_x(x^0) < 0$. Eq. (2.2) reveals the fact that the asymptotic variance depends on the parameter Γ. As a function of Γ, $\Sigma(\Gamma)$ is well behaved. Minimizing Σ w.r.t. Γ leads to the choice of $\Gamma^* = -1/H$ and the optimal variance is given by $\Sigma^* = \Sigma_0/H^2$.

A first glance may make one believe that the problem is completely solved. Nevertheless, H is very unlikely to be known to start with. Therefore, much work has been devoted to the design of efficient algorithms in order to achieve the asymptotic optimality. One of the approaches is the adaptive stochastic approximation method. The essence of such an approach is that in lieu of Γ, a sequence of estimates $\{\Gamma_n\}$ is constructed and (2.1) is replaced by

$$(2.3) \qquad x_{n+1} = x_n + \frac{\Gamma_n}{n}(f(x_n) + \tilde{\zeta}_n).$$

The emphasis is then placed on designing the algorithm such that $\Gamma_n \to -H^{-1}$ and $x_n \to x^0$. Moreover, it is desired to have that $\sqrt{n}(x_n - x^0) \sim N(0, \Sigma^*)$, where $\Sigma^* = H^{-1}\Sigma_0(H^{-1})'$.

The aforementioned approach can be adopted to treat adaptive filtering problems. In this case, the algorithm takes the form

$$\theta_{n+1} = \theta_n + \frac{\Gamma_n}{n}\varphi_n(y_n - \varphi'_n\theta_n).$$

Similar to the argument above, it can be shown that $\Gamma_n \to R^{-1}$, $\theta_n \to \theta$, and $\sqrt{n}(\theta_n - \theta) \sim N(0, \Sigma^*)$, where $\Sigma^* = R^{-1}\Sigma_0 R^{-1}$ and Σ_0 is the covariance of the signals involved. Further discussion on this matter and related problems (with the corresponding approaches in adaptive filtering like algorithms) can be found in [12] and the references therein. While this approach does give us the consistency of $\{\Gamma_n\}$ and $\{x_n\}$ or $\{\theta_n\}$, and the desired optimality, it is computationally intensive for multidimensional problems. If a multidimensional problem is encountered, a sequence of matrix-valued estimates must be constructed, i.e., the estimate of every entry of the gradient matrix or the matrix R must be obtained.

Now, coming back to algorithm (1.2), take $a_n = a/n^\gamma$, for $0 < \gamma \le 1$ and some $a > 0$. A moment of reflection reveals that as far as the scaling factor is concerned, $\gamma = 1$ leads to the best order due to the central limit theorem. In order to implement adaptive filtering procedures, one wishes the iterates move to a neighborhood of the true parameter θ reasonably fast. Rapid decreasing sequence a_n often yields poor results in the initial phase of computation. Therefore, one might wish to choose large step size a_n, i.e., $\gamma < 1$. Nevertheless, larger step size will result in slower rate of convergence. Therefore, there seems to be a dilemma.

Very recently, some new methods were proposed and suggested for stochastic approximation methods in [13], [14] and [15]. In these new developments, arithmetic averaging is used in an essential way. The procedures are multi-step iterative schemes. Two of the notable algorithms are

$$x_{n+1} = x_n + a_n(f(x_n) + \tilde{\zeta}_n)$$

(2.4)
$$\bar{x}_n = \frac{1}{n}\sum_{i=1}^{n} x_i,$$

and

$$x_{n+1} = \bar{x}_n + a_n n\bar{y}_n$$

(2.5)
$$\bar{x}_n = \frac{1}{n}\sum_{i=1}^{n} x_i, \; \bar{y}_n = \frac{1}{n}\sum_{i=1}^{n}(f(x_i) + \tilde{\zeta}_i),$$

where $\{a_n\}$ is a sequence of 'slowly' varying gain (slow with respect to $1/n$). Some amazing things happen. It turns out that for both algorithms, $\{\bar{x}_n\}$ is an asymptotically optimal convergent sequence of estimates.

Algorithm (2.4) was suggested independently in [13] and [14], respectively, whereas (2.5) was initially studied in [15] in the context of application to sequential estimation of LD_{50}, which is a measure of toxicity defined as the dose level that would produce a death rate of 50% in a given population of animals.

In treating algorithm (2.4), independent, identically distributed (i.i.d.) noise and martingale difference type of processes were considered in [13] and [14]. It was shown in [13],

$$E(\bar{x}_n - x^0)(\bar{x}_n - x^0)' = \frac{1}{n}\Sigma^* + o(1/n),$$

whereas asymptotic normality was obtained in [16]. φ-mixing type of noise was dealt with in [17]. Further extensions were provided in [18]. Algorithms with state feedback were proposed in [19]. As for (2.5), some interesting heuristic argument was given in [15]; one dimensional linear problem with i.i.d. random processes was considered in [20], whereas much more general situation was studied in [21].

The use of the averaging approach allows the iterates to get to a vicinity of θ faster, mean while, it keeps the best possible order of rate of convergence and makes the asymptotic covariance to be the optimal one. It produces a "squeezing effect" forcing the iterates get to a vicinity of θ faster without paying the price of increasing the asymptotic covariance matrix or slowing down the convergence speed.

It should be noted that one of the crucial requirements is that the step size a_n is slowly varying with respect to $1/n$. We shall return to this point in Section 5. Motivated and inspired by the approaches mentioned above, two classes of adaptive filtering algorithms will be studied in the sequel.

3. Two classes of algorithms with averaging. In this section, two classes of adaptive filtering type of algorithms with averaging are presented.

Conditions needed in the subsequent study are given. For simplicity, the slowly varying gain is taken to be of the form $a_n = 1/n^\gamma$, $1/2 < \gamma < 1$. Algorithms with more general gain sequences can be treated. For related work in stochastic approximation, we refer to [16], [18] and [19] among others. In 3.1, adaptive algorithm with averaging in the trajectories is given and in 3.2, another algorithm with averaging in both trajectories and observed signals is presented.

3.1. Algorithm I: averaging in the iterates. The following algorithm is inspired by the averaging approach suggested in [13] and [14]. The idea here is to generate a sequence of rough estimates using slowly varying gain first, and then take arithmetic averages of the resulting iterates. Consider the algorithm

(3.1)
$$\theta_{n+1} = \theta_n + \frac{1}{n^\gamma}\varphi_n(y_n - \varphi'_n\theta_n), \ 1/2 < \gamma < 1$$
$$\bar{\theta}_n = \frac{1}{n}\sum_{j=1}^n \theta_j.$$

Notice that the averaging here creates no additional burden since it can be recursively updated as

$$\bar{\theta}_{n+1} = \bar{\theta}_n - \frac{1}{n+1}\bar{\theta}_n + \frac{1}{n+1}\theta_{n+1}.$$

3.2. Algorithm II: averaging in both iterates and observations. Motivated by the work [15] (cf. also [20] and [21]), another class of adaptive filtering algorithm which uses averaging in both trajectories and observations, is suggested in this paper. In addition to the advantages mentioned at the last section, the algorithm with averaging in both iterates and signals appears to be more stable in the initial period, whereas for the algorithms studied in [13], [14], [17] and [18], the averaging normally should be carried out after the iterations have passed the transient period, i.e., in implementing the algorithm given in (3.1), one normally needs to wait for a while until the sequence $\{\theta_n\}$ has 'settled down', then to start the averaging procedure since taking averages in the first a few iterations may result in poor performance and create large errors. Apparently, to improve the initial performance of the algorithm is an important task. This leads us to consider the following algorithm

(3.2)
$$\theta_{n+1} = \bar{\theta}_n + \frac{1}{n^\gamma}\sum_{i=1}^n \varphi_i(y_i - \varphi'_i\theta_i), \ 1/2 < \gamma < 1$$
$$\bar{\theta}_n = \frac{1}{n}\sum_{j=1}^n \theta_j.$$

It appears that this algorithm works better in the initial computation period in that the averaging can be executed from beginning without producing large burst of errors. The reason stems from the fact that it is the averaged signal instead of the signal itself is used in the iteration, i.e., the random processes are smoothed out in this procedure and used in the iteration.

We close this section by making the following remarks regarding to the literature. The two algorithms suggested above fall into the category of multistep algorithms. Early attempts and investigations in this direction can be found in [23], where ideas from numerical analysis for improving approximation to solution of ordinary differential equations were utilized. In addition, the work of [24] and [25] are worth mentioning.

4. Convergence and rates of convergence. This section is concentrated on the asymptotic optimality issues. Algorithm II is analyzed. Section 4.1 states the main conditions and hypotheses; Section 4.2 deals with almost sure convergence and Section 4.3 to Section 4.5 are on asymptotic normality. First, a stability theorem is obtained for θ_n; then some asymptotic equivalency results are established; finally asymptotic distribution is derived.

4.1. Assumptions. The following assumptions will be used throughout.

(A1) $\{\varphi_n, y_n\}$ is a stationary sequence such that (1.1) holds. In addition,

$$(4.1) \qquad E|\varphi_n|^{4+\delta} < \infty, \ E|y_n|^{4+\delta} < \infty \text{ for some } \delta > 0.$$

(A2) $\{\varphi_n\varphi_n' - R\}$ and $\{\varphi_n y_n - q\}$ are moving average sequences of order m, i.e.,

$$(4.2) \qquad \begin{aligned} \varphi_n\varphi_n' - R &= \sum_{i=0}^{m} C_i \kappa_{n-i} \\ \varphi_n y_n - q &= \sum_{i=0}^{m} D_i \nu_{n-i}, \end{aligned}$$

where C_i, D_i, $i \leq m$ are matrices with appropriate dimension, and $\{\kappa_n\}$ and $\{\nu_n\}$ are stationary martingale difference sequences.
Remark: By virtue of (4.1),

$$E|\kappa_n|^{2+\tilde{\delta}} < \infty \text{ and } E|\nu_n|^{2+\tilde{\delta}} < \infty \text{ for some } \tilde{\delta} > 0.$$

Much more general conditions can be incorporated in the problem formulation. We refer to [21] for additional references. Although the assumptions stated here are not the most general one, they do allow us to give a simpler presentation. It seems to be more instructive to present the main idea without going through complicated technical details. Owing to these reasons, we choose these relatively simple conditions.

4.2. Convergence of Algorithm II

THEOREM 4.1. *Suppose that the conditions (A1) and (A2) hold. Then*

$$\sup_n |\theta_n| < \infty \text{ w.p.1, and } \sup_n |\bar{\theta}_n| < \infty \text{ w.p.1;}$$

$$\theta_n \xrightarrow{n} \theta \text{ w.p.1, and } \bar{\theta}_n \xrightarrow{n} \theta \text{ w.p.1.}$$

To obtain the desired convergence property, we make use of the well-known ordinary differential equation methods (cf. [7] and [8]). A comparison technique will be used and an auxiliary sequence for which the convergence is easily established will be constructed. Throughout the rest of the paper, K will denote a generic positive constant. Its values may change for different usage.

Proof. First, notice that by virtue of the local martingale convergence theorem,

(4.3)
$$\sum_i \frac{1}{i^\gamma}(\varphi_i y_i - q) \text{ converges w.p.1 and}$$

$$\sum_i \frac{1}{i^\gamma}(\varphi_i \varphi_i' - R) \text{ converges w.p.1.}$$

Define $\xi_i = \varphi_i y_i - \varphi_i \varphi_i' \theta$ for each i. It follows from (4.3),

(4.4)
$$\sum_i \frac{1}{i^\gamma}\xi_i \text{ converges w.p.1.}$$

Hence by Kronecker's Lemma,

(4.5)
$$\frac{1}{n^\gamma}\sum_{i=1}^n \xi_i \xrightarrow{n} 0 \text{ w.p.1.}$$

Rewrite (3.2) as

(4.6)
$$\theta_{n+1} = \theta_n + \frac{1}{n^\gamma}\varphi_n(y_n - \varphi_n'\theta_n) + \frac{1-\gamma}{(n-1)^\gamma n}\sum_{i=1}^{n-1}\varphi_i(y_i - \varphi_i'\theta_i)$$

$$+\frac{1}{(n-1)^\gamma}\eta_n\sum_{i=1}^n \frac{1}{i}\varphi_i(y_i - \varphi_i'\theta_i), \text{ for } n > 1;$$

$$\theta_2 = \theta_1 + (\varphi_1 y_1 - \varphi_1\varphi_1'\theta_1),$$

where $\eta_n = O\left(\frac{1}{n^2}\right).$

To obtain the desired result, define an auxiliary sequence $\{u_n\}$ as follows.

(4.7)
$$u_{n+1} = u_n + \frac{1}{n^\gamma}\varphi_n(y_n - \varphi_n' u_n), \text{ for } n > 1;$$

$$u_1 = \theta_1, \ u_2 = \theta_2.$$

The sequence $\{u_n\}$ is essentially generated by a standard adaptive filtering algorithm. By virtue of an argument as in [26] Section IV (E), $\sup_n |u_n| < \infty$ w.p.1 and $u_n \xrightarrow{n} \theta$ w.p.1.

To proceed, set $e_n = \theta_n - u_n$. Direct computation yields that

(4.8)
$$
\begin{aligned}
e_{n+1} = e_n &- \frac{1}{n^\gamma}\varphi_n\varphi_n' e_n - \frac{1-\gamma}{(n-1)^\gamma n}\sum_{i=1}^{n-1}\varphi_i\varphi_i' e_i \\
&- \frac{1}{(n-1)^\gamma}\eta_n\sum_{i=1}^{n-1}\varphi_i\varphi_i' e_i \\
&- \frac{1-\gamma}{(n-1)^\gamma n}\sum_{i=1}^{n-1}\varphi_i(y_i - \varphi_i' u_i) \\
&- \frac{1}{(n-1)^\gamma}\eta_n\sum_{i=1}^{n-1}\varphi_i(y_i - \varphi_i' u_i), \quad \text{for } n > 1;
\end{aligned}
$$

$$e_1 = e_2 = 0.$$

Let

$$
\pi_n = -\left(\frac{1-\gamma}{n}\sum_{i=1}^{n-1}\varphi_i(y_i - \varphi_i' u_i) + \eta_n\sum_{i=1}^{n-1}\varphi_i(y_i - \varphi_i' u_i)\right).
$$

In view of the definition of $\{u_n\}$,

$$
u_{n+1} = u_1 + \sum_{i=1}^{n}\frac{1}{i^\gamma}\varphi_i(y_i - \varphi_i' u_i).
$$

Since $\sup_n |u_n| < \infty$ w.p.1,

$$
\sum_{i=1}^{n}\frac{1}{i^\gamma}\varphi_i(y_i - \varphi_i' u_i) \text{ converges w.p.1, and } \frac{1}{n^\gamma}\sum_{i=1}^{n}\varphi_i(y_i - \varphi_i' u_i) \xrightarrow{n} 0 \text{ w.p.1}
$$

by Kronecker's lemma. As a result $\pi_n \xrightarrow{n} 0$ w.p.1.

Let

$$
B_{nk} = \begin{cases} \prod_{i=k+1}^{n}(I - \varphi_i\varphi_i'/i^\gamma), & k < n; \\ I, & k = n. \end{cases}
$$

It follows from (4.8),

$$
\begin{aligned}
e_{n+1} = (\gamma - 1)\sum_{j=2}^{n}\frac{B_{nj}}{(j-1)^\gamma j}\sum_{i=1}^{j-1}\varphi\varphi_i' e_i \\
- \sum_{j=2}^{n-1}\frac{B_{nj}}{(j-1)^\gamma}\eta_j\sum_{i=1}^{j-1}\varphi_i\varphi_i' e_i + \sum_{j=2}^{n-1}\frac{B_{nj}}{(j-1)^\gamma}\pi_j.
\end{aligned}
$$

Consequently, by interchanging the order of summations,

$$(4.9) \qquad |e_{n+1}| \le K \sum_{i=1}^{n} \left(\sum_{j=i}^{n} \frac{|B_{nj}|}{j^{1+\gamma}} \right) |\varphi_i \varphi_i'| |e_i| + K \sum_{i=1}^{n} \frac{|B_{ni}|}{i^\gamma} |\pi_i|.$$

It can be verified that

$$\sum_{j=1}^{n} \frac{|B_{nj}|}{j^\gamma} = O(1) \quad \text{and} \quad \sum_{j=1}^{n} \frac{|B_{nj}|}{j^{1+\gamma}} = O(1/n).$$

Since $\pi_n \xrightarrow{n} 0$ w.p.1,

$$\tilde{\pi}_n = \sum_{i=1}^{n} \frac{|B_{ni}|}{i^\gamma} |\pi_i| \xrightarrow{n} 0 \text{ w.p.1}.$$

Applying the Gronwall's inequality to (4.9), we arrive at

$$(4.10) \qquad |e_{n+1}| \le K \tilde{\pi}_n \exp \left(\frac{1}{n} \sum_{i=1}^{n} |\varphi_i \varphi_i'| \right) \xrightarrow{n} 0 \text{ w.p.1}$$

by the boundedness of $(1/n) \sum_{i=1}^{n} |\varphi_i \varphi_i'|$. Therefore, with probability one, $\{\theta_n\}$ is bounded uniformly in n, $\lim_n \theta_n$ exists and is equal to $\lim_n u_n = \theta$. The convergence of $\{\theta_n\}$ is thus established. Finally, since $\bar{\theta}_n$ is the arithmetic average of θ_i, $i \le n$, it is also bounded w.p.1 and $\bar{\theta}_n \xrightarrow{n} \theta$ w.p.1. $\square$

4.3. A stability result. To carry out the analysis in the sequel, we need to make sure that a scaled sequence of the estimation error $\{\theta_n - \theta\}$ is bounded (tight) in some appropriate sense. As a preparation for further study, first an order of magnitude estimate or a stability result of $\{\theta_n\}$ is proved. In studying dynamical systems, Liapunov functions are very helpful. Let $V(\theta) = (1/2)\theta'\theta$. $V(\cdot)$ is a Liapunov function. A stability result in terms of $V(\cdot)$ will be given below.

PROPOSITION 4.2. *Under conditions of Theorem 4.1,*
- $V(\theta_n - \theta) = O(n^{-\gamma})$ *for n sufficiently large;*
- $\{n^{\gamma/2}(\theta_n - \theta)\}$ *is tight.*

Proof. We derive the order of magnitude estimate first. The argument to follow can be applied to more general correlated signals as well.

Define $\tilde{\theta}_n = \theta_n - \theta$. Owing to Theorem 4.1, (4.6) can be rewritten as

$$(4.11) \qquad \tilde{\theta}_{n+1} = \tilde{\theta}_n - \frac{1}{n^\gamma} R \tilde{\theta}_n + \frac{1}{n^\gamma} (R - \varphi_n \varphi_n') \tilde{\theta}_n + \frac{1}{n^\gamma} \xi_n + \frac{1}{n^\gamma} \zeta_n$$

where $\zeta_n = \hat{\zeta}_n \left(1 + O\left(\frac{1}{n}\right) \right)$ and

$$\hat{\zeta}_n = \frac{1-\gamma}{n} \sum_{i=1}^{n-1} \varphi_i(y_i - \varphi_i'\theta_i) + \eta_n \sum_{i=1}^{n-1} \varphi_i(y_i - \varphi_i'\theta_i),$$

and $\xi_n = \varphi_n y_n - \varphi_n \varphi_n' \theta$. From Theorem 4.1, it can be seen that $\zeta_n \xrightarrow{n} 0$ w.p.1.

To proceed, for some $\Delta > 0$, let

$$p(j, \Delta) = \max \left\{ k; \ \sum_{l=j}^{k} E^{1/2} |\zeta_l|^2 \leq \Delta \right\}.$$

For some $M > 0$ sufficiently large, consider the partition

$$M = \tau_0 < \tau_1 = p(\tau_0, \Delta) < \tau_2 = p(\tau_1, \Delta) < \cdots < \tau_\nu = p(\tau_{\nu-1}, \Delta).$$

It is now clear that

$$\sum_{l=\tau_i}^{\tau_{i+1}} E^{1/2} |\zeta_l|^2 \leq \Delta, \quad \text{for each } i < \nu.$$

For any n satisfying $\tau_i \leq n < \tau_{i+1}$, we have

$$V(\tilde{\theta}_{n+1}) - V(\tilde{\theta}_n) = \tilde{\theta}_n' \left(-\frac{1}{n^\gamma} R \tilde{\theta}_n - \frac{1}{n^\gamma} (\varphi_n \varphi_n' - R) \tilde{\theta}_n + \frac{1}{n^\gamma} \xi_n + \frac{1}{n^\gamma} \zeta_n \right)$$
$$+ \rho_n + O(n^{-2\gamma})(1 + V(\tilde{\theta}_n)),$$

where $E\rho_n = O(n^{-2\gamma})$.

The technique of perturbed Liapunov function method (cf. [27] and the references therein) will be employed. Due to the fact that $(1/n) \sum_{j=1}^{n} \xi_j$ is the 'effective' noise process, direct adoption of the approach in [27] will not work. In the following proof, we first prove the desired result on a subsequence via a perturbed Liapunov function approach. Using the estimate on the subsequence as a bridge, the result then is established for any n large enough.

A number of perturbed Liapunov functions are introduced. These perturbations are small in magnitude and result in desired cancellations.

Define

$$V_1(\theta, n) = \sum_{j=n}^{\tau_{i+1}} E_n \frac{1}{j^\gamma} \theta' \xi_j$$

where E_n denotes conditioning on the data up to n, i.e., conditioning on the σ-algebra $\mathcal{F}_n = \sigma\{(\varphi_j, y_j); \ j \leq n\}$. It can be seen that

(4.12)
$$E|V_1(\theta, n)| = O(n^{-\gamma})(1 + V(\theta)) \quad \text{for each } \theta$$
$$E_n V_1(\tilde{\theta}_{n+1}, n+1) - V_1(\tilde{\theta}_n, n) = \tilde{\rho}_n - \frac{1}{n^\gamma} \tilde{\theta}_n' \xi_n,$$

where $E\tilde{\rho}_n = O(n^{-2\gamma})$.

Define

$$V_2(\theta, n) = \sum_{j=n}^{\tau_{i+1}} \frac{1}{j^\gamma} E_n \theta'(R - \varphi_j \varphi_j')\theta$$

$$V_3(\theta, n) = \sum_{j=n}^{\tau_{i+1}} \frac{1}{j^\gamma} \theta' \zeta_j.$$

Similar as above, it can be shown that

$$(4.13) \quad \begin{aligned} & E|V_2(\theta, n)| = O(n^{-\gamma})(1 + V(\theta)) \text{ for each } \theta \\ & E_n V_2(\tilde{\theta}_{n+1}, n+1) - V_2(\tilde{\theta}_n, n) = \hat{\rho}_n - \frac{1}{n^\gamma} \tilde{\theta}_n'(R - \varphi_n \varphi_n')\tilde{\theta}_n, \end{aligned}$$

where $E\hat{\rho}_n = O(n^{-2\gamma})$;

$$(4.14) \quad \begin{aligned} & E|V_3(\theta, n)| = O(n^{-\gamma})(1 + V(\theta)) \text{ for each } \theta \\ & V_3(\tilde{\theta}_{n+1}, n+1) - V_3(\tilde{\theta}_n, n) = \varpi_n - \frac{1}{n^\gamma} \tilde{\theta}_n' \zeta_n, \end{aligned}$$

where $E\varpi_n = O(n^{-2\gamma})$.

Let

$$\tilde{V}(\theta, n) = V(\theta) + \sum_{j=1}^{3} V_j(\theta, n).$$

Detailed computation leads to

$$E\tilde{V}(\tilde{\theta}_{n+1}, n+1) - E\tilde{V}(\tilde{\theta}_n, n) \le -\frac{\hat{\lambda}}{n^\gamma} EV(\tilde{\theta}_n) + O(n^{-2\gamma})(1 + EV(\tilde{\theta}_n))$$

for some $\hat{\lambda} > 0$. Owing to (4.12), (4.13) and (4.14),

$$E\tilde{V}(\tilde{\theta}_{n+1}, n+1) - E\tilde{V}(\tilde{\theta}_n, n) \le -\frac{\lambda}{n^\gamma} E\tilde{V}(\tilde{\theta}_n, n) + O(n^{-2\gamma})$$

for some $\lambda > 0$ with $\lambda < \hat{\lambda}$. It then follows

$$E\tilde{V}(\tilde{\theta}_{n+1}, n+1) \le \Phi_{n|\tau_i - 1} E\tilde{V}(\tilde{\theta}_{\tau_i}, \tau_i) + K \sum_{j=\tau_i}^{n} \Phi_{n|j} \frac{1}{j^{2\gamma}},$$

where

$$\Phi_{n|k} = \begin{cases} \prod_{j=k+1}^{n}(1 - \lambda/j^\gamma), & k < n; \\ 1, & k = n. \end{cases}$$

By virtue of a summation by parts,

$$\sum_{j=1}^{n} \Phi_{n|j} \frac{1}{j^{2\gamma}} = O(n^{-\gamma}).$$

Thus,

$$E\tilde{V}(\tilde{\theta}_{n+1}, n+1) \leq \Phi_{n|\tau_i-1} E\tilde{V}(\tilde{\theta}_{\tau_i}, \tau_i) + K/n^\gamma.$$

Owing to the estimates in (4.12), (4.13) and (4.14), we also have

$$EV(\tilde{\theta}_{n+1}) \leq \Phi_{n|\tau_i-1} EV(\tilde{\theta}_{\tau_i}) + K/n^\gamma.$$

To proceed, we derive an upper bound when the subsequence $\{\tau_i\}$ is used as the iteration number. Let

$$\alpha_n = (n-1)^\gamma EV(\tilde{\theta}_n).$$

It then follows

$$\alpha_{\tau_{i+1}} \leq \left(\frac{\tau_{i+1}-1}{\tau_i-1}\right)^\gamma \Phi_{\tau_{i+1}-1|\tau_i-1}\alpha_{\tau_i} + K.$$

Notice that

$$0 < \left(\frac{\tau_{i+1}-1}{\tau_i-1}\right)^\gamma \Phi_{\tau_{i+1}-1|\tau_i-1} \leq c < 1,$$

for some $c \in \mathbb{R}$. Solving the difference inequality with c given above yields

$$\alpha_{\tau_{i+1}} \leq c^{i+1}\alpha_{\tau_0} + K\frac{1-c^{i+1}}{1-c}.$$

As a result, $\sup_i \alpha_{\tau_i} \leq K < \infty$. It in turn implies that $EV(\tilde{\theta}_{\tau_i}) = O(\tau_i^{-\gamma})$.

Finally, passing the result from the subsequence $\{\tau_i\}$ to $\{n\}$ leads to that for any $T < \infty$, and any n satisfying $M \leq n < T$, there must exist an integer j, such that $\tau_j \leq n < \tau_{j+1}$. Similar estimates as above yield that

$$\alpha_{n+1} \leq \sup_j \alpha_{\tau_j} + K, \quad \text{and} \quad \sup_n \alpha_n < \infty.$$

To prove the second statement, notice that the Liapunov function is quadratic. For any $\varepsilon > 0$, choose $K_\varepsilon = [1/\varepsilon]$, where $[1/\varepsilon]$ denotes the largest integral part of $1/\varepsilon$. By virtue of the Markov inequality and the first part of the proposition,

$$P\left(\frac{V(\theta_n-\theta)}{n^\gamma} \geq K_\varepsilon\right) \leq \frac{EV(\theta_n-\theta)}{K_\varepsilon n^\gamma} \leq \frac{K}{K_\varepsilon} \leq K\varepsilon.$$

The tightness thus follows, and the proof of the proposition is completed.
□

4.4. Asymptotic equivalency. Noticing that the desired asymptotic properties is on the sequence $\{\bar{\theta}_n\}$, first rewrite Algorithm II in an appropriate form. Since

$$\theta_{n+1} = (n+1)(\bar{\theta}_{n+1} - \bar{\theta}_n) + \bar{\theta}_n,$$

(3.2) yields that

$$(4.15) \qquad \bar{\theta}_{n+1} = \bar{\theta}_n + \frac{1}{n^\gamma(n+1)} \sum_{i=1}^n \varphi_i(y_i - \varphi_i'\theta_i), \ 1/2 < \gamma < 1.$$

Using the definition of ξ_n and putting $\hat{\theta}_n = \bar{\theta}_n - \theta$, (4.15) can be further written as

$$(4.16) \qquad \begin{aligned} \hat{\theta}_{n+1} &= \hat{\theta}_n - \frac{R}{n^\gamma}\hat{\theta}_n + \frac{R}{n^\gamma(n+1)}\hat{\theta}_n \\ &+ \frac{1}{n^\gamma(n+1)} \sum_{i=1}^n (R - \varphi_i\varphi_i')\tilde{\theta}_i + \frac{1}{n^\gamma(n+1)} \sum_{i=1}^n \xi_i. \end{aligned}$$

Define

$$A_{nk} = \begin{cases} \prod_{i=k+1}^n (I - R/i^\gamma), & k < n; \\ I, & k = n. \end{cases}$$

Solution to (4.16) gives us

$$(4.17) \qquad \begin{aligned} \sqrt{n}\hat{\theta}_{n+1} &= \sqrt{n}A_{n0}\hat{\theta}_1 + \sqrt{n} \sum_{k=1}^n \frac{1}{k^\gamma(k+1)} A_{nk} R\hat{\theta}_k \\ &+ \sqrt{n} \sum_{k=1}^n \frac{1}{k^\gamma(k+1)} A_{nk} \sum_{i=1}^k (R - \varphi_i\varphi_i')\tilde{\theta}_i \\ &+ \sqrt{n} \sum_{k=1}^n \frac{1}{k^\gamma(k+1)} A_{nk} \sum_{i=1}^k \xi_i. \end{aligned}$$

Our effort in this subsection is devoted to proving that the first three terms on the right-hand side of the equality above are asymptotically unimportant, and the last term is asymptotically equivalent to $(R^{-1}/\sqrt{n}) \sum_{i=1}^n \xi_i$. More precise statements is provided below.

PROPOSITION 4.3. *Under the conditions of Proposition 4.2,*

$$\sqrt{n}\hat{\theta}_{n+1} = \frac{R^{-1}}{\sqrt{n}} \sum_{i=1}^n \xi_i + o(1),$$

where $o(1) \xrightarrow{n} 0$ *in probability.*

Proof. In what follows, we examine each of the terms in (4.17) separately.

First, since $\sup_n |\hat{\theta}_n| < \infty$ w.p.1, for some $\mu > 0$,

$$(4.18) \quad |\sqrt{n} A_{n0} \hat{\theta}_1| \leq \sqrt{n} |A_{n0}| |\hat{\theta}_1| \leq K \sqrt{n} \exp\left(-\mu \sum_{i=1}^{n} 1/i^\gamma\right) \xrightarrow{n} 0 \text{ w.p.1.}$$

As for the second term, by virtue of Proposition 4.2,

$$
\begin{aligned}
& E\left| \sqrt{n} \sum_{k=1}^{n} \frac{1}{k^\gamma(k+1)} A_{nk} R \hat{\theta}_k \right| \\
(4.19) \quad & \leq K \sqrt{n} \sum_{k=M}^{n} \frac{1}{k^{1+\gamma}} |A_{nk}| E|\hat{\theta}_k| + \sqrt{n} \sum_{k=1}^{M-1} \frac{1}{k^{1+\gamma}} |A_{nk}| |R| E|\hat{\theta}_k| \\
& \leq K n^{-\frac{1+\gamma}{2}} + K \sqrt{n} |A_{nM}| \xrightarrow{n} 0.
\end{aligned}
$$

Therefore, the second term also tends to 0 in probability.

To proceed, we examine the last term in (4.17). By using a partial summation,

$$
\begin{aligned}
(4.20) \quad & \sqrt{n} \sum_{k=1}^{n} \frac{1}{k^\gamma} A_{nk} \frac{1}{k} \sum_{i=1}^{k} \xi_i = \left(\sum_{k=1}^{n} \frac{1}{k^\gamma} A_{nk}\right)\left(\frac{1}{\sqrt{n}} \sum_{i=1}^{n} \xi_i\right) \\
& + \sqrt{n} \sum_{k=1}^{n-1} \left(\sum_{i=1}^{k} \frac{1}{i^\gamma} A_{ni}\right)\left(\frac{1}{k} \sum_{i=1}^{k} \xi_i - \frac{1}{k+1} \sum_{i=1}^{k+1} \xi_i\right).
\end{aligned}
$$

Owing to the property of A_{nk},

$$A_{nk} - A_{n,k-1} = \frac{R}{k^\gamma} A_{nk}.$$

This then yields that

$$\sum_{k=1}^{n} \frac{1}{k^\gamma} A_{nk} = R^{-1}(I - A_{n0}).$$

Due to the fact that $(1/\sqrt{n}) \sum_{i=1}^{n} \xi_i$ is bounded in probability and $A_{n0} \xrightarrow{n} 0$,

$$(4.21) \quad \left(\sum_{k=1}^{n} \frac{1}{k^\gamma} A_{nk}\right)\left(\frac{1}{\sqrt{n}} \sum_{i=1}^{n} \xi_i\right) = \frac{R^{-1}}{\sqrt{n}} \sum_{i=1}^{n} \xi_i + o(1),$$

where $o(1) \xrightarrow{n} 0$ in probability. Likewise,

$$
\begin{aligned}
(4.22) \quad & \sqrt{n} \sum_{k=1}^{n-1} \left(\sum_{i=1}^{k} \frac{1}{i^\gamma} A_{ni}\right)\left(\frac{1}{k} \sum_{i=1}^{k} \xi_i - \frac{1}{k+1} \sum_{i=1}^{k+1} \xi_i\right) \\
& = \sqrt{n} R^{-1} \sum_{k=1}^{n-1} (A_{nk} - A_{n0})\left(\frac{1}{(k+1)k} \sum_{i=1}^{k} \xi_i - \frac{1}{k+1} \xi_{k+1}\right) \\
& \xrightarrow{n} 0 \text{ in probability.}
\end{aligned}
$$

A few details are omitted.

Owing to (4.20)–(4.22) and noticing

$$\sqrt{n} \sum_{k=1}^{n} \frac{1}{k^\gamma(k+1)} A_{nk} \sum_{i=1}^{k} \xi_i$$

$$= \sqrt{n} \sum_{k=1}^{n} \frac{1}{k^{1+\gamma}} A_{nk} \sum_{i=1}^{k} \xi_i - \sqrt{n} \sum_{k=1}^{n} \frac{1}{k^{1+\gamma}(k+1)} A_{nk} \sum_{i=1}^{n} \xi_i.$$

It is easily seen that the last term above goes to 0 in probability. This together with (4.20) yields

(4.23)
$$\sqrt{n} \sum_{k=1}^{n} \frac{1}{k^\gamma(k+1)} A_{nk} \sum_{i=1}^{k} \xi_i$$
$$= \frac{R^{-1}}{\sqrt{n}} \sum_{i=1}^{n} \xi_i + o(1), \quad \text{where } o(1) \xrightarrow{n} 0 \text{ in probability.}$$

Finally, we come back to the next to the last term in (4.17). Using similar arguments as above, it can be shown that

$$\sqrt{n} \sum_{k=1}^{n} \frac{1}{k^\gamma(k+1)} A_{nk} \sum_{i=1}^{k} (R - \varphi_i \varphi_i') \tilde{\theta}_i = \frac{R^{-1}}{\sqrt{n}} \sum_{i=1}^{n} (R - \varphi_i \varphi_i') \tilde{\theta}_i + o(1),$$

where $o(1) \xrightarrow{n} 0$ in probability.

Now,

(4.24)
$$E \left| \frac{1}{\sqrt{n}} \sum_{i=1}^{n} (R - \varphi_i \varphi_i') \tilde{\theta}_i \right|^2$$
$$= \frac{1}{n} \sum_{i=1}^{n} E \tilde{\theta}_i' (R - \varphi_i \varphi_i')' (R - \varphi_i \varphi_i') \tilde{\theta}_i$$
$$+ \frac{2}{n} \sum_{i=1}^{n} \sum_{j>i} E \tilde{\theta}_i' (R - \varphi_i \varphi_i')' (R - \varphi_j \varphi_j') \tilde{\theta}_j.$$

Recall the assumptions (A1) and (A2). For ease of presentation, set $m = 1$. (For more general cases, the proof is the same except more complex notations are needed). The first term on the right side of the equality in (4.24) tends to 0 by the fact $\tilde{\theta}_n \xrightarrow{n} 0$ w.p.1, $\sup_n |\tilde{\theta}_n| < \infty$ w.p.1 and $E|R - \varphi_n \varphi_n'|^2 < \infty$.

As for the second term, notice that the signals involved are m-dependent (with $m = 1$ in this case). First, when $j = i+1$, since $\tilde{\theta}_{i+1}$ is $\mathcal{F}_i$ measurable,

$$E \tilde{\theta}_i' (R - \varphi_i \varphi_i')' E_i (R - \varphi_{i+1} \varphi_{i+1}') \tilde{\theta}_{i+1} = E \tilde{\theta}_i' (R - \varphi_i \varphi_i')' C_1 \kappa_i \tilde{\theta}_{i+1}.$$

For $j \geq i + 2$,

$$
\begin{aligned}
E\tilde{\theta}'_i(R - \varphi_i\varphi'_i)'(R - \varphi_j\varphi'_j)\tilde{\theta}_j \\
= E\tilde{\theta}'_i(R - \varphi_i\varphi'_i)'E_iE_{j-2}E_{j-1}(C_0\kappa_j + C_1\kappa_{j-1})\tilde{\theta}_j \\
= E\tilde{\theta}'_i(R - \varphi_i\varphi'_i)'C_1E_iE_{j-2}\kappa_{j-1} \\
\left\{ \tilde{\theta}_{j-1} + \frac{1}{(j-1)^\gamma}[\varphi_{j-1}(y_{j-1} - \varphi'_{j-1}\theta_{j-1}) + \zeta_{j-1}] \right\} \\
= E\tilde{\theta}'_i(R - \varphi_i\varphi'_i)'C_1E_i\frac{\kappa_{j-1}}{(j-1)^\gamma}[\varphi_{j-1}(y_{j-1} - \varphi'_{j-1}\theta_{j-1}) + \zeta_{j-1}].
\end{aligned}
$$

Thus

(4.25)
$$
\begin{aligned}
\left| \frac{1}{n}\sum_{i=1}^n\sum_{j>i} E\tilde{\theta}'_i(R - \varphi_i\varphi'_i)'(R - \varphi_j\varphi'_j)\tilde{\theta}_j \right| \\
\leq \frac{K}{n}\sum_{i=1}^n E|\tilde{\theta}_i| \|R - \varphi_i\varphi'_i\|E_i \\
\left| \sum_{j>i} \frac{1}{(j-1)^\gamma}[\varphi_{j-1}(y_{j-1} - \varphi'_{j-1}\theta_{j-1}) + \zeta_{j-1}] \right| \\
+ \frac{K}{n}\sum_i E|\tilde{\theta}_i| \|R - \varphi_i\varphi'_i\| \|\kappa_i\| |\tilde{\theta}_{i+1}|.
\end{aligned}
$$

Since $\sup_n |\tilde{\theta}_n| < \infty$ w.p.1, $\tilde{\theta}_n \to 0$ w.p.1, $E|R - \varphi_i\varphi'_i|^2 < \infty$, and $E|\kappa_i|^2 < \infty$,

$$
\frac{K}{n}\sum_i E|\tilde{\theta}_i| \|R - \varphi_i\varphi'_i\| \|\kappa_i\| |\tilde{\theta}_{i+1}| \xrightarrow{n} 0.
$$

We have already demonstrated that

$$
\sum_j \frac{1}{j^\gamma}\varphi_j(y_j - \varphi'_j\theta_j) \quad \text{converges w.p.1}
$$

in the previous section. In view of Proposition 4.2,

$$
\begin{aligned}
\frac{1}{n}\sum_{i=1}^n E|\tilde{\theta}_i| \|R - \varphi_i\varphi'_i\|E_i \left| \sum_{j>i} \frac{1}{(j-1)^\gamma}\varphi_{j-1}(y_{j-1} - \varphi'_{j-1}\theta_{j-1}) \right| \\
\leq Kn^{-\gamma/2} \xrightarrow{n} 0.
\end{aligned}
$$

Notice that since

$$
\sum_i \frac{1}{i^\gamma}\varphi_i(y_i - \varphi'_i\theta_i) \quad \text{converges w.p.1},
$$

$$
\frac{1}{n^\gamma}\sum_{i=1}^{n-1}\varphi_i(y_i - \varphi'_i\theta_i) \xrightarrow{n} 0 \quad \text{w.p.1}.
$$

Owing to the definition of $\{\zeta_n\}$, a very rough estimate gives us

$$\left| \sum_{j>i} \frac{1}{(j-1)^\gamma} \zeta_{j-1} \right| \leq K \ln n \text{ uniform in } i,$$

and as a result, Proposition 4.2 yields

$$\frac{1}{n} \sum_{i=1}^{n} E|\tilde{\theta}_i| \|R - \varphi_i \varphi_i'\| \left| \sum_{j>i} \frac{1}{(j-1)^\gamma} \zeta_{j-1} \right| \leq K n^{-\gamma/2} \ln n \xrightarrow{n} 0.$$

Thus, the right-hand side of (4.25) goes to 0. The asymptotic equivalence is established. □

Remark: For more general random signals, (e.g., certain mixing processes etc.), the asymptotic equivalency still holds. The basic idea and proof of the proposition remain to be the same. Nevertheless, to account the stochastic effect, the idea of 'fixed θ process' (cf. [27] and the references therein), which indicates that for n large enough, the random signals evolves as though θ never changes, needs to be used.

4.5. Limiting distribution. Asymptotic distribution of a suitably scaled sequence of the estimation error $\bar{\theta}_n - \theta$ is considered in this subsection. With the preparation of previous developments, in view of the central limit theorem for martingale difference sequences, and noticing that

$$\sqrt{n}(\bar{\theta}_n - \theta) = \sqrt{n}(\bar{\theta}_{n+1} - \theta) + o(1)$$

where $o(1) \xrightarrow{n} 0$ in probability, we are now in a position to state the following result on asymptotic distribution.

THEOREM 4.4. *Under the conditions of Proposition 4.3,*

$$\sqrt{n}(\bar{\theta}_n - \theta) \sim N(0, R^{-1} \Sigma_0 R^{-1}),$$

where Σ_0 is the covariance of the signals, i.e.,

$$\Sigma_0 = E\xi_1 \xi_1' + \sum_{i=2}^{m} E\xi_1 \xi_i' + \sum_{i=2}^{m} E\xi_i \xi_1'.$$

From the theorem, it follows that Algorithm II is asymptotically optimal in that it has the optimal rate of convergence with the best covariance possible.

5. Further discussions. Algorithm II was analyzed in this paper. Similar approach can be taken to study the asymptotic properties of Algorithm I (cf. [22]). In what follows, several issues are discussed. In Section 5.1, a few remarks are given regarding to the questions of different gain sequences, the noise processes and constrained version of the algorithms etc.;

Section 5.2 is concerned with functional limit theorems; the connection between the averaging algorithms and some singularly perturbed stochastic systems is studied in Section 5.3; continuous parameter problems are treated in Section 5.4.

5.1. A few remarks. By examining the result obtained, one may wonder if the gain sequence is changed to $a_n = a/n^\gamma$ what will change in the outcome. It is certainly interesting to see if there is a contribution in the asymptotic covariance from the constant a. It turns out that the answer is negative. No matter what constant a is placed in a_n, a will be cancelled eventually in the process of averaging. Thus, we conclude that the optimality cannot be improved by placing a constant in the gain.

As was mentioned before, several extensions are possible. Only moving average type of noise processes are treated in this paper. For more general random processes, we refer to [17], [18], [19] and [21] for the stochastic approximation counter part.

Adaptive beam forming algorithms, which is an array processing of the adaptive filtering type with an additional constraint can also be treated in the light of the averaging procedures discussed in this work. Let $\theta \in \mathbb{R}^{r \times o}$, $\varphi_n, C \in \mathbb{R}^{r \times l}$, $y_n, \Phi \in \mathbb{R}^{o \times l}$. The basic problem is to find a recursive algorithm asymptotically converging to θ, the minimizer of

$$E|\theta'\varphi_n - y_n|^2$$

subject to the constraint $\theta'C = \Phi$.

A necessary and sufficient condition for the constraint to hold is

$$\Phi C^\dagger C = \Phi,$$

where $z^\dagger$ denotes the pseudo-inverse of z. Two types of averaging algorithms are devised as follows:

$$\theta_{n+1} = C^{\dagger'}\Phi' + P\left(\theta_n + \frac{1}{n^\gamma}(\varphi_n y_n' - \varphi_n \varphi_n' \theta_n)\right), \quad 1/2 < \gamma < 1$$

(5.1)

$$\bar{\theta}_n = \frac{1}{n}\sum_{i=1}^{n}\theta_i \text{ with } \theta_1 = C^{\dagger'}\Phi',$$

and

$$(5.2)\theta_{n+1} = C^{\dagger'}\Phi' + P\left(\bar{\theta}_n + \frac{1}{n^\gamma}\sum_{i=1}^{n}(\varphi_i y_i' - \varphi_i \varphi_i' \theta_i)\right), \quad 1/2 < \gamma < 1,$$

where $P = I - CC^\dagger$. By considering certain vector spaces, and carrying out appropriate decompositions, these equations can further be written in a more convenient and manageable form (cf. [28] and the references therein); the asymptotic properties can then be studied.

Various projection and truncation algorithms can be designed in conjunction with the averaging approaches. Furthermore, adaptive filtering with averaging can be adopted and used in the framework of using multiprocessors and parallel processing (cf. [29] and the references therein) methods.

5.2. Functional limit theorems. The asymptotic optimality obtained in the previous section can be strengthened. Far reaching functional limit theorems can be established. In lieu of examining $\sqrt{n}(\bar{\theta}_n - \theta)$, let

$$w_n(t) = \frac{[nt]}{\sqrt{n}}(\bar{\theta}_{[nt]+1} - \theta), \text{ for } t \in [0,1]$$

where $[z]$ denotes the largest integral part of z. Then $w_n(\cdot) \in D^r[0,1]$ the space of functions defined on $[0,1]$, that are right continuous, and have left-hand limits endowed with the Skorokhod topology (cf. [30] and the references therein). The pertinent notion of convergence is in the sense of weak convergence (cf. [30]). Using similar arguments as in the previous section, it can be proved that

$$w_n(t) = \frac{R^{-1}}{\sqrt{n}} \sum_{i=1}^{[nt]} \xi_i + o(1),$$

where $o(1) \xrightarrow{n} 0$ in probability uniformly in t. A functional central limit theorem for the sequence $\frac{1}{\sqrt{n}} \sum_{i=1}^{[nt]} \xi_i$ together with the Slutsky's lemma yields that $w_n(\cdot)$ converges weakly to a Brownian motion $w(\cdot)$ with the 'optimal' covariance $t\Sigma$ where $\Sigma = R^{-1}\Sigma_0 R^{-1}$.

5.3. Singularly perturbed systems. In [18], the connection of stochastic approximation algorithms with averaging and some singularly perturbed systems are exploited. This in turn gives clear explanation on why the averaging idea works and why it is important to use slowly varying gains. It will be seen in the sequel that Algorithm II discussed in this work also has a natural connection with a singularly perturbed system.

We begin with the recursion defined by (3.2) and the equation for $\bar{u}_n = \sqrt{n}(\bar{\theta}_n - \theta)$. As in [18], put them in the same time scale, we have

(5.3)
$$\frac{1}{n^{1-\gamma}}(\theta_{n+1} - \theta_n) = \frac{1}{n}(\varphi_n y_n - \varphi_n \varphi_n' \theta_n) + \frac{1}{n}\zeta_n$$
$$\bar{u}_{n+1} - \bar{u}_n = -\frac{1}{2n}\bar{u}_n(1 + O(1/n)) + \frac{1}{\sqrt{n+1}}\theta_{n+1}.$$

Except the extra term ζ_n, the equations above have the same form as that of [18]. Eq. (5.3) can be viewed as a multiple time scale adaptive filtering algorithm, which has a close connection to a singularly perturbed system (cf. [18] Eq. (2.8)) of the form

$$\varepsilon dz^\varepsilon = A_{11}z^\varepsilon dt + dw_1$$
$$dx^\varepsilon = A_{22}x^\varepsilon dt + A_{12}z^\varepsilon dt + dw_2.$$

In addition, notice that the requirement of slowly varying gain is crucial. For example, if $\gamma = 1$, the structure of the singularly perturbed system will be destroyed.

5.4. Continuous time analogue. In addition to the mathematical interest, the reasons for considering continuous version algorithm stem from the fact that the continuous problems are good approximation to discrete ones when the sampling is taken rather frequently. It is important to establish that everything works well if the sampling rate becomes very high.

Continuous time analog of the algorithms discussed here are given below. Corresponding to Algorithm I, we have

(5.4)
$$\dot{\theta}_t = \frac{1}{t^\gamma}\varphi_t(y_t - \varphi_t'\theta_t), \quad 1/2 < \gamma < 1$$
$$\bar{\theta}_t = \frac{1}{t}\int_0^t \theta_s\, ds;$$

corresponding to Algorithm II,

(5.5) $$\theta_t = \frac{1}{t}\int_0^t \theta_s\, ds + \frac{1}{t^\gamma}\int_0^t \varphi_s(y_s - \varphi_s'\theta_s)ds, \ 1/2 < \gamma < 1.$$

To study the asymptotic properties, we first prove $\theta_t \to \theta$ w.p.1. Then define

$$B_t(\tau) = \tau\sqrt{t}(\bar{\theta}_{t\tau} - \theta) \text{ for each } \tau \in [0, 1],$$

and

$$w_t(\tau) = \frac{1}{\sqrt{t}}\int_0^{t\tau} \xi_s\, ds \quad \text{for } \tau \in [0, 1].$$

Then it can be shown as $t \to \infty$,

$w_t(\cdot)$ converges weakly to a Brownian motion $w(\cdot)$ and

$B_t(\cdot)$ converges weakly to a Brownian motion $B(\cdot)$

such that $B(\tau) = R^{-1}w(\tau)$. The discussion on optimality can be carried out similar to that of Section 2. In view of the limiting Brownian motion $B(\cdot)$, the continuous version of the adaptive filtering algorithms are also asymptotic optimal.

REFERENCES

[1] B. WIDROW AND S. STEARNS, *Adaptive Signal Processing*, Prentice-Hell, Englewood Cliffs, NJ, 1985.
[2] A. GERSHO, Adaptive equalization of highly dispersive channels for data transmission, *Bell Syst. Tech. J.* **48** (1969), pp. 55–70.

[3] S. LAKSHMIVARAHAN, *Learning Algorithms and Applications*, Springer-Verlag, New York, 1981.

[4] G.C. GOODWIN AND K.S. SIN, *Adaptive Filtering Prediction and Control*, Prentice-Hall, Englewood Cliffs, NJ, 1984.

[5] K.L. ASTRÖM, *Introduction to Stochastic Control*, Academic Press, New York, 1970.

[6] P.R. KUMAR AND P. VARAIYA, *Stochastic Systems: Estimation, Identification, and Adaptive Control*, Prentice-Hall, New Jersey, 1986.

[7] L. LJUNG, Analysis of recursive stochastic algorithms, *IEEE Trans. Automatic Control*, **AC-22** (1977), pp. 551–575.

[8] H.J. KUSHNER AND D.S. CLARK, *Stochastic Approximation Methods for Constrained and Unconstrained Systems*, Springer-Verlag, 1978.

[9] E. EWEDA AND O. MACCHI, Quadratic and almost sure convergence of unbounded stochastic approximation algorithms with correlated observations, *Ann. Institut. Henri Poincare* **19** (1983), pp. 235–255.

[10] H.J. KUSHNER AND A. SHWARTZ, Weak convergence and asymptotic properties of adaptive filters with constant gains, *IEEE Trans. Inform. Theory* **IT-30** (1984), pp. 177–182.

[11] K.L. CHUNG, On a stochastic approximation method, *Ann. Math. Statist.* **25** (1954), pp. 463–483.

[12] A. BENVENISTE, M. METIVIER AND P. PRIOURET, *Adaptive Algorithms and Stochastic Approximation*, Springer-Verlag, Berlin, 1990.

[13] B.T. POLYAK, New method of stochastic approximation type, *Automat. Remote Control* **51** (1990), pp. 937–946

[14] D. RUPPERT, Stochastic approximation, in *Handbook in Sequential Analysis*, B.K. Ghosh and P.K. Sen Eds., Marcel Dekker, New York, 1991 pp. 503–529.

[15] J.A. BATHER, Stochastic approximation: A generalization of the Robbins-Monro procedure, in *Proc. Fourth Prague Symp. Asymptotic Statist.*, P. Mandl and M. Hušková Eds., 1989, pp. 13–27.

[16] B.T. POLYAK AND A. JUDITSKY, Acceleration of stochastic approximation by averaging, *SIAM J. Control Optim.* **30** (1992), pp. 838–855.

[17] G. YIN, On extensions of Polyak's averaging approach to stochastic approximation, *Stochastics Stochastic Rep.* **36** (1991), pp. 245–264.

[18] H.J. KUSHNER AND J. YANG, Stochastic approximation with averaging of the iterates: optimal asymptotic rate of convergence for general processes, Technical Report, LCDS #91-9, Brown Univ., 1991, also to appear in *SIAM J. Control Optim.*

[19] H.J. KUSHNER AND J. YANG, Stochastic approximation with averaging and feedback: rapidly convergent "on line" algorithms, and applications to adaptive systems, Technical Report, LCDS #92-8, Brown Univ., 1992.

[20] R. SCHWABE, Stability results for smoothed stochastic approximation procedures, Fachbereich Mathematik, Series A, Mathematik, Preprint Nr. A-92-14, Freie Universität Berlin, 1992.

[21] G. YIN AND K. YIN, Asymptotically optimal rate of convergence of smoothed stochastic recursive algorithms, *Stochastics Stochastic Rep.* **47** (1994), pp. 21–46.

[22] G. YIN, Asymptotic optimal rate of convergence for an adaptive estimation procedure, *Lecture Notes in Control Inform.* 184, *Stochastic Theory and Adaptive Control*, T. Duncan and B.Pasik-Duncan Eds., Springer-Verlag, 1992, pp. 480–489.

[23] YA. Z. TSYPKIN, *Adaptation and Learning in Automatic Systems*, Academic Press, New York, 1971.

[24] S.V. SHIL'MAN AND A.I. YASTREBOV, Convergence of a class of multistep stochastic adaptation algorithms, *Avtomatikha i Telemekhanika* **38** (1976), pp. 111–118.

[25] A.P. KOROSTELEV, Multistep procedures of stochastic optimization, *Automat. Remote Control* **43** (1982), pp. 621–627.

[26] M. MÉTIVIER AND P. PRIOURET, Applications of a Kushner and Clark lemma to general classes of stochastic algorithms, *IEEE Trans. Inform.* **IT-30** (1984), pp. 140–150.

[27] H.J. KUSHNER, *Approximation and Weak Convergence Methods for Random Processes, with applications to Stochastic Systems Theory*, MIT Press, 1984.

[28] G. YIN, Asymptotic properties of an adaptive beam former algorithm, *IEEE Trans. Inform. Theory* **IT-35** (1989), pp. 859–867.

[29] H.J. KUSHNER AND G. YIN, Asymptotic properties of distributed and communicating stochastic approximation algorithms, *SIAM J. Control Optim.* **25** (1987), pp. 1266–1290.

[30] S.N. ETHIER AND T.G. KURTZ, *Markov Processes, Characterization and Convergence*, Wiley, New York, 1986.